BASIC ENVIRONMENTAL TECHNOLOGY

BASIC ENVIRONMENTAL

WATER SUPPLY, WASTE DISPOSAL,

TECHNOLOGY

AND POLLUTION CONTROL

Jerry A. Nathanson, P.E.

Union County College
Scotch Plains, New Jersey

JOHN WILEY & SONS

NEW YORK   CHICHESTER   BRISBANE   TORONTO   SINGAPORE

Cover Photo of Water Treatment Plant in Ogden, Utah
by Morton Beebe/The Image Bank

Library of Congress Cataloging in Publication Data:

Nathanson, Jerry A.
   Basic environmental technology.

   Includes index.
   1. Water-supply engineering.   2. Sewage disposal.
3. Pollution   4. Environmental engineering.   I. Title.
TD145.N38   1986   628   86-1682
ISBN 0-471-81886-0

Printed in the United States of America

10 9 8 7 6 5 4 3 2 1

TO GINGER AND ADAM

Sailing up my dirty stream
Still I love it and I'll dream
That some day, though maybe not this year
My Hudson River and my country will run clear

PETE SEEGER

PREFACE

This textbook is an introduction to the fundamentals of water supply, waste disposal, and pollution control. It is particularly useful for students of civil engineering technology and related disciplines in community colleges, technical institutes, and other schools where a basic but practical course of study is required. This book can also be used in an elective course, or for independent study by individuals who wish to explore the underlying principles of environmental quality control and public health protection. Although the focus is primarily on drinking water, stormwater, and wastewater, the topics of solid and hazardous waste management, air pollution, and noise pollution are included.

In addition to offering a pragmatic problem-solving approach to the subject (e.g., how to design a sewer system), this text provides enough descriptive material to enable readers to better understand many current environmental issues, from acid rain to toxic waste. Review or "primer" sections are included so that readers with little or no experience in subjects like chemistry, ecology, and hydraulics can comprehend and use the book. Numerous examples, diagrams, and photographs are used throughout the book to clarify each topic. Review questions and practice problems follow each chapter, with answers to the practice problems given in the Appendix. A comprehensive glossary of environmental terms is also included at the end of the book. In essence, this book is meant to be an effective instructional tool for those who are studying the subject for the first time, rather than an exhaustive or expert technical reference book.

The text is written in a clear and easy-to-read style. Mathematics is kept at a basic level; formulas and example problems are presented without the use of calculus. No attempt is made to discuss chemical stoichiometry. In fact, the emphasis is on applied hydraulics instead of on environmental chemistry. Metric (SI) and customary inch-pound (or "American") units are used alternately, in a series of different examples and practice problems. The ability to use both systems of units, rather than just one or the other, is necessary for all technical personnel during the slow but ongoing period of transition to SI units in the United States.

Environmental technology is so broad a field that it is impracticable to provide rigorous and comprehensive coverage in an introductory textbook such as this. In selecting topics and deciding on the depth of discussion, I was guided by my experience as a civil/environmental engineer and professor of engineering technology. Most of the topics I have included are covered thoroughly in engineering textbooks, but they are presented here in a form that is more readily accessible for technicians, technologists, and many others who may use the book.

This text includes more than enough material for a one-semester course, allowing an instructor to be selective and to concentrate on the topics appropriate for a specific class or program. A list of pertinent technical references is included at the end of each chapter, for those who may wish to explore a particular topic in more detail. In addition to providing an elementary but useful introduction, and to helping prepare the technology student for applications likely to be encountered during future job experiences, an objective of this book is to motivate its readers to pursue the subject at a more advanced level. I hope that most users of this textbook will find that these objectives have been achieved.

For reviewing the manuscript and for offering helpful comments regarding the content of the text, I would like to thank Louis J. Chanin, Hackensack Water Company; Leo A. Ebel, Washington University; Jerry Haimowitz, Boro of North Plainfield; Keith Hancock, Larimer County Vocational-Technical Center; Gayle Huges, Nashville State Technical Institute; Paul H. Klopping, Environmental Training Consultants, Inc.; Paul Mazur, Columbus Technical Institute; Andrew Potter, Monroe Community College; Karl B. Schnelle, Jr., Vanderbilt University; and Paul D. Trotta, Northern Arizona University. Of course, I assume full responsibility for any technical deficiencies, inaccuracies, or errors that may remain in the text, and I welcome any additional suggestions for its improvement.

Jerry A. Nathanson

CONTENTS

BASIC
ENVIRONMENTAL
TECHNOLOGY

1

BASIC CONCEPTS

1.1 OVERVIEW OF ENVIRONMENTAL TECHNOLOGY

WATER SUPPLY
SEWAGE DISPOSAL AND WATER POLLUTION CONTROL
STORMWATER MANAGEMENT
SOLID AND HAZARDOUS WASTE MANAGEMENT
AIR AND NOISE POLLUTION CONTROL
OTHER ENVIRONMENTAL FACTORS

1.2 PUBLIC HEALTH

COMMUNICABLE DISEASES
NONINFECTIOUS DISEASES

1.3 ECOLOGY

FOOD CHAINS AND METABOLISM
AEROBIC AND ANAEROBIC DECOMPOSITION
BIOGEOCHEMICAL CYCLES
STABILITY, DIVERSITY, AND SUCCESSION

1.4 GEOLOGY AND SOILS

TYPES OF ROCK
TYPES OF SOIL
SOIL SURVEY MAPS

1.5 HISTORICAL PERSPECTIVE

AN ERA OF ENVIRONMENTAL AWARENESS
ENVIRONMENTAL LEGISLATION

Environmental technology involves the application of engineering principles to the *planning, design, construction,* and *operation* of systems for:

Drinking water treatment and distribution
Sewage disposal and water pollution control
Stormwater drainage and control
Solid and hazardous waste disposal
Air and noise pollution control
General community sanitation

The structures and facilities that serve these functions, including pipelines, pumping stations, treatment plants, and waste disposal sites, comprise a major portion of society's *infrastructure*—the public and private works that allow human communities to thrive and function productively.

The practice of environmental technology encompasses two fundamental objectives:

1. *Public Health Protection* - to help prevent the transmission of diseases among human beings.

2. *Environmental Health Protection* - to preserve the quality of our natural surroundings, including water, land, air, vegetation, and wildlife.

Actually, there is considerable overlap of these two objectives because of the relationship between the quality of environmental conditions and the health and well-being of the general population. In fact, the terms *public health* and *environmental health* are often used synonymously.

Public health includes more than just the absence of illness. It is a condition of physical, mental, and social well-being and comfort. The cleanliness and aesthetic quality of our surroundings—the atmosphere, rivers, lakes, forests, meadows, as well as towns and cities, have a direct impact on this condition of human well-being and comfort; and the practice of *sanitation,* that is, the promotion of cleanliness, is a basic necessity in the effort to protect public and environmental health.

Environmental technology is usually considered to be a part of the *civil engineering* profession, which has traditionally been called upon to plan, design, build, and operate the facilities required for public and environmental health protection. This particular specialty field within civil engineering is also called

Sanitary engineering

Public health engineering

Pollution control engineering

Environmental health engineering

Whatever the profession is called, a knowledgable and skilled team of engineers, technologists, and technicians is needed to accomplish its fundamental objectives. (A discussion of the role of the technician or technologist as part of the civil/environmental engineering team is included in the appendix.)

Environmental technology can be characterized as an *interdisciplinary field* because it encompasses several different technical subjects. In addition to such traditional civil engineering topics as hydraulics and hydrology, these include biology, ecology, geology, chemistry, as well as others. This variety makes the field an interesting and challenging one.

Fortunately though, it is not necessary to be an expert in all of these subjects to understand and ap-

ply the basic principles of environmental technology. This particular text has been designed so that a student with little academic background in some or all of the supporting subjects can still use it productively.

This chapter is a review of basic and pertinent topics in public health, ecology, and geology. Practical hydraulics is covered in the next chapter, and the fundamentals of hydrology are presented in the third. The essential concepts and terminology from chemistry and microbiology are presented in sections of the fourth chapter, on water quality. The remaining chapters of the book build on these subjects in presenting the principles and applications of environmental technology.

As in any other technical field, this one requires numerical computations in many of its practical applications. Physical properties such as volume, mass or weight, pressure, and flow rate are used in a variety of formulas. For the next several years, technical personnel in the United States will have to be able to use both the SI metric and our traditional inch–pound or "American" units of expression. (Outside of the United States, SI units are used almost exclusively.) A discussion of units, dimensional consistency, and numerical accuracy is presented in the appendix for those who need a review.

1.1 OVERVIEW OF ENVIRONMENTAL TECHNOLOGY

Before beginning a study of the many different topics that comprise environmental technology, it would be helpful to have an understanding of the overall goals, problems, and alternative solutions available to practitioners in this field.

To present an overview of such a broad subject, we can consider a fictitious project involving the subdivision and development of a tract of land into a "new town," which will include residential, commercial, and industrial centers. Whether the "project owner" is a governmental agency or a private developer, a wide spectrum of environmental problems will have to be considered and solved before the construction of the new community can begin. Usually the project owner retains the services of an independent environmental consulting firm to address these problems.

WATER SUPPLY

One of the first problems the project developers and consultants must consider is the provision of a *potable* water supply, one that is clean, wholesome, safe to drink, and available in adequate quantities to meet the anticipated demand in the new community. Some of the questions that must be answered are

1. Is there an existing public water system nearby, with the capacity to connect with and serve the new development? If not,

2. Is it best to build a new centralized treatment and distribution system for the whole community, or would it be better to use individual well supplies? If a centralized treatment facility is selected,

3. What type of water treatment processes will be required to meet federal and state drinking water standards? (Water from a river or a lake usually requires more extensive treatment than groundwater does, to remove suspended particles and bacteria.) Once the source and treatment processes are selected,

4. What would be the optimum hydraulic design of the storage, pumping, and distribution network, to ensure that sufficient quantities of water can be delivered to consumers at adequate pressures?

SEWAGE DISPOSAL AND WATER POLLUTION CONTROL

When running water is delivered into individual homes and businesses, there is an obvious need to provide for the disposal of the used water or *sewage.*

Sewage carries human wastes, wash water, and dishwater, as well as a variety of chemicals if it comes from an industrial or commercial area. It also carries microorganisms that may cause disease and organic material that can damage lakes and streams as it decomposes.

It will be necessary to provide the new community with a means for safely disposing of the sewage, to prevent water pollution and to protect public and environmental health. Some of the technical questions that will have to be addressed include the following:

1. Is there a nearby municipal sewerage system with the capacity to handle the additional flow from the new community? If not,

2. Are the local geologic conditions suitable for on-site subsurface disposal of the wastewater (usually "septic systems"), or is it necessary to provide a centralized sewage treatment plant for the new community and to discharge the treated sewage to a nearby stream? If treatment and surface discharge are required,

3. What is the required degree or level of wastewater treatment to prevent water pollution? Will a "secondary" treatment level, which removes at least 85 percent of the pollutants, be adequate? Or will some form of advanced treatment be required to meet federal and state discharge standards and stream quality criteria? (Some advanced treatment facilities can remove more than 99 percent of the pollutants.)

4. Is it possible to use some type of *land disposal* of the treated sewage, such as spray irrigation, instead of discharging the flow into a stream?

5. What methods will be used to treat and dispose of the *sludge,* or sewage solids, that are removed from the wastewater?

6. What is the optimum layout and hydraulic design of a sewage collection system that will convey the wastewater to the central treatment facility, with a minimum need for pumping?

STORMWATER MANAGEMENT

The development of land for human occupancy and use tends to increase the volume and rate of stormwater runoff from rain or melting snow. Basically, this is due to the construction of roads, pavements, and other impervious surfaces, which prevent the water from seeping into the ground. The increase in surface runoff may cause flooding, soil erosion, and water pollution problems on the site as well as downstream. The following are some of the questions the developer and consultant will have to consider:

1. What is the optimum layout and hydraulic design of a surface drainage system that will prevent local flooding during wet weather periods?

2. What intensity and duration of storm should the system be designed to handle without *surcharging,* or overflowing?

3. Do local municipal land-use ordinances call for facilities that keep post-construction runoff rates equal to or less than the amount of runoff from the undeveloped land? If so,

4. What is the optimum location and design of a stormwater storage basin, to detain the stormwater and reduce the peak runoff flows during wet weather periods?

5. What provisions can be made, during and after construction, to minimize problems related to soil erosion from runoff?

SOLID AND HAZARDOUS WASTE MANAGEMENT

The development of a new community (or growth of an existing community) will certainly lead to the generation of more refuse and industrial waste materials. Ordinarily, the collection and disposal of solid wastes is a responsibility of the local municipality. However, some of the wastes from industrial sources may be particularly dangerous, requiring special handling and disposal methods.

There is a definite relationship between public and environmental health and the proper handling and disposal of solid wastes. Improper garbage disposal practices can lead to the spread of diseases such as *typhus* and *plague* due to the breeding of rats and flies.

If municipal refuse is improperly disposed of on land, in a "garbage dump," it is also very likely that surface and groundwater resources will be polluted with *leachate*. (Leachate is a highly contaminated liquid that seeps through the pile of refuse into nearby streams, as well as into the ground.) On the other hand, incineration of the refuse may cause significant air pollution problems if proper controls are not applied or are ineffective.

Hazardous wastes, such as poisonous or combustible chemicals from industrial processes, must receive special attention with respect to collection, transport, treatment, and final disposal. This is particularly necessary to protect the quality of groundwater, which is the source of water supply for about 100 million people in the United States. In recent years, an increasing number of water supply wells have been found to be contaminated with synthetic organic chemicals, many of which are thought to cause cancer and other illnesses in humans. Improper disposal of these hazardous materials, usually by illegal burial in the ground, is the cause of the contamination.

Some of the general questions related to the disposal of solid and hazardous wastes from the new community include:

1. Is there a suitable *sanitary landfill* serving the area and will it have sufficient capacity to handle the increased amounts of solid waste for a reasonable period of time? (A sanitary landfill provides a means for land disposal of refuse in an environmentally sound manner.) If not,

2. Is there a suitable site for construction and operation of a new sanitary landfill to serve the area, or will it be best to apply modern resource recovery and recycling technology as a means of solid waste management? A sanitary landfill site must meet strict re-

quirements with respect to topography, geology, and hydrology, as well as other environmental conditions.

3. Will commercial or industrial establishments be generating any hazardous waste materials, and, if so, is a *secure landfill* or other means of disposal available for the waste generators to use?

AIR AND NOISE POLLUTION CONTROL

Major sources of air pollution include fuel combustion for power generation, certain industrial and manufacturing processes, and automotive traffic. Project developers can exercise the most control over traffic. Private industry will have to apply appropriate air pollution control technology at their individual facilities to meet federal and state standards.

The volume of traffic in the area will obviously increase, leading to an increase in exhaust fumes from cars and other vehicles. But proper layout of roads and traffic-flow patterns can minimize the amount of stop-and-go traffic, thus reducing the amount of air pollution in the development.

Usually, the developer's consultant will have to prepare an *environmental impact statement (EIS),* which will describe the traffic plan and estimate the expected levels of air pollutants. It will have to be shown that air quality standards will not be violated, for the project to gain approval from regulatory agencies. (In addition to air pollution, the completed EIS will address all other environmental effects related to the proposed project.)

Noise can be considered to be a type of air pollution in the form of waste energy—sound vibrations. Noise pollution will result from the construction activity, causing a temporary or *short-term impact.* The builders may have to observe limitations on the types of construction equipment and the hours of operation, to minimize this negative effect on the environment. A *long-term impact,* with respect to the generation of noise, will be caused by the increased amount of vehicular traffic. This is another environmental factor the consultants will have to address in the EIS.

OTHER ENVIRONMENTAL FACTORS

Not to be overlooked as an environmental factor in any land development project is the potential impact on local vegetation and wildlife. The destruction of woodlands and meadows to make room for new buildings and roads can lead to significant ecological problems, particularly if there are any rare or endangered species in the area. Cutting down trees and paving over meadows can cause short-term impacts related to soil erosion and stream sedimentation. On a long-term basis, it will cause the displacement of wildlife to other suitable habitats, presuming, of course, that they are available nearby. Otherwise, several species may disappear from the area entirely.

Human activity in wetland areas, including marshes and swamps, is often very damaging to the environment. Coastal wetlands are habitats for many different species of organisms, and the tremendous biological productivity of these wetland environments is a very important factor in the "food chain" (see Section 1.3) for many animals. When wetlands are drained, filled in, or dredged, to accommodate building and land development projects, the life cycle of many organisms is disrupted. Many species may be destroyed as a result of habitat loss or loss of a staple food source. There is a definite need to control or restrict construction activities in wetland environments and to implement a nationwide wetlands protection program.

Environmental concerns related to general sanitation in a new community include food and beverage protection, insect and rodent control, radiological health protection, industrial hygiene and occupational safety, and the cleanliness of recreation areas such as public swimming pools. These concerns are generally the responsibility of local health departments.

In the preceding overview of environmental technology, we have briefly considered many factors that are very much interrelated and overlapping, as illustrated in Figure 1.1. In a textbook, it is necessary to organize these factors into chapters and sections. But this is only for academic convenience. The interrelationships should always be kept in

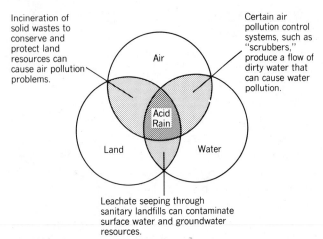

Incineration of solid wastes to conserve and protect land resources can cause air pollution problems.

Certain air pollution control systems, such as "scrubbers," produce a flow of dirty water that can cause water pollution.

Leachate seeping through sanitary landfills can contaminate surface water and groundwater resources.

FIGURE 1.1 Most environmental problems pertaining to air, water, and land quality are interrelated. A problem called "acid rain," for example, is caused by air pollution, and it damages both aquatic and terrestrial ecosystems.

mind. Water, land, and air pollution are part of a single problem.

1.2 PUBLIC HEALTH

Preventing the spread of disease and thereby protecting the health of the general population is a fundamental goal of environmental technology. Public health protection is, of course, a primary concern of doctors and other medical professionals. But engineering technology also plays a significant role in this effort. In fact, the high standard of health enjoyed by citizens of the United States and other developed nations is largely due to the construction and operation of modern water treatment and pollution control systems. The spread of diseases in countries with inadequate sanitary facilities is a major problem for millions of people.

Diseases are classified into two broad groups: *communicable diseases* and *noninfectious diseases*. Communicable diseases are those that can be transmitted from person to person; we commonly refer to these diseases as being infectious or contagious. Noninfectious diseases, as the name implies,

are not contagious; they cannot be transmitted from one person to another by any means. The kinds of noninfectious diseases of concern in environmental technology are associated with contaminated water, air, or food. The contaminants are usually toxic chemicals from industrial sources, although biological toxins can also cause disease.

COMMUNICABLE DISEASES

Communicable diseases are usually caused by *microbes*. These microscopic organisms include bacteria, protozoa, and viruses. Most microbes are essential components of our environment and do not cause disease. Those that do are called pathogenic organisms, or simply *pathogens*.

The ways in which diseases are spread from one person to another vary considerably. They are called *modes of transmission* of disease, and are summarized in Figure 1.2. It is important to make distinctions among the various modes of transmission to be able to apply suitable methods of control. *Direct transmission* involves an immediate transfer of pathogens from a carrier (infected person) to a susceptible contact, that is, a person who has had direct contact with the carrier and is liable to acquire the disease. Clearly, control of this mode of transmission is not within the scope of environmen-

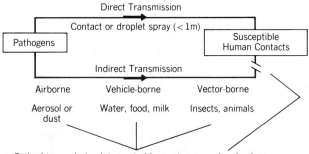

Modes of Disease Transmission

Direct Transmission
Contact or droplet spray (<1m)

Pathogens

Susceptible Human Contacts

Indirect Transmission

| Airborne | Vehicle-borne | Vector-borne |
| Aerosol or dust | Water, food, milk | Insects, animals |

Path of transmission intercepted by *environmental technology:* Water purification, wastewater treatment, air pollution control, solid waste management, and general community sanitation.

FIGURE 1.2 Communicable diseases are spread in several ways, many of which can be controlled or intercepted by applications of modern environmental technology.

tal technology; it is in the province of personal hygiene and the medical profession (who provide immunization and quarantine infected persons).

Environmental technology can be applied to intercept many of the modes of *indirect transmission.* There are three indirect modes of disease transmission, including *airborne, vector borne,* and *vehicle borne.* Airborne transmission involves the spread of microbes from carrier to contact in contaminated mists or dust particles suspended in air. It is the least common of the indirect modes. (This should not be confused with the noninfectious public health problems associated with chemical air pollution, which will be discussed later.)

Vectors of disease include insects, rodents, or other animals that can transport the pathogens to susceptible human contacts. The animals that carry the pathogenic microbes are also called *intermediate hosts* if the microbes have to develop and grow in the vector's body before becoming infective to humans. Vector-borne disease can be controlled to some extent by proper sanitation measures.

A *vehicle* of disease transmission is any nonliving object or substance that is contaminated with pathogens. For example, forks and spoons, handkerchiefs, soiled clothes, or even children's toys, are potential vehicles of transmission. They can physically transport and transfer the pathogens from carrier to contact.

Water, food, and milk are also potential vehicles of disease transmission; these are perhaps the most significant with regard to environmental technology and sanitation. Water, in particular, plays a major role in the transmission of communicable diseases, but it is most amenable to engineering and technological controls. Water and wastewater treatment facilities effectively block the pathway of waterborne disease.

Types of Communicable Diseases

Waterborne and food-borne diseases are perhaps the most preventable types of communicable diseases. The application of basic sanitary principles and environmental technology have virtually eliminated serious outbreaks of these diseases in technologically-developed countries.

Water- and food-borne diseases are also called *intestinal diseases* because they affect the intestinal tract of humans. The pathogens are excreted in the feces of infected people. If these pathogens are inadvertently ingested by others in contaminated food or water, the cycle of disease can continue, possibly in *epidemic* proportions, that is, when the number of occurrences of a disease in a community is far above normal.

Symptoms of intestinal disease include diarrhea, vomiting, nausea, and fever. Intestinal diseases can incapacitate large numbers of people in an epidemic, and sometimes result in the deaths of many infected individuals. Water contaminated with untreated sewage (domestic wastewater) is generally the most common cause of this type of disease.

The most prevalent waterborne diseases include *typhoid fever, dysentery, cholera, infectious hepatitis,* and *gastroenteritis* (common diarrhea and cramps). These can also be transmitted by contaminated food or milk products. (Diseases caused by bacterial toxins include *botulism* and *staphylococcus* food poisoning. Refrigeration, as well as proper cooking and sanitation at food-processing facilities and restaurants, are important for control of these food-borne diseases.)

Although cholera and dysentery have not generally been a problem in the United States, they are prevalent diseases in India, Pakistan, and in many of the technologically underdeveloped countries of southeast Asia. In fact, they are considered to be *endemic* (habitually present) in these areas. Typhoid fever is more universal in occurrence than cholera or dysentery. Until the beginning of the 1900s, typhoid mortality rates in some urban areas of the United States were as high as 650 deaths per 100 000 population. The beginning of modern water purification technology at about that time helped to lower the typhoid death rates to considerably less than 1 per 100 000 people per year. (Immunization and improvements in food and milk sanitation also played a role in reducing the incidence of typhoid.)

Amoebic dysentery, caused by a single-cell microscopic animal called an amoeba, occurred in epidemic proportions in Chicago during the early 1930s. About 100 out of the approximately 1000 people who contracted the disease died from it. The cause of this epidemic was traced to sewage that contaminated the water supplies of two hotels in the city. Although epidemics of intestinal disease like this one are not at all common in the United States, when they do occur they are usually very localized and can be traced to contaminated water supplies in hotels, restaurants, schools, or camps. Generally, the contamination is caused by *cross-connections* in the water distribution system, which may allow backflow of wastewater into the drinking water supply.

Insect-borne diseases include those transmitted by the bites of mosquitoes, lice, and ticks. *Malaria, yellow fever,* and *encephalitis* are typical diseases spread by certain species of mosquitoes. Flies also transmit disease, but not by biting; the contact of their germ-laden bodies, wings, and legs with food consumed by humans spreads diseases such as typhoid fever and gastroenteritis.

The elimination of the breeding places of insects is one of the most important control measures. Proper garbage disposal reduces fly-breeding places and drainage of wet or swampy land is one of the methods available to eliminate mosquito breeding areas. Chemical control with insecticides is usually a last resort because of the environmental and potential health problems associated with use of toxic substances.

In addition to insects, other vectors of disease transmission are vertebrate animals such as dogs and rats. Rabies is a familiar example of a disease spread by the bite of an infected dog or other mammal, but it is not generally related to environmental conditions. Rodent-borne diseases, such as *typhus* and *bubonic plague,* are more readily controlled by applications of environmental technology. Rat populations can be controlled by good community sanitation practices; rodent access to garbage and water should be prevented. Modern building codes include specifications for rodent-proof building construction.

NONINFECTIOUS DISEASES

It is a well-documented fact that the overall death rate for people residing in heavily polluted urban areas is significantly higher than the mortality rate in areas that are relatively pollution free. This is not necessarily because of the incidence of sewage pollution and the spread of infectious diseases. In fact, most current public health problems related to environmental pollution are considered to be the result of contamination of water, food, and air with toxic chemicals. The resulting diseases are noninfectious.

Some noninfectious illnesses associated with toxic chemical pollution have a relatively sudden and severe onset, and the acute or immediate health effects can be readily traced to a specific contaminant. A group of substances known as the *heavy metals* are particularly notorious in this regard. Other noninfectious diseases may take years to develop and can involve chronic or long-lasting health problems. Generally, various synthetic organic substances cause this type of problem, even in extremely small concentrations. Some organics are considered to be *carcinogenic,* having the potential to cause cancer in humans.

Lead is one of the heavy metals involved in noninfectious disease. The public health problems related to lead poisoning have long been associated primarily with ingestion by children of peeling lead-based paint. But the combustion of leaded gasoline fuel in automobiles is a serious threat to air quality and public health. The lead is discharged into the air, inhaled by people, and eventually winds up in the blood. Research has shown that there is a direct correlation between the level of lead in gasoline and the level of lead in the blood of the population as a whole. Lead poisoning can lead to blindness, kidney disease, and mental retardation (particularly in children).

The evidence against lead as a dangerous environmental pollutant is overwhelming. It is a cumulative poison; that is, it accumulates in human tissue and can build up to toxic levels over time. As a result, environmental agencies in Europe as well as in the United States are making an effort to enforce a total ban on lead in gasoline by the end of the 1980s. In the meantime, it is expected that the maximum allowable lead level will soon be reduced about tenfold, to one tenth of a gram per gallon of gasoline.

Mercury is another heavy metal associated with environmental pollution and noninfectious illness. It was first noted as such when it afflicted large numbers of people living in the Minamata Bay region of Japan, in the 1950s. Mercury compounds, discharged into the Bay in wastewater from a local factory, were ingested by people who ate contaminated fish. A severe epidemic of disease resulting in blindness, paralysis, and many deaths, was the result. Less severe symptoms included hand tremors, irritability, and depression.

At the time of the Minamata Bay incident, mercury vapor was known to be harmful, although metallic mercury itself was not considered hazardous (it has long been used in dental fillings). But research after the poisoning episode in Japan led to the discovery that certain microorganisms can cause the metallic mercury to combine with other substances in the water, forming harmful mercury compounds such as *methylmercury*. This substance was ingested by microscopic organisms in the water, called plankton; little fish ate the plankton, bigger fish ate the smaller fish, and finally people ate the contaminated fish and were made ill by the toxic methylmercury.

The episode of mercury poisoning in Japan is one example of a relatively sudden and acute illness related to environmental pollution. The concentration of the pollutant was relatively high and the harmful effects were noticed within a short time. But questions still remain as to the chronic or long-term effects of lower concentrations of mercury compounds. It is common to detect small amounts of mercury in fish and wildlife even in rivers and lakes far from industrial centers.

Discarded batteries and dry cells are a major source of mercury. Ironically, this is becoming a very serious problem in Japan, where it is difficult to properly dispose of the many batteries generated by the growing electronics industry and the use of calculators, cameras, portable stereos, and watches. Almost three billion batteries are produced in Japan yearly. A possible solution to this problem is the further development and use of mercury-free batteries.

Unfortunately, mercury and lead are not the only harmful chemical substances that become environmental pollutants when poorly managed or controlled. For example, the pesticide *kepone* has seriously polluted the James River and Chesapeake Bay. The Hudson River is known to be contaminated with the toxic industrial chemical called *PCB* (polychlorinated biphenyl). This oily substance was widely used in electrical transformer fluids, coolants, paints, and other products. It persists in the environment because it is nonbiodegradable; that is, it does not readily decompose and dissipate by natural processes. PCBs have accumulated in the bottom deposits of the river, and many species of fish are contaminated with it.

Like the pesticide DDT, PCB has been banned from use. But because these substances are extremely persistent in the environment, they remain potential dangers to public health for many years after their initial discharge. Traces of DDT and PCB are still found in the body tissues of animals far removed from the sources of pollution. Both of these chemicals are considered potential human carcinogens.

Environmental pollution with harmful chemicals and the resulting incidences of noninfectious disease are part of a problem now commonly referred to as *hazardous waste disposal*. This will be discussed in more detail in a later chapter.

Perhaps one of the most publicized environmental disasters in recent years that was related to improper disposal of hazardous wastes occurred in the late 1970s at Love Canal in Niagara Falls, New York. Briefly, waste chemicals in steel drums were buried in the unused canal over a period of several years. The land was sold and many homes were built on top of the site. Eventually the chemicals leaked out of the drums and into the soil, water, and air in the vicinity of the old dump site. Soon it was evident that residents in the area of Love Canal were suffering from unusually high rates of cancer, miscarriage, birth defects, kidney disease, and other illnesses. In incidents like this, it is difficult to tie a particular chemical to a specific health problem. But the fact that the noninfectious illnesses suffered by the residents of the area were associated with environmental contamination with chemical wastes is beyond question. Research is now being conducted by many universities and governmental agencies to determine some of the actual long-term effects of heavy metals and synthetic organic chemicals on human health.

Finally, there are several noninfectious diseases specifically associated with air pollution. Air pollution and its control will be discussed in a later chapter. Briefly, common diseases related to air pollution include *bronchial asthma, bronchitis,* and *emphysema. Lung cancer* also occurs more frequently in congested industrial and urban areas, and poor air quality is considered to play a role in this. Again, it is difficult to prove a direct cause-and-effect relationship between a specific pollutant and these illnesses. But the overall negative effect of dirty air on public health is obvious: The incidence of respiratory ailments and increased mortality rates is directly related to the severity of air pollution.

1.3 ECOLOGY

Ecology is a branch of biological science that is concerned with the relationships and interactions between living organisms and their physical surroundings or environment. Living organisms and the environment with which they exchange materials and energy together make up an *ecosystem,* which is the basic unit of ecology. An ecosystem includes *biotic components*—the living plants and animals—and *abiotic components*—the air, water, minerals, and soil that comprise the environment. A third and essential component of most natural ecosystems is *energy,* usually in the form of sunlight.

Familiar examples of land-based or *terrestrial* ecosystems include forests, deserts, jungles, and meadows. Water-based or *aquatic* ecosystems include streams, rivers, lakes, marshes, and estuaries. There is no specific limitation on the size or boundaries of an ecosystem. A small pond can be studied as a separate ecosystem, as can a desert comprising hundreds of square kilometers. Even the entire surface of the earth can be viewed as an ecosystem; the term *biosphere* is often used in this context.

If the earth is imagined to be about the size of an apple, then the layer of air surrounding it would not be much thicker than the skin of that apple. This thin envelope of air and the shallow crust of land and water just beneath it provide the abiotic components that support life in the biosphere. It is a "closed ecosystem" because there is essentially no transfer of material into or out of it. Only the constant flow of energy from the sun provides power to sustain the life cycles within the biosphere. Nutrients are continually recycled and reused.

The biosphere seems so big, it is sometimes difficult to believe that humans can affect or disrupt its natural balances. But global problems related to environmental pollution, such as *acid rain* and the *greenhouse effect,* are significant and must be controlled before irreversible environmental changes occur. These and other pollution problems will be discussed later in the text.

In addition to natural ecosystems, such as lakes or forests, there are several types of artificial ecosystems of particular importance in environmental technology. For example, one of the most common methods of wastewater treatment is based on a biological system called the *activated sludge process.* Briefly, this is an "engineered" ecosystem comprising a steel or concrete tank, a suspended population of microorganisms in wastewater, and a constant input of air. The microbes are the biotic component; the tank, wastewater, and air are the abiotic components. The system removes organic pollutants from the wastewater. This will be discussed in more detail in a later chapter.

FOOD CHAINS AND METABOLISM

There are two basic principles or "laws" of ecology: *one-way flow of energy* and *circulation of materials.*

Energy is the capacity to do work. It can be transformed from one form to another, such as from mechanical to electrical energy, or from energy in the form of sunlight to potential energy stored in food molecules. But it cannot be created or destroyed. No energy transformation is 100-percent

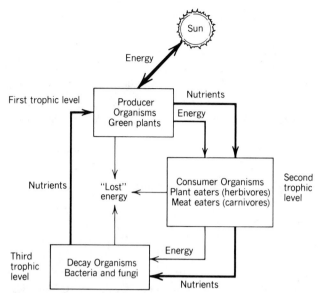

FIGURE 1.3 A simplified diagram of the "food chain." Nutrients are recycled, but energy must be continuously supplied by the sun. The efficiency of energy transfer from one trophic level to the next is less than 10 percent.

phosphorus, potassium, iodine, nitrogen, sulfur, calcium, iron, magnesium, as well as other elements in smaller amounts. For animals, some of these elements must be in the form of organic molecules such as carbohydrates or proteins. (A brief review of basic inorganic and organic chemistry is given in Chapter 4.)

Photosynthesis and Respiration

The food chain shown schematically in Figure 1.3 begins with what ecologists call the first *trophic level* of organisms—the producers. These are the green plants. Green plants are *autotrophic,* which simply means that they are self-nourishing. They have the unique ability to convert carbon dioxide, water, and some basic nutrients into organic compounds that store the sun's energy.

This natural process, called *photosynthesis,* is illustrated in Figure 1.4. The plants utilize solar energy to form carbohydrates from carbon dioxide and water. The carbohydrates can also combine with nitrogen, phosphorus, sulfur, and other elements, forming other organic compounds that are the building blocks of living organisms. *Chlorophyll,* the pigment that gives plants their characteristic green color, plays a key role in trapping solar energy and converting it into chemical energy. The plants manufacture more organic material than they need for their own sustenance. The excess "energy-rich" organic compounds stored in the plant tissue are then available for use by other organisms in the next trophic level that consume the plants.

During the process of photosynthesis, gaseous

efficient; some is always lost to the environment. Because of this, energy cannot be "recycled" in an ecosystem; it can only "flow" one way.

On the other hand, nutrient materials needed to sustain life can be reused over and over again. They are constantly recycled or circulated through the ecosystem. The one-way flow of energy and the circulation of nutrients is illustrated in Figure 1.3. This is a very simplified diagram of a *food chain,* showing three broad groups or types of organisms— the *producers,* the *consumers,* and the *decomposers.*

The biological and chemical process by which an organism sustains its life is called *metabolism.* Two fundamental metabolic processes of living organisms are *photosynthesis* and *respiration,* which will be discussed shortly. Living organisms require energy and, as shown in Figure 1.3, the original or primary source of energy for all natural ecosystems is the sun.

In addition to energy, living organisms need certain chemicals from the environment, called *nutrients,* in sufficient quantities. All organisms need water and most require gaseous oxygen. In addition, plants and animals require carbon, hydrogen,

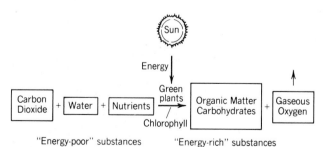

FIGURE 1.4 A schematic diagram of *photosynthesis.* Energy from the sun is stored in organic molecules and is available for use by the next trophic level.

oxygen is released into the atmosphere. Oxygen, of course, is essential for the metabolism of the next trophic level in the food chain—the consumers. Actually, the consumer organisms include several intermediate trophic levels, including the *herbivores,* the *carnivores,* and the *omnivores.* Herbivores are plant-eating animals, carnivores are meat-eaters, and omnivores eat both plants and animals.

The consumer organisms are *heterotrophic.* Unlike the autotrophic plants which manufacture their own food from simple inorganic chemicals, the herbivores must utilize the energy-rich compounds synthesized by the plants. In turn, the carnivores obtain energy for their metabolism when they consume the herbivores. The process by which the consumers obtain energy from the organic material stored in the plants and animals they eat is called *respiration.*

Respiration, illustrated in Figure 1.5, may be viewed as a process of slow combustion or *oxidation* of organic material, in which energy is released. Essentially, respiration is the opposite of photosynthesis. Photosynthesis builds energy-rich organic substances and gives off oxygen; respiration breaks down the organics and gives off carbon dioxide. Photosynthesis requires carbon dioxide, and respiration requires oxygen. This is an example of one of the most fundamental balances in nature.

The simplified food chain shown in Figure 1.3 is completed or closed by the decomposers, or *decay organisms.* These are primarily microscopic organisms such as bacteria and fungi. During their own metabolism, microorganisms break down the waste products and the remains of dead organisms into simpler inorganic substances that are then readily usable by the autotrophs. For example, nitrogen in ammonia is not available to plants as a nutrient until it is broken down and converted to inorganic nitrates by certain bacteria. The nitrates can be absorbed by the plants. Decomposers are essential not only for all natural ecosystems, they are the "work horses" of engineered water pollution control systems.

AEROBIC AND ANAEROBIC DECOMPOSITION

Decomposition that occurs in the presence of free oxygen is called *aerobic decomposition,* and the microorganisms that thrive in oxygen are called *aerobes.* Aerobic decomposition results in the oxidation of the carbon, hydrogen, sulfur, nitrogen, and phosphorus that are tied up in complex organic molecules. These elements become combined with oxygen, forming carbon dioxide, water, sulfates, nitrates, and other simple substances that can be taken up by green plants for photosynthesis. The energy released from the organic molecules in this process is used by the microbes for growth and reproduction. Aerobic decomposition is an efficient and "clean" biochemical process that does not produce the offensive odors often associated with decay.

Certain species of microorganisms are able to decompose organic material in the absence of freely available oxygen. These organisms are called *anaerobes,* and the process is called *anaerobic decomposition.* As illustrated in Figure 1.6, the end products of anaerobic decomposition include methane, ammonia, hydrogen sulfide, and volatile organic acids, many of which are responsible for the unpleasant odors associated with *putrefaction* (the anaerobic decay of proteins). Hydrogen sulfide, with the chemical formula H_2S, causes the familiar "rotten-egg" odor. (See Section 4.1 for a review of chemical symbols and formulas.)

Anaerobic decomposition is an inefficient biochemical process. Although the anaerobes get energy from it for their growth and reproduction, the end products are still relatively unstable and can decompose further. In effect, anaerobic decay is

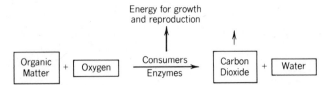

FIGURE 1.5 A schematic diagram of *respiration,* the opposite of photosynthesis. Organic matter is metabolized or "burned," thereby releasing the stored energy for use by consumer organisms. Enzymes are chemicals that help the metabolism reactions occur in the living cell.

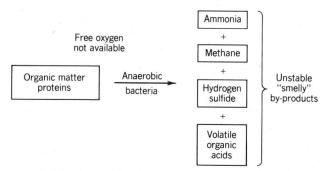

FIGURE 1.6 Anaerobic decomposition of proteins in the absence of free or molecular oxygen is called *putrefaction.*

similar to incomplete combustion. But it plays a key role in some wastewater treatment processes. Methane, CH_4, one of the few odorless products of anaerobic decomposition, has a high enough energy value to be useful as a fuel; it is collected for that purpose at some sewage treatment plants or sanitary landfills used for garbage disposal.

A type of anaerobic decomposition that is useful in producing certain foods and beverages is called *fermentation,* the decomposition of carbohydrates by microbes without free oxygen. Although it is used to produce cheese and alcohol, for example, some kinds of fermentation are not desirable, such as those that sour milk or produce acid in wine.

BIOGEOCHEMICAL CYCLES

We have already discussed that although an ecosystem needs a constant source of energy from outside, the nutrients upon which life depends can be recycled indefinitely. The pathways in which the chemical nutrients move through the biotic and abiotic components of the ecosystem are called *biogeochemical* or *nutrient cycles.*

One of particular concern in the field of environmental technology is the *hydrologic (water) cycle.* This will be discussed in some detail in a separate chapter. Three other important nutrient cycles will be considered here—the *carbon cycle,* the *nitrogen cycle,* and the *phosphorus cycle.*

Carbon, nitrogen, and phosphorus are considered to be among the *macronutrients* because they are

needed in relatively large amounts in *protoplasm,* the fundamental substance of which a living cell is made. Other macronutrients essential to life include hydrogen, oxygen, potassium, calcium, magnesium, and sulfur. A few of the many *micronutrients,* required only in very small quantities, are iron, manganese, copper, zinc, and sodium.

Carbon Cycle

Carbon dioxide, CO_2, in the air and dissolved in water, is the primary source of the element carbon. Through the process of photosynthesis, the carbon is removed from the CO_2 and incorporated with other chemical elements in complex organic molecules. The CO_2 eventually finds its way back into the atmosphere when the organics are broken down during respiration. A schematic diagram of this cycle is shown in Figure 1.7.

The combustion of fossil fuels (oil and gasoline) for energy is a human activity that tends to increase the concentration of CO_2 in the atmosphere. Carbon dioxide plays a role in absorbing radiated heat and in regulating global atmospheric temperatures. A rise in CO_2 levels in the atmosphere will

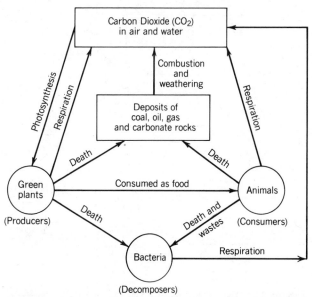

FIGURE 1.7 A simplified diagram of the carbon cycle. The arrows show the various directions of carbon transfer through the biosphere.

tend to cause the average temperature to increase. This problem, called the *greenhouse effect,* will be considered again in the chapter on air pollution.

Nitrogen Cycle

About 78 percent of the atmosphere is nitrogen gas, N_2, but in this molecular form it is not active in biological systems. The nitrogen must first be "fixed" in the form of nitrates, NO_3^-, in which form it can be utilized by plants during photosynthesis. Eventually it is combined with other substances and converted into proteins, consumed by heterotrophs, and broken down again in the process of decay. This cycle is illustrated in Figure 1.8. *Nitrification,* the process in which nitrogen in the form of ammonia, NH_3, is converted to nitrate nitrogen, is of particular significance in water pollution control.

Phosphorus Cycle

Phosphorus is another nutrient that plays a central role in aquatic ecosystems and water quality. Un-like carbon and nitrogen, which come primarily from the atmosphere, phosphorus occurs in large amounts as a mineral in phosphate rocks and enters the cycle from erosion and mining activities. As will be discussed in the chapter on water pollution, this is the nutrient considered to be the main cause of excessive growth of rooted and free-floating microscopic plants in lakes.

STABILITY, DIVERSITY, AND SUCCESSION

Each species of living organism occupies a particular habitat and serves a particular function in an ecosystem. The function and habitat comprise the organism's *ecological niche.* A basic characteristic of a healthy or well-balanced ecosystem is an overlapping of niches occupied by different species. The more complex the ecosystem is, in terms of the numbers and interrelationships among different species, the more stable it will be. A stable ecosystem can withstand some external stress, like pollution, construction, or hunting, without being completely disrupted or damaged.

In a stable ecosystem, if any one species disappears because of natural or artificial causes, other species are available to occupy its niche and take over its role in the food chain. Actually, the term *food web* is more appropriate for a healthy ecosystem because of the overlapping nature and complexity of the eat-and-be-eaten-by relationships. A jungle is a good example of a stable ecosystem because of the tremendous number of plant and animal species thriving in it. The loss of one species of tree or one species of animal is not likely to have a significant impact on the whole ecosystem.

In an ecosystem with little diversity, that is, only a few different species of organisms, the situation is more unstable and susceptible to the effects of stress. The disappearance of a group of organisms from the food web is more likely to break the chain of trophic levels and severely disrupt the ecosystem. Diversity of species, then, provides a factor-of-safety or buffer against ecological disruptions by increasing the likelihood of adaptation to changing environmental conditions. *The greater the diversity of species, the healthier is the ecosystem.*

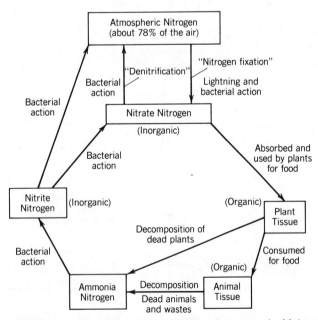

FIGURE 1.8 A simplified diagram of the nitrogen cycle. Molecular nitrogen must first be "fixed" (combined with oxygen) into the form of nitrate nitrogen before it can be used by plants as a nutrient.

Although aquatic ecosystems such as streams and lakes are generally stable, they are sensitive to disruption from human activity. A diagram of an aquatic ecosystem is shown in Figure 1.9. Most desirable organisms, from the fish down to the microscopic plankton and bacteria, need oxygen to survive.

As will be discussed later on, one of the effects of water pollution is the reduction of the dissolved oxygen level in the water. This type of pollution changes the ecological balance, favoring a smaller number of species of organisms which are tolerant of low oxygen levels. In heavily polluted water, only maggots and sludge worms may survive.

In studying the health or quality of a stream or lake, ecologists may use a formula to compute a *diversity index* for the ecosystem. In a field survey the number of different species is counted and the population of each species is estimated from sampling limited areas. These data are used in the diversity index formula, and a single number or index is determined to characterize the condition of the ecosystem.

Generally, a low diversity index is indicative of a polluted ecosystem, and the pollution-tolerant species are readily identified. In a clean stream, for example, many different species of fish may be found, including trout. But in a polluted stream, only a few species of more tolerant organisms, such as catfish, may be found.

It is important to realize that even healthy or well-balanced ecosystems change over time in a process called *natural succession.* For example, a lake will eventually become shallower as silt and organic material accumulate in bottom sediments. As time goes on, the lake will eventually turn into a marsh, and finally a meadow. These natural changes in a lake, called *eutrophication,* will be discussed in more detail in a later chapter. As we shall see, this natural process can be affected by human activity and pollution.

Although the lake, marsh, and meadow may be stable and healthy ecosystems during their individual "lifetimes," natural geological and biological processes will cause the succession from one stage to another. If geologic and weather conditions are suitable, the process of natural succession will continue until a *climax stage* is reached. For example, the meadow, once a lake, will eventually become a hardwood forest. Natural succession, though, takes

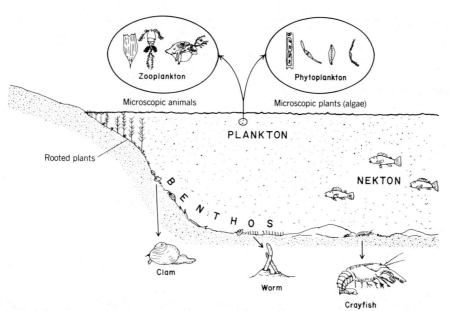

FIGURE 1.9 An aquatic ecosystem showing the various biological components of a fresh-water or marine habitat. (EPA)

place over very long periods of time, and the changes are not ordinarily visible during a human lifespan.

1.4 GEOLOGY AND SOILS

To a large extent, both liquid and solid wastes are disposed of on top of or below the ground surface. One of the most important concerns in environmental technology is the interaction between such waste materials and naturally occurring bodies of water. The protection of groundwater quality is of particular concern. Later in the text, the topics of groundwater and land disposal of wastes will be discussed. This will involve terminology related to soil types and behavior. Since soil consists of unconsolidated rock particles, it will be helpful for the student who has little or no background in the subject of geology or soils to first review the brief discussion in this section.

TYPES OF ROCK

Rocks are composed of inorganic substances called *minerals.* Some common minerals are quartz, mica, feldspar, calcite, magnetite, and kaolinite. The fundamental chemical elements making up these minerals include silicon, potassium, aluminum, calcium, iron, oxygen, and many others.

There are three major types of rocks: igneous, sedimentary, and metamorphic. *Igneous rocks,* which comprise most of the solid crust of the planet, have cooled and solidified from a hot molten state. *Granite,* composed primarily of the minerals quartz and feldspar, is a common type of igneous rock.

Even the hardest and most durable igneous rocks that are exposed at the earth's surface are subject to physical disintegration and chemical decay. Changes in the composition and structure of the rock are constantly occurring because of the action of wind, water, temperature changes, carbon dioxide, and oxygen. This gradual process is called

weathering. The solid rock made up of consolidated minerals is broken down into relatively small unconsolidated fragments called *soil.*

When the soil particles are moved by wind or water and deposited elsewhere, they form *sediments.* These sediments may be covered under additional deposits of material; eventually, they are compacted and consolidated under the load of overlying layers. With time, the minerals may recrystalize and the rock fragments can become cemented together, forming a second type of rock called *sedimentary rock. Sandstone* and *shale* are common sedimentary rocks. *Limestone,* consisting primarily of the mineral calcite (calcium carbonate, $CaCO_3$), is also a widely occurring sedimentary rock type.

Under conditions of excessive heat and pressure caused by environmental conditions, both the igneous and sedimentary rock types can be changed from their original forms and mineral structures. The newly formed rock is called *metamorphic rock. Marble* and *slate* are familiar examples of this third fundamental type of rock.

Both sedimentary and metamorphic rocks can again be subject to the weathering, transportation, and deposition process in a continual cycle of rock formation. This is illustrated in Figure 1.10.

The physical properties of rocks that are of primary interest in environmental technology include those that are related to the underground storage and flow of water. Two terms are of significance in this regard: *porosity* and *permeability.* Rock is not entirely solid through and through. Porosity refers

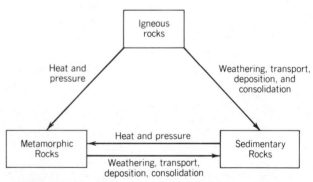

FIGURE 1.10 A schematic diagram of the "rock cycle."

to the percentage of total rock volume that is occupied by *voids* or *pore spaces*. Permeability refers to the ability of the rock to allow the flow of water through the pore spaces. It should be noted that all porous rocks are not necessarily highly permeable, particularly if the pore spaces are very small or are not interconnected.

Sedimentary rocks generally are porous and relatively permeable, whereas most igneous and metamorphic rocks are impermeable. Carbonate sedimentary rocks, such as limestone, are relatively soluble. In this type of rock, *solution cavities* may form as water slowly flows through the pore spaces and dissolves the mineral calcite. The solution cavities increase the rock's permeability even more.

Rock formations have characteristic structural features that may also affect permeability. For example, a feature called *layering* or *stratification* is usually present in sedimentary rocks. This is because the unconsolidated soil particles are deposited as sediments by water or wind in horizontal layers; these layers may differ in particle size or mineral composition. As a consequence, permeability is generally higher in the direction parallel to the layers in sedimentary rocks.

Rock masses can gradually bend and fold because of environmental changes and earth movements. When the rocks shatter and crack from excessive stresses, fracture lines or fissures are formed; these are called *joints*. Joints in which one side of the rock mass moves or is displaced relative to the other side, as happens during an earthquake, are called *faults*. Both joints and faults serve as convenient pathways for the flow of water. In soluble rocks like limestone, the joints can be enlarged by solution to form caves.

TYPES OF SOIL

Soil, the unconsolidated rock fragments formed from weathering, may be classified or grouped on the basis of *texture* or the size and shape of the soil particles. There are four major textural classifications of soil:

1. *Gravel* - Rock fragments between 2 and 75 mm in size.
2. *Sand* - Rock particles larger than 0.05 but less than 2 mm in size.
3. *Silt* - Fine, powderlike particles larger than 0.002 but less than 0.05 mm.
4. *Clay* - Very small particles, less than 0.002 mm in size.

The term *loam* is also used for a combination of silt and sand that also contains organic material, and that is suitable for the growth of plants or crops. Clay differs from gravel, sand, and silt not only in size, but in shape and mineral composition. Gravel, sand, and silt consist of relatively coarse-grained, bulky particles; the very finely divided clay particles are platelike in shape and have a strong affinity for water.

The permeability of soil decreases as the particle size decreases. Gravel and sand are porous and highly permeable, readily allowing the flow of water through the spaces between the soil grains. Silt is considerably less permeable because of the small particle and void size; and clay, although a porous material, is virtually impervious to the flow of water. This is basically because the water in the clay is held by molecular forces on the platelike clay particles. The impermeability of clay soil can be used to advantage in building a sanitary landfill for solid waste disposal, but it is a serious disadvantage for subsurface disposal of wastewater. This will be discussed in more detail in subsequent chapters.

In addition to *porosity* and *permeability,* other terms related to the flow of water in soil include *infiltration* and *percolation.* Infiltration refers to the penetration of the water through the ground surface layer of soil or rock, and percolation refers to the continuing movement or flow of the water through the pore spaces, under the force of gravity.

Generally, the so-called tight soils, which contain a significant percentage of silt and clay, have low infiltration and percolation rates. However, coarse textured soils containing mostly sand or gravel have high infiltration and percolation rates.

Tight soils are not suitable for on-site subsurface disposal of sewage. The percolation or "perc" test, used when designing on-site septic systems, will be discussed in the chapter on wastewater disposal.

Soil Gradation

Although soils can be classified according to particle size as either gravel, sand, silt, or clay, naturally occurring soil deposits are not usually found in such distinctive groupings. They are generally mixed, containing a variety of soil particle sizes. The distribution and percentage of different particle sizes in the soil is referred to as *soil gradation*.

Soil gradation can be easily determined mechanically, by separating the different soil particles in sieves or screens with varying sizes of mesh openings. The results of this mechanical analysis are usually expressed graphically, in a *gradation curve*. A typical gradation curve is shown in Figure 1.11 where the vertical axis shows the percentage of the soil particles that are finer than a specific particle size, and particle size is shown on the horizontal axis. This curve shows 60 percent of the soil is less than 0.7 mm in size and 10 percent of the soil is less than about 0.12 mm.

The particle size for which 10 percent of the soil is smaller is called the *effective size* of the soil because there is a reasonably consistent relationship between this size and soil permeability. The effective size, designated D_{10}, is used as a factor when specifying sand for filters used in water or waste-

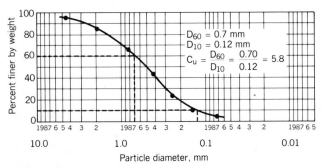

FIGURE 1.11 A typical soil gradation curve. The uniformity coefficient, C_u, can be used to classify the soil. (From McCarthy, *Essentials of Soil Mechanics and Foundations,* with permission of the Reston Publishing Company).

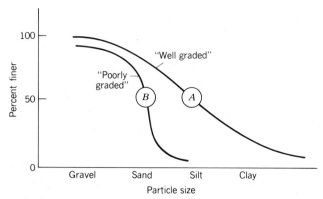

FIGURE 1.12 A well-graded soil (curve A) has lower porosity and permeability than a poorly graded soil (curve B).

water treatment systems. Another gradation factor, called the *uniformity coefficient,* is also used in this regard. The uniformity coefficient is the ratio of two particle sizes: the 60 percent finer size and the effective size, or D_{60}/D_{10}. The soil depicted in the gradation curve in Figure 1.11 has an effective size of 0.12 mm and a uniformity coefficient of 0.7 mm/0.12 mm = 5.8.

Mixed soils can be described as being well graded or poorly graded, depending on the distribution of particle sizes. Good gradation is represented by curve *A* in Figure 1.12. There is a wide variation of particle sizes in this kind of soil; the smaller particles fit into and fill up most of the pore spaces between the larger particles. As a result, a well-graded soil tends to have relatively low porosity and permeability. It is also resistant to erosion and scour.

Gradation curve *B* is typical of a poorly graded soil that consists primarily of soil grains in a very narrow range of sizes. In a soil of this type, the porosity and permeability are relatively high. This is because there are not enough small particles to fill the voids between the larger particles. Poorly graded soils generally have uniformity coefficients less than 10.

Mixed soils are ordinarily classified or grouped to help predict their behavior and characteristics. There are several classification systems used in soil mechanics technology, most of which are based upon sieve analysis and other laboratory tests. One of the simplest methods for classifying soil makes

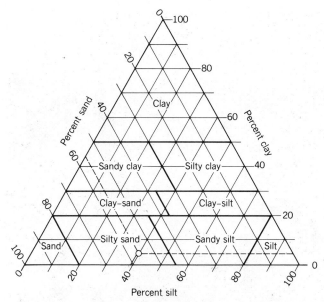

FIGURE 1.13 A triangular soil classification chart (From Hough, *Basic Soils Engineering,* with permission of the Ronald Press.)

use of a *triangular classification chart,* as illustrated in Figure 1.13.

The triangular classification chart relies only upon the relative amounts of sand, silt, and clay in the soil sample. For example, a soil that contained 25 percent sand, 50 percent silt, and 25 percent clay would be characterized as *clay–silt* on the chart. Generally, when the proportion of clay exceeds 20 percent of the total soil sample, the clay will tend to dominate the soil characteristics and behavior, and the primary soil type or designation will be clay. A classification scheme more common than the triangular chart, called the *unified system,* makes use of additional lab test data to classify the soil.

SOIL SURVEY MAPS

A group of related soils that have developed from similar "parent" rock formations is called a *soil series.* Soils within a specific soil series are essentially alike in all basic characteristics. They are designated on the basis of their textural classification and the name of the geographical location that is particularly representative of the soil type.

The Soil Conservation Service (SCS) of the United States Department of Agriculture has surveyed local areas of the country and has prepared countywide *soil survey maps.* Soil survey maps show the location of different soil series in an area and are readily available from most county SCS offices. A part of a typical SCS soil survey map is illustrated in Figure 1.14.

Soil survey maps are superimposed on top of aerial photographs of the area. In addition to the soil series distribution, they show lakes, roads, and other physical land features. They are particularly useful as an aid for good land-use planning and development. The letter symbols such as HnB and RrD designate the soil series. For example, HnB stands for "Hibernia very stony loam, 3- to 8-percent slopes." RrD stands for "Rockaway rock outcrop association, sloping and moderately steep." Hibernia and Rockaway are the geographic locations in Sussex County, New Jersey where these soils have been identified.

The slope of the ground is of importance in environmental planning and land development. Slopes expressed in percent (%) represent the change in elevation of the ground per 100 m or 100 ft. of horizontal distance. For example, a hill with a 20-percent slope changes 20 m in elevation in a distance of 100 m. This is illustrated in Figure 1.15.

Slopes less than 5 percent are usually considered to be "gentle," whereas those over 15 percent are considered to be steep. Steep slopes are much more susceptible to high rates of stormwater runoff and soil erosion than are gentle slopes. On the other hand, areas of very flat topography (0- to 2-percent slopes) may suffer from poor stormwater drainage and flooding problems. Modern municipal land-use ordinances may limit the extent of home building and other development allowed on steep slopes, and proper hydraulic design of storm drainage systems is important for the very flat areas.

The descriptive material in the SCS soil survey reports contains a good deal of information regarding soil characteristics and behavior. Data on depth to *bedrock* and depth to the *water table* are given. Bedrock is the unweathered rock formation underlying the surface soils. The water table represents

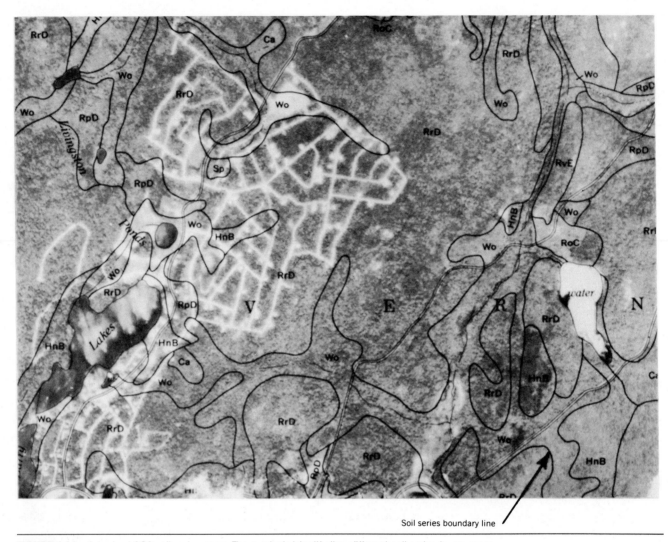

FIGURE 1.14 A typical SCS soil series map. The symbols identify the different soil series in an area. The characteristics of each soil series are described in the SCS publications. (Soil Conservation Service, USDA)

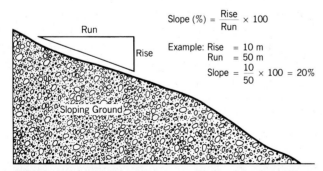

$$\text{Slope (\%)} = \frac{\text{Rise}}{\text{Run}} \times 100$$

Example: Rise = 10 m
         Run = 50 m

$$\text{Slope} = \frac{10}{50} \times 100 = 20\%$$

FIGURE 1.15 The slope of the ground surface is important in environmental planning. Slopes of over 15 percent are generally considered to be "steep."

the depth at which the soil is saturated with water, as will be discussed in more detail in a later chapter. Depth to bedrock and the water table are important with respect to designing solid or liquid waste disposal facilities. Other SCS data include soil index properties, gradation, permeability, erosion potential, and suitability of the land for various types of development. For example, the HnB series is described as having a seasonal high water table that severely limits use of the ground for a sanitary landfill or septic tank absorption fields.

1.5 HISTORICAL PERSPECTIVE

Piped water supply systems have been in existence for well over two thousand years. Archeological findings show that pipes were made out of hollow logs, clay, and other materials that could not withstand much in the way of pressure. Even the most sophisticated of the ancient systems, the masonry aqueducts and tunnels constructed by the early Romans (many of which are still standing), were not capable of carrying water under pressure.

Ancient Rome had sewers as well as aqueducts, but these were used primarily as storm drains rather than for sanitary disposal of human wastes. At that time in history there was no knowledge of the relationship between improper waste disposal and the spread of disease.

Not until the middle of the 1800s did people realize that there was a direct connection between contaminated drinking water and disease. The cause of a localized cholera epidemic in London was traced to a polluted community well, known as the Broad Street well. People who drew their water from the well were affected by the epidemic, whereas people with other water sources were not affected at all. When the well was taken out of service by simply removing the pump handle, the outbreak of cholera subsided.

The Broad Street well incident of 1849 pointed to the need for proper disposal of human wastes and for clean water supplies. Today, it is common knowledge that sewage carries "germs" that can spread disease, but at that time the germ theory of disease had not even been postulated. The only evidence of the need for sanitation was the coincidence of the dirty well water and the epidemic. It was not until the late 1800s that Robert Koch proved that the presence of certain microscopic living organisms caused disease in humans.

In the United States, a piped water supply system was used for the first time in 1801 when a steam-powered water works station was put into operation in Philadelphia. By the end of the 1800s, about 25 percent of all urban households in the country had running water. During the mid 1800s,

sewers began to replace cesspools for waste disposal. This was because of the rapid growth of urban populations and the provision of running water in individual homes. Construction of the first large sewage collection systems in the United States was completed in Chicago and Brooklyn in the late 1850s.

By the turn of the century, newly applied water purification techniques, mostly disinfection with chlorine, drastically reduced the incidence of typhoid and other waterborne diseases in the United States. The relationship between the number of water supplies being chlorinated and the decline of waterborne disease is illustrated in Figure 1.16.

Chlorine was used for the first time on a large scale for disinfecting the Jersey City, New Jersey water supply in 1908. Water filtration also plays an important role in disease prevention. The first water supply filters were built in Poughkeepsie, New York in 1872.

In the first half of the twentieth century, many public health authorities believed that "the solution to pollution is dilution." They felt that the best and most economical way to protect public health was to purify drinking water, not wastewater. Most sewage was discharged untreated into streams and

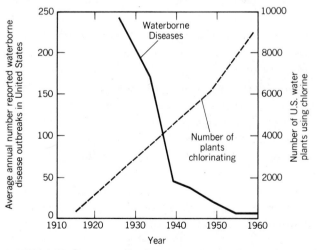

FIGURE 1.16 Fewer outbreaks of waterborne disease are the result of increasing use of chlorine for disinfecting water supplies. (Figure 6.1 from *New Concepts in Water Purification*, by Culp and Culp Copyright © 1971, by Litton Educational Pub. Co. Reprinted by permission of Van Nostrand Reinhold Company, Inc.)

rivers. The first large scale municipal sewage treatment plant was built in Milwaukee, Wisconsin, in 1919. But it was not until the late 1950s that it really became apparent that wastewater treatment was important for water pollution control and public health protection.

AN ERA OF ENVIRONMENTAL AWARENESS

In the 1960s, a broad awareness of environmental pollution problems developed among the general public. Many people came to realize the value and importance of protecting environmental quality. Clean air and clean water were worth rallying about, and public demonstrations were held. People wanted streams and lakes that could be used for swimming and fishing, as well as for safe drinking water supplies.

The words *ecosystem* and *biosphere* became popular buzzwords, and newspaper articles about local pollution problems became more common. Educational programs that focused on environmental issues were developed for grade school through the university level and grew in popularity.

In addition to stopping air and water pollution, solving problems related to garbage disposal, radiation, noise, pesticides, and wildlife preservation became important in the modern quest for environmental quality. Although infectious diseases like typhoid and cholera had virtually disappeared in the United States, people became aware of other types of problems caused by human and ecosystem exposure to industrial toxic chemical substances.

Silent Spring, a popular book by the scientist Rachel Carson, published in the early 1960s, focused attention on the environmental damage caused by improper use of pesticides. DDT and other chemicals used to protect agricultural crops were seriously disrupting the natural balance of ecosystems on a wide scale. After DDT was finally banned for most uses in the United States, the concentrations of this chemical found in human and animal tissues was observed to decline significantly, and several endangered species, such as the bald eagle, began to thrive once again.

The emergence of an environmental awareness on the part of the general public in the 1960s was apparently more than just a passing fad. It was a genuine concern that served to focus the attention of politicians, lawmakers, and governmental agencies on the need for an appropriate legal and regulatory framework for environmental quality control.

ENVIRONMENTAL LEGISLATION

In 1970 the *National Environmental Policy Act (NEPA)* was signed into law in the United States. NEPA established a national policy to "maintain conditions under which humans and nature can exist in productive harmony, and fulfill the social, economic, and other requirements of present and future generations of Americans." The concept of using *environmental impact statements (EIS)* as a planning tool to minimize harmful effects of land development and urban growth was developed for the first time under NEPA.

In addition to formally establishing a coherent nationwide policy on environmental issues, NEPA created the *President's Council on Environmental Quality (CEQ).* The CEQ, made up of experts in the various fields of environmental science and technology, advises the president on environmental concerns and makes recommendations for policy decisions. In addition to the CEQ, the 1970 act also formed the *Environmental Protection Agency (EPA),* consolidating into one independent federal department a means for establishing and enforcing pollution control and environmental quality standards.

The EPA is basically a regulatory agency, but it has other responsibilities in addition to establishing and enforcing environmental standards. It is a research organization with a scientific and technical staff that collects environmental data and studies the causes, effects, and methods of control for all types of pollution. The EPA also provides technical and financial assistance to state governments. Individual states have set up their own environmental protection agencies to provide similar assistance to county and municipal governments. Federal funds allocated to the states by the EPA have fueled the fight against water pollution and supported the largest public works construction effort in the United States.

Federal laws for environmental quality control began even before the environmental movement of the 1960s. But in the years since 1970, most of the old laws were amended and many new ones were passed. One of the first attempts to legislate clean water was the *Rivers and Harbors Refuse Act* of 1899. This law was intended to be applied primarily to control the dumping of debris that might obstruct navigation, but in recent times it has even been invoked against sewage pollution.

The following list of recent federal laws pertaining to environmental technology illustrates the widespread involvement of the government in this field. Together these laws provide a framework for guiding efforts to protect public health and to preserve environmental quality in the United States. Some are quite new, but most of them were written and passed in the 1970s, sometimes referred to as the "environmental decade." Many of them have been reauthorized and amended over a period of more than 10 or 15 years. A few will be discussed in more detail later on in the text.

National Environmental Policy Act

Clean Water Act

Clean Air Act

Safe Drinking Water Act

Toxic Substances Control Act

Resource Conservation and Recovery Act

Comprehensive Environmental Response, Compensation and Liability Act

Marine Protection, Research, and Sanctuaries Act

Occupational Safety and Health Act

Endangered Species Act

National Parks and Recreation Act

Surface Mining Control and Reclamation Act

Environmental Pesticide Control Act

Noise Control Act

Hazardous Materials Transportation Act

Coastal Zone Management Act

Many of these laws are commonly referred to with abbreviations, such as *CWA* for the Clean Water Act, *SDWA* for the Safe Drinking Water Act, *RCRA* (pronounced "recra") for the Resource Conservation and Recovery Act, *OSHA* (pronounced "ōsha") for the Occupational Safety and Health Act, and so on. The Marine Protection, Research, and Sanctuaries Act is sometimes referred to as the *Ocean Dumping Act.* The Comprehensive Environmental Response, Compensation, and Liability Act, often called the *Superfund Law,* is a recent piece of legislation to clean up the thousands of hazardous waste dumps in the nation.

The long list of environmental laws points to a commitment to improve and maintain the quality of life in America. Environmental awareness and action was not, however, confined solely to the United States in those years. For example, in 1972 a United Nations Conference on the Human Environment was held in Stockholm. International agreements were made to control ocean pollution from ships, to protect the ozone layer in the stratosphere, and to avoid complete destruction of whales and other wildlife. The problems of environmental quality were recognized to exist on a global scale rather than just as concerns for individual nations.

In 1981, the United Nations declared the 1980s to be the *International Drinking Water and Sanitation Decade.* The UN World Health Organization estimates that almost one third of the present world population is without safe drinking water. The goal is to reduce this number and to provide as much of the world as possible with adequate environmental sanitation and wholesome water supplies.

In the United States the environmental focus for the 1980s is primarily on the problems of hazardous waste disposal and groundwater pollution. About half of the nation's population relies upon groundwater for drinking supplies. Improper land disposal of hazardous chemical wastes has caused many public and private wells to become contaminated. In the past, the protection of groundwater quality did not receive as much attention as streams and lakes. Although there still is not any federal legislation that deals exclusively with the country's groundwater resources, many of the existing laws, including the CWA, SDWA, and RCRA, do provide rules and regulations to protect that important natural resource.

Another current environmental problem receiving much attention in the mid 1980s is referred to as *acid rain*. This is basically an air pollution problem related to the emission of sulfur and nitrogen oxides from coal-fired power plants and from automobiles. These oxides are converted to sulfuric and nitric acid in the atmosphere and reach the earth's surface, often hundreds of miles away from the source, in the form of precipitation. Many lake and forest ecosystems are believed to have been seriously damaged by acidic rain and snow. Research on this problem is still being conducted, and immediate action to alleviate the damage is strongly resisted by industry.

Unfortunately, the objectives of environmental protection often seem to be at odds with those of industrial growth and economic "progress." In the long run however, the "external diseconomies"

caused by environmental degradation will be far greater than the costs of regulation and control. This simply means that if our environment is damaged now in the interest of maximizing profits, we will have to "pay the piper" in the future, either in the form of expensive clean-up operations and increased health problems and medical costs, or in a generally lower quality of life with respect to our natural surroundings.

Although there is often controversy on just what steps to take and how much to spend for environmental protection, most people agree that the problems cannot be ignored altogether. A certain amount of governmental regulation will always be necessary. Eventually, a balance will be reached in which environmental, economic, energy, and social problems will be solved without one preempting or overshadowing the others. Meanwhile, there will be a need for technical personnel at all levels of training and education to plan, design, build, and operate environmental control systems.

REVIEW QUESTIONS

1. Give a brief definition of *environmental technology,* including mention of basic activities and objectives.

2. List at least fifteen technical factors or options related to the environmental aspects of a land development project.

3. Define the term *pathogen.*

4. Briefly discuss the modes of transmission of communicable disease.

5. List five common intestinal diseases. What is the most common mode of transmission of these diseases?

6. Briefly discuss the environmental aspects of noninfectious disease with reference to some specific illnesses.

7. What comprises an *ecosystem?* Give examples of five different types and sizes of ecosystems.

8. Make a sketch that illustrates the two basic principles of ecology.

9. Briefly describe two fundamental metabolic processes of living organisms.

10. What is the difference between a *heterotrophic* and an *autotrophic* organism?

11. What is the difference between *aerobic* and *anaerobic* decay?

12. What is the difference between *putrefaction* and *fermentation?*

13. Sketch a biogeochemical cycle for two different macronutrients.

14. What are *plankton,* and what role do they play in the aquatic food web?

15. Why is species diversity important for an ecosystem?

16. What is meant by *natural succession* of an ecosystem?

17. Briefly describe three types of rocks and give one example of each type. Briefly compare their permeabilities. What role do structural features such as layering, joints, and faults play in permeability?

18.   What is the difference between the terms *porosity* and *permeability?* What is the difference between *infiltration* and *percolation?*

19.   What is *soil?* List four basic types of soil and compare their permeability characteristics.

20.   What is meant by *soil gradation?* What is meant by the *effective size* and the *uniformity coefficient* of soil? Why is soil gradation important in environmental technology?

21.   What is the difference between a well-graded soil and a poorly graded soil? Illustrate the difference by sketching typical gradation curve shapes for each. If a soil sample had a uniformity coefficient of 50, would it be considered well-graded, or poorly graded?

22.   A soil sample is determined to contain 25 percent sand, 20 percent silt, and 55 percent clay. Using the triangular classification chart in Figure 1.13, how would you classify this soil? What do you think its permeability will be, high or low?

23.   What information does an SCS soil survey map convey, and what other information is given in SCS publications?

24.   During what period of history did people first begin to recognize the connection between contaminated drinking water and disease? About when was the Germ Theory of Disease proven?

25.   Approximately when were filtration and chlorination applied to drinking water supplies in the United States? What was the effect on public health?

26.   The first major challenge for environmental technology was the control of communicable disease. Has this challenge been met? How? What would you say is the present day challenge in this field?

27.   What major piece of environmental legislation was passed by Congress in 1970? What did it accomplish?

28.   List at least five other federal environmental laws. What do SDWA and RCRA stand for?

SELECTED REFERENCES

1. Salvato, J. A., *Environmental Engineering and Sanitation,* 2nd ed., Wiley-Interscience, New York, 1972.

2. Andrews, W. A., et al., *Environmental Pollution,* Prentice–Hall, Englewood Cliffs, N.J., 1972.

3. Steel, E. W., and T. J. McGhee, *Water Supply and Sewerage,* 5th ed., McGraw-Hill, New York, 1979.

4. Ehlers, U. M., and E. W. Steel, *Municipal and Rural Sanitation,* 6th ed., McGraw–Hill, New York, 1965.

5. Odum, E. P., *Ecology,* Holt, Rinehart & Winston, New York, 1963.

6. Turk, A., Turk, J., and Wittes, J. T., *Ecology, Pollution, Environment,* W. B. Saunders, Philadelphia, 1972.

7. Trefethen, J. M., *Geology for Engineers,* 2nd ed., D. Van Nostrand, New York, 1959.

8. Hough, B. K., *Basic Soils Engineering,* Ronald Press, New York, 1957.

9. Culp and Culp, *New Concepts in Water Purification,* Van Nostrand Rheinhold, New York, 1971.

2

HYDRAULICS

2.1 PRESSURE

HYDROSTATIC PRESSURE
PRESSURE HEAD
MEASUREMENT OF PRESSURE

2.2 FLOW

CONTINUITY OF FLOW
CONSERVATION OF ENERGY

2.3 FLOW IN PIPES UNDER PRESSURE

HAZEN–WILLIAMS EQUATION
HAZEN–WILLIAMS NOMOGRAPH
FLOW MEASUREMENT

2.4 GRAVITY FLOW IN PIPES

MANNING'S FORMULA
PARTIAL FLOW IN PIPES
FLOW MEASUREMENT

The study of water at rest and in motion is called *hydraulics*. Applied hydraulics is concerned primarily with the computation of flow rates, pressures, and forces in water or wastewater storage and conveyance systems. For practical purposes, water and wastewater are considered to be incompressible liquids, each with a unit weight of 62.4 pounds per cubic foot (62.4 lb/ft^3). The physical and hydraulic behavior of wastewater (sewage) is so similar to that of clean water that there is generally no difference in the design or analysis of systems involving those two liquids.

In the SI metric system, the term for weight (force due to gravity) is called a newton (N), and the unit weight of water is 9800 newtons per cubic meter (9800 N/m^3). More appropriately, this is expressed as 9.8 kilonewtons per cubic meter (9.8 kN/m^3), where the prefix *kilo* stands for 1000.

Hydraulics is a very important aspect of environmental technology. A knowledge of basic hydraulic principles is particularly necessary for technical personnel working on the design or analysis of water supply, drainage, and water pollution control systems. The purpose of this chapter is to present only the fundamental concepts of hydraulics, which will be necessary for an understanding of the environmental topics covered subsequently in the text. It can serve as a primer for students who have not yet been exposed to the subject. For those who have previously studied hydraulics or fluid mechanics in other courses, this chapter may be useful for a quick review.

2.1 PRESSURE

Water or wastewater exerts forces against the walls of its container, whether it is stored in a tank or flowing in a pipeline. We can also say that it exerts a *pressure*. But there is a difference between force and pressure, although they are closely related. Specifically, pressure is defined as a *force per unit area*. In equation form, this can be expressed as follows:

$$P = \frac{F}{A} \qquad (2\text{-}1)$$

where P = pressure
F = force
A = area over which the force is distributed

In American units, pressure is usually expressed in terms of pounds per square inch (lb/in.^{2} or psi). In SI units pressure is expressed in terms of newtons per square meter (N/m^2). For convenience, the unit N/m^2 is called a *pascal,* abbreviated as Pa. Since a pressure of 1 Pa is a relatively small pressure (1 Pa = 0.000 145 psi), the term *kilopascal* (kPa) will be used in most practical hydraulics applications. 1 kPa = 1000 Pa = 0.145 psi.

HYDROSTATIC PRESSURE

The pressure water at rest exerts is called *hydrostatic pressure*. Some very important principles always apply for hydrostatic pressure.

1. The pressure depends only on the depth of water above the point in question (not on the water surface area).

2. The pressure increases in direct proportion to the depth.

3. The pressure in a continuous volume of water is the same at all points that are at the same depth.

4. The pressure at any point in the water acts in all directions at the same magnitude.

Consider the two tanks connected by a horizontal pipe shown in Figure 2.1. The water surfaces in both tanks are at the same elevation. We can consider the pressure at the water surface to be equal to zero. Actually, there is some pressure at the free surface because of the weight of the column of air above. This pressure is called *atmospheric* or *barometric pressure*.

Atmospheric pressure at sea level is approximately 101 kPa or 14.7 psi. For most practical applications, we neglect the atmospheric pressure in hydraulic computations. In other words, we consider atmospheric pressure to be a zero reference or starting point. When we do this, we are working in terms of *gage pressure* as opposed to *absolute pressure*.

A total vacuum would have a pressure of absolute zero. Pressures less than atmospheric, but greater than absolute zero, are called *partial vacuums*. Partial vacuums expressed in terms of gage pressure would have a negative sign; absolute pressures are always positive. For example, an absolute pressure of 61 kPa (8.9 psi) is equivalent to a gage pressure of −40 kPa (−5.8 psi). This is illustrated in Figure 2.2.

The second principle of hydrostatic pressure listed is one that you can appreciate from personal experience if you ever dove under water in a pool or lake. You can feel the pressure on your body (especially your eardrums) increase as you descend deeper into the water. We can say that the pressure at point B near the bottom of tank 1 (Figure 2.1) is greater than the pressure at point A; likewise, we know that the pressure at point A is greater than zero. Actually, if point B was exactly twice as

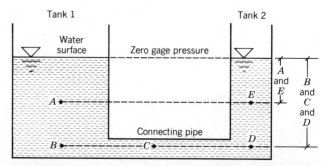

FIGURE 2.1 The pressure at point A equals the pressure at point E, since those points are at the same depth in the water. Likewise, the hydrostatic pressures at points B, C, and D are equivalent.

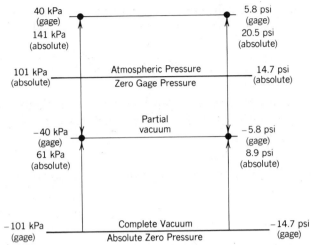

40 kPa
(gage)
141 kPa
(absolute)

5.8 psi
(gage)
20.5 psi
(absolute)

101 kPa
(absolute)

Atmospheric Pressure
Zero Gage Pressure

14.7 psi
(absolute)

Partial
vacuum

−40 kPa
(gage)
61 kPa
(absolute)

−5.8 psi
(gage)
8.9 psi
(absolute)

−101 kPa
(gage)

Complete Vacuum
Absolute Zero Pressure

−14.7 psi
(gage)

FIGURE 2.2 Pressure measured with reference to standard or normal atmospheric pressure is called *gage pressure.* Gage pressure can have a negative sign when it is less than atmospheric pressure. Absolute pressure is always positive.

deep as point A, the gage pressure at point B would be exactly twice the pressure at point A. This is because the pressure varies in direct proportion to the depth.

Consider point E in tank 2. Since point E is at the same depth below the water surface as point A, the pressure at point E is the same as the pressure at point A. It makes no difference that tank 2 is narrower than tank 1. Hydrostatic pressure depends only on the height of water above the points and not on the volume or surface area of the water.

Even though point C in the connecting pipe does not have water directly above it, it still has the same pressure as points B and D. This is in accordance with the third principle of hydrostatics listed. Another way of expressing this is to say that *pressure in a continuous fluid at rest is transmitted undiminished at the same depth throughout the fluid.*

Computation of Pressure

Consider the tank shown in Figure 2.3a, with a bottom area of 1 m^2. If the tank is filled with water to a depth of 1 m, the volume of water would be 1 m^3, and its weight would be 9.8 kN. The pressure at the bottom of the tank can be computed from Equation 2–1 as $P = F/A = 9.8$ kN/1 m^2 = 9.8 kN/m^2 = 9.8 kPa.

If the depth of water in the tank were increased to 2 m, the total weight of water would be 2 × 9.8 = 19.6 kN, and the pressure at the bottom would be $P = 19.6$ kN/1 m^2 = 19.6 kPa, as shown in Figure 2.3b. It can also be seen, in Figure 2.3c, that for a tank with a bottom area of 4 m^2 filled with water to a depth of 2 m, the bottom pressure is still 19.6 kPa. This is because the additional weight of water is spread over a proportionally greater area. This demonstrates one of the basic principles of hydrostatics: Pressure at a point in water depends only on the depth of water above the point. Expressed in the form of an equation, this becomes

$$P = 9.8 \times h \qquad (2\text{-}2a)$$

where P = hydrostatic pressure, kPa
h = water depth from surface, m

In American units, one cubic foot of water weighs 62.4 lb. Considering a 1-ft depth of water on an area of 1 ft^2, or 144 square in. the pressure at the bottom would be $P = 62.4$ lb/144 in.^{2} = 0.43 psi. Following the same reasoning as given earlier, we can say that in water

$$P = 0.43 \times h \qquad (2\text{-}2b)$$

where P = hydrostatic pressure, psi
h = water depth from surface, ft

The following examples illustrate the use of Equation 2-2.

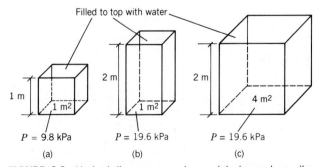

Filled to top with water

1 m   2 m   2 m
1 m^2 1 m^2 4 m^2
$P = 9.8$ kPa $P = 19.6$ kPa $P = 19.6$ kPa
(a)   (b)   (c)

FIGURE 2.3 Hydrostatic pressure at a point depends on the depth of water above the point, but not on the water surface area or volume. Note that the pressure at the bottom of tank (b) is the same as the pressure at the bottom of (c).

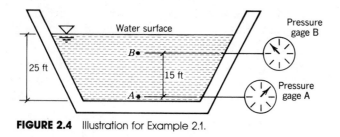

FIGURE 2.4 Illustration for Example 2.1.

EXAMPLE 2.1

The tank shown in Figure 2.4 has a total depth of 25 ft of water in it. What pressure would be recorded on gage A at the tank bottom? What pressure would gage B, at a height of 15 ft from the bottom, indicate?

Solution:

Using Equation 2-2b, we can compute the pressure at the tank bottom as follows:

$$P_A = 0.43 \times 25 = 11 \text{ psi} \qquad \text{(rounded to two significant figures)}$$

To compute the pressure that would be recorded by gage B, it is first necessary to determine the depth of point B below the water surface. This is $25 - 15 = 10$ ft. Do not use the 15-ft height above the tank bottom! The pressure is computed as

$$P = 0.43 \times 10 = 4.3 \text{ psi}$$

EXAMPLE 2.2

An elevated water storage tank and connecting pipeline are shown in Figure 2.5. Compute the hydrostatic pressures at points $A, B, C, D,$ and E.

Solution:

Point *A:* The depth of water above point A in the tank is equal to the difference in elevation between the water surface and the tank bottom, or $100.00 - 95.00 = 5.00$ m. Using Equation 2-2a, we get

$$P_A = 9.8 \times 5 = 49 \text{ kPa}$$

Point *B:* The total depth of water above point B is $100.00 - 70.00 = 30.00$ m. The pressure at that point is

$$P_B = 9.8 \times 30.00 = 290 \text{ kPa} \qquad \text{(rounded to two significant figures)}$$

Point *C:* The pressure at point C is equal to the pressure at point B because those points are at the same elevation.

$$P_C = P_B = 290 \text{ kPa}$$

Point *D:* The total depth of water above point D is $100.00 - 55.00 = 45.00$ m. The pressure at that point is

$$P_D = 9.8 \times 45.00 = 440 \text{ kPa} \qquad \text{(rounded to two significant figures)}$$

Point *E:* We do not have enough information to determine the pressure at point E because it is isolated from the system above it by the closed valve. Remember, pressures are transmitted only in a continuous fluid; the water at point E is not continuous with the water on the other side of the valve.

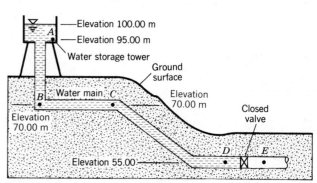

FIGURE 2.5 Illustration for Example 2.2.

PRESSURE HEAD

It is often convenient to express pressure in terms of the height of a column of water, in meters or feet, instead of in terms of kPa or psi. This *pressure head,* as it is called, is the actual, or equivalent, height of water above the point in question.

For example, in the tank shown in Figure 2.4, we could simply say that the pressure head at point A is 25 ft of water instead of 11 psi. Likewise, the pressure head at point B is 10 ft of water. In Figure 2.5, the pressure head at point D is 45 m of water. If a tall vertical tube was inserted into the pipe at point D, the water in the tube would rise exactly 45 m, to the original water surface elevation of 100.00 m.

In some instances, we might know the pressure in terms of kPa or psi, but would like to use units of pressure head instead. This is done, for example, when evaluating water distribution systems. Rearranging the terms in Equation 2-2, we get:

$$h = \frac{P}{9.8} = 0.1 \times P \qquad (2\text{-}3a)$$

or

$$h = \frac{P}{0.43} = 2.3 \times P \qquad (2\text{-}3b)$$

where Equation 2-3a is for SI units and Equation 2-3b is for American units.

EXAMPLE 2.3

A pressure gage on an open tank of water at a point 5 ft above the tank bottom registers a pressure of 35 psi. What is the pressure head at that point? What is the total depth of water in the tank?

Solution:

Using Equation 2-3b, we get

$$h = 2.3 \times 35 = 81 \text{ ft} \qquad \text{(rounded to two significant figures)}$$

The pressure head is equivalent to the depth of water above the gage. The total depth of water in the tank, therefore, is $81 + 5 = 86$ ft.

EXAMPLE 2.4

A sealed tank, shown in Figure 2.6, has a pocket of air trapped above the water, which is 1 m deep. A pressure gage at the bottom of the tank reads 30 kPa. Determine (a) the pressure head of water at the tank bottom, (b) the height that water would rise in the vertical tube if the valve is opened, and (c) the pressure in the trapped air.

Solution:

(a) Using Equation 2-3a, we can compute the pressure head as

$$h = 0.10 \times 30 = 3 \text{ m}$$

Notice that the pressure head is more than the depth of water in the tank. This means that the air in the tank must be exerting additional pressure, pushing downward on the water.

(b) The water would rise 3 m in the vertical tube, a height equal to the pressure head at the bottom of the tank.

(c) If the tank was open to the atmosphere, then 1 m of water depth would cause a pressure of only 9.8 kPa to register on the gage. The difference, or

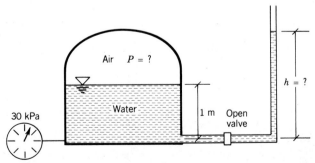

FIGURE 2.6 Illustration for Example 2.4. When the valve is open, the water will rise in the vertical pipe to a certain height h that depends on the pressure in the tank.

30 − 9.8 = 20 kPa (rounded to two significant figures), must be exerted by the pressurized air in the sealed tank. Therefore we can say the air pressure in the tank is 20 kPa. (The pressure in a small volume of gas is considered to be uniform, and does not depend on the height or depth of gas.)

MEASUREMENT OF PRESSURE

Pressure measurement is important in the operation of environmental control facilities. The operating pressure of pumps in water and sewage treatment plants must be monitored, and the pressures throughout a water distribution system must be determined to ensure adequate service. Often pressure measurements are made and recorded automatically with electromechanical instrumentation. But it is sometimes necessary for technical personnel to measure pressures in the field with other devices.

The simplest way to determine pressure is to use a *piezometer tube*. For example, if a narrow transparent tube were attached to a pipeline under pressure, as shown in Figure 2.7, the water in the pipe would rise in the tube until the pressure head caused by the column of water was equal to the pressure in the pipe. By measuring the height of the column in meters or feet and using a simple computation (Equation 2-2) one can find the pressure, in kPa or psi.

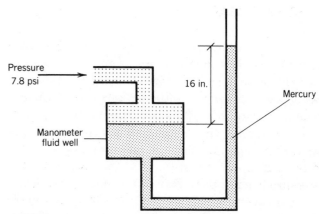

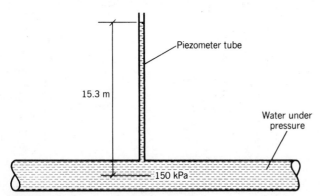

FIGURE 2.7 A *piezometer tube* offers a simple means for determining pressure by direct measurement of the corresponding pressure head.

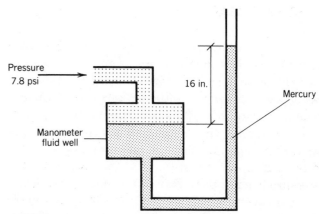

FIGURE 2.8 A well-type mercury manometer is more practical than a piezometer tube for measuring pressures in most hydraulic systems.

Although they are simple, piezometer tubes are not very practical for field use. As seen in Figure 2.7, for a pressure of 150 kPa (22 psi), the piezometer tube would have to be about 15 m (50 ft) high. They are used primarily in laboratory situations to measure very low pressures. In Section 3.7, on groundwater, we use the term *piezometric surface* to indicate pressure in water that is confined under the ground.

The *manometer* is a somewhat more practical device for measuring pressure using the height of a column of liquid. But the liquid in the manometer tube is different from the liquid in the system being measured. A well-type manometer, using mercury as the manometer fluid, is illustrated in Figure 2.8. Mercury is a heavy metal that is liquid at room temperature; it is 13.6 times as heavy as water. In a well-type mercury manometer, the equivalent pressure head of water in the system is 13.6 times the measured height of the column of mercury. For example, if the column of mercury shown in Figure 2.8 is 16 in. high, the pressure in the system is $P = 13.6 \times 0.43 \times (16/12)$ ft = 7.8 psi.

One of the most commonly used pressure measuring devices is the *Bourdon tube gage*. It works on the principle that a flattened hollow metal tube, curved in the form of a spiral or circular arc, will tend to uncurl as pressure is applied inside the tube. As the tube uncurls, a pointer linked to it will indicate the pressure on a calibrated scale. This is

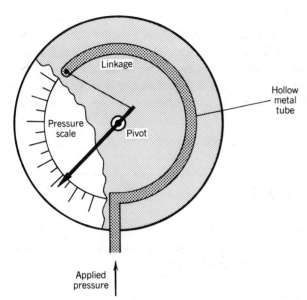

FIGURE 2.9 A simplified cutaway view of a Bourdon pressure gage; the internal hollow tube uncurls as pressure is applied, thereby moving a pointer on the scale.

illustrated schematically in Figure 2.9. Bourdon tubes are calibrated to indicate "gage" pressure, that is, pressure above atmospheric pressure, which is taken as 0 psi.

Pressure transducers, devices that sense changes in pressure and convert them to pneumatic or electrical signals, are installed in water and wastewater treatment plants and in pumping stations. They transmit signals to a central control panel where the operator can see the pressure readings conveniently displayed. One example of a pressure transducer is illustrated in Figure 2.10. In this device, a change in pressure causes the vertical cantilever

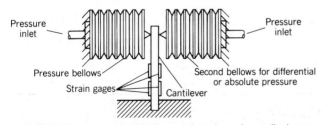

FIGURE 2.10 An example of a pressure transducer, that converts pressure into a proportional electrical signal. (From *Operation of Wastewater Treatment Plants*.) Manual of Practice No. 11, Water Pollution Control Federation, © 1976)

arm to bend. The displacement changes the wire resistance in attached strain gages, and an appropriate signal proportional to the pressure change is transmitted.

2.2 FLOW

Most applications of hydraulics in environmental technology involve water in motion—in pipes under pressure, or in open channels under the force of gravity. The volume of water flowing past any given point in the pipe or channel per unit time is called the *flow rate* or *discharge*.

In the SI system, the basic unit for flow rate is cubic meters per second (m^3/s). The term *liters per second* (L/s) is also used, particularly for relatively small flow rates. Another expression for flow rate is megaliters per day (ML/d). (*Note:* 1 m^3 = 1000 L and 1 ML = 10^6 L.)

In American units, flow rate may be expressed as cubic feet per second (ft^3/s or cfs), gallons per minute (gpm), or million gallons per day (mgd). An approximate but convenient conversion to remember is that 1 mgd = 1.55 cfs = 700 gpm.

EXAMPLE 2.5

Convert a flow of 50 m^3/s to its equivalent value expressed in terms of L/s and ML/d.

Solution:

50 m^3/s × 1000 L/m^3 = 50 000 L/s
50 000 L/s × 3600 s/h × 24 h/d × 1 ML/10^6L
= 4320 ML/d

EXAMPLE 2.6

Convert a flow rate of 50 ft^3/s to its equivalent value expressed as gpm and mgd.

Solution:

Using the conversion 1 ft^3 = 7.48 gal (see appendix), we get

50 ft^3/s × 7.48 gal/ft^3 × 60 s/min
$$= 22\ 440\ \text{gal/min or } 22\ 440\ \text{gpm},$$

and

22 440 gal/min × 60 min/hr × 24 hr/d
$$= 32.3 × 10^6\ \text{gal/d or } 32\ \text{mgd}$$

Alternate Solution: Using the approximate conversions and setting up appropriate ratios, we get

$$\frac{\text{flow}}{50\ \text{cfs}} = \frac{700\ \text{gpm}}{1.55\ \text{cfs}} \quad \text{and flow} = 50 × \frac{700}{1.55}$$
$$= 22\ 580\ \text{gpm}$$

To the nearest 500 gpm, this is for practical purposes the same as the first answer, or 22 500 gpm, and, for mgd we get

$$\frac{\text{flow}}{50\ \text{cfs}} = \frac{1\ \text{mgd}}{1.55\ \text{cfs}} \quad \text{and flow} = 50 × \frac{1}{1.55}$$
$$= 32\ \text{mgd}$$

Many students confuse flow rate with velocity of flow, but there is a distinct difference between those two terms. Flow rate represents volume per unit time, whereas velocity represents distance per unit time. There is a relationship among flow rate, flow velocity, and the flow area, expressed by the following formula:

$$Q = A × V \qquad (2\text{-}4)$$

where Q = flow rate or discharge
A = cross-sectional flow area
V = velocity of flow

In SI units, area A would be expressed in terms of m^2 and velocity V would be expressed in terms

of m/s, resulting in units of m^3/s for Q. In American units, V is usually expressed in terms of ft/s, therefore A should be in units of ft^2 and the units for Q would be ft^3/s. It is important to use the appropriate units in Equation 2-4 so that the results are dimensionally correct. The following examples illustrate the use of the basic flow equation, $Q = A × V$.

EXAMPLE 2.7

Water is flowing with an average velocity of 4.0 ft/s in an 18-in.-diameter storm drain. The pipe is flowing full. Compute the flow rate in cfs.

Solution:

Since the pipe is flowing full, the flow area is the same as the cross-sectional area of the pipe. The formula for the area of a circle is $A = \pi D^2/4$, where D is the diameter and π can be approximated as 3.14.

To keep units consistent in applying Equation 2-4, the area must be expressed in terms of ft^2. First convert D from inches to feet.

$$18\ \text{in.} × 1\ \text{ft}/12\ \text{in.} = 1.5\ \text{ft.}$$

Now compute the flow area.

$$A = \frac{(\pi × 1.5^2)}{4} = 1.77\ \text{ft}^2$$

Now, applying Equation 2-4, we get

$$Q = A × V = 1.77\ \text{ft}^2 × 4\ \text{ft/s}$$
$$= 7.1\ \text{ft}^3/\text{s or } 7.1\ \text{cfs}$$

EXAMPLE 2.8

Determine the required diameter of a pipe that will carry a discharge of 50 ML/d of water at a velocity of 3 m/s.

Solution:

We would be able to determine the pipe diameter D if we knew the required flow area A. Rearranging the terms in Equation 2-4 to solve for A, we get $A = Q/V$. In this problem, both Q and V are given, but we must be careful to use the proper units so that the equation is dimensionally correct. First convert the flow rate of 50 ML/d to an equivalent value in terms of m^3/s, as follows:

$$Q = 50 \times 10^6 \text{ L/d} \times 1 \text{ d/24 h} \times 1 \text{ h/3600 s}$$
$$\times 1 \text{ m}^3/1000 \text{L} = 0.58 \text{ m}^3/\text{s}$$

Now we can apply Equation 2-4 to get

$$A = \frac{0.58 \text{ m}^3/\text{s}}{3 \text{ m/s}} = 0.2 \text{ m}^2$$

Rearranging the terms in the formula $A = \pi D^2/4$, we get

$$D = \left[\frac{(4 \times A)}{\pi}\right]^{1/2} = \left[\frac{(4 \times 0.2)}{\pi}\right]^{1/2} = 0.5 \text{ m}$$

or

$$D = 0.5 \text{ m} \times 1000 \text{ mm/m} = 500 \text{ mm}$$

CONTINUITY OF FLOW

As previously mentioned, we consider water to be an incompressible fluid. In other words, its volume does not change significantly with changing pressure. This means that for a steady discharge in a pipe, the flow rate Q must be constant at any section in the pipe, no matter how the flow area or velocity may change.

Referring to Figure 2.11, we can say that the flow rate Q_1 at section 1 must equal the flow rate Q_2 at section 2, since water is neither added nor removed from the pipe between those two sections. But the path of flow is constricted at section 2 of the pipe. Common sense tells us that something

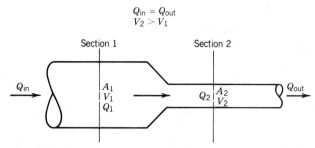

FIGURE 2.11 For an incompressible fluid, such as water or sewage, the volume flow rate Q is constant at any section of the pipeline. Since Q = A × V, when the flow area A is constricted, the velocity V of flow must increase.

must be happening to the water because of the reduced area of flow.

One of the things that is happening is that the flow velocity is increasing as the water moves from section 1 to section 2 of the pipe. Since Q is constant and $Q = A \times V$, when A gets smaller V must get larger; the product $A \times V$ must always equal Q. Conversely, if the area of flow increased, the velocity of flow would have to decrease.

This principle is sometimes referred to as *continuity of flow*. A common formula used to express this is

$$Q = A_1 \times V_1 = A_2 \times V_2 \qquad (2\text{-}5)$$

Equation 2-5 is sometimes called the *continuity equation*. The product of area and velocity is constant anywhere in the pipeline. The following examples illustrate the concept of continuity of flow.

EXAMPLE 2.9

In the pipeline shown in Figure 2.11, the area at section 1 is 0.5 m^2 and the area at section 2 is 0.25 m^2. For $Q_{in} = 1000$ L/s, determine the velocities at sections 1 and 2.

Solution:

First convert 1000 L/s to 1.0 m^3/s for use in Equation 2–5.

$$Q = A_1 \times V_1 = A_2 \times V_2$$
$$1 \text{ m}^3/\text{s} = 0.50 \times V_1 = 0.25 \times V_2$$

and

$$V_1 = \frac{(1 \text{ m}^3/\text{s})}{(0.50 \text{ m}^2)} = 2.0 \text{ m/s}$$

$$V_2 = \frac{(1 \text{ m}^3/\text{s})}{(0.25 \text{ m}^2)} = 4.0 \text{ m/s}$$

Note that although the area decreased by a factor of ½, the velocity increased by a factor of 2. The velocity is inversely proportional to the area. Also, the velocity is inversely proportional to the square of the diameter. If the diameter of a pipe was reduced by a factor of 3, for example, the velocity would increase by a factor of 3^2 or 9.

EXAMPLE 2.10

For the branching pipe section shown in Figure 2.12, compute the velocity of flow at section C in the 6-in. diameter branch. The velocity in the 12-in. branch at point A is 1.0 ft/s, and the velocity at point B in the 4-in. branch is 5.0 ft/s.

Solution:

Since water is incompressible, the total volume of flow entering the system at branch A must equal the total volume of flow leaving the system in branches B and C. This can be stated mathematically as $Q_A = Q_B + Q_C$.

First compute the cross-sectional flow areas using the unit of feet for diameter.

$$A_A = \frac{\pi(1 \text{ ft})^2}{4} = 0.785 \text{ ft}^2$$

$$A_B = \frac{\pi(1/3 \text{ ft})^2}{4} = 0.087 \text{ ft}^2$$

$$A_C = \frac{\pi(1/2 \text{ ft})^2}{4} = 0.196 \text{ ft}^2$$

Now we can compute the flow rate in branch A as

$$Q_A = A_A \times V_A = 0.785 \text{ ft}^2 \times 1.0 \text{ ft/s} = 0.79 \text{ ft}^3/\text{s}$$
$$Q_B = A_B \times V_B = 0.087 \text{ ft}^2 \times 5.0 \text{ ft/s} = 0.44 \text{ ft}^3/\text{s}$$

The flow rate in branch C is the difference between that in branch A and branch B, or $Q_C = Q_A - Q_B$.

$$Q_C = 0.79 \text{ cfs} - 0.44 \text{ cfs} = 0.35 \text{ cfs}$$

and

$$V_C = \frac{Q_C}{A_C} = \frac{0.35}{0.196} = 1.8 \text{ ft/s}$$

CONSERVATION OF ENERGY

It is a basic principle in physics that energy can neither be created nor destroyed, but it can be converted from one form to another. In a given closed system, the total energy is constant. This is the *law of conservation of energy.* Applied to problems involving the flow of water, it proves to be a most useful principle.

In hydraulic systems, there exist three forms of mechanical energy: potential energy due to elevation, potential energy due to pressure, and kinetic energy due to velocity. Energy has the units of foot pounds (ft-lb), or newton-meters (N-m). It is

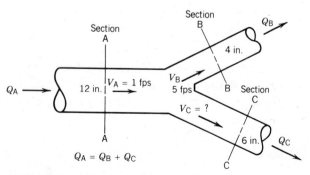

FIGURE 2.12 Illustration for Example 2.10.

convenient to express hydraulic energy in terms of *energy head,* in meters or feet of water. This is equivalent to foot-pounds per pound of water (ft-lb/lb = ft), or newton-meters per newton of water (N-m/N = m).

In a hydraulic system, then, there are elevation head, pressure head, and velocity head. (Pressure head has already been discussed in Section 2.1.) The total energy head in a hydraulic system is equal to the sum of these individual energy heads. This can be expressed mathematically as follows:

$$E = z + \frac{p}{w} + \frac{v^2}{2g} \qquad (2\text{-}6)$$

$$\text{TOTAL} \atop \text{HEAD} = \text{ELEVATION} \atop \text{HEAD} + \text{PRESSURE} \atop \text{HEAD} + \text{VELOCITY} \atop \text{HEAD}$$

where E = total energy head
  z = height of the water above a reference plane, m (ft)
  p = pressure, kPa (psi)
  w = unit weight of water, 9.8 kN/m^3 (62.4 lb/ft^3)
  v = flow velocity, m/s (ft/s)
  g = acceleration due to gravity, 9.8 m/s^2 (32.2 ft/s)^2

Consider the constricted section of pipe shown in Figure 2.13. From the law of energy conservation, we know that the total energy head at section 1, E_1, must equal the total energy head at section 2, E_2. Setting $E_1 = E_2$, and using Equation 2-6, we get

$$z_1 + \frac{p_1}{w} + \frac{v_1{}^2}{2g} = z_2 + \frac{p_2}{w} + \frac{v_2{}^2}{2g} \qquad (2\text{-}7)$$

This equation is called Bernoulli's equation, and is one of the most useful formulas in hydraulics. As written here, it applies to so-called ideal fluids because viscosity and energy loss due to friction are neglected.

For the system shown in Figure 2.13, we can simplify Bernoulli's equation because the pipeline is horizontal, and $z_1 = z_2$. Since they are equal, the elevation heads cancel out from both sides, leaving

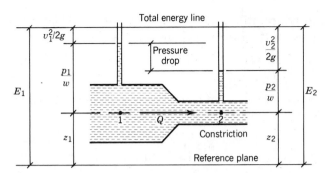

FIGURE 2.13 Since the velocity and kinetic energy of the water flowing in the constricted section must increase, the potential energy must decrease (from the law of energy conservation). This is observed as a pressure drop in the constriction.

$$\frac{p_1}{w} + \frac{v_1{}^2}{2g} = \frac{p_2}{w} + \frac{v_2{}^2}{2g} \qquad (2\text{-}8)$$

Now consider what happens as water passes through the constricted section of the pipe, section 2. We know from continuity of flow that the velocity at section 2 must be greater than the velocity at section 1, because of the smaller flow area at section 2. This means that the velocity head in the system increases as the water flows into the constricted section.

But the total energy must remain constant. For this to happen, the pressure head, and therefore the pressure, must drop. In effect, pressure energy is converted into kinetic energy in the constriction. The fact that the pressure in the narrower pipe section is less than the pressure in the bigger section contradicts what many beginning students often "feel" about the system. But it follows logically from continuity of flow and conservation of energy. As we shall see in the next section, the fact that there is a pressure difference will allow measurement of flow rate in the closed pipe.

EXAMPLE 2.11

For the system illustrated in Figure 2.13, the diameter at section 1 is 12 in. and at section 2 it is 4 in. The flow rate through the pipe is 2.0 cfs and the pressure at section 1 is 100 psi. What is the pressure in the constriction at Section 2?

Solution:

First compute the flow area at each section, as follows:

$$A_1 = \frac{\pi(1 \text{ ft})^2}{4} = 0.785 \text{ ft}^2$$

$$A_2 = \frac{\pi(0.333 \text{ ft})^2}{4} = 0.087 \text{ ft}^2$$

Now from $Q = A \times V$ or $V = Q/A$, we get

$$V_1 = \frac{2.0 \text{ ft}^3/\text{sec}}{0.785 \text{ ft}^2} = 2.5 \text{ ft/s}$$

$$V_2 = \frac{2.0}{0.087} = 23 \text{ ft/s}$$

Now applying Equation 2-8, we get

$$\frac{100 \times 144}{62.4} + \frac{2.5^2}{2 \times 32.2} = \frac{p_2 \times 144}{62.4} + \frac{23^2}{2 \times 32.2}$$

Note that the pressures are multiplied by 144 in.^{2}/ft^2 to convert from psi to lb/ft^2 to be consistent with the units for w; the energy head terms are in feet of head. Continuing, we get

$$231 + 0.1 = 2.3p_2 + 8.2$$

and

$$p_2 = \frac{231.1 - 8.2}{2.3} = \frac{222.9}{2.3} = 97 \text{ psi}$$

2.3 FLOW IN PIPES UNDER PRESSURE

When water flows in a pipeline, there is friction acting between the flowing water and the pipe wall, and between layers of the water moving at different

velocities in the pipe. This is because of the *viscosity* of the water. The flow velocity is actually zero at the pipe wall and maximum along the center line of the pipe. When we use the term *velocity of flow* in this text, we really mean the average velocity over the cross-section of flow.

The frictional resistance to flow causes a loss of energy in the system. This loss of energy is manifested as a continuous pressure drop along the path of flow. It is often necessary to be able to compute the expected pressure drop in a given system, or to design a new system with a specified maximum pressure loss.

In Figure 2.14a, a straight section of pipe filled with water under pressure is shown attached to a tank. There is no flow in the system and therefore no pressure loss when the valve in the pipe is closed. It can be seen that the pressure head at section 1 equals the pressure head at section 2.

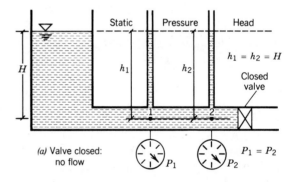

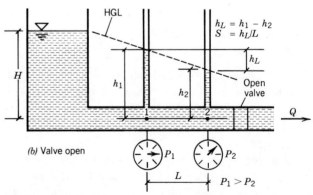

FIGURE 2.14   The hydraulic grade line, or HGL, is a graph of the pressure head above the pipe centerline. Its downward slope in the direction of flow shows pressure loss due to friction.

When the valve is opened, flow begins to occur with corresponding energy loss due to friction. We can see this loss by measuring the pressures along the pipeline. In Figure 2.14*b,* the difference in pressure heads between sections 1 and 2 can be seen in the piezometer tubes attached to the pipe. A line connecting the water surface in the tank with the water levels at sections 1 and 2 shows the pattern of continuous pressure loss along the pipeline. This is called the *hydraulic grade line (HGL)* or *hydraulic gradient* of the system. It is a very useful graphical aid when analyzing pipe flow problems.

The HGL is actually a graph of the pressure head along the pipe, plotted above the pipe centerline. It is not necessary to draw the piezometer tubes, as in Figure 2.14. *The HGL always slopes downward in the direction of flow,* unless additional energy is added to the system by a pump. The vertical drop in the HGL between two sections separated by a distance L is called the *head loss (h_L).* The ratio of h_L to L is the slope *(S)* of the HGL or *hydraulic gradient.* In equation form, $S = h_L/L$.

The HGL always passes through the free water surface of any storage tank in the system, since that elevation is equivalent to the system's pressure head at that point. The greater the flow rate is in a given pipeline, the greater the rate of pressure loss is, and the steeper the slope of the HGL is.

HAZEN–WILLIAMS EQUATION

To be able to design new water distribution pipelines or sewage force mains, or to analyze existing pipe networks, we must be able to compute head losses, pressures, and flows throughout the system. There are several formulas in hydraulics to do this, but one of those most commonly used is the Hazen–Williams equation:

$$Q = 0.28 \times C \times D^{2.63} \times S^{0.54} \quad (2\text{-}9)$$

where Q = flow rate, m³/s or gpm
C = pipe roughness coefficient
D = pipe diameter, m (in.)
S = slope of HGL, dimensionless

With the constant 0.28, this equation can be used with sufficient accuracy for both SI units and American units, as long as the appropriate units noted here for Q and D are used. For practical purposes, computed values may be rounded off to two significant figures.

HAZEN–WILLIAMS NOMOGRAPH

In actual practice, charts, tables, or graphs that "solve" the Hazen–Williams equation are generally used. The *nomograph* shown in Figure 2.15 is an example of one such commonly used design aid. A nomograph is a specially prepared chart that is set up to help solve a specific equation; a straightedge is used to line up data and intersect the axes at appropriate values.

The Hazen–Williams nomograph presented in Figure 2.15 has been prepared for a pipe roughness coefficient of 100. A value of $C = 100$ could represent the friction of an unlined iron pipe that is about 20 years old. Most designers would use this value when preparing plans for a new unlined pipeline, to account for the inevitable aging and deterioration of the pipe. Newer pipes would be smoother and allow more flow ($C > 100$); older pipes would offer more resistance to flow ($C < 100$). Concrete pipes generally have a value of $C = 130$, and smooth plastic pipes may have a C value as high as 150. In this text, all problems will have an assumed value $C = 100$.

In order to use the Hazen–Williams nomograph, two out of the three variables *(Q, D,* and *S)* must be known; the unknown variable can then be determined. A straightedge is placed across the axes so that the two known variables are intersected by a straight line. The "solution" for the third variable is found where the line crosses its corresponding nomograph axis. The following examples illustrate this procedure.

EXAMPLE 2.12

A 12-in.-diameter pipe carries water with a head loss of 10 ft per 1000 ft of pipeline. Determine the

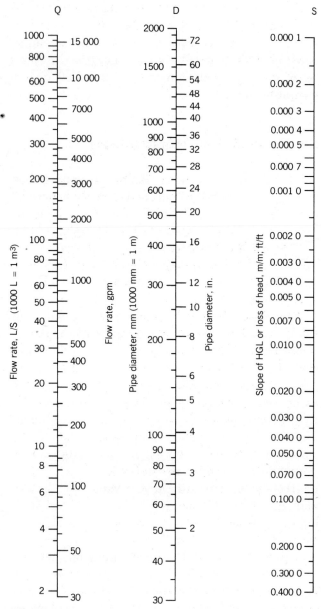

FIGURE 2.15 A nomograph that provides a graphical solution to the Hazen–Williams equation, for water flowing in circular pipes under pressure, with C = 100.

flow rate in the pipe using the nomograph in Figure 2.15. Check the solution using Equation 2-9.

Solution:

Compute the value of $S = h_L/L = 10$ ft/1000 ft = 0.010.

Place a straightedge across the nomograph intersecting 12 in. on the D line and 0.010 on the S line. Read a flow rate of 1600 gpm on the Q line.

Applying Equation 2-9, we get:

$$Q = 0.28 \times 100 \times 12^{2.63} \times 0.010^{0.54} = 1605 \text{ gpm}$$

This checks very well with the nomograph solution.

EXAMPLE 2.13

An 8-in.-diameter pipe carries a flow of 1.0 cfs. Compute the pressure drop per mile of pipeline, in psi.

Solution:

First, it is necessary to convert the flow to units of gpm in order to use the nomograph.

$$1.0 \text{ ft}^3/\text{s} \times 7.48 \text{ gal/ft}^3 \times 60 \text{ s/min} = 450 \text{ gpm}$$

Using a straightedge to connect $Q = 450$ gpm and $D = 8$ in. on the nomograph, we read a corresponding value of 0.0064 on the S axis.

Since $S = h_L/L$, we also can write $h_L = S \times L$. In this problem, L is specified to be 1 mile or 5280 ft. Therefore we get

$$h_L = 0.0064 \times 5280 = 34 \text{ ft}$$

In other words, the HGL would drop a vertical distance of 34 ft over each mile of pipeline. To convert this pressure head loss to its equivalent value in terms of psi, we must use Equation 2-2b:

$$\text{pressure drop } P = 0.43 \times h_L = 0.43 \times 34$$
$$= 15 \text{ psi per mile}$$

EXAMPLE 2.14

What minimum pipe diameter is required to carry a flow of 30 L/s without causing the pressure to drop more than 10 kPa per kilometer of pipeline?

Solution:

Convert the pressure drop to an equivalent pressure head, using Equation 2-3a, as follows:

$$h_L = 0.10 \times P = 0.10 \times 10 = 1.0 \text{ m/km}$$

Since $S = h_L/L$ and $L = 1$ km $= 1000$ m, we get $S = 1.0$ m/1000 m $= 0.001$.

Entering the nomograph with $Q = 30$ L/s and $S = 0.001$, we read 305 mm on the D or diameter axis.

The energy loss due to viscosity and friction along the straight length of the pipeline accounts for most of the pressure drop. This loss is called the *major loss*. As the water flows through valves, bends, and other pipe fittings, there are additional losses due to turbulence. These losses are called *minor losses*.

In large water distribution systems, the combined minor losses are very small compared to the major loss, and they are neglected. In pumping stations, where there may be many valves and bends in a confined space, the minor losses must be accounted for by using appropriate loss coefficients or equivalent length factors.

FLOW MEASUREMENT

The rate at which water is pumped into a distribution system or sewage is pumped in a force main must be known for proper control and operation of the system. One of the most common types of flow meters used to measure the discharge in a closed pipe under pressure is the *venturi meter*. This is actually a differential pressure meter; the flow rate is

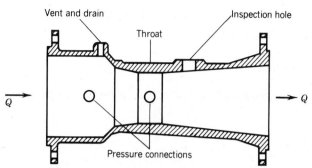

FIGURE 2.16 A *venturi meter* can be installed in a pipeline to measure the flow rate. The difference between the pressure in the throat of the venturi and in the upstream section can be converted to discharge, using the Bernoulli Equation. (From *Operation of Wastewater Treatment Plants*. Manual of Practice No. 11, Water Pollution Control Federation, © 1976)

related to the pressure difference caused by the meter, using the formulas from continuity of flow and the Bernoulli equation.

A section through a venturi meter is shown in Figure 2.16. As we have seen in the discussion on energy conservation in Section 2.2, the pressure in the constricted section, called the *throat* of the venturi tube, must be lower than the pressure just upstream of the converging section.

Using the continuity equation and Bernoulli's equation, the following formula can be derived to relate the discharge Q to the measured pressure difference $p_1 - p_2$.

$$Q = C \times A_2 \times \left[\frac{2g(p_1 - p_2)/w}{1 - (A_2/A_1)^2} \right]^{1/2} \quad (2\text{-}10)$$

In Equation 2-10, C is a discharge coefficient that accounts for a small amount of head loss in the venturi meter; it is usually about 0.98. All the other terms are as previously defined in the continuity and Bernoulli equations. Care must be taken to use appropriate units so that the equation is dimensionally consistent.

EXAMPLE 2.15

A venturi meter in a 100-mm pipe has a throat diameter of 50 mm. A pressure difference of 75 kPa is measured in the meter. What is the flow rate under these conditions?

Solution:

First compute the flow areas in terms of m², as follows:

$$A_1 = \frac{\pi \times (0.1 \text{ m})^2}{4} = 0.00785 \text{ m}^2$$

$$A_2 = \frac{\pi \times (0.05 \text{ m})^2}{4} = 0.00196 \text{ m}^2$$

The ratio $A_2/A_1 = 0.00196/0.00785 = 0.25$, and $1 - (0.25)^2 = 0.9375$. Applying Equation 2-10 we get

$$Q = 0.98 \times 0.00196 \times \left[\frac{2 \times 9.8 \times 75/9.8}{0.9375}\right]^{1/2}$$
$$= 0.024 \text{ m}^3/\text{s} = 24 \text{ L/s}$$

In most systems where flow rates must be monitored continuously, the pressure difference in the venturi tube is sensed by pressure transducers and the flow rate is recorded automatically on a rotating chart in the control room. Venturi meters are available in a wide range of sizes from several manufacturers. They must have the correct shape and proportions in the converging section, throat, and diverging section in order to maintain streamline flow and accurate measurements. Other pressure differential meters, such as orifice or nozzle meters, are available. They are shorter in overall length, but they obstruct the flow and cause greater head loss than the venturi.

Another device that is used to measure flow rates in closed pipelines is the *magnetic flow meter.* It has the advantage of not causing any constriction at all in the path of flow. The operating principle is based on the fact that water is a slight conductor of electricity, and when it moves through a magnetic field it induces a voltage. The meter produces the magnetic field around the pipe, and also senses the induced voltage. The greater the flow rate, the greater the voltage. The voltage signals are transmitted to a recording chart calibrated in units of flow rate.

A device called a *pitot tube,* illustrated schematically in Figure 2.17, can be used to measure the flow discharged by an open hydrant in a water distribution system. It consists basically of a tube open at both ends and bent so that one end can be pointed into the flowing water while the other end is vertical. The vertical part of the tube fills with water to a height that is proportional to the flow velocity. If the area of the hydrant opening is known, the discharge can be computed by a formula derived from the continuity and Bernoulli equations.

The common household water meter serves to record the total volume of water that passes through it. This provides a means for the water utility company to bill customers on the basis of actual water use instead of at a flat rate. Water conservation is encouraged when users must pay for actual metered consumption.

A common type of meter for small water service connections is a positive displacement or *nutating-disk meter.* In this device the water must pass through a small chamber of known volume. An inclined hard-rubber disk undergoes a wobbling ro-

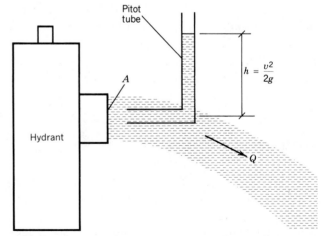

FIGURE 2.17 The *pilot tube* is a simple device which can be used to measure flow velocity. The velocity is proportional to the square root of the height of water in the tube.

tation as the water flows through the chamber. The number of rotations of the disk, which is proportional to the volume of water, is transmitted to a recording register. In modern installations, a digital register can be mounted outside the customer's house to facilitate meter reading by the utility company. Computerized billing systems based on automatic remote meter readings are also being used.

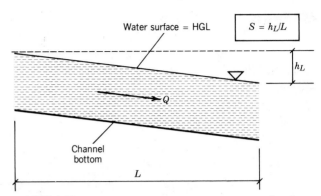

FIGURE 2.18 In *steady uniform open channel flow*, the slope of the water surface or HGL is equal to the slope of the channel bottom.

2.4 GRAVITY FLOW IN PIPES

When water flows in a pipe or channel with a "free surface" exposed to the atmosphere, it is called *open channel* or *gravity flow*. Gravity provides the moving force while friction resists the motion and causes energy loss. Stream or river flow is open channel flow. Flow in storm or sanitary sewers is also open channel flow, except when the water is pumped through a pipe under pressure (a *force main*).

In most routine problems in the design or analysis of storm or sanitary sewer systems, a condition called *steady uniform flow* is assumed. Steady flow means that the discharge is constant with time. Uniform flow means that the slope of the water surface and the cross-sectional flow area are also constant. A length of a stream, channel, or pipeline that has a relatively constant slope and cross-section may be called a *reach*.

Under steady uniform flow conditions, the slope of the water surface is the same as the slope of the channel bottom. The hydraulic grade line lies along the water surface, and as in pressure flow in pipes, the HGL slopes downward in the direction of flow. Energy loss is manifested as a drop in elevation of the water surface. A typical profile view of uniform steady flow is shown in Figure 2.18. The slope of the water surface represents the rate of energy loss. It may be expressed as the ratio of the drop in elevation of the surface in the reach to the length of the reach.

Typical cross-sections of open channel flow are

shown in Figure 2.19. In Figure 2.19a, the pipe is only partially filled with water and there is a free surface at atmospheric pressure. It is still open channel flow, even though the pipe is a "closed" conduit, under ground. The important factor is that gravity, not a pump, is moving the water.

The top of the inside pipe wall is called the *crown* and the bottom of the pipe wall is called the *invert*. One of the basic objectives of sewer design is to establish appropriate invert elevations along the pipeline. The length of wetted surface on the pipe or stream cross-section is called the *wetted perimeter*. The size of the channel, as well as its slope and wetted perimeter, are important factors related to its discharge capacity.

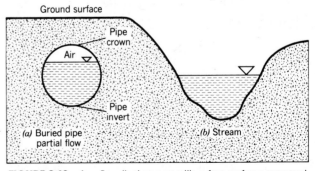

FIGURE 2.19 Any flow that occurs with a free surface exposed to atmospheric pressure is open channel flow, whether it occurs in a surface stream or in an underground pipe.

MANNING'S FORMULA

A common formula for solving open channel flow problems is called *Manning's formula,* written as follows:

$$Q = \frac{1.0 \text{ or } 1.5}{n} \times A \times R^{2/3} \times S^{1/2} \quad (2\text{-}11)$$

It is an empirical or experimentally derived equation,

where Q = channel discharge capacity, m³/s (ft³/s)

1.0 = a constant for SI metric units

1.5 = a constant for American units

n = channel roughness coefficient

A = cross-sectional flow area, m² (ft²)

R = hydraulic radius of the channel, m (ft)

S = slope of the channel bottom

The hydraulic radius of a channel is defined as the ratio of the flow area to the wetted perimeter P. In formula form, $R = A/P$. The roughness coefficient n depends on material and age for a pipe or lined channel and on topographic features for a natural stream bed. It can range from a value of 0.01 for a smooth clay pipe to 0.1 for a small natural stream. A value of n commonly assumed for concrete pipes or lined channels is 0.013.

The following example illustrates the application of Manning's formula for a channel with a rectangular cross-section.

EXAMPLE 2.16

A rectangular drainage channel is 3 ft wide and is lined with concrete, as illustrated in Figure 2.20. The bottom of the channel drops in elevation at a rate of 0.5 ft per 100 ft. What is the discharge in the channel when the depth of water is 1.5 ft? Assume $n = 0.013$.

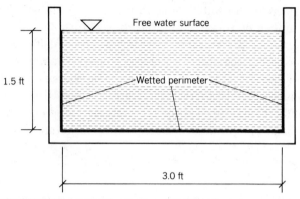

FIGURE 2.20   Illustration for Example 2.16.

Solution:

Since the data use American units, we will use the constant 1.5 in Equation 2-11. Referring to Figure 2.20, we see that the cross-sectional flow area $A = 3.0 \text{ ft} \times 1.5 \text{ ft} = 4.5 \text{ ft}^2$, and the wetted perimeter $P = 1.5 \text{ ft} + 3.0 \text{ ft} + 1.5 \text{ ft} = 6.0 \text{ ft}$. The hydraulic radius $R = A/P = 4.5 \text{ ft}^2/6.0 \text{ ft} = 0.75 \text{ ft}$. The slope $S = 0.5/100 = 0.005$.

Applying Manning's formula, we get:

$$Q = \frac{1.5}{0.013} \times 4.5 \times 0.75^{2/3} \times 0.005^{1/2} = 30 \text{ cfs}$$

(rounded off)

Circular Pipes Flowing Full

Most sanitary and storm sewer systems are built with sections of circular pipe. In addition to the roughness coefficient, the important factors related to the design or analysis of these pipelines are discharge Q, flow velocity V, pipe diameter D, and pipe slope S. Limitations on slope and pipe velocity will be discussed in Chapters 8 and 9.

In a circular pipe carrying water such that the pipe is just full to the crown (but still under atmospheric pressure and gravity flow), the flow area A would be $\pi D^2/4$, the area of the pipe. The wetted perimeter P is the perimeter of the pipe, or πD. Since hydraulic radius R is defined as A divided by P, we get

$$R = \frac{\dfrac{\pi D^2}{4}}{\pi D} = \frac{\pi D^2}{4} \times \frac{1}{\pi D} = \frac{D}{4}$$

For circular pipes flowing full, Manning's formula then takes the following form:

$$Q = \frac{1.0 \text{ or } 1.5}{n} \times \frac{\pi D^2}{4} \times \left(\frac{D}{4}\right)^{2/3} \times S^{1/2}$$

For a given value of n, only the pipe diameter and slope are needed to solve for discharge in a circular pipe flowing full. To facilitate the application of Manning's formula, particularly for routine problems with circular pipes, charts or nomographs are usually used. A nomograph for Manning's formula for circular pipes flowing full and $n = 0.013$ is shown in Figure 2.21. For pipes with values of n not equal to 0.013, simply multiply the Q and V values obtained from the nomograph by the ratio of 0.013 to the actual value of n.

To use the nomograph, two of the four variables (D, S, Q, and V) must be known. A straightedge lined up across the two known variables will intersect the "solution" for the other two variables on their respective axes. Applications of the Manning nomograph are illustrated in the following examples.

EXAMPLE 2.17

A 12-in.-diameter pipeline is built on a slope of 1 percent. Assuming $n = 0.013$, determine the discharge capacity of the pipeline with full flow. What is the flow velocity?

Solution:

Line up the values $D = 12$ and $S = 0.01$ on the nomograph in Figure 2.21. Read a value of 1580 gpm for flow rate or discharge and a value of 4.6 ft/s for velocity on the appropriate axes.

Using Manning's formula as a check, and to illustrate the accuracy of the nomograph, we get the following:

$$Q = \frac{1.5}{0.013} \times \frac{\pi \times 1^2}{4} \times \left(\frac{1}{4}\right)^{2/3} \times 0.01^{1/2} = 3.6 \text{ cfs}$$

$$Q = 3.6 \text{ ft}^3/\text{s} \times 7.48 \text{ gal/ft}^3 \times 60 \text{ s/min} = 1600 \text{ gpm}$$

and

$$V = \frac{Q}{A} = \frac{3.6}{(\pi \times 1^2)/4} = 4.6 \text{ ft/s}$$

These values check very well with the nomograph solution.

EXAMPLE 2.18

A 450-mm-diameter storm sewer is built on a grade of 2.0 percent. What is the discharge capacity and velocity of flow when the pipe is full?

Solution:

Using the Manning nomograph, line up 45 cm (1 cm = 10 mm) with 2.0 percent, or 0.02, on the appropriate axes. The nomograph solution is $Q = 0.4$ m³/s and $V = 2.6$ m/s.

Checking with Manning's formula, we get

$$Q = \frac{1.0}{0.013} \times \frac{\pi \times 0.45^2}{4} \times \left(\frac{0.45}{4}\right)^{2/3} \times 0.02^{1/2}$$
$$= 0.4 \text{ m}^3/\text{s}$$

and

$$V = \frac{Q}{A} = \frac{0.4}{(\pi \times 0.45^2/4)} = 2.5 \text{ m/s}$$

EXAMPLE 2.19

What diameter pipe is needed to carry a peak flow of at least 500 L/s on a 0.25-percent grade?

Manning Formula
Pipe Flow Chart
American/SI units

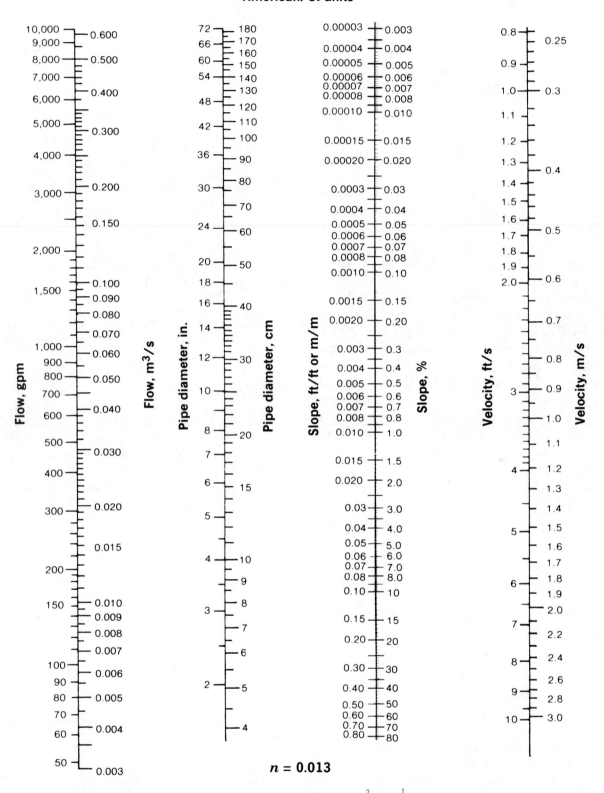

Chart based on the formula $Q = \dfrac{1.0 \text{ or } 1.5}{n} \times AR^{\frac{2}{3}} \times S^{\frac{1}{2}}$ for pipe flowing full.

FIGURE 2.21 *Manning's nomograph* for circular pipes flowing full, with *n* = 0.013. Manning's equation is used for open channel or gravity flow, while the Hazen–Williams equation is used for flow under pressure. (Reprinted with permission from the U.S. Pipe and Foundry Company and Scranton Gillette Communications, Inc.)

Solution:

First convert the flow to cubic meters per second.

$$Q = 500 \text{ L/s} \times 1 \text{ m}^3/1000 \text{ L} = 0.5 \text{ m}^3/\text{s}$$

Line up $Q = 0.5$ m³/s and $S = 0.25$ percent or 0.0025 on the appropriate axes on the nomograph in Figure 2.21. The solution read on the diameter axis is about 73 cm, or 730 mm. In practice, the next largest standard pipe size that is manufactured would be selected.

EXAMPLE 2.20

On what slope should a 16-in.-diameter sanitary sewer be built if it is to carry at least 750 gpm of sewage at a velocity not less than 2 ft/s?

Solution:

First line up $D = 16$ in. and $Q = 750$ gpm on the Manning nomograph, and read $S = 0.00048$. But at that slope the velocity would be about $V = 1.2$ ft/s, which is less than the minimum allowable velocity of 2 ft/s. It is necessary to increase the slope to increase the flow velocity.

Line up $D = 16$ and $V = 2$ on the nomograph. Read $S = 0.0014$ or 0.14 percent. Note that the actual discharge capacity of a full 16-in. pipe at that slope is $Q = 1250$ gpm. In this problem, the required minimum velocity is the controlling factor. If the flow rate is 750 gpm, the pipe would be flowing partially full.

PARTIAL FLOW IN PIPES

Most of the time, gravity sewers flow only partially full. The free water surface is usually below the crown of the pipe. This condition is depicted in Figure 2.19a. Partial flow hydraulics can be analyzed directly, using the Manning formula, but it

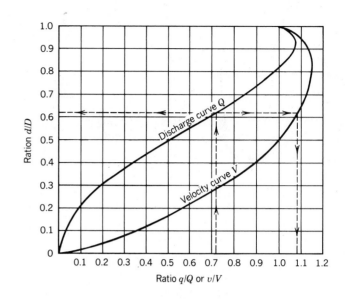

Nomenclature:
d = partial depth
D = full depth or pipe diameter
q = partial discharge
Q = full-flow discharge
v = velocity, partially full
V = velocity, full

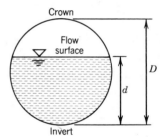

FIGURE 2.22 A *partial-flow diagram* for a circular pipe that carries flow with the water surface below the pipe crown.

is much more convenient to use the *partial flow diagram* shown in Figure 2.22. This diagram takes into account the variation of hydraulic radius with depth, which otherwise would require tedious computations. The following examples illustrate its use.

EXAMPLE 2.21

A 300-mm-diameter pipe is constructed on a slope $S = 0.02$. What would be the depth of flow in the pipe when it carries a flow of 96 L/s? What would be the velocity of flow at that depth?

Solution:

First, using the Manning nomograph, we find that a 300-mm or 30-cm pipe on a 0.02 slope would carry $Q = 0.135$ m^3/s $= 135$ L/s when flowing full (which is the condition the nomograph is set up for). Its full-flow velocity would be $V = 1.9$ m/s.

The discharge under partial flow conditions is $q = 96$ L/s, and we get the ratio of partial flow to full flow to be $q/Q = 96/135 = 0.71$.

Now enter the partial flow diagram, Figure 2.22, on the horizontal or *x*-axis with the value $q/Q = 0.71$. Move straight up to an intersection with the "Discharge curve Q"; from that point on the Q curve move horizontally to the left and read $d/D = 0.62$. Since $D = 300$ mm, we can solve for the partial depth d as follows:

$$d/300 = 0.62 \quad \text{and} \quad d = 300 \times 0.62 = 186 \text{ mm}$$

To compute the velocity at that depth of flow, reenter Figure 2.22 on the vertical or *y*-axis with $d/D = 0.62$ and move horizontally to the right to an intersection with the "Velocity curve V"; from that point on the V curve move straight down and read $v/V = 1.08$ on the horizontal axis. Since the full-flow velocity $V = 2$ m/s, we can compute the partial flow velocity as follows:

$$v/1.9 = 1.08 \quad \text{and} \quad v = 1.9 \times 1.08 = 2.1 \text{ m/s}$$

Notice that the partial-flow velocity is actually greater than the velocity under full-flow conditions. As can be seen from Figure 2.22, there is a range of depths at which discharge as well as velocity exceeds full depth values. The maximum velocity in the pipe occurs when the depth is 82 percent of the diameter, and the maximum discharge occurs when the depth is 93 percent of the diameter of the pipe. The basic reason for this is the reduction in friction at the pipe wall. But at shallower depths, the smaller flow area outweighs the effect of the reduced friction, and the discharge ratio drops below 1.0.

EXAMPLE 2.22

What is the maximum possible discharge capacity of a 900-mm-diameter pipe built on a slope of 0.1 percent?

Solution:

The Manning nomograph gives the full-flow capacity as $Q = 550$ L/s. From the partial-flow diagram, we see that the maximum discharge occurs when $d/D = 0.93$; at that depth $q/Q = 1.08$. Therefore the maximum discharge $q = 1.08 \times 550 = 590$ L/s. It will occur at a depth of $d = 900 \times 0.93 = 840$ mm.

EXAMPLE 2.23

An 18-in.-diameter sewer line drops 1.6 ft in elevation over a 400-ft distance. Determine the discharge and velocity in the pipe when the depth of flow is 6 in.

Solution:

The slope $S = 1.6/400 = 0.004$. Using the Manning nomograph, we get $Q = 2900$ gpm and $V = 3.8$ fps. The partial-flow depth ratio is $6/18 = 0.33$, and from the partial-flow diagram we read $q/Q = 0.22$ and $v/V = 0.82$. Therefore, we get $q = Q \times 0.22 = 2900 \times 0.22 = 640$ gpm, and $v = V \times 0.82 = 3.8 \times 0.82 = 3.1$ ft/s.

FLOW MEASUREMENT

An approximate but very simple method to determine open channel discharge is to measure the velocity of a floating object moving in a straight uniform reach of the channel. If the cross-sectional geometry of the channel is known, and the depth of flow is measured, then the flow area can be computed. From the relationship $Q = A \times V$, the discharge Q can be estimated.

This is a useful way to get a "ballpark" estimate for the flow rate as part of a preliminary field study, but it would not be suitable for routine measurements. The average velocity of flow in a reach is approximated by timing the passage of the floating object along a measured length of the channel.

EXAMPLE 2.24

A floating object is placed on the surface of water flowing in a stormwater drainage ditch and is observed to travel a distance of 10 m downstream in 20 s. The ditch is 1.5 m wide and the average depth of flow is estimated to be 0.5 m. Estimate the discharge under these conditions.

Solution:

The flow velocity is computed as distance over time, or

$$V = D/T = 10 \text{ m}/20 \text{ s} = 0.5 \text{ m/s}$$

The channel area is $A = 1.5 \text{ m} \times 0.5 \text{ m} = 0.75$ m^2.

The discharge $Q = A \times V = 0.75 \text{ m}^2 \times 0.5 \text{ m/s} = 0.375 \text{ m}^3/\text{s}$.

Weirs

A widely used device to measure open channel flow is the *sharp-crested weir*. A weir is simply a dam or obstruction placed in the channel so that the water backs up behind it and then flows over it. The sharp crest or edge allows the water to spring clear of the weir plate and to fall freely in the form of a *nappe*, as shown in Figure 2.23.

When the nappe discharges freely into the air, there is a hydraulic relationship between the height or depth of water flowing over the weir crest and the flow rate. This height, the vertical distance between the crest and the water surface, is called the *head on the weir;* it can be measured directly with a meter or yard stick or automatically by float-operated recording devices (see Section 3.4 on stream gaging stations). The head on the weir should be measured a short distance upstream of

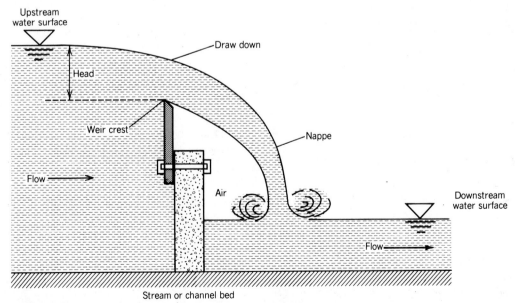

FIGURE 2.23 Side view of a *sharp-crested weir,* a simple device used for measuring open channel or stream flow.

the weir plate to avoid the effect of the surface drawdown as the flow passes over the crest.

The part of the weir plate over which the water flows can have one of several different shapes, depending on the particular application. The most common shapes are rectangular, triangular, or trapezoidal. A *contracted rectangular weir* is one that does not extend across the full width of the channel, as shown in Figure 2.24*a*. Triangular weirs, also called V-notch weirs, can have notch angles ranging from 22.5 to 90°, but right-angle notches are the most common; this is illustrated in Figure 2.24*b*. (A trapezoidal or *Cipolletti*-shaped weir is shown in Figure 3.15, installed as part of a stream gaging station.)

V-notch weirs allow accurate measurement of much lower discharges than rectangular or trapezoidal weirs. They are commonly used in small sewage treatment plants to monitor the sewage effluent flow rate. V-notch weirs can also be inserted into sewer lines to measure sewage discharges, and they are used frequently during infiltration–inflow

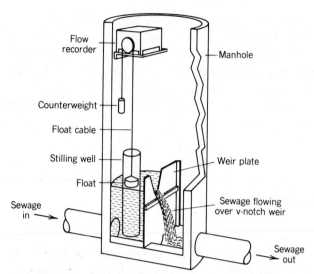

FIGURE 2.25 A temporary weir installation used to measure wastewater flow rates in a sewer manhole. (Leupold & Stevens, Inc.)

studies (see Section 9.4). These weirs, manufactured to fit a range of different pipe sizes, can be held in the end of a pipe in a manhole; the flow rate can be read directly from a calibrated scale on the weir plate.

For larger sewer lines, or when continuous flow data over 24 hours are needed, a temporary weir installation can be set up in the manhole. An installation that includes an automatic level recorder is shown in Figure 2.25. The disadvantage of this type of flow metering installation is that sewage solids settle out and accumulate behind the weir plate.

There are many equations, tables, and charts in hydraulics textbooks and handbooks that relate discharge to the head on a weir. Some of the formulas account for end contractions, approach velocities, and other factors. A simple formula for a 90° V-notch weir is as follows:

$$Q = 2.5 \times H^{2.5} \qquad (2\text{-}12)$$

where Q = discharge, ft^3/s
H = head on weir, ft

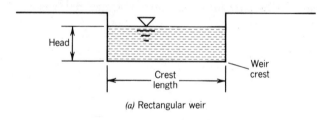

(a) Rectangular weir

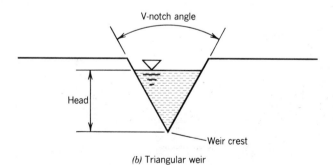

(b) Triangular weir

FIGURE 2.24 Commonly used weir shapes include *(a)* the contracted rectangular weir and *(b)* the triangular V-notch weir.

EXAMPLE 2.25

Estimate the discharge in L/s over a 90° V-notch weir when the head on the weir $H = 100$ mm.

Solution:

It is necessary to use some conversion factors in this problem since Equation 2-12 is given only for American units.

100 mm × 1 in./25.4 mm × 1 ft/12 in. = 0.33 ft

Applying Equation 2–12, we get

$$Q = 2.5 \times 0.33^{2.5} = 0.156 \text{ ft}^3/\text{s}$$

and

$$Q = 0.156 \text{ ft}^3/\text{s} \times 28.32 \text{ L/ft}^3 = 4.4 \text{ L/s}$$

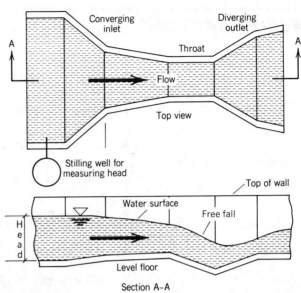

FIGURE 2.26 A Parshall flume is often used to measure flow rate in an open channel that carries sewage; discharge is related to the head or depth of water just upstream of the constricted flume section.

Flumes

A flume is a specially shaped constricted section in an open channel (similar to the venturi tube in a pressure conduit). The geometry of the flume causes a "free-fall" condition in the channel which allows a correlation to be made between the discharge and the depth of flow. Although they are more expensive to install than weirs, flumes offer the advantage of a "self-cleansing" action that prevents deposits of sewage solids. Also, there is little head loss and no significant backup of sewage upstream of the meter.

The most common flume used for a permanent sewage flow metering installation is called the *Parshall flume,* shown in Figure 2.26. For small channels, prefabricated fiberglass flumes can be installed, but for larger systems, the flumes are built of cast-in-place concrete.

A set of tables or a nomograph can be used to relate the water level in the flume to the flow rate.

Usually automatic recording devices provide a continuous record of discharge once the instrument has been calibrated with the flume and level sensing device. Another type of flume, called a *Palmer–Bowlus flume,* can be placed in existing circular channels or sewer pipes, as illustrated in Figure 2.27.

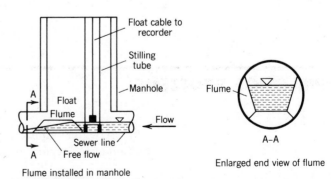

FIGURE 2.27 A typical Palmer–Bowlus flume installation.

REVIEW QUESTIONS

1. Why is hydraulics an important aspect of environmental technology?

2. What is the definition of *pressure?*

3. List four important characteristics of hydrostatic pressure.

4. What is the difference between *absolute pressure* and *gage pressure?*

5. What is *pressure head?*

6. Briefly describe three different ways to measure pressure.

7. What is the simple formula that relates flow rate to flow velocity and area?

8. Briefly describe what is meant by *continuity of flow.*

9. Briefly describe the principle of conservation of energy as it relates to hydraulic systems.

10. Explain why the pressure in a pipeline drops in a constricted section of the pipe.

11. What is a *hydraulic grade line?*

12. Briefly describe the operating principle of a venturi meter.

13. Give a definition of *uniform, steady open-channel flow.*

14. Briefly describe two methods for measuring discharge in an open channel.

PRACTICE PROBLEMS

1. A water supply reservoir is 50 ft deep. Compute the pressure at the bottom of the reservoir and at a point 30 ft above the bottom.

2. The pressure at the bottom of an open tank is measured to be 50 kPa. How deep is the water in the tank?

3. A water storage tank is situated on a hill, as shown in Figure 2.28. The water main is 2 m below the ground surface at all points. The elevations above sea level of the ground and water surface are given. Compute the hydrostatic pressure at the closed valve at the bottom of the hill.

4. The pressure in a water main is 50 psi. What is the pressure head in the main? What would be the hydrostatic pressure at a customer's tap which is 40 ft above the main?

5. If the air pressure in the sealed tank of Figure 2.6 was 30 kPa, what would the pressure gage read at the tank bottom and how high would the water rise in the vertical tube?

6. A 300-mm-diameter pipe carries a flow of 100 L/s. Compute the velocity of flow.

7. What diameter pipe would be needed to carry a flow of 500 gpm at a velocity of 1.4 fps?

8. Water is flowing at a velocity of 2 m/s in a 200-mm-diameter pipe. The pipe diameter is reduced to 100 mm at a constriction in the pipe. Determine the flow velocity in the constricted pipe section.

9. For the branching pipe section shown in Figure 2.29, determine the flow velocity in pipe A if the velocity of flow in pipe B is 2 m/s and in pipe C it is 1 m/s.

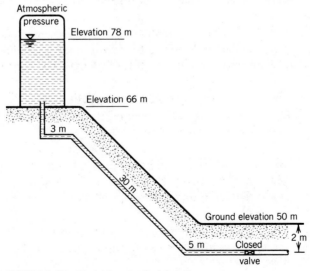

FIGURE 2.28 Illustration for Problem 3.

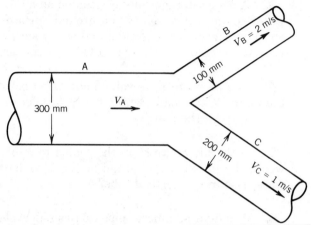

FIGURE 2.29   Illustration for Problem 9.

10.   For the system illustrated in Figure 2.13, the diameter at section 1 is 16 in. and at section 2 it is 8 in. The flow rate through the pipe is 6.0 cfs, and the pressure at section 1 is 50 psi. What is the pressure in the constriction at section 2?

11.   For the system illustrated in Figure 2.13, the diameter at section 1 is 300 mm and at section 2 it is 100 mm. The flow rate through the pipe is 50 L/s, and the pressure at section 1 is 700 kPa. What is the pressure in the constriction at section 2?

12.   A 600-mm-diameter pipe carries a flow of 0.20 m³/s. Compute the pressure drop per kilometer of pipeline.

13.   What minimum pipe diameter would be needed to carry a flow of 1000 gpm without exceeding a pressure loss of 20 psi per mile of pipeline?

14.   A 300-mm-diameter pipe carries water with a head loss of 10 m/km of pipeline. Determine the flow rate in the pipe using the Hazen–Williams nomograph and check the solution using Equation 2-9.

15.   A venturi meter has a pipe diameter of 6 in. and a throat diameter of 3 in. A pressure difference of 10 psi is measured in the meter. What is the flow rate under these conditions?

16.   A venturi meter has a pipe diameter of 150 mm and a throat diameter of 75 mm. A pressure difference of 100 kPa is measured in the meter. What is the flow rate under these conditions?

17.   An 800-mm-diameter sanitary sewer is built on a slope of 0.2 percent. What is the full-flow discharge capacity of the sewer and what would be the flow velocity?

18. An 18-in.-diameter sewer is placed on a grade of 1.5 ft/1000 ft. What would be the discharge and flow velocity when the pipe was full?

19. What diameter pipe is needed to carry a peak flow of 200 L/s on a grade of 0.007?

20. What grade is required for a 36-in.-diameter pipe to carry at least 7 mgd of sewage at a velocity not less than 2 ft/s?

21. A 300-mm-diameter sewer is built at a grade of 2 percent. What would be the depth of sewage flowing in the pipe when it carried a flow of 50 L/s? What would the velocity be?

22. What is the highest discharge capacity of an 18-in.-diameter pipe on a slope of 0.15 percent, and what would be the depth of flow in the pipe?

23. What would be the highest flow velocity in a 900-mm pipe on a 0.1-percent slope? What would be the discharge and depth of flow?

24. The invert elevation of a 600-mm sewer drops 0.5 m over a 100-m distance. Determine the discharge and flow velocity in the sewer when the depth of flow is 200 mm.

25. A float placed on the surface of water flowing in a stormwater drainage ditch is seen to move a distance of 25 m downstream in 75 s. The ditch is 2 m wide and the average depth of flow is estimated to be 0.75 m. Estimate the channel discharge.

SELECTED REFERENCES

1. King, H. W., and E. F. Brater, *Handbook of Hydraulics,* McGraw–Hill, New York, 1963.

2. Mott, R. L., *Applied Fluid Mechanics,* Charles E. Merrill, Columbus, Ohio, 1979.

3. Simon, A. L., *Practical Hydraulics,* John Wiley & Sons, 1981.

4. Hammer, M. J., *Water and Waste-Water Technology,* John Wiley & Sons, New York, 1977.

5. *Operation of Wastewater Treatment Plants,* Manual of Practice No. 11, Water Pollution Control Federation.

HYDROLOGY

Hydrology is a branch of earth science that is concerned with the distribution and movement of water on and under the earth's surface. The science of hydrology is of great importance in environmental technology for many reasons. Two extreme hydrologic conditions, droughts (not enough water where needed) and floods (too much water in the wrong place), are well known for the environmental problems they cause. But droughts and floods are not the only aspects of hydrology that are important. In general, the presence and quantity of water must be estimated in order to plan, design, and operate water supply, pollution control, and stormwater management facilities.

The purpose of this chapter is to present fundamental hydrologic concepts for measuring present conditions and estimating future variations in water availability. Applications of these concepts will be discussed in further detail in the chapters on stormwater management and water supply and in other sections of the book.

3.1 WATER USE AND AVAILABILITY

Everyone knows that water is essential to sustain life. It also plays a central role in the growth and environmental health of cities and towns. We depend on water for more than just drinking, cooking, and personal hygiene. Vast quantities are often required for industrial and commercial uses. In some parts of the country, large quantities of water for irrigation are necessary to support argiculture. We also rely on our water resources for power generation, recreation, fish and wildlife conservation, and navigation.

Water use refers to the withdrawal of water from its source, which may be a river, lake, or well, and the transport of that water to a specific location. For example, water used for cooling purposes in a power plant may be diverted from a nearby river, passed through the power plant, and then discharged back into the river without significant loss in quantity. (The water would have to be cooled down before discharge to prevent "thermal pollution," which is discussed in Chapter 5.) Navigation and recreation are other examples of *nonwithdrawal use*. However it is necessary to make a distinction between water *use* and water *consumption*. Water that is used for drinking or combined with a product, and is not available for use again, is consumed water.

More than 100 million cubic meters of water per day are withdrawn for public water supplies in the United States. More than 500 million cubic meters are withdrawn each day for irrigation. Industrial use accounts for the largest share of water demand, almost 1 billion cubic meters per day. Most of this, though, is used as cooling water at electric power utilities. These approximate figures are presented to give you an appreciation of the tremendous quantities of water needed.

Water is present in abundant quantities on and under the earth's surface. But only less than 1 percent of the earth's water is actually available for use in economically satisfying the needs mentioned here. Most of the earth's water is salt water or is frozen in the polar ice caps.

Many of our freshwater lakes and rivers have been deteriorating in quality because of land development and pollution, limiting the availability of water for use, particularly for public water supplies. Even groundwater is affected by pollution in some areas, although much of it is just too deep to pump out of wells economically.

THE DISTRIBUTION OF WATER

In addition to the limited availability of usable water, another basic problem in managing water resources is the fact that it is not evenly distributed geographically. In some regions there is ample precipitation, including rain, snow, hail, sleet, and dew, and water is readily available for use. On the average, about one third of this precipitation becomes available in lakes and rivers and some makes its way into the groundwater. But where there is little precipitation, water is scarce. The fact that there is a close relationship between the amount of rain or snow and the amount of water available for use should be self-evident. This will be discussed in more detail later in this chapter.

Figure 3.1 illustrates the different annual precipitation amounts across the United States. Except for the extreme northwestern corner of the country, where the total annual amount of rainfall may exceed 2500 mm (100 in.), it can be seen that the eastern half of the country gets significantly more rainfall than the western half. In some areas of the southwest, less than 100 mm (4 in.) of rain may fall in any one year. In the northeast, an annual rainfall of about 1000 mm (40 in.) is moderate compared to the two previously mentioned extremes.

The amount of rainfall and the availability of water can vary considerably even within a relatively small area. California is an example of a state with a very uneven distribution of water. Although southern California is very dry, the growing population there generates a large demand for water. Most of the needed water must be transported to the south from the northern part of the state where water is more readily available. A huge system of reservoirs, open channels, pumping stations, and tunnels is used to accomplish this transfer of water.

Part of the system, called the California Aqueduct, can convey about 2800 m^3 (100 000 ft^3) of water per second. The aqueduct is an open channel, about 40 m (130 ft) wide at the surface and about 9 m (30 ft) deep. At one point in the system, the water is pumped up about 600 m (2000 ft) to get it over a mountain, quite an engineering undertaking.

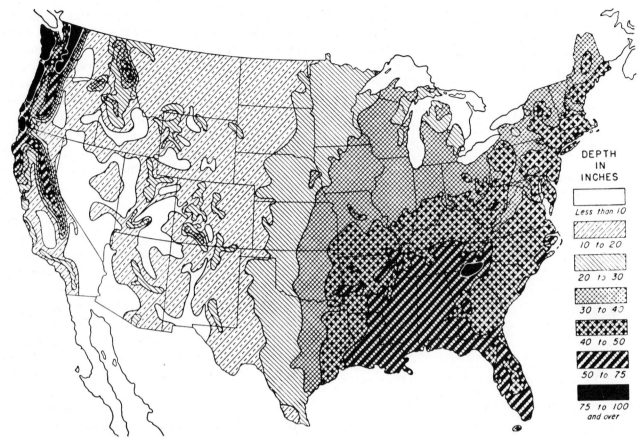

FIGURE 3.1 The distribution of annual precipitation, and therefore the availability of water, is not uniform over the continental United States. (From *Hydrology Handbook,* figure 3 with permission of the American Society of Civil Engineers.)

The uneven distribution of water from one geographic location to another is only part of the problem in hydrology and water resources management. The occurrence and availability of water also varies with time. In any given location there may be occasional periods of little rainfall or drought and severe water shortages may result as water in storage reservoirs is used up during these dry periods.

On the other hand, the same area may sometimes experience periods of above average rainfall. Serious flooding problems may result, with accompanying loss of lives and property, as well as environmental pollution problems. In any given area, then, there can be too little water or too much water, depending on natural climatic conditions.

3.2 THE HYDROLOGIC CYCLE

Water is in constant motion on, under, and above the earth's surface. Even in what appears to be a stagnant pond, the water is evaporating or changing into a vapor and moving into the atmosphere. Powered with energy from the sun and from gravity, there is a constant circulation of water and water vapor. This natural process is called the *hydrologic cycle.* It is illustrated in schematic form in Figure 3.2.

Suprisingly enough, there was a time when people did not have an understanding of the cyclical

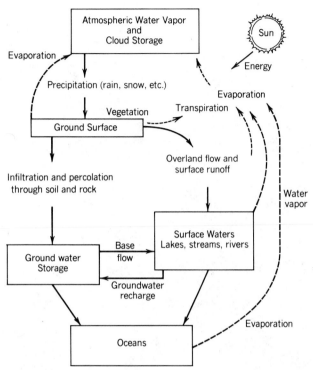

FIGURE 3.2 A schematic diagram of the natural *hydrologic cycle*. The constant circulation of water is powered by energy from the sun and by gravity.

motion of water through the environment and had misconceptions about the origin of water in streams or lakes. Even today, some people still have misconceptions, particularly with respect to groundwater.

Although the hydrologic cycle looks simple when sketched in schematic form as in Figure 3.2, there is more to it than initially meets the eye. The science of hydrology gets quite complicated, applying a good deal of statistics and higher mathematics. The basic objective is to measure and analyze the relationships controlling the form, quantity, and distribution of water. When these relationships are understood, reliable predictions may be made concerning the occurrence of future floods or droughts. It is important that technicians involved in environmental control have an appreciation of the basic structure of the hydrologic cycle.

Precipitation begins when atmospheric moisture (water vapor) is cooled and condensed into water

droplets. There are three different paths the precipitation can follow after it reaches the ground. First some of it may be intercepted by vegetation or small surface depressions. In other words, it is temporarily stuck on the surfaces of leaves or grass or it is retained in puddles. Second, a portion of the water can *infiltrate* through the earth's surface and seep (or *percolate,* as it is called) downward into the ground. Third, a portion of the water can flow over the earth's surface. Measuring and predicting the relative amounts of water that follow each of these paths is of importance in hydrology.

Some of the intercepted water soon evaporates and some of it is absorbed by the vegetation. A process called *transpiration* takes place as water is used by the vegetation and passes through the leaves of grass, plants, and trees, returning to the atmosphere as vapor. The combined process of evaporation and transpiration is called *evapotranspiration*. Overall, more than half of the precipitation that reaches the ground is returned to the atmosphere by this process, before reaching the oceans.

Overland flow and surface runoff occur when the rate of precipitation exceeds the combined rates of infiltration and evapotranspiration. Eventually, the overland flow finds its way into stream channels, rivers and lakes, and finally the oceans. The ocean can be thought of as the final "sink" to which the water flows. As previously mentioned, about one third of the average annual rainfall in the United States becomes surface runoff in streams and rivers. This, of course, varies from region to region. In some areas of the southwest, for example, there is no runoff for years at a time since the rate of precipitation does not often exceed the rate of infiltration and evapotranspiration in that area.

The water that infiltrates the ground surface will percolate into saturated soil and porous rock layers, forming vast groundwater reservoirs. A groundwater "reservoir" should not be visualized as an underground lake—the water fills the tiny voids in the soil or cracks in the rock. The groundwater may later seep out onto the ground surface in springs or into streams. Eventually, the groundwater makes its way to the ocean, either directly or via surface streams. Evaporation from the ocean surface substantially replenishes the water vapor in the atmosphere, winds carry the moist air over land, and the hydrologic cycle continues.

Synthetic Hydrologic Cycle

This description of the hydrologic cycle is only a brief summary of a complex natural phenomenon. Some of the details of this natural cycle are discussed in the following sections of this chapter. But one water cycle should be mentioned here—the "synthetic" or "urban" water cycle, illustrated in Figure 3.3.

In human communities there is a constant circulation of water. Water is withdrawn from its source in the natural hydrologic cycle—surface waters or groundwater—and is pumped through treatment and distribution systems. After use, the wastewater is collected in sewer systems, treated to reduce the effect of pollution, and finally disposed of back into surface or groundwaters. A most significant aspect of environmental technology is the maintenance of this urban water cycle while protecting public health and environmental quality. Much of this textbook focuses on this topic.

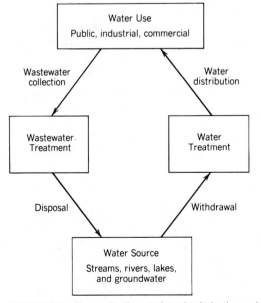

FIGURE 3.3 The synthetic or urban hydrologic cycle.

3.3 RAINFALL

Water in streams, rivers, and lakes, as well as water in the ground, is the residue of precipitation. It is possible, and often necessary, to refer to records of rainfall in order to estimate the quantity of water that will be found on and under the ground. Other factors, such as topography and land-use, play a role in the relationship between rainfall and water availability. These will be considered later on. In this section, basic concepts related to rainfall intensity and volume will be discussed.

DEPTH, VOLUME, AND INTENSITY

The collection of rainfall data is the responsibility of the United States National Weather Service, a government agency that maintains rain gage stations throughout the country. Rainfall amounts are expressed in terms of the depth of water accumulated in the rain gage during a storm. The units can be expressed in millimeters or inches. It is usually necessary to compute weighted averages of rainfall amounts over a region, using the data from several rain gages. The data may be weighted in proportion to the area covered by each gage.

Sometimes it is necessary to compute the total volume of water that falls on an area during a storm. The volume is computed by multiplying the land area by the rainfall depth, as follows:

$$\text{volume} = \text{depth} \times \text{area} \qquad (3\text{-}1)$$

In SI metric units, the volume is usually expressed in terms of cubic meters, but rainfall depth is expressed in terms of millimeters. To keep the units consistent when applying Equation 3-1, area should be expressed in square meters and rainfall depth should be expressed in meters. Relatively large areas that are expressed in units of hectares (ha) should first be converted to m^2 (1 ha = 10 000 m^2).

64 **CHAPTER 3 HYDROLOGY**

EXAMPLE 3.1

During a 20-min rain storm, a depth of 25 mm of rainfall was recorded for an area of 2.5 ha. Compute the total volume of water that fell on that area during the storm.

Solution:

In this problem, we do not need the data regarding storm duration. But we will use that information in a subsequent example.

First convert the rainfall depth from mm to m.

$$25 \text{ mm} \times \frac{1 \text{ m}}{1000 \text{ mm}} = 0.025 \text{ m}$$

Now convert ha to m^2 as follows:

$$2.5 \text{ ha} \times \frac{10\ 000 \text{ m}^2}{1 \text{ ha}} = 25\ 000 \text{ m}^2$$

Applying Equation 3-1, we get

volume = depth × area
volume = 0.025 m × 25 000 m^2 = 625 m^3

We can round this up to 630 m^3 since the given data had only two significant figures.

More important than total volume of rain is the rate at which the rain falls. This is called *rainfall intensity*. As you will see later on, the rainfall intensity that occurs during a storm is of particular significance in civil and environmental technology, particularly when designing urban drainage facilities.

Rainfall intensity is expressed in terms of depth per unit time, as either in./h or mm/h. The National Weather Service gathers this kind of data using automatic rain gages that record rainfall duration as well as depth; a continuous record of rainfall

amount and intensity is plotted on a revolving drum. It is generally observed that short duration storms have higher rainfall intensities than longer duration storms. This will be of significance when we consider problems in stormwater control.

EXAMPLE 3.2

For the storm described in Example 3.1, compute the rainfall intensity.

Solution:

Even though the storm lasted only 20 min, we can still compute its intensity in terms of mm/h. In effect, we are computing how much rain would have fallen if the storm had lasted for 1 h at a steady intensity.

$$\text{intensity} = \frac{25 \text{ mm}}{20 \text{ min}} \times \frac{60 \text{ min}}{1 \text{ h}} = 75 \frac{\text{mm}}{\text{h}}$$

When using American units, cubic feet is a common term for volume. But in hydrologic applications, large volumes of water are usually expressed in terms of *acre–feet* (ac–ft). This may seem to be a strange term at first, but as illustrated in Figure 3.4, 1 ac–ft can easily be visualized as the volume required to cover 1 ac of land to a depth of 1 ft. Since 1 ac is equivalent to 43 560 ft^2, one ac–ft is equal to 43 560 ft^2 × 1 ft or, 43 560 ft^3.

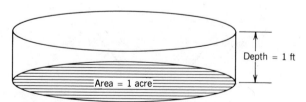

FIGURE 3.4 One *acre–foot* of water is a volume equivalent to the volume that would cover 1 acre of land at a depth of 1 ft, or 43 560 cubic feet (ft^3) of water.

EXAMPLE 3.3

During a storm, a total of 4.0 in. of rain fell on an area of 120 ac. The storm duration was 8.0 h. What was the average rainfall intensity during the storm? Determine the total volume of rain that fell on the area in the 8-h period. Express the volume in acre-feet and cubic feet.

Solution:

The average intensity is determined by dividing the total depth of rain by the storm duration, as follows:

$$\text{intensity} = \frac{4 \text{ in.}}{8 \text{ h}} = 0.5 \, \frac{\text{in.}}{\text{h}}$$

To apply Equation 3-1, the depth should be expressed in feet. The computation can be done in one step as follows:

volume = 120 ac × 4.0 in. × 1 ft/12 in. = 40 acre–feet

To convert from acre–feet to ft³,

volume = 40 ac–ft × 43 560 ft³/ac–ft = 1 700 000 ft³

(The answer is rounded off to two significant figures.)

RECURRENCE INTERVAL

Common experience shows us that hydrologic events, such as rain storms, do not occur with any definite regularity. The time span or period between storms is not constant. The occurrence of rainfall, its intensity, and its duration are random natural events. Consider, for instance, the storm described in Example 3.1. It dropped 25 mm of rain in 20 min. Despite the random nature of precipitation events, we can determine average frequencies of occurrence of storms having specific intensities and durations. It would be convenient if we could predict the exact dates that identical storms would occur in the future, but obviously that is not possible. For instance, even though we can not determine the date of the next 20-min, 25-mm storm, we can at least predict how many times we should expect a similar storm over the next year, or several years. In addition, we can predict the likelihood or probability of observing that storm again in any given period.

By examining many years of rainfall records and applying statistical analyses, we can determine the average number of years between storms of specific intensities and durations. This time span between identical storms is called the *recurrence interval* or *return period* of the storm. These return periods are determined and reported by the National Weather Service, and designers of environmental facilities must know how to interpret and use the data.

When applying these data, we use the expression "*N*-year storm," where *N* stands for the recurrence interval in years. For example, a storm with a return period of 5 years would be called a "5-year storm." This means that over a long period of time, the average time span between storms of that particular intensity and duration will be 5 years. It does **not** mean that a similar storm will occur once exactly every 5 years. In fact, it is possible that more than one of these so-called 5-year storms could occur within a shorter time span, even within a single year, but the chances for this are slim. It should also be noted here that the probability of the 5-year storm occurring in any given 5-year period is not quite 100 percent. In other words, no one can say for sure that what is called a 5-year storm will actually take place within, say, the next 5 years. But over a long time span, 500 years for example, it is a good bet that there will be about 100 of those 5-year storms.

PROBABILITY OF OCCURRENCE

Data on storm intensity, duration, and return period are important in the design of urban drainage structures and for predicting peak flows in rivers. On the other end of the hydrologic spectrum, the severity of droughts, and their frequency of occurrence, are of importance in designing water supply reservoirs.

Because of the uncertain and irregular nature of hydrologic events, there is always some risk of failure when designing a structure or facility involving water resources. For example, a river used for water supply may not provide enough water for a growing community during dry periods. Even if a small reservoir were built to overcome this deficiency, there would remain the risk that a more severe (though less frequent) drought would cause the reservoir to run dry. This risk can be reduced by building a larger reservoir, but this would be more expensive. Designers must be able to balance the economics and the risks, using probability concepts.

The probability or chance that a given event will occur can be expressed either as a fraction, a decimal, or a percent. For example, the probability of a tossed coin coming up "heads" is one chance out of two, or $\frac{1}{2} = 0.5 = 50$ percent. In the long run, 50 tosses out of 100 tosses can be expected to come up "heads." A probability of 1 or 100 percent represents a certainty, and a probability of 0 represents an impossibility.

There is a simple relationship between the return period of a hydrologic event and the probability of occurrence of that event. If N is the recurrence interval of the event (in years), then the probability P of that event being equalled or exceeded in any given year is the reciprocal of N. Expressed as a formula, we get

$$P = \frac{1}{N} \qquad (3\text{-}2)$$

For example, the probability of a 5-year storm occurring in any single year is $P = \frac{1}{5} = 0.2$ or 20 percent. In effect, this also means that there is less than a 20-percent chance that a "worse" or more intense storm will occur in any given year.

Relying on common experience again, we recognize that the really intense storms are few and far between. In other words, the more extreme the hydrologic event is, the larger is its recurrence interval. And the larger the recurrence interval N is, the lower the probability of occurrence P is because of the inverse relationship between the two. For example, there is only a 1 percent chance that a "100-year storm" will occur in a given year. It is much less likely to observe a severe 100-year storm than a 5-year storm. (Although in many regions of the country rainfall records do not go as far back as 100 years, statistics and probability theory can be used to "extrapolate" or extend the existing data beyond the actual period of record.)

To summarize, the larger the recurrence interval N, the less likely it is for a hydrologic event to be equalled or exceeded in a given year. This is an important concept. Generally, the more critical a project is in terms of potential loss of life, economic damage, or adverse environmental effects, the larger is the value of N used in design computations.

A dam, for instance, may be designed to accommodate a 100-year flood; whereas a local storm drain may be designed to handle only the flow from a 2-year storm. In the former case, designing the dam for the big flow will reduce the chance of failure or breach of the dam, and ensure the protection of human lives and property downstream. In the latter case, a trade-off is made between saving money and taking a chance on the stormdrain's backing-up or overflowing once every two years or so.

INTENSITY–DURATION–FREQUENCY RELATIONSHIPS

In these discussions, we have been examining terms such as *storm intensity, storm duration,* and *recurrence interval* as if they were independent quantities. But these three factors are related to each other and must be considered together. The term *frequency* is often used instead of return period. The frequency of a storm or other hydrologic event varies inversely with its return period. A 10-

year storm, for example, would occur less frequently than a 5-year storm.

The rainfall data that are collected by the National Weather Service are compiled, analyzed, and published in various forms. The relationships among rainfall intensity, duration, and frequency may be shown graphically in curves or maps, or they may be expressed as formulas. As will be shown in the chapter on stormwater control, these data are used by designers to estimate storm runoff and peak streamflow or discharge.

Rainfall Curves

A typical set of rainfall intensity–duration–frequency curves is illustrated in Figure 3.5. Rainfall patterns vary significantly with geographic location and climate. For an actual application of rainfall data to a real design problem, the appropriate rainfall curves for the specific location under study should be obtained from the National Weather Service or from appropriate state or county agencies.

Rainfall curves of this form are generally used by entering the horizontal axis with a preselected storm duration, moving vertically to an intersection with a specific storm return period (the curved lines), and then moving horizontally to the vertical axis where an expected rainfall intensity is read. For example, it can be seen in Figure 3.5 that a 10-year storm with a 30-min duration would have an intensity of 100 mm/h (or about 4 in./h). The shape of these rainfall curves reflects the fact that storms of shorter durations have higher average intensities than do longer storms. Also, for a given duration, the higher intensities correspond to storms with longer recurrence intervals.

EXAMPLE 3.4

A storm of 40-min duration drops 50 mm (2 in.) of rain. Using the rainfall curves in Figure 3.5, estimate the probability of observing a similar storm in the next year.

Solution:

First compute the storm intensity, as follows:

$$\text{intensity} = \frac{50 \text{ mm}}{40 \text{ min}} \times \frac{60 \text{ min}}{1 \text{ h}} = 75 \text{ mm/h}$$

Now enter Figure 3.4 on the horizontal axis with 40 min, and on the vertical axis with 75 mm/h. The intersection of horizontal and vertical lines extended from those points falls about halfway between the 5-year storm and the 10-year storm. From this we can conclude that the return period for the storm in question is about 7.5 years. The probability of observing a similar or more intense storm in the next year is computed using Equation 3-2, as follows:

$$P = \frac{1}{7.5} = 0.13 \text{ or } 13 \text{ percent}$$

Rainfall Formulas

Rainfall intensity–duration–frequency relationships may be expressed in equation form instead of

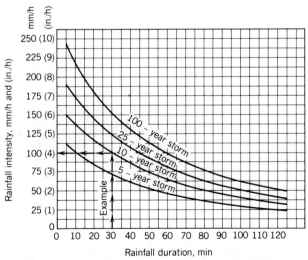

FIGURE 3.5 Typical rainfall *intensity–duration–frequency* curves. Curves like these are prepared from rainfall statistics by the U.S. National Climatic Data Center.

in the form of curves on a graph. One of the equations that may be used is the following:

$$i = \frac{A}{t + B} \qquad (3-3)$$

where i = rainfall intensity, mm/h (in./h)
t = rainfall duration, min
A and B = constants that depend on the recurrence interval and geographic locale.

Values of the constants A and B have been derived from data for various sections of the country. For example, for 10-year storms in the eastern Middle–Atlantic States, A and B are reported as 5840 and 29, respectively. A and B are 1520 and 13, respectively, for the western states. (These values of A and B are for use in the SI metric system; intensity, i, will be in mm/h.)

EXAMPLE 3.5

Using the rainfall formula given by Equation 3-3, determine the expected rainfall intensity for a 10-year storm of 60-min duration (a) in California, and (b) in Delaware.

Solution:

(a) For California, use the constants $A = 1520$ and $B = 13$, for the western states. Applying Equation 3-3 with $T = 60$, we get

$$i = \frac{1520}{60 + 13} = \frac{1520}{73} = 21 \text{ mm/h}$$

(b) For Delaware, use the constants $A = 5840$ and $B = 29$, for the Middle–Atlantic States. Applying Equation 3-3 with $T = 60$, we get

$$i = \frac{5840}{60 + 29} = \frac{5840}{89} = 66 \text{ mm/h}$$

The rainfall formula presented in Equation 3-3 covers large regions of the country for given values of A and B and can be rather insensitive to more local variations in rainfall patterns. It is preferable to use more local information regarding rainfall data if they are available.

Rainfall Maps

Rainfall data can be depicted on a map, as illustrated in Figure 3.6. The lines seen crisscrossing the United States may look like ground contour lines at first glance. But instead of contours, they represent lines of equal rainfall depth for a 24-hour storm with a 100-year return period. It can be seen, for example, that a 100-year, 24-hour storm would cause 5 in. of rain in northern Maine, western Texas, and many other locations in between. A storm with the same frequency and duration would cause 14 in. of rain in southern Florida. Maps of this nature are available for a wide range of durations and frequencies from the National Weather Service.

3.4 SURFACE WATER

Water that flows over the ground is often called *runoff*. Runoff that has not yet reached a definite stream channel is called *overland flow* or *sheetflow* (on a smooth surface, such as pavement). This type of surface water will be of importance when we discuss stormwater drainage systems. For the most part, the term *surface water* refers to water flowing in streams and rivers, as well as water stored in natural or artificial lakes.

WATERSHEDS

As mentioned in Section 3.2, runoff occurs when the rate of precipitation exceeds the rate of interception and evapotranspiration. The total land area that contributes runoff to a stream or river is called

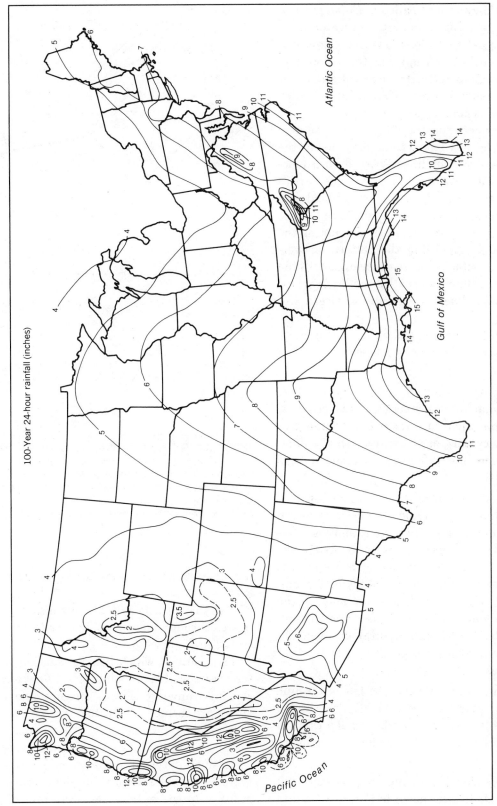

FIGURE 3.6 Rainfall maps, which show lines of equal rainfall depth across the nation, are available for a range of recurrence intervals and storm durations. The lines of equal rainfall depth are called *isohyetal* lines. (National Weather Service)

a *watershed*. It may also be called a *drainage basin* or *catchment area*, particularly if the water flows toward or in an urban drainage system. Generally, we are interested in determining the amount of runoff at a specific point in the natural stream or engineered drainage system. That point is called the *basin outlet* or *point of concentration*.

The boundary or perimeter of the watershed may be determined from a topographic map, using the ground elevation contour lines. Viewed on a topographic map, water flowing freely over the earth's surface would move in a direction perpendicular to the contour lines, which is the direction of the steepest slope at any given point. By examining the contour map and visualizing the pattern of overland flow, it is possible to locate the boundary of the watershed. This boundary is called the *drainage divide line* or *ridge line;* it separates adjacent watersheds.

To draw a drainage divide line on a topo map, the following procedure may be followed:

1. Start at the point of concentration. This might be at the intersection of two streams, at a point where a stream flows through a highway culvert, or at the location of a dam. The divide line will begin and end at this point.

2. Examine the contours to determine flow patterns. You can imagine a drop of water on the ground at any given point and visualize which way it will flow. Start to sketch sections of the divide line that clearly separate the watershed that you are studying from an adjacent watershed. These sections of the line will follow ridges and pass through "saddles." Remember, the natural drainage divide line is always perpendicular to the contour lines.

3. Fill in any gaps that may be left in the line you are sketching. Occasionally, the divide line will turn sharply on the top of a ridge to pass through one of the saddles on the line.

A perspective sketch showing flow patterns and a drainage divide line is shown in Figure 3.7*a*, and

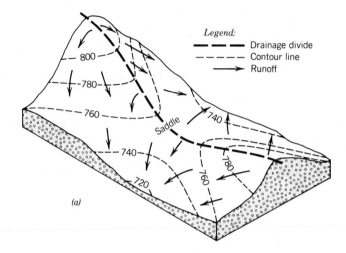

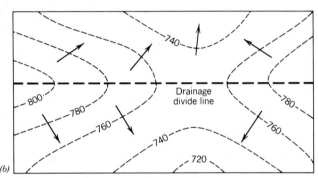

FIGURE 3.7 *(a)* A perspective view of runoff patterns; arrows show the direction of sheet flow. The *drainage divide line* passes through the ridges and the saddle, separating two adjacent watersheds. The direction of sheetflow is perpendicular to the contour lines. *(b)* A plan view topographic map that shows the same drainage divide line and contours as depicted in *(a)*.

a plan view of the same area is shown in Figure 3.7*b*.

The sharp turns a divide line may comprise as it passes through adjacent ridges and saddles is illustrated in Figure 3.8.

The point at which two streams converge or intersect is called a point of *confluence*. As small streams converge, larger streams and eventually rivers are formed. The catchment area for a particular stream may be only a part of a larger watershed; the smaller area is called a *subbasin* of the watershed. A typical drainage network is shown in Figure 3.9. The streams may be classified by their position in the overall network. Typical classifica-

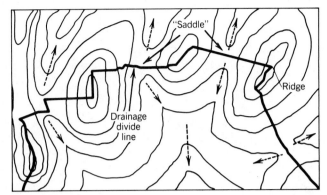

FIGURE 3.8 A drainage divide line will sometimes turn sharply on a ridge or saddle, as shown here. The dashed arrows show the direction of overland flow.

tions are "first-order" streams, "second-order" streams, and so on. A first-order stream does not have any *tributaries,* or smaller streams flowing into it.

A watershed for a large river may encompass thousands of square miles and include many smaller tributaries. These large watersheds are

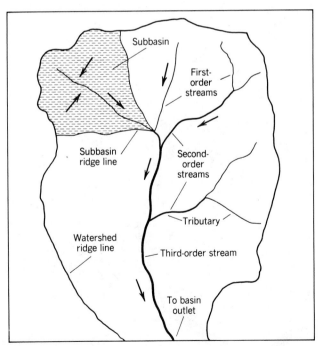

FIGURE 3.9 A large watershed usually comprises several smaller catchment areas or *subbasins.*

also called *river basins.* The Raritan River Basin in New Jersey, for example, encompasses about 2850 km³ (1100 square miles). The Raritan basin is illustrated in Figure 3.10.

The size of a drainage basin refers to its total horizontal surface area. Relatively small basins may be expressed in terms of acres or hectares. A mechanical device called a planimeter is often used to measure the area by simply tracing the boundary of the watershed. Modern electronic planimeters can be calibrated to display the area digitally, based on the scale of the map being used.

The volume and rate of runoff in a watershed is a function of many variables. The basin area, as well as the intensity and duration of rainfall, have a direct effect on the amount and rate of runoff. Other factors include the slope of the ground, the type of soil and vegetative cover, and the type of land use. For example, a flat area with sandy soil would produce less runoff than a sloping area with clay soil. More of the water would infiltrate the ground surface through the porous sand, in the former case, leaving a smaller fraction of the rain to become surface flow. Also, densely populated urban areas generate more runoff than suburban or rural areas. The relationships among these various factors and runoff will be discussed in more detail in the chapter on stormwater management.

STREAMFLOW

The amount or volume of water that flows in a stream is called the flow rate or *discharge* of the stream. The discharge is expressed in terms of volume per unit time passing any given point in the stream. The SI units for discharge are usually cubic meters per second (m³/s), cubic meters per hour (m³/h), or megaliters per day (ML/d). In American units, discharge may be expressed as cubic feet per second (ft³/s), gallons per minute (gpm), million gallons per day (mgd), or acre–feet per day (ac–ft/d). A section of a stream that has a relatively constant slope, cross-section, and discharge may be called a *reach* of the stream.

Stream discharge varies with time. Generally, higher flow rates are observed in the spring and

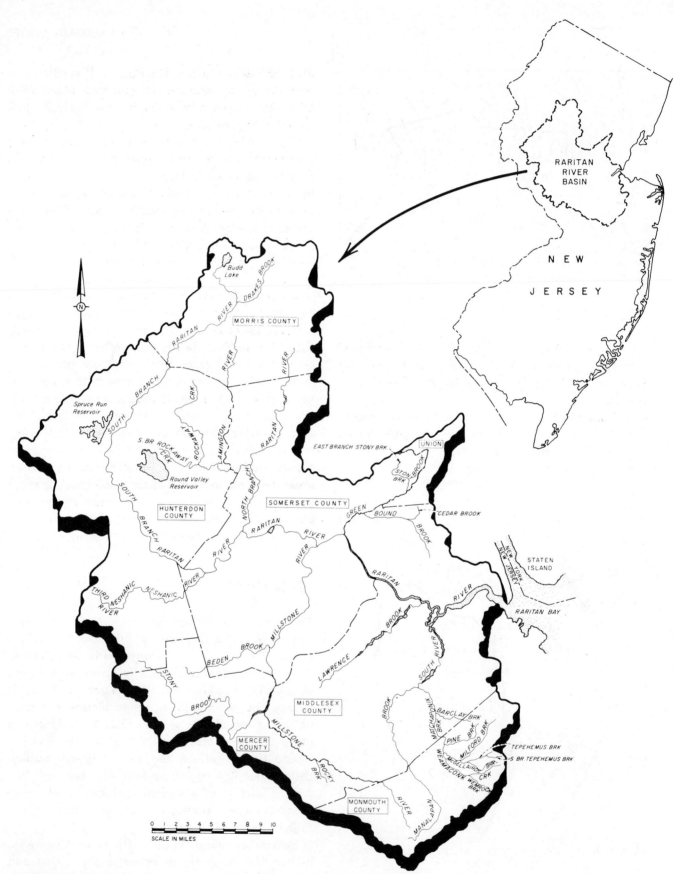

FIGURE 3.10 The Raritan River Basin. This 1100 mile² watershed includes a total of 333 miles of streams. (Division of Water Resources/New Jersey Department of Environmental Protection)

summer months, whereas lower discharges occur in the fall and winter. Snowmelt can contribute significantly to the streamflow. Variations in discharge that occur on a weekly, daily, or even hourly basis are directly associated with rainfall events. In some streams, an extremely wide variation in discharge can occur, from a raging torrent during wet weather to hardly a trickle of water during dry periods.

Low flow rates can cause environmental problems in streams receiving discharges from wastewater treatment plants. This is because there is less water in the stream to dilute the wastewater. The chapter on water pollution discusses the relationship between the discharge and the *waste-assimilative capacity* of the stream. (Under certain conditions, a stream can *assimilate* or absorb sanitary wastes without excessive environmental damage.)

Low stream discharges also cause problems if the stream is used as a source for water supply. On the other end of the spectrum, excessively high discharges usually necessitate the construction of flood control facilities.

Hydrographs

A graph of discharge versus time is called a *hydrograph*. The vertical axis represents stream discharge. The horizontal axis represents time. Time intervals may span periods of several years or several hours, depending on the type of hydrograph and its use.

Streamflow over a span of 1 year may be depicted in an *annual hydrograph*, as illustrated in Figure 3.11. The isolated peaks on the hydrograph correspond to periods of heavy rainfall.

A *flood hydrograph* or *storm hydrograph* represents the flow response of a stream to one particular rainfall event. The time interval on the horizontal axis is generally hours or days, rather than months as in the annual hydrograph. A typical storm hydrograph is shown in Figure 3.12.

Shortly after the rain begins and overland flow reaches the stream channel, the discharge in the stream starts to increase. This is depicted as the rising limb of the hydrograph. After a time interval

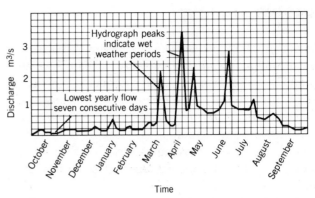

FIGURE 3.11 A typical *annual hydrograph* for a small stream.

called the *lag time,* which depends on the physical characteristics of the watershed, the maximum discharge occurs. This peak streamflow can occur many hours after the rain stops. After the peak is reached, the streamflow gradually recedes toward the *base flow.* The base flow is the normal dry weather flow in the stream; it is sustained by groundwater seeping out of the ground into the stream channel. A stream that has a base flow throughout the year is called a *perennial stream.* A stream that dries up completely during periods of very little rainfall is called an *intermittent* or *ephemeral stream.* Perennial stream channels penetrate the groundwater table, whereas ephemeral streams lie above the water table. Groundwater is discussed in more detail in Section 3.7.

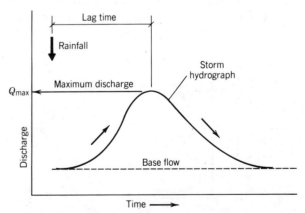

FIGURE 3.12 A *storm or flood hydrograph* shows the direct effect of a specific rainfall event on streamflow. A sharply rising limb, a peak flow after a lag time, and a gradually receding limb are typical characteristics.

GAGING STATIONS

The United States Geological Survey (USGS) is a government agency that measures streamflows and publishes records of discharges for most of the large streams or rivers in the United States as one of its responsibilities. In many cases, a permanent structure called a gaging station is constructed along the river to provide a continuous record of flow versus time. A sketch of a typical gaging station is shown in Figure 3.13.

The basic measurement made at a gaging station is of the depth of water in the stream or river. The elevation or height of the water surface above a reference level is called the *stage* of the stream. The stage changes as the discharge changes; as you would expect, higher stages correspond to higher discharges.

The stage is measured and recorded graphically on a rotating chart by a float-operated device. A cable with a float on one end and a counterweight on the other end is hung over a pulley, as shown. The float moves up or down as the stage changes, rotating the pulley and thus changing the position of a pen on the chart. The stilling well, connected to the stream channel by a pipe, prevents excessive fluctuations of the water level due to wind or other disturbances.

Before a gaging station can provide data on streamflow, it is necessary to determine the actual relationship between the stage and the discharge. This relationship is often expressed graphically in a rating curve or stage-discharge curve, as illus-

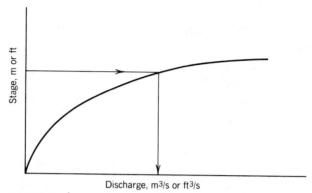

FIGURE 3.14   A typical *stage-discharge curve* for a stream or river, showing the relationship between flow rate and depth of water in that particular river.

trated in Figure 3.14. Once a rating curve is established for the stream, it is necessary only to measure the stage in order to know what the discharge is in terms of a volume flow rate.

One of the methods used to correlate stage with discharge is to construct a low dam or *weir* in the stream channel. As discussed in more detail in Section 2.4, a weir is an obstruction in the stream over which the water must flow. The height of water flowing over the weir, called the head on the weir, is related hydraulically to the volume flow rate. Figure 3.15 shows the installation of a weir and unsheltered gaging station in a small stream.

In larger streams or rivers, it may be impractical to obstruct the flow of water with a weir. The increase in water depth behind the weir can cause excessive flooding upstream. Instead, devices called current meters can be submerged at various points in the river to measure velocity of flow at various stages. A typical current meter includes a small propeller that rotates in the water in proportion to the water velocity. Knowing the depth and cross-sectional area of the stream where the current measurement is made, one can compute the discharge. (The hydraulic relationship between velocity and area of flow is explained in Section 2.2.) Since the depth and shape of the stream bed may change gradually because of erosion or sedimentation, the rating curve must be checked and revised from time to time.

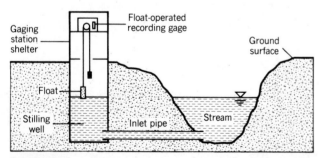

FIGURE 3.13   A cross-section of a typical stream gaging station. The water elevation in the stilling well is the same as the *stage* of the river or stream.

FIGURE 3.15 An unsheltered gaging station and a trapezoidal weir for a small stream. (Leupold & Stevens, Inc.)

3.5 DROUGHTS

In everyday terms, a *drought* is a long period of dry weather that causes a lack of available water. At the other extreme, a flood is what happens when a stream or river overflows its banks, shortly after periods of excessive rainfall or snowmelt. Both of these events are hydrologic extremes that are notorious for the environmental problems, in addition to possible loss of life and property damage they cause.

In order to reduce or mitigate the problems caused by floods or droughts, designers of hydraulic structures and water management facilities must be able to quantify the severity and frequency of these events. The magnitude of the "*N*-year flood" for a particular watershed must be determined if flood control efforts are to be effective. The low flow in a stream due to a drought must be estimated if the problems associated with prolonged periods of dry weather are to be avoided.

To a large extent, the occurrence and severity of floods or droughts can be related to precipitation.

Since precipitation records are more commonly available than streamflow data, designers often have little choice than to make estimates on the occurrence of droughts or floods from correlations with rainfall data. The computation of peak streamflows from rainfall data will be discussed in more detail in the chapter on stormwater management. It will be assumed that the return period of the maximum stream discharge is the same as the return period of the storm from which the discharge is computed.

The low flows that occur in perennial streams during droughts are of importance for two reasons. If a stream is to be used for water supply, we must determine whether or not a storage reservoir must be built to ensure adequate supply during a drought; and if the stream is receiving wastewater discharges from a sewage treatment plant, we must determine if the low streamflow will still be adequate to dilute the sewage, or if some type of advanced wastewater treatment is necessary.

"MA7CD10" Flow

In water pollution studies, a drought flow is commonly defined as the lowest average discharge over a period of 1 week with a recurrence interval of 10 years. This is called the *minimum average 7 consecutive day 10 year flow*, or, the *MA7CD10 flow*. Since the value of N for the MA7CD10 flow is 10 years, there is only a 10 percent probability that there will be a more severe drought in any given year. Or in other words, the probability is 90 percent that the minimum weekly discharge in the stream will be greater than the MA7CD10 flow. This is generally considered to provide an acceptable risk for water pollution control projects, and the MA7CD10 flow is used for design computations.

When many years of records of stream discharge are available, a statistical procedure called *frequency analysis* may be used to estimate return periods or frequencies of droughts. The same method may be used to determine flood frequencies, or recurrence intervals of storms from precipitation records. It is a good idea for designers, or others who make use of return period data, to have some understanding of how the data are determined.

To illustrate the procedure, a simplified example for determining a drought flow in a stream is presented here. In Example 3.6, only 5 years' worth of discharge records are used. In practical applications, much longer periods of record would be required for meaningful results. But for the sake of clarity, as well as to illustrate extrapolation beyond the period of record, this is a useful example.

EXAMPLE 3.6

Given the following record of streamflow data, estimate the MA7CD10 flow for the stream:

Year	Lowest Seven-Day Average Flow, m^3/s
1980	4.4
1981	2.8
1982	4.0
1983	3.4
1984	5.2

Solution:

First rearrange the flow data in decreasing order of magnitude and assign a "rank" or m value to each flow, beginning with 1 and increasing by 1 sequentially. The probability of observing an equal or higher flow in any given year is estimated by dividing the rank m by the number of years of record plus 1 $(n + 1)$; in this example $n = 5$. In formula form, the probability $P = m/(n + 1)$.

Low flow, m^3/s	Rank	Probability
5.2	1	1/6 = 0.167
4.4	2	2/6 = 0.333
4.0	3	3/6 = 0.500
3.4	4	4/6 = 0.667
2.8	5	5/6 = 0.833

Hydrologic data are often plotted on a special type of graph paper called logarithmic probability paper. The points usually plot as a straight line, or close to it. The low flows and their corresponding probabilities in this problem have been plotted, as shown in Figure 3.16. A straight line of best fit has been drawn through the plotted points and extended, or extrapolated, to the 90-percent probability value. This identifies a flow rate on the vertical axis of the graph that would be exceeded 9 times out of 10 in any given subsequent year. Conversely, the probability of observing a lower flow (a more severe drought) is 10 percent. This flow, therefore, represents the MA7CD10 flow. As seen in Figure 3.16, the MA7CD10 flow for this stream (based on the very limited 5 years of record) is estimated at 2.4 m^3/s.

3.6   RESERVOIRS

When streamflow is not sufficient for a dependable water supply, particularly during dry spells, a reservoir can be built to overcome this problem. A reservoir equalizes the flow in a stream, storing excessive wet weather flows for use during periods of low streamflow. A reservoir serving primarily for water supply is called a *conservation reservoir*. A reservoir of this type would be built on a natural site with suitable topography by blocking the stream with a dam, allowing an artificial lake to be formed.

Conservation reservoirs are usually large and provide capacity for long periods of dry weather. The flooding of land by the artificial lake can have significant environmental as well as social effects; these must be considered in addition to the purely technical or economic aspects of the project.

Because of economic and environmental factors, it is often not feasible to build a dam and reservoir for only one purpose, such as water supply. Reservoirs that simultaneously serve this and other needs, such as for flood control, hydroelectric power, and recreation, are called *multipurpose reservoirs*. Other types of reservoirs include *distribution storage reservoirs* for water distribution, and *detention*

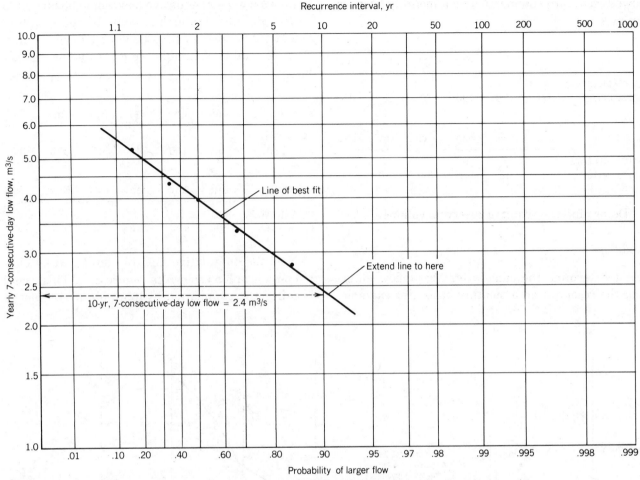

FIGURE 3.16 Logarithmic *probability paper* is used to estimate the MA7CD10 drought flow in a stream or river.

reservoirs for stormwater control. These will be discussed in subsequent chapters.

The storage capacity of a large reservoir is usually expressed in units of megaliters or acre–feet. The *yield* of a reservoir represents the amount of water the reservoir can supply in a specific time interval without going dry. The relationship between reservoir yield and storage capacity is a key factor in its design.

SUMMATION HYDROGRAPH

In order to determine the required volume of a conservation reservoir, records of streamflow spanning many years must be used. Conservation reservoirs are often designed to provide the needed yield during a drought equal to the worst drought on record. A *summation hydrograph*, also called a *mass diagram*, is a convenient graphical tool for determining the required storage volume. This technique is illustrated in Example 3.7.

EXAMPLE 3.7

A conservation reservoir is to provide a uniform withdrawal or yield of 60 ML/month without being depleted. The streamflow records for the year of lowest flows are summarized on a monthly basis as follows:

MONTH	Jan	Feb	Mar	Apr	May	Jun
STREAMFLOW, ML/MONTH:	60	100	180	20	15	15

MONTH:	Jul	Aug	Sep	Oct	Nov	Dec
STREAMFLOW, ML/MONTH:	5	15	115	200	180	100

Determine the required reservoir volume.

Solution:

First determine the cumulative streamflow entering the reservoir on a monthly basis. For example,

in February the cumulative flow would be 60 + 100 = 160 ML, and in March it would be 160 + 180 = 340 ML. Just keep adding the flows for each month. Prepare a table of cumulative monthly flows as follows:

MONTH:	Jan	Feb	Mar	Apr	May	Jun
CUMULATIVE FLOW, ML:	60	160	340	360	375	390

MONTH:	Jul	Aug	Sep	Oct	Nov	Dec
CUMULATIVE FLOW, ML:	395	410	525	725	905	1005

Now the cumulative monthly flows can be plotted on a graph as shown in Figure 3.17. This graph is called the summation hydrograph; it is a plot of

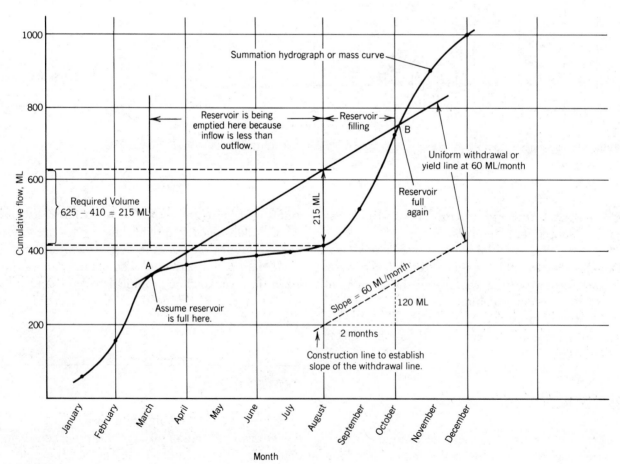

FIGURE 3.17 A *summation hydrograph* (for Example 3.7).

flow versus time, but the flows are cumulative with time.

The slope of the summation hydrograph or mass curve represents the rate of inflow into the reservoir. Notice that the slope is very flat during the summer months because of the low streamflows during that period.

A uniform yield or withdrawal can be represented as a straight line on the graph; in this case the withdrawal line has a slope of 60 ML/month, as shown. Where the mass curve slope is flatter than the withdrawal line slope, more water is leaving the reservoir than is flowing in; the reservoir is emptying. Where the mass curve is steeper than the withdrawal line, more water is flowing in than is flowing out of the reservoir; the reservoir is filling up.

Draw a line parallel to the withdrawal line and tangent to the mass curve at a point labeled A in Figure 3.17. Point A, in general, represents a peak of the mass curve where it is concave downward. Assume that the reservoir is just full at this point. It would immediately start to decrease in water volume since just past point A the rate of withdrawal exceeds the rate of inflow. But after several months, the slope of the mass curve increases, and the rate of inflow surpasses the rate of withdrawal; the water volume would begin to increase. At point B, where the line crosses the mass curve, the reservoir would be at full capacity once again.

The vertical distance between the yield line AB and the mass curve represents the volume of water taken out of storage to satisfy the yield or withdrawal. In this example, the maximum vertical distance is measured to be 215 ML, as shown in Figure 3.17. This is the minimum storage volume needed to ensure that the specified yield can be satisfied.

Since this volume of 215 ML was determined for the worst drought year on record, it is reasonable to assume that during years of normal precipitation and streamflow, the reservoir will be more than adequate to provide the required yield. But it is still possible that a more severe drought could occur. A frequency analysis could be done to allow estimates of recurrence intervals and probabilities of more serious droughts.

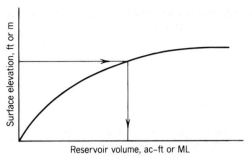

FIGURE 3.18 A typical *reservoir capacity curve.*

Reservoir Capacity

The maximum volume of water that can be stored in a reservoir depends on the elevation of the spillway of the dam forming the reservoir and on ground topography upstream of the dam. In addition to this total volume, it is important to know the relationship between the volume and the elevation of the reservoir surface. A graph of water elevation versus volume is called a *reservoir capacity curve* or *elevation-storage curve*. A typical capacity curve is illustrated in Figure 3.18. Using a curve like this, one can determine the volume of water in the reservoir at any given time by simply measuring the elevation of the water surface.

All streams and rivers carry suspended soil particles to some degree. These particles tend to settle out by gravity in the reservoir, forming deposits of *sediment*. All reservoirs ultimately become filled with sediment and therefore have limited design lives or periods during which they can fulfill their intended purposes.

Figure 3.19 illustrates the accumulation of sediment behind a dam. Although reservoir sedimen-

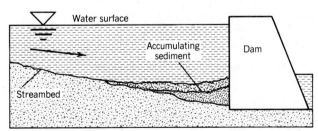

FIGURE 3.19 Sedimentation in a reservoir reduces its capacity to store water. Over time, a reservoir can eventually become completely filled in with sediment.

tation cannot be prevented, it can be controlled and slowed down. Sluice gates below the crest of the dam permit the occasional discharge of sediment before it has a chance to settle to the bottom.

3.7 GROUNDWATER

As discussed in Section 3.2, part of the precipitation that falls upon the land may infiltrate the surface, percolate downward through the soil under the force of gravity, and become what is known as *groundwater*. The groundwater is an extremely important part of the hydrologic cycle. Almost half of the people in the United States obtain their public water supply from groundwater. Overall, there is more groundwater than surface water in the United States, including the water in the Great Lakes. But it sometimes is uneconomical to pump it to the surface for use, and in recent years, pollution of groundwater supplies from improper disposal of wastes has become a significant problem.

After water first infiltrates the ground surface, it seeps downward through a layer of soil called the *zone of aeration.* This is a layer of soil in which the small spaces between the solid soil particles are partially filled with air as well as with water. As the water continues to percolate downward, it eventually reaches the *zone of saturation,* a layer of soil or rock in which all the pore spaces or rock fissures are completely filled with water. Even though the individual pore spaces and rock crevices are relatively small, the total volume of groundwater is large because the geologic formations that can hold water are so vast. Groundwater can be considered to be a huge subsurface reservoir.

The dividing line between the zone of aeration and the zone of saturation is called the *water table.* An excavation or a well that is deep enough to penetrate the zone of saturation would fill up with water to the height or elevation of the water table. This is illustrated in Figure 3.20.

The elevation of the water table is not constant. It depends on weather conditions and varies sea-

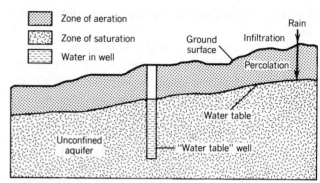

FIGURE 3.20 Soil below the *water table* is saturated with water.

sonally. The water table is generally closer to the ground surface in the spring or during rainy periods and deeper during dry spells. The water table can also be lowered by pumping, as will be seen in the discussion on wells later on in this section.

AQUIFERS

An *aquifer* is a layer of soil or rock in which groundwater can move relatively freely. It is, in other words, a geologic stratum that can transmit water in sufficient quantities to permit economical use of the groundwater for supply purposes. Porous sand and gravel aquifers yield more water than do more impermeable silt or clay deposits. Rock formations may contain enough cracks or fissures to yield significant quantities of water.

An aquifer that is "sandwiched" between two impermeable layers that block the flow of water is called a *confined* or *artesian* aquifer. The water in a confined aquifer is under hydrostatic pressure. It does not have a "free" water table. An imaginary line, representing the *piezometric surface,* can be used to represent the height to which water would rise in a well that penetrated the aquifer. This is illustrated in Figure 3.21.

The *recharge area* for an aquifer is where precipitation infiltrates the ground to replenish the water flowing through the aquifer. As seen in Figure 3.21, the recharge area may be remote from the point of actual water use. This is an important factor in

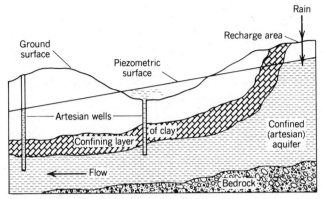

FIGURE 3.21 Water in a confined or *artesian aquifer* is under hydrostatic pressure. Water in an artesian well rises to the level of the *piezometric surface.*

land-use planning and urban development. Covering recharge areas with pavements and parking lots blocks the infiltration process and reduces the amount of water that can be withdrawn from the aquifer.

GROUNDWATER FLOW

Groundwater is in a constant state of motion through the pores and crevices of the aquifer in which it occurs. The water table is rarely level; it generally follows the shape of the ground surface. The groundwater flows in the downhill direction of the sloping water table, as illustrated in Figure 3.22.

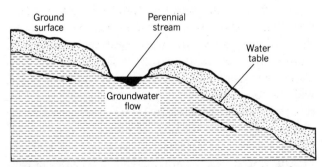

FIGURE 3.22 Groundwater flows slowly through the soil. Sometimes it seeps out of the ground in a *spring,* or into the channel of a *perennial stream,* resulting in a relatively stable dry-weather base flow in the stream.

The water table sometimes intersects low points of the ground where it seeps out into springs, lakes, or streams. As previously discussed, the base flow of a perennial stream is actually sustained by groundwater flow. Such streams are also called *influent* streams.

The rate of movement of groundwater due to gravity is usually very slow. It is limited by the frictional resistance to flow in the soil and rock openings. A velocity of about 18 m/d (60 ft/d) would be considered high, even in porous sand and gravel deposits. In more impervious clay soils, the velocity of flow may be as low as a fraction of a meter per year.

The velocity of groundwater flow is a function of the slope of the water table and the permeability of the soil. This relationship is expressed in a formula known as *Darcy's law,* as follows:

$$V = K \times S \qquad (3\text{-}4)$$

where V = flow velocity, mm/s
K = permeability coefficient, mm/s
S = slope of the water table

Permeability is a characteristic of a porous material that allows it to transmit water; it has the same units as velocity. The slope of the water table is the drop in elevation divided by the horizontal distance; it is a dimensionless number.

Darcy's law is the basis of more complicated mathematical analyses of groundwater hydraulics. If aquifer conditions are known, it is possible to predict such things as how much water can be pumped out of a well. But the limitations of mathematics, including Darcy's law, for evaluating groundwater conditions should be mentioned. To apply the formulas, we must assume that the aquifer is a uniform or homogeneous material that can be described by a single coefficient K. In reality, natural soil deposits or rock aquifers are rarely uniform over large areas. Nevertheless, Darcy's law is of value in making initial estimates of the rate of groundwater movement. Typical values of K are presented in Table 3.1 and the application of Darcy's law is illustrated in Example 3.8.

TABLE 3.1 TYPICAL PERMEABILITY COEFFICIENTS

Soil Type	K, mm/s
Gravel	10 to 40
Sand	0.01 to 10
Silty sand	0.001 to 0.02
Silt	0.0001 to 0.005
Clay	10^{-6} to 10^{-8}

EXAMPLE 3.8

Compute the velocity of groundwater flow in an aquifer that has a coefficient of permeability of K = 0.1 mm/s. The water table slopes at a rate of 1 m over a distance of 200 m.

Solution:

The slope of the water table is computed as

$$S = \frac{1 \text{ m}}{200 \text{ m}} = 0.005$$

and

$$V = K \times S = 0.1 \times 0.005 = 0.0005 \text{ mm/s}$$

or

$$V = 0.0005 \text{ mm/s} \times 3600 \text{ s/h} \times 24 \text{ h/d} = 432 \text{ mm/d}$$

WELLS

The most common method for withdrawing groundwater is to penetrate the aquifer with a vertical well and then pump the water up to the surface. Other methods include using natural springs or infiltration galleries. An infiltration gallery consists basically of several horizontal perforated pipes radiating outward from a large-diameter central shaft.

Wells may be constructed in a variety of ways,

depending on the depth and nature of the aquifer. A *dug well* is a shallow excavation, up to about 10 m (30 ft) deep, that penetrates an unconfined aquifer. It is generally lined with stone or masonry to support the side walls. Dug wells are not dependable sources of water because of the seasonal variation in the depth of the water table and the well's susceptibility to pollution. They still may be observed in use in some agricultural and rural areas, but modern environmental sanitation standards generally prohibit their construction for public water supply.

Wells up to about 20 m (65 ft) deep may be constructed in soft soils by driving a *well point* into the ground. A well point is a section of perforated pipe with an internal screen and a point on the lower end. The required depth is reached by coupling additional sections of pipe to the well point as it is driven down. This type of well is more commonly used to dewater construction excavations rather than to provide a water supply.

Deep wells, those more than 30 m (100 ft) deep, are most commonly used for public water supplies. They can penetrate extensive aquifers with more dependable yields of water and better water quality than shallower wells can. Deep wells are typically 100 to 300 mm (4 to 6 in.) in diameter. They are drilled using percussion or rotary drilling techniques.

Deep wells are permanently lined with a metal pipe, called a casing. The annular space around the casing is filled with cement grout. The casing and the grout serve to seal off poor quality water coming from the surface and upper soil layer, protecting the well from contamination. A sanitary seal is installed at the top of the casing to further protect the water quality. In unconsolidated aquifers, a slotted well screen is usually attached to the bottom of the casing to strain silt and sand out of the well water. These basic features of deep well construction are illustrated in Figure 3.23.

Multistage vertical turbine pumps (a type of centrifugal pump) are commonly used in deep wells to lift the water. (The operating characteristics of centrifugal pumps will be discussed in the chapter on water distribution systems.) The well pump may be driven by an electric motor at ground level con-

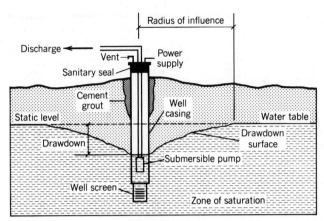

FIGURE 3.23 A schematic diagram of a water table well showing the *drawdown* that occurs during water withdrawal by pumping.

nected to the submerged pump by a shaft, or by a special submersible motor directly connected to the pump.

The elevation of the water table in the well before pumping begins is called the *static level*. When the well is pumped, the water level in the well drops below the static level, as seen in Figure 3.23. The elevation difference between the static level and the pumping level is called the *drawdown*. A *drawdown surface* of the water table, or *cone of depression,* as it is also called, is formed around the well during pumping.

As the distance from the well increases, the slope of the drawdown curve flattens out, eventually merging with the undisturbed static water table. The horizontal distance from the well to the area where the water table has not been appreciably affected by the pumping is called the *radius of influence* of the well. These terms are illustrated in Figure 3.23.

The drawdown and the circle of influence will increase as the pumping rate is increased. From the discussion of groundwater flow, it should be clear that a very permeable aquifer would have a smaller drawdown than an aquifer with less permeability,

for the same rate of pumping. Likewise, the radius of influence would be larger in the more porous aquifer. It should be noted that there are mathematical formulas that allow for the computation of all the terms defined here. But a full discussion of groundwater hydraulics is beyond the scope of this book.

The *safe yield* of a well is the rate at which water can be withdrawn without pumping the well dry. The larger the drawdown is, the greater the yield is. The relationship between the yield and drawdown of a well is called its *specific capacity*. For example, if the drawdown in a well is 50 m when the withdrawal rate is 500 m^3/h, the specific capacity would be expressed as 500/50 or 10 m^3/h per meter of drawdown. If the drawdown was 25 m instead of 50 m, the output from the well could be estimated to be 10 m^3/h/m $\times$ 25 m = 250 m^3/h. The diameter of a well has very little effect on the yield; doubling the diameter would increase the yield by only about 10 percent.

Wells must be "developed" before they are put into use in order to remove silt and fine sand adjacent to the well screen. Well development unplugs the aquifer, producing a natural filter of coarser particles around the well screen and allowing silt-free water to flow freely into the well. In one method of developing a well, called *surging,* a plunger is moved rapidly up and down in the well. In aquifers consisting of very fine uniform sand and silt, a filter must be constructed around the well screen. This is called *gravel-packing* and is accomplished by filling the annular space around the well screen with gravel.

After a well is developed, a pump test is conducted to determine if it can supply the required amount of water. The well is generally pumped for at least 6 hours at a rate equal to or greater than the desired yield. A stabilized drawdown should be obtained at that rate, and the original static level should be recovered within 24 hours after pumping stops. During this test period, samples are taken and tested for bacteriological and chemical quality.

REVIEW QUESTIONS

1. Why is the science of hydrology of importance in environmental technology?

2. List two uses of water other than for public supplies. Which use requires the greatest amount of water?

3. Is there a difference between water withdrawal and water consumption? Name two nonwithdrawal uses of water.

4. Briefly discuss the relative availability of water across the United States.

5. Briefly outline the basic features of the hydrologic cycle.

6. What is the origin of subsurface water (groundwater)?

7. What is the meaning of *rainfall intensity?* How is it measured?

8. What is an "acre–foot"?

9. Briefly discuss the meaning of "*N*-year storm."

10. If a so-called 5-year storm occurred today, when would you next expect to observe a similar storm? Explain.

11. Is a 50-year storm more likely to be observed than a 20-year storm? Explain.

12. What is the difference between the expressions *storm recurrence interval* and *storm frequency?*

13. On average, are the intensities of long-duration storms less than, equal to, or greater than the intensities of short-duration storms? Sketch a graph of intensity versus duration to illustrate your answer.

14. List three basic characteristics of a drainage divide line.

15. List three general characteristics of a watershed that may affect the volume and rate of runoff.

16. What is a hydrograph?

17. What is meant by "base flow" of a perennial stream?

18. Briefly explain the basic operation of a stream gaging station.

19. What is a stage-discharge curve?

20. What is the MA7CD10 flow of a stream?

21. What is a summation hydrograph and how is it used?

22. Sketch a typical reservoir capacity curve. Why do you think it is shaped concave downward?

23. What is the difference between the zone of aeration and the zone of saturation?

24. What is a water table? What may cause it to change position?

25. What is an aquifer?

26. What is the difference between a water table well and an artesian well?

27. Does groundwater constantly flow through the ground or is it stationary? Explain.

28. What is the most common method of groundwater withdrawal for public water supply? Briefly discuss construction and operation details.

29. For a given rate of groundwater withdrawal, would the drawdown occurring in a sand and gravel aquifer be any different than that in a less permeable, silty aquifer? Would there be any difference in the radius of influence?

PRACTICE PROBLEMS

1. A total of 500 mm of rain fell on a 75-ha watershed in a 10-h period. Compute the average rainfall intensity and the total volume of rain that fell on the watershed.

2. A total of 1 in. of rain fell on a 96-ac area in 30 min. Compute the average rainfall intensity and the total volume of rain that fell.

3. Using the rainfall curves in Figure 3.5, determine the expected rainfall intensity for (a) a 5-year storm of 10-min duration, (b) a 10-year storm of 1.5-h duration, and (c) a 100-year storm of 2-h duration.

4. A storm of 30-min duration causes 75 mm of rainfall. Using the rainfall curves in Figure 3.5, estimate the probability of observing a similar storm in the next year.

5. Using Equation 3-3, determine the expected rainfall intensity for a storm of 1.5-h duration if $A = 3000$ and $B = 20$.

6. What is the probability of a 20-year storm being equalled or exceeded in any given year?

7. Based on the following record of weekly low flows in a river, determine the MA7CD10 drought flow. Use the probability graph of Figure 3.16; multiply vertical axis values by 10.

Year	Discharge, m^3/s	Year	Discharge, m^3/s
1970	50	1978	40
1971	47	1979	45
1972	57	1980	50
1973	42	1981	33
1974	36	1982	45
1975	39	1983	48
1976	53	1984	50
1977	44	1985	41

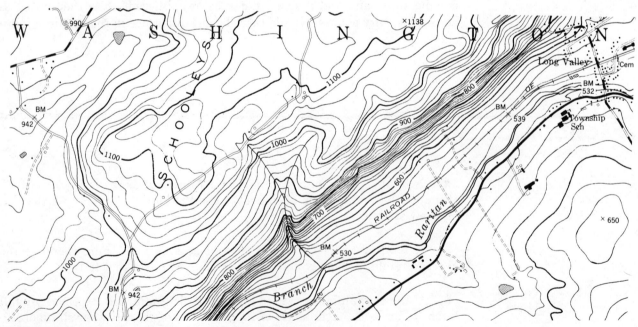

FIGURE 3.24   Illustration for Problem 8*(a)*. (U.S. Geological Survey)

FIGURE 3.25 Illustration for Problem 8*(b)*. (U.S. Geological Survey)

8. (a) In Figure 3.24, a stream tributary to the south branch of the Raritan River is shown on a USGS topographic map. (The point of confluence is near BM 530 on the RR line.) Sketch the drainage divide line for the stream, beginning at its point of confluence.

(b) In Figure 3.25, Clyde Potts Reservoir is shown on a USGS topographic map. Sketch the drainage basin boundary for the reservoir.

9. A conservation reservoir is needed to provide a uniform withdrawal or yield of 0.5 mgd without being depleted. The streamflow records for the year of lowest flows are summarized on a monthly basis as follows:

MONTH	January	February	March	April	May	June
STREAMFLOW MIL. GAL/MONTH	20	20	10	2	10	38

MONTH	July	August	September	October	November	December
STREAMFLOW, MIL. GAL/MONTH	8	3	2	6	30	50

Determine the required reservoir volume.

10. Compute the velocity of groundwater flow in soil that has a coefficient of permeability of 0.05 mm/s if the water table drops 0.5 m in elevation over a distance of 100 m.

11. The slope of the water table is determined to be 0.035 and the average velocity of groundwater flow is determined to be 0.5 m/h. In what type of soil deposit is the flow probably occurring?

12. The specific capacity of a 100-mm-diameter well is 2 $m^3/h/m$. What is the yield of the well when the drawdown is 15 m? If the well diameter were doubled to 200 mm, what would you expect the yield to be at the same drawdown of 15 m?

SELECTED REFERENCES

1. Chow, V. T. (ed.), *Handbook of Applied Hydrology,* McGraw–Hill, New York, 1964.

2. *Hydrology Handbook,* American Society of Civil Engineers, MOP 28, 1955.

3. *Stevens Water Resources Data Book,* 3rd ed., Leupold & Stevens, Beaverton, Oregon, 1978.

4. Linsley, R. K. and J. B. Franzini, *Water Resources Engineering,* 2nd ed., McGraw–Hill, New York, 1972.

5. Steel, E. W. and T. J. McGhee, *Water Supply and Sewerage,* 5th ed., McGraw–Hill, New York, 1979.

6. Fair, G. M. et al., *Elements of Water Supply and Wastewater Disposal,* 2nd ed., John Wiley & Sons, New York, 1971.

7. Clark, J. W. et al., *Water Supply and Pollution Control,* 3rd ed., Harper and Row, New York, 1977.

4

WATER QUALITY

4.4 BIOLOGICAL PARAMETERS OF WATER QUALITY

MICROORGANISMS
INDICATOR ORGANISMS
TESTING FOR COLIFORMS

4.5 WATER SAMPLING PROCEDURES

GRAB SAMPLES
COMPOSITE SAMPLES
GENERAL REQUIREMENTS

The topic of water quality focuses on the presence of "foreign" substances in water and their effects on people or the aquatic environment. Water of "good quality" for one purpose may be considered to be of "poor quality" for some other use. For example, water suitable for swimming may not be of good enough quality for drinking. But even drinking water may not be suitable for certain industrial or manufacturing purposes that require "pure" water.

What exactly is pure water? Just how pure does it have to be for drinking, or for other uses? Obviously, it is not enough for us to simply describe water quality as being "good" or "poor." We need some quantitative measures in order to determine and describe the condition of the water. It is necessary to determine what substances are in the water and in what concentrations they are present. We also must have some knowledge of the effects of those substances on public and environmental health. Finally, we need to have some "yardsticks" or standards against which we can compare the results of our analysis and thereby judge the suitability of the water for a particular use.

Water has a remarkable tendency to dissolve other substances. Because of this, it is rarely found in nature in a pure condition. Even water in a mountain stream, far from civilization, contains some natural impurities in solution and suspension.

Changes in water quality begin with precipitation. As rain falls through the atmosphere, it picks up dust particles and such gases as oxygen and carbon dioxide. In some industrialized regions, the quality of rain water is altered significantly before it ever touches the ground. (The topic of "acid rain," a prime example of this, will be discussed in the chapter on air pollution.)

Surface runoff picks up silt particles, bacteria, organic material, and dissolved minerals. Groundwater usually contains more dissolved minerals than surface water because of its longer contact with soil and rock. Finally, water quality is very much affected by human activities, including land use (like agriculture) and the direct discharge of municipal or industrial wastewaters to the environment.

Protecting water quality and modifying it for a particular purpose are major objectives in the field of environmental technology. We will therefore have to make use of technical terms in discussing the various aspects of water quality and pollution. In particular, we will be referring to the different *parameters* of physical, chemical, and biological quality.

This chapter begins with an overview of chemical concepts and terminology. In the discussion of specific water quality parameters that follows, only brief reference is made to actual laboratory analysis procedures. These are thoroughly described in *Standard Methods for the Examination of Water and Wastewater* and other professional publications (see References). Discussion of federal water quality standards is included in subsequent chapters, as they relate to drinking water, surface water, or wastewater treatment plant effluents.

Portable field test kits, as illustrated in Figure 4.1, are particularly useful for conducting preliminary water quality surveys. But for most water quality analyses to be "official" and able to stand up to legal scrutiny if challenged, they must be done by qualified personnel in certified laboratories, following *Standard Methods*.

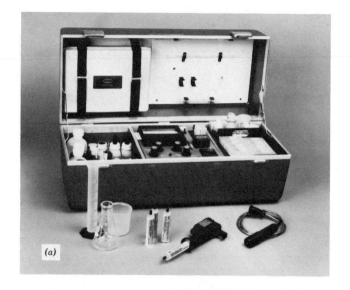

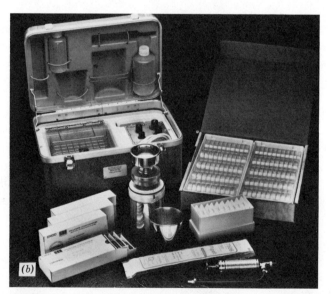

FIGURE 4.1 Portable water testing kits are used in the field to measure (a) chemical and physical quality (HACH Company, Loveland, Colorado), and (b) microbiological or sanitary quality (Millipore Corporation, Bedford, Massachusetts).

4.1 FUNDAMENTAL CONCEPTS IN CHEMISTRY

The study of water quality and pollution control requires a basic knowledge of *chemistry,* a science that focuses on the composition and properties of substances. For those students with little or no previous training in chemistry, this section will provide a foundation for an understanding of the environmental topics covered later. It may serve as a quick review for others.

ELEMENTS AND COMPOUNDS

All matter is composed of basic substances called *elements* that cannot be subdivided or broken down into simpler substances by ordinary chemical change. The smallest part of an element that can exist and still retain the same chemical characteristics of that substance is called an *atom.*

There are over 100 known elements. Some of the more common elements along with the symbols used to represent them in chemical formulas and equations are listed in Table 4.1.

TABLE 4.1 COMMON ELEMENTS AND THEIR SYMBOLS

Element	Symbol	Element	Symbol
Aluminum	Al	Magnesium	Mg
Arsenic	As	Manganese	Mn
Barium	Ba	Mercury	Hg
Cadmium	Cd	Nitrogen	N
Calcium	Ca	Oxygen	O
Carbon	C	Phosphorus	P
Chlorine	Cl	Potassium	K
Chromium	Cr	Selenium	Se
Copper	Cu	Silicon	Si
Fluorine	F	Silver	Ag
Hydrogen	H	Sodium	Na
Iron	Fe	Sulfur	S
Lead	Pb	Zinc	Zn

The science of chemistry is concerned with how the elements react and combine with each other, forming *compounds.* Compounds are substances made up of various combinations of the basic elements. The smallest part of a chemical compound that can exist and still retain the same chemical properties of that compound is called a *molecule.*

Molecules can be represented using combinations of the symbols for the atoms in the molecule; such a combination is called a *chemical formula.* For example, a single molecule of water is composed of two atoms of hydrogen H and one atom of oxygen O. Its chemical formula is H_2O, pronounced "H-two-O." The subscript *2* after the *H* indicates that two atoms of hydrogen are in a water molecule. The formula for iron oxide, commonly called rust, is Fe_2O_3, indicating that there are two atoms of iron and three atoms of oxygen in one molecule of this compound.

There are hundreds of thousands of known compounds. Chemists have traditionally separated them into two broad groups, called *organic* compounds and *inorganic* compounds. Organic compounds are typically complex molecules of carbon in combination with other elements such as hydrogen and oxygen. We will discuss this type of substance in more detail later in this section.

Inorganic compounds usually do not contain carbon, although there are exceptions to this. In a very general sense, organic compounds are closely related to living organisms, whereas inorganic compounds are more a part of the inanimate world. A list of common inorganic compounds is presented in Table 4.2. Many of these compounds are used in water or wastewater treatment operations and will be mentioned again later. The list includes the common names and formulas of each compound as well as its physical state (solid, liquid, or gas) at ordinary room temperature and pressure.

Atomic Structure

The way in which elements combine with each other to form compounds depends on their atomic structure. A simplified model of atomic structure includes a dense center or *nucleus* of positively charged particles called *protons* and uncharged or

TABLE 4.2   COMMON INORGANIC COMPOUNDS

Chemical Name	Common Name	State	Formula
Aluminum sulfate	Alum	Solid	$Al_2(SO_4)_2$
Ammonia	—	Gas	NH_3
Calcium carbonate	Limestone	Solid	$CaCO_3$
Calcium hydroxide	Slaked lime	Solid	$Ca(OH)_2$
Calcium hypochlorite	—	Solid	$Ca(ClO)_2$
Calcium oxide	Lime	Solid	CaO
Carbon dioxide	—	Gas	CO_2
Carbon monoxide	—	Gas	CO
Chlorine	—	Gas	Cl_2
Copper sulfate	Blue vitriol	Solid	$CuSO_4$
Iron oxide	Rust	Solid	Fe_2O_3
Hydrogen	—	Gas	H_2
Hydrogen sulfide	—	Gas	H_2S
Hydrochloric acid	Muriatic acid	Liquid	HCl
Hypochlorous acid	—	Liquid	$HOCl$
Nitric acid	—	Liquid	HNO_3
Nitrogen	—	Gas	N_2
Nitrogen oxide	—	Gas	NO
Nitrogen dioxide	—	Gas	NO_2
Oxygen	—	Gas	O_2
Ozone	—	Gas	O_3
Sodium carbonate	Soda ash	Solid	Na_2CO_3
Sodium chloride	Table salt	Solid	$NaCl$
Sodium hydroxide	Lye	Solid	$NaOH$
Sodium hypochlorite	—	Solid	$NaClO$
Sulfur dioxide	—	Gas	SO_2
Sulfuric acid	Oil of vitriol	Liquid	H_2SO_4

neutral particles called *neutrons*. Very light, negatively charged particles called *electrons* spin around the atomic nucleus in concentric shells or orbitals. To illustrate this basic model of atomic structure, schematic diagrams of a hydrogen atom and an oxygen atom are shown in Figure 4.2.

The unique identity of an element is established by the number of protons in its nucleus, called its *atomic number*. Each of the known elements has a different atomic number. For example, the atomic number of hydrogen is 1 and the atomic number of oxygen is 8. The atoms themselves are electrically neutral since the number of negatively charged electrons orbiting the nucleus is the same as the

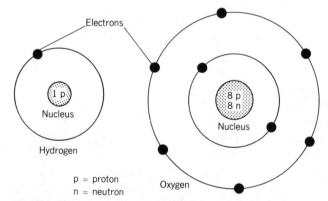

FIGURE 4.2 The "solar system" model of atomic structure, for hydrogen and oxygen. Hydrogen has one proton in its nucleus; oxygen has eight protons and eight neutrons.

number of positively charged protons inside the nucleus; the opposite charges cancel or balance each other.

The total number of protons plus neutrons in a nucleus is called the mass number of the element, and it is approximately equal to its *atomic weight.* The electrons themselves have very little mass or weight. For example, the atomic weight of hydrogen is 1 (since it has no neutrons in its nucleus), and the atomic weight of oxygen is 16 (eight protons plus eight neutrons).

The way an element behaves chemically depends primarily on the number of electrons in the atom's outermost shell or orbital. The orbital closest to the nucleus is most stable when it has two electrons, which is the maximum it can contain. The second orbital is most stable when it contains its maximum of eight electrons. Larger atoms have additional electron orbitals.

Formation of Molecules

Compounds are formed by either the transfer or the sharing of electrons among two or more atoms. For example, sodium chloride, NaCl, is formed by the transfer of one electron from the outermost shell of the Na atom to the outermost shell of the Cl atom. As a result, the Na has a positive charge and the Cl has a negative charge. Since unlike charges attract each other, like magnets, the NaCl molecule is formed as the Na and Cl atoms "stick" together. This is an example of what is called *ionic bonding* between atoms.

In the case of the water molecule, H_2O, the two atoms of hydrogen tend to "share" their electrons with the oxygen atom, as illustrated in Figure 4.3. This is called *covalent bonding.* In ionic bonding there is a complete transfer of electrons. But in covalent bonding, the outer orbitals are stabilized by a sharing of electrons.

The atoms in a water molecule are arranged at an angle instead of along a straight line, and the shared electrons are pulled closer to the oxygen atom than to the hydrogen atoms. This results in what is called a *polar molecule,* in which the positive and negative charges are not evenly distributed. The oxygen end of the molecule is negatively

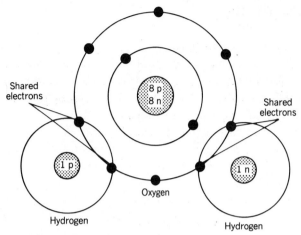

FIGURE 4.3 The water molecule, written as H_2O, is formed by the sharing of electrons between a hydrogen and an oxygen atom. This is an example of *covalent bonding.*

charged and the hydrogen ends are positively charged. It is this polarity of the water molecule that accounts for most of its properties, including its ability to dissolve many other substances. A schematic drawing of polar water molecules and the so-called hydrogen bonds that hold them together in a volume of water is shown in Figure 4.4.

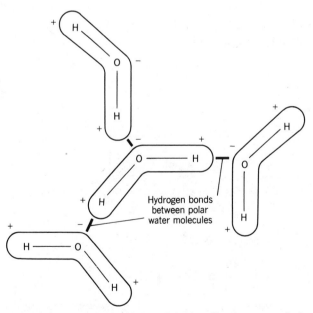

FIGURE 4.4 Water molecules "stick" together because of attractive forces in what are called *hydrogen bonds.*

SOLUTIONS

A *solution* is a uniform mixture of two or more substances existing in a single phase, that is, as either a solid, a liquid, or a gas. Solutions in water are called *aqueous solutions* and are the most familiar to us. But we can also consider carbon monoxide, CO, to be "dissolved" in the air; this is a gaseous solution. Also, carbon, C, and manganese, Mn, are "dissolved" in iron, forming steel, an example of a solid solution.

In these mixtures, the substance present in the largest amount is called the *solvent* and the substances present in smaller amounts are called *solutes*. The properties of solutions differ from the properties of the solvent. For example, although water freezes at 0°C, the presence of a dissolved salt, such as NaCl, in water, lowers the freezing point of the solution to below 0°C. Most chemical changes or reactions take place in solution.

Consider the aqueous solution of sugar shown schematically in Figure 4.5. The individual sugar molecules are uniformly dispersed throughout the water and do not settle out to the bottom. The solute molecules will always remain evenly mixed in the solvent because of the kinetic energy and constant motion of all the molecules.

If more sugar were added, the solution would eventually reach a state where the sugar molecules would no longer dissolve. At this point, the solution is called a *saturated solution*. Temperature has a great effect on the saturation point, that is, on the amount of solute a solution can hold before it becomes saturated. Solid substances are more soluble in warm water than in cold water (e.g., increasing the temperature of the sugar-water solution would allow more sugar molecules to dissolve).

Some substances can dissolve in water without limit. Alcohol, for example, can be mixed with water in all proportions. If there is more alcohol than water, the alcohol would then be considered to be the solvent and water would be the solute.

The solubility of various gases in water, such as oxygen, carbon dioxide, and chlorine, is of particular concern in environmental technology. As with solids, the solubility of gases depends on temperature, but the relationship is just the opposite of that for solids. *The solubility of a gas decreases with increasing temperature.*

The solubility of oxygen is of particular importance in water quality, as we will discuss in more detail later in this chapter and in the following chapter. Typical saturation values of dissolved oxygen in fresh water, at selected temperatures, are summarized in Table 4.3. This illustrates the very limited solubility of oxygen, as well as the pronounced effect of temperature on solubility.

The term mg/L used in the table for the concentration of oxygen in the water is an abbreviation for *milligrams per liter*. We will discuss this and other ways of expressing concentrations later.

Factors other than temperature also affect the solubility of gases. At higher altitudes and lower atmospheric pressures, the solubility of a gas is less than it is at sea level. Also, increasing the salinity, or salt content, decreases the solubility of gases.

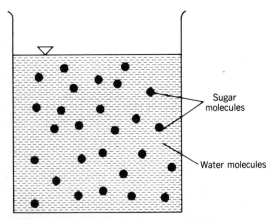

FIGURE 4.5 A schematic representation of an aqueous sugar solution. The sugar molecules remain uniformly dispersed in the volume of water.

TABLE 4.3 SOLUBILITY OF OXYGEN IN FRESH WATER

Temperature, °C	Saturation Solubility, mg/L
0	14.6
10	11.3
20	9.2
30	7.6

For example, less oxygen can be dissolved in sea water than in fresh water, under the same conditions of temperature and pressure.

Ionization

In the previous illustration of an aqueous sugar solution, the sugar molecules retained their identity. In other words, they did not break apart into fragments smaller than sugar molecules. The uncharged or neutral sugar molecules remained dispersed in the solution, surrounded by the water molecules.

There are many substances, however, that *dissociate* or break apart as they dissolve, forming smaller electrically charged particles called *ions*. This process is called *ionization*.

We previously described sodium chloride, NaCl, as an ionic compound because of the nature of the chemical bond between the Na and Cl atoms. Sodium chloride ionizes in water, as shown in the following chemical equation:

$$\text{NaCl} \rightarrow \text{Na}^+ + \text{Cl}^-$$

sodium	positive	negative
chloride	sodium ion	chloride ion

The sodium ion, Na^+, has a positive charge because it has given up its outermost electron to the chlorine atom, and therefore it has more protons than electrons. The chloride ion, Cl^-, is negative because it has more electrons than protons; its outer shell is stable with the extra electron from the sodium atom. Ionic solutions are neutral; the total positive charges must equal the total negative charges in the solution. Figure 4.6 illustrates schematically the solution of NaCl in water.

Radicals

In some cases, molecules dissociate into charged particles consisting of groups of atoms that act together as a unit. These charged fragments are called *radicals* and they have special names that are used quite frequently in water chemistry. For example, the water molecule itself can dissociate

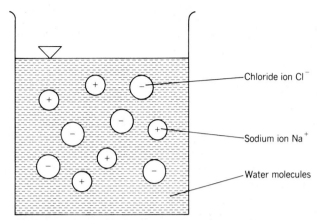

FIGURE 4.6 A schematic representation of an *ionic solution* of sodium chloride, NaCl, in water.

into a hydrogen ion, H^+, and a *hydroxyl radical*, OH^-, as follows:

$$\text{H}_2\text{O} \rightarrow \text{H}^+ + \text{OH}^-$$

Sulfuric acid is an example of a compound that dissociates readily in water, as follows:

$$\text{H}_2\text{SO}_4 \rightarrow 2\text{H}^+ + \text{SO}_4^{2-}$$

The SO_4^{2-} is called a *sulfate radical;* it has a double negative charge, balancing the total positive charge of the two hydrogen ions.

Other radicals of interest in environmental applications include *nitrate,* NO_3^-, *phosphate,* PO_4^{2-}, *ammonium,* NH_4^+, and *hypochlorite,* OCl^-. There are many others, some of which we will come across in later discussions of water treatment and pollution.

SUSPENSIONS AND COLLOIDS

Although many substances occur in solution in molecular or ionic form, some substances may be suspended in the mixture in fragments larger than the size of molecules. The properties of these mixtures differ from those of true solutions and are of particular significance in environmental technology.

Perhaps one of the most significant characteristics of a true solution is that the solute particles do not settle out or separate from the mixture, no matter how long the solution remains under quiescent or still conditions. Further, solutes can not be physically separated from the solvent by filters.

In contrast to this, suspended particles will settle out from the water if allowed enough time. Also, suspended particles can be removed from the water by filters. Because of this, suspensions of silt, organic material, and even microbes are among the first of the impurities to be removed in conventional water and sewage treatment systems.

A suspension of relatively large particles is called a *coarse suspension*. In quiescent or still water, large particles will settle out of the water in a matter of minutes. Finer particles, however, will take many hours to settle out. Even some bacteria, about 0.001 mm or 1 μm in size, will eventually settle to the bottom. (A μm, pronounced "micrometer," is one-millionth or 10^{-6} of a meter; see Appendix C.1 for further discussion of metric terms and symbols.) It can be predicted mathematically that bacteria will settle at a rate of about 1 m in 175 h. But this is hardly an efficient way to remove bacteria from water.

Extremely fine particles, those less than about 0.1 μm, are generally too small to settle out because of the force of gravity or to be removed by most filters. These particles, smaller than those in coarse suspensions but larger than those in true solutions, are called *colloids*. Colloids occur in air as well as in water. In water, for example, clay particles or tiny fragments of decaying vegetation and organic wastes may form colloidal suspensions. In air, colloidal suspensions of tiny solid particles (smoke) or liquid particles (fog) are often encountered.

Like the particles in a true solution, the particles in a colloidal suspension cannot be seen under a microscope. In a true solution, a beam of light can pass through the solution without any scattering of the light. However in a colloidal suspension, the colloidal particles will scatter the light, allowing the beam to be seen. This is called the *Tyndall effect,* and it is one of the characteristics distinguishing colloids from solutions. The Tyndall effect is illustrated in Figure 4.7.

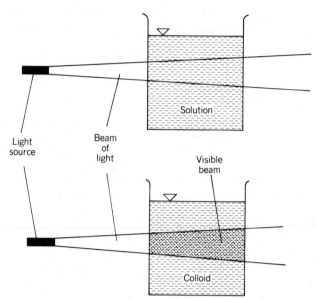

FIGURE 4.7 The *Tyndall effect*: Light is scattered by the colloidal particles, and the light beam is visible in the liquid.

Colloidal particles may have either all positive or all negative electrical charges of variuos magnitudes, depending on the nature of the substance. Since like charges repel each other, colloidal particles keep their distance from each other. There is a force of repulsion between them, similar to the force between the same poles of two bar magnets. Because colloidal particles repel each other, they very rarely collide, so they have no chance to stick together to form larger, heavier particles. In addition to their small size, this is a basic reason that colloids are stable and do not settle out of suspension. A very common water treatment process involves the addition of certain chemicals to neutralize the effect of the colloidal charges. This allows the particles and chemicals to collide and form *flocs* that can settle out or be separated from the water by filters. This will be discussed in more detail in the chapter on water treatment.

EXPRESSING CONCENTRATIONS

The properties of solutions, suspensions, and colloids depend to a large extent on their concentrations. A *dilute* or weak solution has a relatively small amount of solute dissolved in the solvent. It would have characteristics different from those of a *concentrated* or strong solution of the same substance, in which a relatively large amount of solute is present. Since we need to express concentrations quantitatively, instead of using qualitative terms like *dilute* or *strong,* concentrations are usually expressed in terms of mass per unit volume, parts per million or billion, or percent.

Mass per Unit Volume

One of the most common terms for concentration is *milligrams per liter* (mg/L). For example, if a mass of 10 mg of oxygen is dissolved in a volume of 1 L of water, the concentration of that solution would be expressed simply as 10 mg/L. If 0.3 g of salt were dissolved in 1500 mL of water, then the concentration would be expressed as 300 mg ÷ 1.5 L = 200 mg/L, where 0.3 g = 300 mg and 1500 mL = 1.5 L.

Very dilute solutions are more conveniently expressed in terms of *micrograms per liter* (μg/L). For example, a concentration of 0.004 mg/L is preferably written as its equivalent 4 μg/L. Since 1000 μg = 1 mg, simply move the decimal point three places to the right when converting from mg/L to μg/L. Move the decimal three places to the left when converting from μg/L to mg/L. For example, a concentration of 1250 μg/L is equivalent to 1.25 mg/L.

In air, concentrations of particulate matter or gases are commonly expressed in terms of *micrograms per cubic meter* (μg/m^3).

Parts per Million

One liter of water has a mass of 1 kg. But 1 kg is equivalent to 1000 g or 1 000 000 mg. Therefore, if 1 mg of a substance is dissolved in 1 L of water, we can say there is 1 mg of solute per 1 million mg of water. Or, in other words, there is "one part per million."

Neglecting the small change in the density of water as substances are dissolved in it, we can say that, in general, a concentration of one milligram per liter is equivalent to one part per million: *1 mg/L = 1 ppm.* Conversions are very simple: For example, a concentration of 17.5 mg/L is identical to 17.5 ppm.

The expression *parts per million* is useful in conveying a picture of just how small most of the concentrations we encounter in environmental technology actually are. If 1 lb of salt were dissolved in a 50 ft × 50 ft × 50 ft tank of water, the concentration of salt would be about 1 ppm. One part per million is the same as 1 in. in about 16 miles, or 1 s in about 12 d.

Very dilute concentrations can be expressed in terms of *parts per billion* (ppb), instead of parts per million. For example, a concentration of 0.005 ppm is better written as its equivalent of 5 ppb. Even such tiny concentrations of some substances can significantly affect environmental quality and human health.

The expression *mg/L* is preferred over *ppm*, just as the expression *μg/L* is preferred over its equivalent of ppb. But both types of units are still used and the student should be familiar with each.

Percentage Concentration

Concentrations in excess of 10 000 mg/L are generally expressed in terms of percent, for convenience. For practical purposes, we can use the conversion of 1 percent = 10 000 mg/L, even though the densities of the solutions are slightly more than that of pure water. (10 000 mg/L = 10 000 mg/1 000 000 mg = 1 mg/100 mg = 1 percent.)

The concentration of salts in sea water is about 35 000 mg/L. To convert to percent salts, divide mg/L by 10 000, or 3.5 percent. The concentration of wastewater sludge may be about 3 percent solids. To convert this to mg/L, multiply by 10 000, getting 30 000 mg/L solids.

A concentration expressed in terms of percent may also be computed from the following equation:

$$\text{percent} = \frac{\text{mass of solute (mg)}}{\text{mass of solvent (mg)}} \times 100 \quad (4\text{-}1)$$

American Units

The expression *grains per gallon* (gpg) is sometimes used for the concentrations of certain substances in water. One grain per gallon is equivalent to a concentration of 17.1 milligrams per liter: *1 gpg = 17.1 mg/L.*

The expression *pounds per million gallons* is also used in American units of concentration for water treatment applications. Since 1 gallon of water weighs 8.34 pounds, 1 gallon per million gallons is the same as 8.34 pounds per million gallons. Or, we can say that *1 mg/L = 8.34 lb/mil gal.* To convert from mg/L to lb/mil gal, simply multiply by 8.34; to go from lb/mil gal to mg/L, divide by 8.34.

EXAMPLE 4.1

A 500-mL aqueous salt solution has 125 mg of salt dissolved in it. Express the concentration of this solution in terms of *(a)* mg/L, *(b)* ppm, *(c)* gpg, *(d)* percent, and *(e)* lb/mil gal.

Solution:

(a) 125 mg/500 mL × 1000 mL/L = 250 mg/L
(b) 250 mg/L = 250 ppm
(c) 250 mg/L × 1 gpg/17.1 mg/L = 14.6 gpg
(d) Applying Equation 4-1, and the fact that 500 mL of water has a mass of 500 g, we get,

percent = 0.125 g/500 g × 100 = 0.025 percent

Or, divide 250 mg/L by 10 000 to get 0.025 percent
(e) 250 mg/L × 8.34 = 2090 lb/mil gal

EXAMPLE 4.2

How many pounds of chlorine gas should be dissolved in 8 mil gal of water to result in a concentration of 0.2 mg/L?

Solution:

$$0.2 \text{ mg/L} \times 8.34 = 1.67 \text{ lb/mil gal}$$

and

$$1.67 \text{ lb/mil gal} \times 8 \text{ mil gal} \approx 13 \text{ lb}$$

ACIDS, BASES, AND pH

Technically, an *acid* is a substance that causes an increase of the hydrogen ion, H^+, concentration in solution. A substance that causes the hydroxyl, OH^-, concentration to increase is called a *base*. Acids and bases may be characterized as "strong" or "weak," depending on the degree to which they increase the relative concentrations of H^+ or OH^-.

Hydrochloric acid, HCl, is an example of a strong acid that readily dissociates in water forming H^+ and Cl^- ions; the H^+ concentration in the water is greatly increased. Sodium hydroxide, NaOH, is an example of a strong base that dissociates into Na^+ and OH^-, greatly increasing the OH^- concentration. A substance that is basic, like NaOH, is also called an *alkaline* substance.

The chemical reaction between an acid and a base is called *neutralization*. The two products of a neutralization reaction are water and a salt. Table salt, NaCl, for example, is a product of the neutralization reaction between hydrochloric acid, HCl, and sodium hydroxide, NaOH, as follows:

$$HCl + NaOH \rightarrow NaCl + H_2O$$

The pH Scale

The pH is a dimensionless number that indicates the "strength" of an acidic or basic solution. The pH scale ranges from 0 to 14. The middle of the range, pH = 7, represents a neutral solution, or one that is neither acidic nor basic. Pure water is neutral because it contains the same number of hydrogen ions, H^+, as hydroxyl radicals, OH^-.

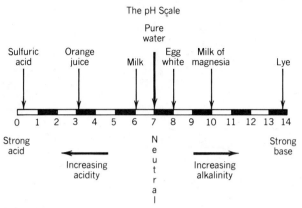

FIGURE 4.8 The pH scale is used to indicate the intensity or strength of an acidic or basic solution. A pH of 7 is neutral—neither acidic nor basic.

In more technical terms, pH is defined as the negative logarithm of the hydrogen ion concentration. For example, in pure water the numerical value of the hydrogen ion concentration is 10^{-7}. The logarithm (or exponent) is -7, and the negative of that is 7.

Because the pH scale is based on logarithms to the base 10, each unit change in pH actually represents a ten-fold change in the degree of acidity or alkalinity of a solution. For instance, a solution with a pH = 5 is 10 times more acidic than a solution with a pH = 6. Likewise, a solution with a pH = 4 is 100 times more acidic than the solution with pH = 6.

Solutions with pH values less than 7 are acidic; those with pH values greater than 7 are basic or alkaline. Figure 4.8 illustrates this scale, along with the relative positions of some familiar substances.

ORGANIC SUBSTANCES

The defining characteristic of an organic compound is that it contains carbon in combination with other elements, such as hydrogen, oxygen, nitrogen,

NAME	CHEMICAL FORMULA	MOLECULAR STRUCTURE
METHANE	CH_4	$\begin{array}{c} H \\ | \\ H-C-H \\ | \\ H \end{array}$
BENZENE	C_6H_6	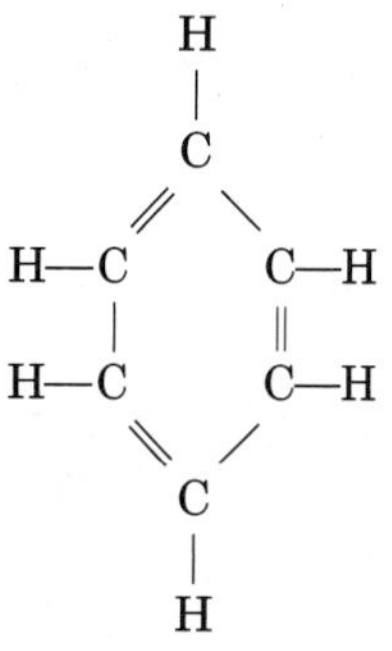 (ring structure of benzene)
BUTANE	C_4H_{10}	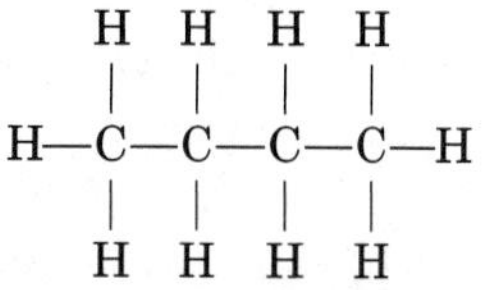 $H-C-C-C-C-H$ (with H atoms)

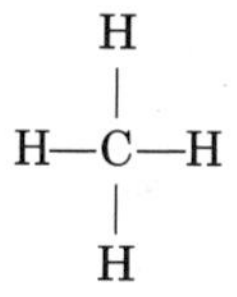

FIGURE 4.9 Schematic representations of the molecular structures for three common organic compounds, called *hydrocarbons*.

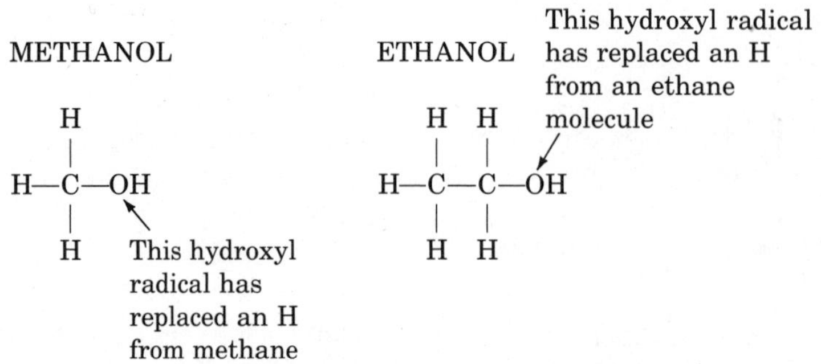

METHANOL   ETHANOL   This hydroxyl radical
                                       has replaced an H
                                       from an ethane
         H   molecule
         | H   H
   H—C—OH | |
         | H—C—C—OH
         H   This hydroxyl | |
              radical has   H   H
              replaced an H
              from methane

FIGURE 4.10 Molecular structure for two different types of alcohol: methanol and ethanol. The OH$^-$ radical replaces an H$^+$ ion in both types.

phosphorus, and sulfur. All living organisms are composed of organic compounds, some of which are so complex in molecular structure that they are still not fully understood by scientists.

There are hundreds of thousands of organic chemicals known to exist. Many of these occur naturally, in animal and plant tissues and wastes. Other organic compounds are synthetic substances that never occur in nature outside of the chemist's test tube. The basic reason for there being so many organic compounds, both natural and synthetic, is that the carbon atom is very "reactive." It combines readily with other carbon atoms and other elements, linking together in long chains or rings.

Organic compounds that contain only carbon and hydrogen are called *hydrocarbons*. The simplest hydrocarbon is methane, CH_4, which is a gas at ordinary temperature and pressure. It is produced naturally during the decay of other organic compounds, such as those found in sewage sludge or garbage.

The simplest ring hydrocarbon is benzene, C_6H_6, in which the carbon atoms link together to form a hexagon-shaped ring. Butane is an example of a straight-chain hydrocarbon molecule. Schematic diagrams showing the molecular structure of methane, benzene, and butane are shown in Figure 4.9.

In addition to the ability of carbon atoms to bond to each other forming rings and chains of various lengths, different groups of atoms can readily replace the hydrogen atoms in the hydrocarbons. This is another reason for the existence of the extremely large number of organic substances.

The classification of different organic chemicals depends on which particular group of atoms replaces the hydrogen. *Alcohols,* for example, are formed when hydrogen atoms are replaced by hydroxyl radicals. Methanol, CH_3OH, is an example of an alcohol used in solvents and fuel additives. Its molecular structure is illustrated schematically in Figure 4.10. Ethanol, C_2H_5OH, is an alcohol resulting from the fermentation of sugar and is used in alcoholic beverages. Oxidation of alcohol may result in organic compounds called *aldehydes* and *ketones*.

Organic acids are formed when the carboxyl group, —COOH, replaces hydrogen in a hydrocarbon. Acetic acid, commonly called vinegar, is an example of an organic acid. It is illustrated in Figure 4.11.

Many organic substances are *biodegradable*. This is a popular term used in reference to substances that can be used by microbes as food; biodegradable organic molecules are readily broken down into smaller, simpler molecules by biological action.

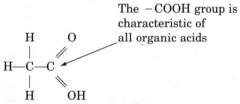

FIGURE 4.11 The molecular structure of acetic acid, commonly known as vinegar.

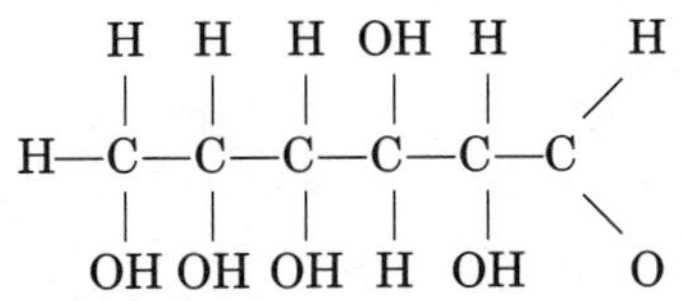

$$
\begin{array}{ccccccc}
\text{H} & \text{H} & \text{H} & \text{OH} & \text{H} & & \text{H} \\
| & | & | & | & | & & / \\
\text{H}-\text{C}-&\text{C}-&\text{C}-&\text{C}-&\text{C}-&\text{C} \\
| & | & | & | & | & & \backslash \\
\text{OH} & \text{OH} & \text{OH} & \text{H} & \text{OH} & & \text{O}
\end{array}
$$

FIGURE 4.12 The molecular structure of the sugar called glucose, $C_6H_{12}O_6$.

Carbohydrates are the most abundant group of biodegradable organic compounds in existence and are sometimes called the "fuel of life." They are the basic products of photosynthesis in green plants. Photosynthesis is the process by which the sun's energy is converted into a form that can be used by living organisms (see Section 1.3).

Carbohydrate molecules are formed from the elements carbon, hydrogen, and oxygen; the hydrogen and oxygen always occur in the same proportion as in water, that is, two to one. The sugar called *glucose* is an example of a simple carbohydrate. Its molecular structure is illustrated schematically in Figure 4.12. *Sucrose,* common table sugar, is a carbohydrate formed by the combination of glucose and another sugar called fructose. Starch and cellulose are larger and more complex carbohydrates that are not as biodegradable as the simpler sugars. Cellulose is the primary material in plants.

Fats are also biodegradable organic compounds, made up principally of carbon, oxygen, and hydrogen atoms. Although they are important energy storage molecules in living organisms, they are not very soluble in water and decompose at a slow rate. *Proteins* are much more complex than carbohydrates or fats, and they form the primary substance of animal tissue. In addition to carbon, oxygen, and hydrogen, proteins contain nitrogen and sulfur.

4.2 PHYSICAL PARAMETERS OF WATER QUALITY

The parameters that are commonly used to describe the physical quality of water include *turbidity, temperature, color, taste,* and *odor.*

TURBIDITY

When small particles are suspended in water, they tend to scatter and absorb light rays. This gives the water a murky or *turbid* appearance, and this effect is called *turbidity*. Clay, silt, tiny fragments of organic matter, and microscopic organisms are some of the substances that cause turbidity. They occur in water naturally or because of human activities and pollution.

Turbidity is a particularly important parameter of drinking water quality. Suspended particles can provide "hiding places" for harmful microorganisms and thereby shield them from the disinfection process in a water treatment plant. Because of this shielding effect, the microbes can be consumed by people who drink the water, and the spread of disease may result.

Turbidity in drinking water is also unacceptable for esthetic reasons—it makes the water look very unappetizing. Most people find even a slight degree of turbidity in their water objectionable. Even when told the water is safe to drink in spite of its turbidity, people tend to seek alternative water supplies (which could possibly be of poorer quality).

Turbidity is measured in units that relate the clarity of the water sample to that of a standardized suspension of silica. The interference in the passage of light caused by a suspension of 1 mg/L of silica is equivalent to one *turbidity unit* (TU). For example, a water sample that has the same degree of cloudiness as a 10 mg/L suspension of silica would be said to have a turbidity of 10 TU.

To interpret turbidity data, it is useful to be familiar with the typical ranges that occur. Turbidity in excess of 5 TU is just noticeable to the average person; most people do not complain about the clarity of the water at TU values less than 5. Turbidity in what most people would consider to be a relatively clear lake may be as high as 25 TU. In "muddy" water, turbidity generally exceeds 100 TU. Modern water treatment plants can routinely produce crystal-clear water with turbidities less than 1 TU.

Groundwater normally has very low turbidity because of the natural filtration that occurs as the water percolates through the soil. Most streams and rivers, though, have relatively high turbidities. This is particularly true during and just after rain storms which cause soil erosion. The treatment of turbid stream water for drinking supplies can be an expensive process; the greater the turbidity is, the greater is the amount of chemicals needed and the more frequently the filters must be cleaned.

For drinking water, instruments called *nephelometers* are used to measure the turbidity after purification. These devices measure the amount of scattered light electronically and do not depend on human vision or judgment in making comparisons to standard suspensions. Measurements made with nephelometric turbidimeters may be expressed in terms of *NTU* instead of just TU, to indicate how the measurement was made.

A conventional *Jackson Candle Turbidimeter* is illustrated in Figure 4.13. It may be used to measure raw (untreated) water turbidities. The water is added to a vertical glass tube until the candle flame is just obscured from view. The glass tube is graduated with turbidity units; the higher the water column required to obscure the flame, the less

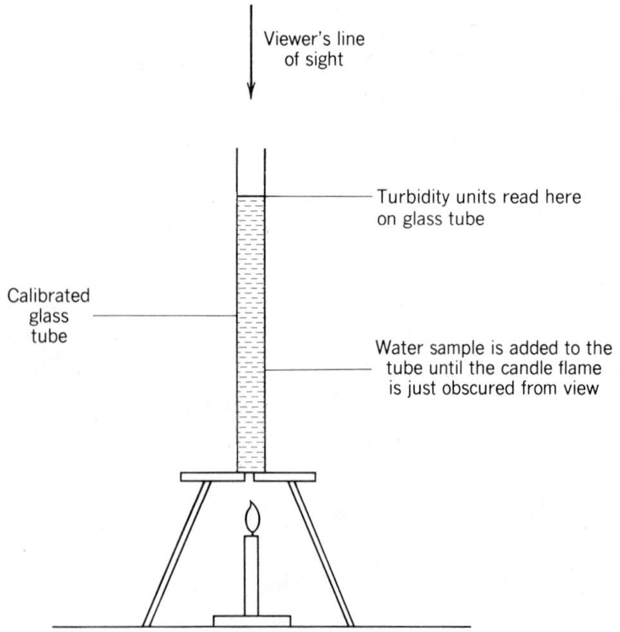

Viewer's line of sight

Turbidity units read here on glass tube

Calibrated glass tube

Water sample is added to the tube until the candle flame is just obscured from view

FIGURE 4.13 The candle turbidimeter.

the turbidity. Turbidity values obtained using the candle turbidimeter may be expressed as *JTU*.

Excessive turbidity in a lake reduces the depth to which sunlight penetrates the water. This has an effect on the photosynthesis of microscopic plants or algae and on the overall environmental balance of the lake. In field surveys, small white *Secchi disks* may be lowered into the water on a line marked off in meters, until the disk disappears from view. The depth of the disk at that point can be correlated with the turbidity of the lake water.

TEMPERATURE

Fish and other aquatic organisms require certain conditions of temperature in order to live and reproduce. The optimum temperature for trout, for example, is 15°C. A temperature of about 24°C is best for perch, and carp do very well at a cozy 32°C, which is more than twice the preferred temperature for trout (a "cold-water" fish).

Most species can adapt to a moderate change from their optimum temperature, but if the change is excessive the organisms will perish or migrate to a new location. Generally, a change of about 5°C can significantly alter the balance and health of an aquatic environment. Sudden drops in temperature can be harmful, but usually an increase in temperature will cause more damage than a decrease. Rivers must be protected from warm water discharges from power plants.

A basic reason for this, as discussed previously, is that the solubility of oxygen in water decreases markedly as the temperature of the water goes up. Fish and other organisms need the oxygen to survive, and higher temperatures increase their rate of metabolism. In other words, the rate at which the organisms use oxygen to "burn" food for energy increases at higher temperatures. The combined effect of there being less available oxygen and the organisms having faster metabolism rates can eventually be very damaging.

Other than the fact that most people prefer cold drinking water, temperature is of little direct significance in public water supplies. Temperature plays a more important role, however, in wastewater treatment and water pollution control. Biological wastewater treatment systems are more efficient at higher temperatures. In colder regions of the country, treatment plants may be sheltered in heated enclosures to maintain optimum temperature ranges.

COLOR, TASTE, AND ODOR

Color, taste, and odor are physical characteristics of drinking water that are important for aesthetic reasons. They do not cause any direct harmful effect on health. But no matter how safe the water may be to drink, most people object strongly to water that offends their sense of sight, taste, or smell.

Color may be caused by dissolved or suspended colloidal particles, primarily from decaying leaves or microscopic plants. This tends to give the water a brownish–yellow hue. Streams or rivers with tributaries in swampy areas may have this problem. Color is measured by comparing the water sample with standard color solutions or colored glass disks. One color unit is equivalent to the color produced by a 1 mg/L solution of platinum. It is not practical to isolate and identify specific chemicals that cause the color.

Hydrogen sulfide gas, H_2S, is a common cause of odor in water supplies. The "rotten-egg" smell of this gas may be encountered in water that has been in contact with naturally occurring deposits of decaying organic matter. Groundwater supplies sometimes have this problem; the wells are called "sulfur wells."

Odor is measured and expressed in terms of a *threshold odor number*. The threshold odor number is the ratio by which the sample has to be diluted for the odor to become virtually unnoticeable. For example, if a 50-mL volume of water sample has to be diluted to a volume of 200 mL for the odor to be just barely detectable, the threshold number would be 200/50 = 4. A similar technique may be applied in measuring the taste of the water. Taste and odor measurements are very subjective and depend on the sensitivity of the person conducting the test.

4.3 CHEMICAL PARAMETERS OF WATER QUALITY

There are many organic and inorganic chemicals that affect water quality. In drinking water, these effects may be related to public health or to aesthetics and economics. In surface waters, chemical quality can affect the aquatic environment. Several chemical parameters are also of concern in wastewater. In this section, the most common chemical parameters of water quality are discussed.

DISSOLVED OXYGEN

Dissolved oxygen is generally considered to be one of the most important parameters of water quality in streams, rivers, and lakes. It is usually abbreviated simply as *D.O.* Just as we need oxygen in the air we breathe, fish and other aquatic organisms need D.O. in the water to survive. With most other substances, the less there is in the water, the better is the quality. But the situation is reversed for D.O. *The higher the concentration of dissolved oxygen is, the better the water quality is.*

Oxygen is only slightly soluble in water. For example, the saturation concentration at 20°C is about 9 mg/L or 9 ppm. (Remember, that is equivalent to the relationship between 9 in. and 16 miles.) Because of this very slight solubility, there is usually quite a bit of competition among aquatic organisms, including bacteria, for the available dissolved oxygen. As we will discuss in some detail later on, bacteria will use up the D.O. very rapidly if there is much organic material in the water. Trout and other fish soon perish when the D.O. level drops. Another factor to remember is that oxygen solubility is very sensitive to temperature. Changes in water temperature have a significant effect on D.O. concentrations.

Dissolved oxygen has no direct effect on public health. But drinking water with very little or no oxygen tastes "flat" and may be objectionable to some people. Dissolved oxygen does play a part in the corrosion or rusting of metal pipes; it is an important factor in the operation and maintenance of water distribution networks.

Dissolved oxygen is used extensively in biological wastewater treatment facilities. Air, or sometimes pure oxygen, is mixed with sewage to promote the aerobic decomposition of the organic wastes. The role of dissolved oxygen in water pollution and wastewater treatment will be discussed in subsequent chapters.

The D.O. concentration can be determined using standard "wet chemistry" methods of analysis or by using membrane electrode meters in the lab or in the field. Field instruments are available having probes that can be lowered directly into a stream or treatment tank. The electrode probe senses small electric currents that are proportional to the dissolved oxygen level in the water.

BIOCHEMICAL OXYGEN DEMAND

Bacteria and other microorganisms use organic substances for food. As they metabolize organic material, they consume oxygen. The organics are broken down into simpler compounds, such as CO_2 and H_2O, and the microbes use the energy released for growth and reproduction.

When this process occurs in water, the oxygen consumed is the D.O. If oxygen is not continually replaced in the water by artificial or natural means, then the D.O. level will decrease as the organics are decomposed by the microbes. This need for oxygen is called the *biochemical oxygen demand*. In effect, the microbes "demand" the oxygen for use in the biochemical reactions that sustain them. The abbreviation for biochemical oxygen demand is *BOD;* this is one of the most commonly used terms in water quality and pollution control technology.

As we will discuss in the next chapter, organic waste in sewage is one of the major types of water pollutants. It is impractical to isolate and to identify each specific organic chemical in these wastes and to determine its concentration. Instead, we use the BOD as an indirect measure of the total amount of biodegradable organics in the water. *The more organic material there is in the water, the higher the BOD exerted by the microbes will be.*

In addition to being used as a measure of the amount of organic pollution in streams or lakes, the BOD is used as a meaure of the "strength" of sewage. As we will see in the chapter on wastewater treatment, this is one of the most important parameters for the design and operation of a water pollution control plant. A "strong" sewage has a high concentration of organic material and a correspondingly high BOD. A "weak" sewage, with a low BOD, may not require as much treatment.

The complete decomposition of organic material by microorganisms takes time, usually 20 days or more under ordinary circumstances. The amount of oxygen used to completely decompose or "stabilize" all of the biodegradable organics in a given volume of water is called the *ultimate BOD*, or BOD_L. For example, if a 1-L volume of municipal sewage requires 300 mg of oxygen for complete decomposition of the organics, the BOD_L would be expressed as 300 mg/L. One liter of wastewater from an industrial or food processing plant may require as much as 1500 mg of oxygen for complete stabilization of the waste. In this case, the BOD_L would be 1500 mg/L, indicating a much stronger waste than ordinary municipal or domestic sewage. In general, then, the BOD is expressed in terms of mg/L of oxygen.

The BOD is a function of time. At the very beginning of a BOD test, or time = 0, no oxygen will have been consumed and the BOD = 0. As each day goes by, oxygen is used by the microbes and the BOD increases. Ultimately, the BOD_L is reached and the organics are completely decomposed. A graph of the BOD versus time has the characteristic shape illustrated in Figure 4.14. This is called the *BOD curve*.

The BOD curve can be expressed mathematically by the following equation:

$$BOD_t = BOD_L \times (1 - 10^{-kt}) \qquad (4\text{-}2)$$

where BOD_t = BOD at any time t, mg/L
  BOD_L = ultimate BOD, mg/L
    k = a constant representing the rate of the BOD reaction
    t = time, d

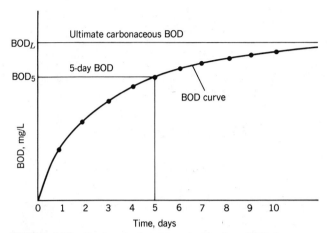

FIGURE 4.14 *Biochemical oxygen demand, or BOD, increases over time until all the organic pollutants are stabilized. The value of the BOD after 5 days, or BOD₅, is used for routine measurement and analysis.*

The rate at which oxygen is consumed is expressed by the constant k. The value of this rate constant depends on the temperature, on the type of organic material, and on the type of microbes exerting the BOD. For ordinary domestic sewage, at a temperature of 20°C, the value of k is usually about 0.15/d.

EXAMPLE 4.3

A sample of sewage from a town is found to have a BOD after 5 days (BOD₅) of 180 mg/L. Estimate the ultimate BOD (BOD$_L$) of the sewage. Assume $k = 0.1$/d for this wastewater.

Solution:

Applying Equation 4-2, we get

$$180 = BOD_L \times (1 - 10^{-0.1 \times 5}) = BOD_L \times (1 - 10^{-0.5})$$
$$= BOD_L \times (1 - 0.316) = BOD_L \times 0.684$$

Rearranging terms to solve for BOD$_L$, we get

$$BOD_L = 180/0.684 = 260 \text{ mg/L} \quad \text{(rounded-off)}$$

The effect of different temperatures on the rate of the BOD reaction and on the shape of the BOD curve is shown in Figure 4.15. At higher temperatures, the organics decompose at a faster rate, but the BOD$_L$ remains the same.

The 20 days or so required for the ultimate BOD to be developed is much too long a time to wait for lab results. This is particularly true when the BOD data are used to monitor the efficiency of a water pollution control plant. It has been found that more than two thirds of the BOD$_L$ is usually exerted within the first five days of decomposition. For instance, in the preceding example, the 5-day BOD is 180/260 = 0.69 or 69 percent of the ultimate BOD. For practical purposes, the 5-day BOD, or *BOD₅*, has been chosen as a representation of the organic content of water or wastewater. For standardization of results, the test must be conducted at a temperature of 20°C.

In summary, the parameter BOD₅ is the amount of dissolved oxygen used by microbes in 5 days to decompose organic substances at 20°C.

Measurement of BOD₅

The traditional BOD test is conducted in standard 300 mL glass "BOD bottles." The test for the 5-day BOD of a water sample involves taking two D.O. measurements: an initial measurement when the test begins at time $t = 0$ and a second measurement at $t = 5$, after the sample has been incubated

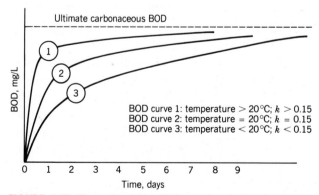

FIGURE 4.15 *The rate of the BOD reaction is directly proportional to temperature. But the total amount of organics in the sample, and therefore the ultimate BOD, does not change.*

in the dark for 5 days at 20°C. The BOD_5 is simply the difference between the two D.O. measurements.

For example, consider that a sample of water from a stream is found to have an initial D.O. of 8.0 mg/L. It is placed directly into a BOD bottle and incubated for 5 days at 20°C. After the 5 days, the D.O. is determined to be 4.5 mg/L. The BOD is the amount of oxygen consumed, or the difference between the two D.O. readings. That is, BOD_5 = 8.0 − 4.5 = 3.5 mg/L.

Very clean bodies of surface water usually have a BOD_5 of about 1 mg/L, due to naturally occurring organics from decaying leaves and animal wastes. BOD_5 values in excess of 10 mg/L, however, usually indicate the presence of sewage pollution.

When measuring the BOD_5 of sewage, it is necessary to first dilute the sample in the BOD bottle. Domestic sewage generally has a BOD_5 value of about 200 mg/L. If the sample were not diluted, all the D.O. would be very quickly depleted, and it would not be possible to get a D.O. reading on the fifth day. Computation of the BOD_5, using the so-called *dilution method* in a 300-mL BOD bottle, is done using the following equation:

$$BOD_5 = \frac{(D.O._{.0} - D.O._{.5}) \times 300}{V} \qquad (4\text{-}3)$$

where $D.O._{.0}$ = initial D.O. at $t = 0$
$\qquad\quad D.O._{.5}$ = D.O. at $t = 5$ d
$\qquad\quad V$ = sample volume, mL

EXAMPLE 4.4

A 6.0-mL sample of wastewater is diluted to 300 mL with distilled water in a standard BOD bottle. The initial D.O. in the bottle is determined to be 8.5 mg/L, and the D.O. after 5 days at 20°C is found to be 5.0 mg/L. Determine the BOD_5 of the wastewater and compute its BOD_L. Assume $k = 0.1$/d.

Solution:

Applying Equation 4-3, we get

$$BOD_5 = \frac{(8.5 - 5.0) \times 300}{6.0} = \frac{3.5 \times 300}{6.0}$$
$$= 180 \text{ mg/L}$$

Now applying Equation 4-2, we get

$$180 = BOD_L \times (1 - 10^{-0.1 \times 5})$$

and

$$BOD_L = \frac{180}{0.684} = 260 \text{ mg/L}$$

In some cases, particularly when analyzing industrial or food-processing wastewater that does not contain bacteria, the dilution water must be *seeded* with sewage. This provides a suitable population of microorganisms for the BOD reaction to take place. Remember, even though there may be a lot of organic material present in the water, if there are no microbes to use oxygen and stabilize the organics, a measurement of the BOD cannot be obtained. When seeded dilution water is used, Equation 4-3 must be modified to account for the BOD added by the dilution water.

Nitrification

In Figure 4.14, the BOD curve "flattens out" after about 8 days, as it approaches the ultimate BOD. That BOD_L is called the ultimate *carbonaceous* BOD because during the first week or so of decomposition, the bacteria act primarily on the carbon-containing substances.

But as time goes on and the carbonaceous material is depleted, another group of bacteria become active. These are called *nitrifying bacteria*. This group of microorganisms thrives on the noncarbonaceous ammonia, NH_3, in the wastewater, metabolizing it for energy. In this process, called *nitrification*, the ammonia is converted into the more stable nitrite, NO_2^-, and nitrate, NO_3^-, forms of nitrogen.

During this process, the nitrifying bacteria consume additional oxygen, causing a rise in the BOD curve after the first 8 to 10 days of decomposition.

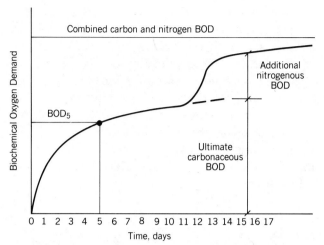

FIGURE 4.16 A complete BOD curve, showing the delayed effect of *nitrification* on the total oxygen demand. Additional oxygen is consumed by the nitrifying bacteria as they convert ammonia to nitrate.

This is illustrated in Figure 4.16. Most sewage treatment plants remove only the carbonaceous BOD. But it is possible that the discharge from a conventional treatment plant could still deplete the D.O. in a small receiving stream because of nitrogenous BOD that is exerted. Sometimes advanced treatment systems must be built to also remove the ammonia and protect sensitive aquatic environments from D.O. depletion due to nitrification.

Chemical Oxygen Demand

The BOD test provides a measure of the biodegradable organic material in water, that is, of the substances that microbes can readily use for food. But there also might be nonbiodegradable or slowly biodegradable substances that would not be detected by the conventional BOD test.

The *chemical oxygen demand,* or *COD,* is another parameter of water quality that measures all organics, including the nonbiodegradable substances. It is a chemical test using a strong oxidizing agent (potassium dichromate), sulfuric acid, and heat. The results of the COD test can be available in just 2 hours, a definite advantage over the 5 days required for the standard BOD test.

COD values are always higher than BOD values for the same sample, but there is generally no consistent correlation between the two tests for different wastewaters. In other words, it is not feasible to simply measure the COD and then predict the BOD. Because most wastewater treatment plants are biological in their mode of operation, the BOD test is more representative of the treatment process and remains a more commonly used parameter than the COD.

SOLIDS

Solids occur in water either in solution or in suspension. These two types of solids are distinguished by passing the water sample through a glass-fiber filter. By definition, the *suspended solids* are retained on top of the filter and the *dissolved solids* pass through the filter with the water.

If the filtered portion of the water sample were placed in a small dish and then evaporated, the solids in the water would remain as a residue in the evaporating dish. This material is usually called *total dissolved solids* or *TDS.* The concentration of TDS is expressed in terms of mg/L. It can be calculated as follows:

$$TDS = \frac{(A - B) \times 1000}{C} \qquad (4\text{-}4)$$

where A = weight of dish plus residue, mg
B = weight of empty dish, mg
C = volume of sample filtered, mL

EXAMPLE 4.5

The weight of an empty evaporating dish is determined to be 40.525 g. After a water sample is filtered, 100 mL of the sample is evaporated from the dish. The weight of the dish plus the dried residue is found to be 40.545 g. Compute the TDS concentration.

Solution:

Applying Equation 4-4, we get

$$TDS = \frac{(40\ 545\ mg - 40\ 525\ mg) \times 1000}{100\ mL}$$

$$= \frac{20 \times 1000}{100}$$

$$= 200\ mg/L$$

In drinking water, dissolved solids may cause taste problems. Hardness, corrosion, or aesthetic problems may also accompany excessive TDS concentrations. In wastewater analysis and water pollution control, the suspended solids retained on the filter are of primary importance and are referred to as *total suspended solids* or *TSS*.

The TSS concentration can be computed using Equation 4-4, where *A* would represent the weight of the filter plus retained solids, *B* would represent the weight of the clean filter, and *C* would represent the volume of sample filtered.

One of the routine solids tests used in wastewater treatment plants to determine the efficiency of the treatment process is the measurement of *settleable solids*. Settleable solids are the coarser fraction of the suspended solids that readily settle out because of gravity. A 1-L volume *Imhoff Cone,* illustrated in Figure 4.17, is filled with the sewage sample. After 1 h of quiescent settling, the solids accumulate at the bottom of the cone; the cone is graduated in milliliters and the amount of settleable solids is expressed in terms of mL/L.

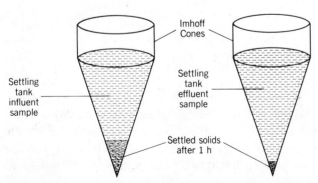

FIGURE 4.17 Imhoff cones are used to measure the amount of settleable solids in raw and in treated sanitary sewage.

Another classification of solids that is of particular significance in wastewater treatment is called *volatile solids.* These are organic substances that can be burned off or volatilized at 550°C in a furnace. The residues remaining after burning at that temperature are the "fixed" or nonvolatile solids. The concentration of volatile suspended solids gives an indication of the organic loading on biological treatment units. It can be determined by measuring the loss in weight of the glass-fiber filter plus solids after burning.

HARDNESS

Hardness is a term used to express the properties of certain highly mineralized waters (high TDS concentrations). The dissolved minerals cause problems such as scale deposits in hot water pipes and difficulty in producing lather with soap. The economic aspects of these problems, rather than any adverse health effect, are what make "hard" water generally unacceptable to the consumer.

Calcium, Ca^{2+}, and magnesium, Mg^{2+}, ions cause the greatest portion of hardness in naturally occurring waters. These minerals enter the water primarily from contact with soil and rock, especially limestone deposits. In general, groundwater is "harder" than surface water because it is in contact with mineral deposits for long periods.

Hardness is usually expressed in terms of milligrams per liter of calcium carbonate, $CaCO_3$; grains per gallon is also used to express hardness concentrations. Water with more than 300 mg/L of hardness is generally considered to be "hard," and water with less than 75 mg/L is considered to be "soft." Very soft water is undesirable in public supplies because it tends to increase corrosion problems in metal pipes; also some health officials believe it to be associated with the incidence of heart disease.

IRON, MANGANESE, COPPER, AND ZINC

Although iron, Fe, and manganese, Mn, do not cause health problems, they do impart a noticeable bitter taste to drinking water, even at very low concentrations. These metals usually occur in groundwater in solution as ferrous, Fe^{2+}, and manganous, Mn^{2+}, ions. When exposed to air, they form the insoluble ferric, Fe^{3+}, and manganic, Mn^{3+}, forms, making the water turbid and unacceptable to most people. They also cause brown or black stains on laundry and on plumbing fixtures.

Copper, Cu, and zinc, Zn, are nontoxic in small concentrations, and in fact they are both beneficial and essential for human health. But they cause undesirable tastes in drinking water and at high concentrations zinc imparts a milky appearance to the water.

FLUORIDES

A moderate amount of fluoride ions, F^-, in drinking water contributes to good dental health. Extensive research over many years has demonstrated that a fluoride concentration of about 1 mg/L is effective in preventing tooth decay, particularly in children, without any harmful side effects. Fluoridation, the intentional addition of compounds containing fluoride to drinking water, is practiced by many cities and towns throughout the United States.

Fluorides occur naturally in water in some areas. When the concentrations are excessive, either alternate water supplies must be used or treatment to reduce the fluoride concentration must be applied. This is because excessive amounts of fluoride cause mottled or discolored teeth, a condition called *dental fluorosis*. There is a very small margin of error between beneficial levels of fluoride and levels that cause fluorosis. The maximum allowable levels of fluoride in public water supplies depends on local climate because climate affects the amount of water consumption. In the warmer regions of the country, the maximum allowable concentration of fluoride is 1.4 mg/L; in colder climates, up to 2.4 mg/L is allowed.

CHLORIDES

Chlorides, Cl^-, in drinking water do not cause any harmful effects on public health, but high concentrations can cause a salty taste that most people find objectionable. Chloride levels are, of course, very high in ocean waters—about 35 000 mg/L.

Chlorides do occur naturally in groundwater, streams, and lakes. But the presence of relatively high chloride concentrations in fresh water (about 500 mg/L or more) may be an indication of sewage pollution. Salt (NaCl), used in foods, is excreted with body wastes; sanitary sewage would carry these chlorides into the receiving waters. Also, chlorides from roadway de-icing salts may enter the groundwater as well as streams and lakes. And saltwater intrusion into wells is a problem in some coastal areas.

CHLORINE RESIDUAL

It is important to make a distinction between chloride and chlorine in water; many beginning students confuse these two. Chlorine, Cl_2, does not occur naturally in water. It is, however, one of the most common chemicals added to water and wastewater, primarily for disinfection. This will be discussed in more detail in subsequent chapters.

Although chlorine itself is a toxic gas, in dilute aqueous solutions it is not harmful to human health. One of the advantages of chlorine as a disinfectant is that a leftover or *residual* concentration can be maintained in the water distribution system, ensuring good sanitary quality of the water. In drinking water, a residual of about 0.2 mg/L is optimal. The measurement of chlorine residual in a water sample can be made using a DPD color comparator test kit, illustrated in Figure 4.18.

One of the problems with chlorination of water supplies is that the chlorine can react with organics in the water, forming toxic compounds. The naturally occurring organics, primarily from vegetation, are called *precursors*. By themselves they are harmless. The toxic compounds, called *trihalomethanes*, or THM, have been identified as potential *carcinogens*, or cancer causing substances. One of the most common THM compounds formed is chloroform, $CHCl_3$. The chlorine replaces three of the hydrogen atoms in the methane molecule (see Figure 4.9) to form chloroform.

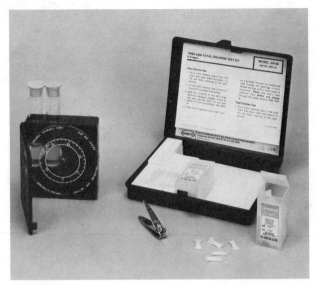

FIGURE 4.18 A DPD color comparator test kit for measuring chlorine residual (HACH Company, Loveland, Colorado).

It is ironic that the disinfection process using chlorine, designed to destroy microbes that cause disease, may be the source of a different public health hazard. Methods of control, including removal of the precursors before chlorination, are being studied.

SULFATES

Sulfates, SO_4^{2-}, occur in natural waters and in wastewater. If high concentrations are consumed in drinking water, there may be objectionable tastes or unwanted laxative effects, but there is no significant danger to public health from sulfates.

Sulfates in sewage can result in offensive odors from the formation of hydrogen sulfide gas, H_2S, with its characteristic rotten-egg odor. It also leads to a problem in sewer systems called *crown corrosion*. This will be discussed in more detail in the chapter on sanitary sewer systems.

NITROGEN

Nitrogen, N_2, occurs in many forms in the environment and takes part in many biochemical reactions. The four forms of nitrogen that are of particular significance in environmental technology are organic nitrogen, ammonia nitrogen, nitrite nitrogen, and nitrate nitrogen. The circulation of nitrogen in its various forms through the environment is illustrated in the nitrogen cycle, in Figure 1.8.

In water contaminated with sewage, most of the nitrogen is originally present in the form of complex organic molecules (proteins) and ammonia, NH_3. These substances are eventually broken down by microbes to form nitrites and nitrates. This process of *nitrification* has been discussed in the section on biochemical oxygen demand.

Nitrogen is a basic nutrient that is essential to the growth of plants, particularly in the nitrate form. Excessive nitrate concentrations in surface waters encourage the rapid growth of microscopic plants called *algae;* excessive growth of algae degrades water quality. This problem, referred to as *eutrophication,* is discussed in more detail in the next chapter, on water pollution.

Nitrates can enter the groundwater from chemical fertilizers used in agricultural areas. Excessive nitrate concentrations in drinking water pose an immediate and serious health threat to infants under three months of age. The nitrate ions react with blood hemoglobin, reducing the blood's ability to carry oxygen; this produces a disease called *blue baby* or methemoglobinemia.

PHOSPHORUS

Like nitrogen, phosphorus, P, is an essential nutrient that contributes to the growth of algae and the eutrophication of lakes, although its presence in drinking water has little effect on health. Phosphorus can enter water from sewage or from agricultural runoff containing fertilizers and animal wastes. Phosphate, PO_4^{3-}, the inorganic form of phosphorus, had been commonly used in detergents in the past. But even with the ban on phosphate-based detergents, the amount of phosphorus occurring in water from other sources poses a significant environmental problem.

ACIDITY, ALKALINITY, AND pH

Very high levels of either acidity or alkalinity in water may indicate the presence of industrial or chemical pollution. But acidity and alkalinity also occur naturally. Carbon dioxide from the atmosphere, or from the respiration of aquatic organisms, causes acidity when dissolved in water by forming carbonic acid, H_2CO_3. Dissolved carbonates, CO_3^{2-}, or bicarbonates, HCO_3^-, of sodium, calcium, or magnesium, cause natural alkalinity. Contact between the water and minerals in the ground is the major source of these substances.

Acids are substances that dissociate to yield H^+ ions in water, and alkaline substances yield OH^- radicals. The pH is a measure of the intensity of the acidity or alkalinity, as discussed in Section 4.1. The primary reason for measuring acidity, alkalinity, and pH of water is to be able to control the water treatment process in a water purification facility. The required doses of various chemicals depend on the concentrations of acidity or alkalinity, or on the pH of the water.

Drinking water with moderate amounts of acidity or alkalinity can be consumed without adverse health effects, but excessive concentrations would cause objectionable tastes; acids are sour and alkaline solutions are bitter.

The acidity and alkalinity in natural waters provide a buffering action that protects fish and other organisms from sudden changes in pH. For example, if an acidic chemical somehow contaminated a lake that had natural alkalinity, there would be a neutralization reaction between the acid and alkaline substances; the pH of the lakewater would remain essentially unchanged. Most aquatic organisms can survive in a pH range of about 6 to 9.5.

TOXIC AND RADIOACTIVE SUBSTANCES

A wide variety of *toxic* inorganic and organic substances are found in water in very small or trace amounts. Some of these substances are from natural sources, but many come from industrial activities. Even in trace amounts, they can be a danger to public health.

A toxic chemical may be a poison, causing death, or it may cause disease that is not noticeable until many years after exposure. A carcinogenic substance is one that causes cancer; substances that are mutagenic cause harmful effects in the offspring of exposed people.

Some *heavy metals* that are toxic are cadmium, Cd, chromium, Cr, lead, Pb, mercury, Hg, and silver, Ag. Arsenic, As, barium, Ba, and selenium, Se, are also poisonous inorganic elements that must be monitored in drinking water.

Many toxic organic chemicals have been identified. Those that are currently monitored in public water supplies are the *chlorinated hydrocarbons.* Among these are the trihalomethanes formed after chlorination, as previously discussed in the section on chlorine residual. Pesticides such as endrin and toxaphene are toxic chlorinated hydrocarbons that are monitored; DDT and chlordane are not routinely checked for in drinking water because they have been banned from use.

Relatively expensive and sophisticated instruments are required to analyze water samples for trace contaminants. Atomic absorption spectrophotometers for detecting heavy metals, and gas chromotography/mass spectrometry (GC/MS) instrumentation for detecting organics are now commonly found in water-quality labs. These instruments are capable of detecting substances in extremely dilute concentrations, in the parts-per-billion or micrograms-per-liter range.

To illustrate the low concentrations involved, consider that the maximum allowable concentration of the pesticide endrin in drinking water is 0.2 μg/L or 0.2 ppb; this is equivalent to the presence of only one endrin molecule among several billion water molecules. It is quite a technical achievement to make measurements in this range—sort of like finding a needle in a haystack as big as a house.

Radiation

The emission of subatomic particles or energy from the unstable nuclei of certain atoms, referred to as radiation, poses a serious public health hazard. Obviously the consumption of radioactive substances in water is undesirable, and maximum allowable concentrations of radioactive materials have been established for public water supplies. Potential sources of radioactive pollutants in water include wastes from nuclear power plants, from industrial or medical research using radioactive chemicals, and from refining of uranium ores. The unit of radioactivity used in water quality applications is the picocurie per liter (pCi/L); 1 pCi is equivalent to about two atoms disintegrating per minute.

4.4 BIOLOGICAL PARAMETERS OF WATER QUALITY

The presence or absence of living organisms in water can be one of the most useful indicators of its quality. In streams, rivers, and lakes, the diversity of fish and insect species provides a measure of the biological balance or "health" of the aquatic environment. A wide variety of different species of organisms usually indicates that the stream or lake is unpolluted. The disappearance of certain species and overabundance of other groups of organisms is generally one of the effects of pollution. Trout, for example, will soon disappear from a polluted stream, whereas catfish and other scavenger organisms will thrive. If the pollution is very severe, fish life will vanish altogether. Biologists can survey the fish and insect life of natural waters and assess the water quality on the basis of a computed *species diversity index*. (See Section 1.3.)

Microscopic plants and animals are also important in assessing the quality of water, particularly drinking water and sewage. In this section, some basic facts about bacteria and other microbes is discussed. The main focus is on a group of organisms called *coliforms*, which are perhaps the most important of the biological parameters of water quality.

MICROORGANISMS

Microscopic plants and animals play an essential role in the life processes of all living organisms, including humans. Contrary to a popular misconception that microbes are harmful, the fact is that most of them are beneficial, particularly in their role as decomposers in the food chain (see the discussion on ecology in Section 1.3). Only a relatively small number of species of microbes cause disease in humans or otherwise harm our environment.

Microorganisms are ubiquitous in nature; that is, they occur everywhere. There are millions of bacteria and molds living in a single gram of rich garden soil, for example. They serve to decompose organic materials, converting them into simpler nutrients that can be absorbed through the roots of plants. Foods also contain microorganisms, such as yeasts, which cause fermentation, producing CO_2 and alcohol from sugars.

Since the foods we eat are not sterile, our bodies acquire a normal population of microbes in the intestinal tract; the *coliform* group of bacteria makes up a large part of this population. Animal wastes consist primarily of microorganisms from the intestines. Although sewage contains millions of microbes per milliliter, most of them are harmless. It is only when sewage contains wastes from people infected with disease that the presence of harmful organisms in the sewage is likely.

Bacteria

Bacteria are considered to be single-celled plants because of their cell structure and the way they take in food. They utilize soluble food taken in through a rigid cell wall. But unlike green plants, using photosynthesis, they do not produce their own food.

Bacteria are extremely small, typically about 2 μm in size, and can be seen only with the aid of a microscope. They occur in three basic cell shapes: rod shaped or *bacillus,* sphere shaped or *coccus,* and spiral shaped or *spirellus.* In some cases, the individual cells grow together in larger groups or chains. *Sphaerotilus natans* is an example of a species of bacteria that grows in a chain or filament enclosed within a long sheath or tube. Excessive growth of these filamentous organisms is known to be one of the causes of reduced treatment efficiency in biological sewage treatment plants.

In less than 30 min, a single bacterial cell can mature and divide into two new cells. This process of reproduction is called *binary fission.* Under favorable conditions of food supply, temperature, and pH, bacteria can reproduce so rapidly that a bacterial culture may contain as many as 20 million individual cells per milliliter after just one day of growth. This rapid growth of visible "colonies" of bacteria on a suitable nutrient media makes it possible to detect and count the number of bacteria in water. This will be discussed in more detail in the section on coliform bacteria.

There are several distinctions among the various species of bacteria. One depends on how they metabolize their food. Bacteria that require oxygen for their metabolism are called *aerobic bacteria* or *aerobes*. Those that can live only in an oxygen-free environment are called *anaerobic bacteria* or *anaerobes*. The distinction between aerobes and anaerobes is of great significance in water pollution and wastewater treatment.

Another distinction among species of bacteria is a function of the type of food they require. Those that utilize simple inorganic compounds for nourishment are called *autotrophic* bacteria; those that require complex organic substances are called *heterotrophic* bacteria. The nitrifying bacteria, for example, which use ammonia as food and convert it to nitrate, are among the autotrophs. Other examples of autotrophs include the so-called *iron bacteria* and the *sulfur bacteria*. Iron bacteria thrive in some water pipelines and often cause taste and odor problems in drinking water. The sulfur bacteria, which are also anaerobes, are active in sewers and speed the deterioration of concrete pipes by converting hydrogen sulfide gas to sulfuric acid.

One of the most important factors affecting the growth and reproduction of bacteria is temperature. At low temperatures, bacteria grow and reproduce slowly. As the temperature increases, the rate of growth and reproduction just about doubles for every additional 10°C (up to the optimum temperature for the species). The majority of species of bacteria are classified as *mesophilic*, having an optimum temperature of about 35°C. Those that do best at elevated temperatures of about 60°C are called *thermophilic* bacteria.

Algae

Algae are microscopic plants that contain photosynthetic pigments, such as chlorophyll. They are autotrophic organisms that support themselves by converting inorganic materials into organic matter using energy from the sun. During the process of photosynthesis, they take in carbon dioxide from the air and give off oxygen.

A basic characteristic of these simple plants is their lack of roots, stems, and leaves. Free-floating algae are also called *phytoplankton*. Even though most species of algae are microscopic, they can be easily noticed when their numbers proliferate in the water. Excessive growths of algae, called *algal blooms*, are often unsightly. Some algal species are multicellular, growing as filaments that sometimes appear as a green slime in the water.

Common species include the blue-green algae such as *Anabaena*, green algae such as *Spirogyra*, yellow-green algae such as *Botrydium*, and red algae such as *Gelidium*. Another important group of algae, called *diatoms*, produce hard shells of silica. Deposits of these shells, from dead diatoms, that accumulated over many hundreds of years form *diatomaceous earth*, a material sometimes used for filtering water.

Algae play a role in the aging or eutrophication of lakes. They also are important in wastewater treatment stabilization ponds. Algae are generally nuisance organisms in public water supplies because of the taste and odor problems they cause and because of the extra expense required to filter them out of the water.

Protozoa

Protozoa are the simplest of animal species. These single-celled microscopic animals consume solid organic particles, bacteria, and algae for food. They are, in turn, ingested as food by higher-level multicellular animals. Floating freely in water, these *zooplankton*, as they are sometimes called, are a vital part of the natural aquatic food chain. They are also of significance in biological wastewater treatment systems.

A group of protozoa called *flagellates* move around in water by means of a long threadlike strand, called a *flagella*, that propels them with its whiplike action. One such organism, *Giardia lamblia*, is an intestinal parasite that causes a form of dysentery in humans. Another type of protozoa has hundreds of short hairs called *cilia* that propel the organism through the water and that serve to direct food particles into its digestive system. The *paramecia*, for example, are ciliated protozoa commonly found in fresh water ponds and lakes.

Amoebae are protozoa that move by projecting sections of their bodies; this mobile protoplasm of the amoebae is also used to surround and engulf food particles. Amoebae are commonly found in slimes formed in certain types of sewage treatment processes.

Several types of protozoa, as well as some common forms of algae and bacteria, are illustrated in Figure 4.19.

INDICATOR ORGANISMS

One of the most important attributes of good quality water is that it be free of disease causing organisms—*pathogenic* bacteria, viruses, protozoa, or parasitic worms. Water contaminated with sewage may contain such organisms because they are excreted in the feces of infected individuals. If contaminated water is consumed by others before it is properly treated, the cycle of disease can continue in epidemic proportions.

Testing water for the presence of individual pathogens such as the *Salmonella typhosa* bacteria, which causes typhoid fever, or *Entamoeba histolytica*, the protozoa that causes dysentery, is a very difficult, time consuming, and impractical task. The concentrations of these organisms in a contaminated water sample may be small enough to elude detection, making it necessary to test large volumes of water. And it would be necessary to test for a wide variety of different organisms before the water could be considered safe.

An approach more practical and reliable than testing for individual pathogens is to test for a single species that would signal the possible presence of sewage contamination. If sewage is present in the water, it can be assumed that the water may also contain pathogenic organisms and is a threat to public health. A species of organisms that serves this purpose is called an *indicator organism*.

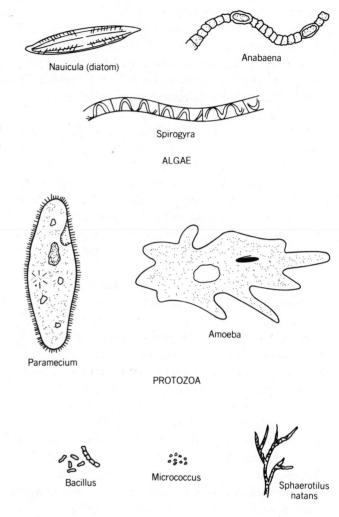

FIGURE 4.19 Sketches of some typical microorganisms found in water and/or in sewage.

The most important biological indicator of water quality and pollution used in public health technology is the group of bacteria called *coliforms*. Not pathogenic, coliforms are always present in the intestinal tract of humans, and millions are excreted with body wastes. Consequently, water that has been recently contaminated with sewage will always contain coliforms.

A particular species of coliforms found in domestic sewage is called *Escherichia coli* or *E. coli*. Even if the water is only slightly polluted, they are very likely to be found; there are roughly three million

E. coli bacteria in a 100-mL volume of untreated sewage. The *E. coli* are generally harmless, but infected individuals also excrete pathogens along with the coliforms.

Coliform bacteria are hardy organisms that survive in water longer than most pathogens. They are also relatively easy to detect. In general, it can be stated that if a sample of water is found not to contain coliforms, then there has not been recent sewage pollution, and therefore the presence of pathogens is extremely unlikely. On the other hand, if coliforms are detected, then there is a possibility of recent sewage pollution. However additional tests would be required to prove that the coliforms are from sewage and not from other sources.

In summary, we can say:

No coliforms → No sewage → No pathogens

Coliforms are actually a broadly defined group of microorganisms. They occur naturally in the soil as well as in the digestive tract of warm-blooded animals, including humans. It is necessary to make a distinction between two groups: *total coliforms* and *fecal coliforms*. *Total coliforms* refers to all the members of the group regardless of origin. Fecal coliforms are those from the intestines of warm-blooded animals. *E. coli* are fecal coliforms from humans.

A total coliform test is particularly applicable to the analysis of drinking water, to determine its sanitary quality. Drinking water must be free of coliforms of any kind. On the other hand, a fecal coliform test is more appropriate for monitoring pollution of natural surface- or groundwaters, since a total coliform count would be inconclusive in this case. Municipalities and industries are required to test for fecal coliforms in their wastewater treatment plant discharges to make sure the disinfection process is working properly.

Fecal-Coliform-to-Fecal-Strep Ratio

It is sometimes necessary to determine whether or not fecal coliforms in a tested sample originated from human wastes or animal wastes. The presence of another type of bacteria, called *fecal streptococci,*

can provide the necessary clue. Fecal strep bacteria are also intestinal bacteria, but they predominate over coliforms in animals other than humans. When the ratio of the number of fecal coliform bacteria to fecal strep bacteria is more than 2, the contamination is likely to be of human origin. When this ratio, abbreviated FC/FS, is less than 1, then animal wastes rather than sewage are more likely to be the source of pollution. FC/FS ratios between 1 and 2 are inconclusive. In addition to the FC/FS ratio, an investigation called a *sanitary survey* is usually required to determine the source of water pollution. In a sanitary survey, factors such as the extent of agricultural activity, the location and condition of residences, and the prevalence of individual on-site sewage disposal systems are studied.

TESTING FOR COLIFORMS

There are two different testing procedures that can be used for detecting and measuring coliforms in water—the *membrane filter method* and the *multiple-tube fermentation method*. The membrane filter method takes less time and provides more of a direct count of the coliforms than the multiple-tube method does. It also requires less laboratory equipment. Although the membrane filter method is gaining in use, the multiple-tube procedure is still practiced in many labs. It is necessary to understand the essential differences between these two tests.

Membrane Filter Method

In this procedure, a measured volume of sample is drawn through a special membrane filter by applying a partial vacuum. The filter, a flat paper-like disk about the size of a silver dollar, has uniform microscopic pores small enough to retain the bacteria on its surface while allowing the water to pass through.

After the sample is drawn through, the filter is placed in a sterile container called a *petri dish*. The petri dish also contains a special *culture media* that the bacteria use as a food source. This nutrient media is usually available in small glass containers called *ampoules,* from which it is readily transferred into the petri dish. Its composition is such that it promotes the growth of coliforms while inhibiting the growth of other bacteria caught on the filter.

A membrane filter apparatus is shown in Figure 4.20, and the filter is shown being placed in a petri dish in Figure 4.21.

The petri dish holding the filter and nutrient media is usually placed in an incubator, which keeps the temperature at 35°C, for 24 h. After incubation, colonies of coliform bacteria, each containing millions of organisms, will be visible. The colonies form by the reproductive process of binary fission. They appear as specks or dots, with a characteristic green metallic sheen. This is illustrated in Figure 4.22.

The coliform concentration is obtained by counting the number of colonies on the filter. A basic premise for the membrane filter test is that each colony started growing from one single organism. From this it can be assumed that *each colony counted represents only one coliform in the original sample.*

The filter has a grid printed on it to facilitate counting colonies; a magnifying glass is helpful to obtain accurate results. Small samples of polluted

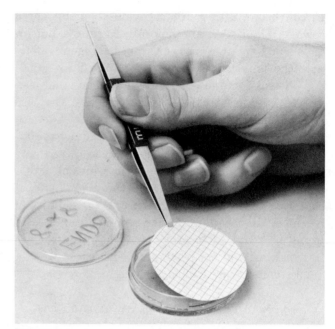

FIGURE 4.21 After filtration, the membrane filter is placed in a petri dish that contains a nutrient media. The trapped bacteria on the filter will grow into visible colonies. (Millipore Corporation, Bedford, Massachusetts)

water or wastewater must be diluted with sterile water before filtering, so that the filter is not overgrown with colonies, making it impossible to get an accurate count.

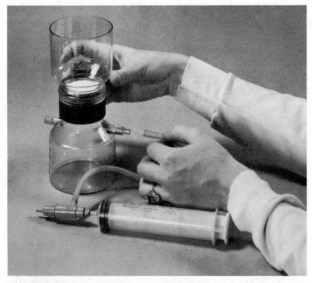

FIGURE 4.20 A membrane filter apparatus, for detecting and counting bacteria in water or sewage. (Millipore Corporation, Bedford, Massachusetts)

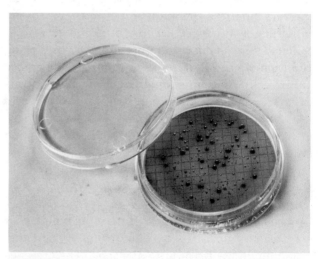

FIGURE 4.22 The visible colonies have a characteristic green metalic sheen that is readily identifiable. (Millipore Corporation, Bedford, Massachusetts)

Coliform concentrations are expressed in terms of the number of organisms per 100 mL of water. The following formula can be used to express the results of samples of various sizes:

coliforms per 100 mL

$$= \frac{\text{number of colonies} \times 100}{\text{mL of sample}} \quad (4\text{-}5)$$

EXAMPLE 4.6

A 4-mL volume of a water sample from a stream was drawn through a membrane filter. The filter was first covered with sterile water to dilute and spread the sample evenly over the filter. Sixteen coliform colonies were counted on the filter after incubation for 24 h at 35°C. Determine the coliform count per 100 mL.

Solution:

Applying Equation 4-5, we get

$$\text{coliforms per 100 mL} = \frac{(16 \times 100)}{4} = 400$$

The basic procedure described here for the membrane filter test can be applied to test for total coliforms or fecal coliforms. But different nutrient media are used, and the fecal coliform test is conducted at 44.5°C rather than at 35°C. A special water bath incubator is used to accurately maintain the higher temperature for the fecal coliform test. The membrane filter technique can also be used to test for fecal streptococci.

Multiple-Tube Fermentation Method

This technique is based on the fact that coliform organisms can use lactose, the sugar occurring in milk, as food, and produce gas in the process. A measured volume of water sample is added to a tube that contains lactose broth nutrient media. A small inverted vial in the lactose broth traps some of the gas that is produced as the coliform bacteria

grow and reproduce. The gas bubble in the inverted vial along with a cloudy appearance of the broth provide visual evidence that coliforms may be present in the sample. But if gas is not produced within 48 h of incubation at 35°C, it can be concluded that coliforms were not present in the sample volume injected into the broth. This is illustrated in Figure 4.23.

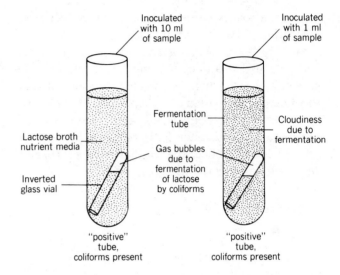

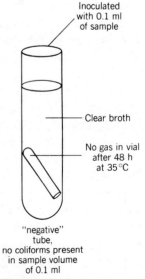

FIGURE 4.23 A "positive test" in a fermentation tube (a trapped gas bubble and cloudiness in the lactose broth) signals the possible presence of coliform bacteria in the sample. The larger the sample volume, the more likely it is that the tube will test positive. And if the number of positive tubes that occur in a series of sample dilutions is high, then the MPN of coliforms is high.

The failure of gas to form after incubation is called a *negative* test. The appearance of gas and the accompanying cloudiness in the broth is called a *positive presumptive test*. There are some bacteria other than coliforms that occasionally produce gas in lactose. Because of this, it is usually necessary to perform another test, called the *confirmed test,* to prove that it was really the coliform bacteria that produced the gas in the positive presumptive tube.

The confirmed test involves transferring the media from a positive presumptive tube to another fermentation tube that contains a different nutrient media, called brilliant green bile. Now, if gas is again formed within 48 h of incubation at 35°C, the presence of coliforms is confirmed. In some cases, a third procedure, called the *completed test,* may have to be performed. The fermentation tube procedure can be used to test for fecal coliforms as well as total coliforms, but the higher temperature of 44.5°C is used for the fecal organisms.

Most Probable Number

The production of gas in a single fermentation tube may indicate the presence of coliforms, but it gives no clue as to the concentration of bacteria in the sample. A coliform count cannot be obtained directly. The gas bubble could have been caused by one bacterium or by thousands. In order to estimate the actual number of organisms, a multiple series of fermentation tubes with different sample volumes must be used.

As the size of the sample volume placed into a fermentation tube is increased, the probability of coliforms being present in the tube increases. Using statistics and probability theory, it is possible to analyze the combinations of positive and negative results in a multiple tube series, and to determine the *Most Probable Number* of coliforms in the original sample; this is referred to as the *MPN* for the sample.

The MPN is expressed in terms of the number of coliforms per 100 mL, but as its name implies, it is more of an educated guess based on probability formulas than a direct count of organisms. Fortunately, the statistical analyses have been worked out for a variety of tube dilutions and combinations

and summarized in references such as *Standard Methods*. To illustrate this, MPN values using a series of nine tubes are presented in Table 4.4. Three of the nine tubes are inoculated with 10 mL of sample, three are inoculated with 1 mL of sample, and three are inoculated with 0.1 mL. Example 4.7 illustrates the use of the table.

TABLE 4.4   MPN INDEX AND 95 PERCENT CONFIDENCE LIMITS FOR VARIOUS COMBINATIONS OF POSITIVE AND NEGATIVE RESULTS WHEN THREE 10-mL PORTIONS, THREE 1-mL PORTIONS, AND THREE 0.1-mL PORTIONS ARE USED

Number of Tubes Giving Positive Reaction Out of			MPN Index per 100 mL	95% Confidence Limits	
3 of 10 mL Each	3 of 1 mL Each	3 of 0.1 mL Each		Lower	Upper
0	0	1	3	<0.5	9
0	1	0	3	<0.5	13
1	0	0	4	<0.5	20
1	0	1	7	1	21
1	1	0	7	1	23
1	1	1	11	3	36
1	2	0	11	3	36
2	0	0	9	1	36
2	0	1	14	3	37
2	1	0	15	3	44
2	1	1	20	7	89
2	2	0	21	4	47
2	2	1	28	10	150
3	0	0	23	4	120
3	0	1	39	7	130
3	0	2	64	15	380
3	1	0	43	7	210
3	1	1	75	14	230
3	1	2	120	30	380
3	2	0	93	15	380
3	2	1	150	30	440
3	2	2	210	35	470
3	3	0	240	36	1300
3	3	1	460	71	2400
3	3	2	1100	150	4800

Source: Standard Methods

EXAMPLE 4.7

The results of a multiple-tube fermentation test on a sample of river water are as follows:

DILUTION SERIES	RESULTS AFTER INCUBATION
10 mL	2 positive
	1 negative
1 mL	1 positive
	2 negative
0.1 mL	0 positive
	3 negative

Determine the MPN of coliforms from these data.

Solution:

Entering Table 4.4, locate the row with 2, 1, and 0 in the first three columns, respectively. These numbers represent the number of positive tubes in each dilution series. Under the column headed MPN Index, we read an MPN of 15 coliforms per 100 mL. The last two columns of the table point out the statistical nature of the MPN. We see that the probability is 95 percent that the actual coliform concentration is at least 3 but no more than 44 per 100 mL.

4.5 WATER SAMPLING PROCEDURES

Proper sampling procedures are an important part of any survey to assess water (or wastewater) quality and to check compliance with water quality standards. A sample which has been improperly collected, preserved, transported, or identified will result in invalid and useless test results, despite the precision of the analytical lab procedure. Since the results of water quality tests are the basis for decisions that affect public health, good sampling procedures must be followed. There are two basic sampling methods: *grab sampling* and *composite sampling*.

GRAB SAMPLES

As its name implies, a grab sample is a single sample collected over a very short period of time. Most people envision this as a quick "scoop," but technically it can take up to 15 min to fill the sample container and still be considered a grab sample.

It is important to note that the test results from a grab sample only represent the conditions of the water or wastewater at the particular time and lo-

FIGURE 4.24 A typical grab sampler device, which allows safe access to the water or wastewater from which a sample is to be collected. (Wheaton Instruments)

cation of sample collection. Grab samples are most suitable when testing for chlorine residual, pH, coliforms, and dissolved oxygen. They are usually collected manually.

For stream or wastewater grab sampling, devices that provide easy access to the flow channel from boats, spillways, or docks are available. This is illustrated in Figure 4.24. Special containers that allow samples to be collected at specific depths below the surface, without mixing with air, are also available. This is particularly important for D.O. sampling.

COMPOSITE SAMPLES

In many instances, grab samples alone are not enough to adequately characterize water or wastewater quality. This is particularly true for wastewater collection and treatment systems in which quality as well as quantity changes from hour to hour. Composite sampling is more appropriate when it is necessary to determine overall or average conditions over a certain period of time.

Composite samples are obtained by mixing individual grab samples taken at regular intervals over the sampling period. For example, a composite sample may consist of a mixture of smaller samples taken every 20 min over an 8 h period.

In wastewater studies, the volumes of the smaller grab samples that comprise the composite are generally taken in proportion to the flow rate, for more meaningful results. For example, if a 100-mL grab sample is taken when the flow rate is 5 L/s, then a 200 mL sample would be taken when the flow increased to 10 L/s.

Automatic sampling equipment is usually used for composite sampling. The cost of the equipment is balanced by the savings on the labor involved in manual collection and compositing. An automatic

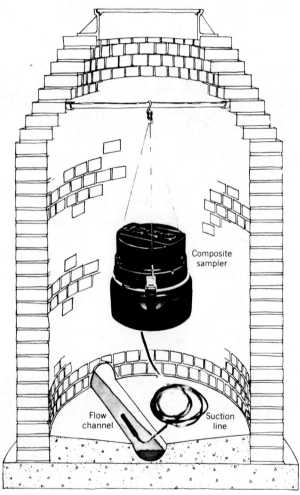

FIGURE 4.25 Automatic composite samplers can be left unattended in sewer manholes, providing a sample that is representative of average flow conditions. (ISCO, Inc., Environmental Division, Lincoln, Nebraska 68501)

FIGURE 4.26 Automatic composite samplers can be used to collect stream samples in order to determine the average water quality conditions over a period of time. (ISCO, Inc. Environmental Division, Lincoln, Nebraska 68501)

sampler installed in a sewer manhole is shown in Figure 4.25, and a device being set up for composite stream sampling is shown in Figure 4.26.

GENERAL REQUIREMENTS

The methods for taking and preserving samples vary, depending on the specific water quality parameter and analysis to be made. The sampling frequencies and locations are stipulated in the NPDES permit for wastewater effluent standards and by the SDWA for drinking water. A summary of four general considerations that apply for any type of sample follows:

1. *The sample must be truly representative of the existing conditions* - for instance, collecting a water sample from a faucet without first letting the water run for a while will not give results representative of conditions in the water main, but only of the water that was stagnant in the service line for an unknown period of time.

2. *The time between collection and analysis should be as short as possible* - for the most reliable results. Certain tests, however, such as chlorine residual or temperature tests, must be determined immediately. Dissolved oxygen is another parameter that needs immediate analysis, although it is possible to add a chemical that "fixes" the D.O. concentration, allowing later testing in the lab.

3. *Appropriate preservation techniques should be applied* - to slow down the biological or chemical changes that may occur in the time between sample collection and sample analysis. This usually involves refrigeration to cool the sample, or chemical fixing (as for D.O.).

4. *Accurate and thorough sampling records must be kept* - to avoid any confusion as to the "what, when, and where" of the sample, as well as to satisfy legal requirements. Figure 4.27 illustrates a sample bottle label that facilitates thorough record keeping.

Example Coliform Bottle Label

COLIFORM BOTTLE No._____	LAB RESULTS
Sample Taken: Date _____ Time _____	Date Tested _____
Sample Source ☐ Consumer's Faucet ☐ Raw Water Supply	MF _____ per 100 ml
Sample Location: _____	LB ____ 24 ____ 48
Sampled By:_____	BGB ____ 24 ____ 48
Type of Sample ☐ Initial Sample Chlorine Residual mg/l ____	SPC_____
☐ Routine Sample WaterTemp. _____	MPN _____
☐ Check Sample	Sample Is:
☐ Special Purpose Sample	☐ Safe ☐ Unsafe
Water System Name _____	☐ Unsatisfactory for test
Address _____	Please Resubmit
City _____ State _____ Zip _____	

FIGURE 4.27 All sample bottles should be properly identified, and records of the date, time, place, and type of sampling should be kept. (Reprinted from *Safe Drinking Water Act Self-Study Handbook,* by permission. Copyright © 1978, The American Water Works Association.)

REVIEW QUESTIONS

1. What is the difference between an element and a compound? What is the difference between an atom and a molecule?

2. Briefly describe the difference between ionic bonding and covalent bonding. Give one example compound for each.

3. Briefly describe the difference between a suspension and a solution and between a colloid and a solution.

4. How does the solubility of solids in water change with increasing temperature? Is it the same for gases?

5. Match the following symbols or formulas for certain chemical substances on the left with the appropriate descriptive term(s) from the list on the right.

_____ Mg	(a)	atom
_____ Pb	(b)	ion
_____ Na^+	(c)	molecule
_____ Cl^-	(d)	radical
_____ O_3	(e)	organic
_____ CaO	(f)	manganese
_____ OH^-	(g)	magnesium
_____ H^+	(h)	silver
_____ NO_3^-	(i)	lead
_____ CH_4	(j)	oxygen
_____ $-COOH$	(k)	lime
_____ HCl	(l)	hydroxyl
_____ Mn	(m)	proton
_____ Cl_2	(n)	nitrate
	(o)	chlorine
	(p)	acid

6. What is the approximate saturation concentration of oxygen in fresh water at a temperature just above freezing?

7. Give one example of an ionic solution.

8. What is a radical? Give three examples.

9. What do *ppm*, *ppb*, and *gpg* stand for?

10. What is a neutralization reaction?

11. A solution has a pH of 8.5. Is it acidic or basic? Which substance is likely to cause that pH in water, HCl or NaOH? Why?

12. The pH of a solution changes from 6 to 3. By what factor did the strength of its acidic condition increase?

13. What is an organic compound? List three different groups or types of organic substances.

14. Match the following parameters of water quality on the left with the possible effects listed on the right.

_____ turbidity	(a) causes dysentery
_____ TDS	(b) interferes with disinfection
_____ D.O.	(c) suffocates fish
_____ iron	(d) causes algal blooms
_____ fluoride	(e) is toxic to humans
_____ phosphorus	(f) prevents tooth decay
_____ fecal coliforms	(g) may cause cancer
_____ lead	(h) increases corrosion
_____ THM	(i) indicates sewage pollution
	(j) causes taste problems in water

15. What is the meaning of the terms *JTU* and *NTU*?

16. Briefly describe the significance of temperature in water quality.

17. Briefly describe how odor or taste is measured in water?

18. What is D.O.? Why is it a significant parameter of water quality?

19. What does BOD stand for? What does this parameter indicate about water or wastewater quality?

20. What is ultimate BOD? Why is BOD_5 used for the standard BOD test?

21. Briefly describe how the BOD of a sewage sample would be determined.

22. Briefly explain the effect of nitrification on oxygen demand.

23. What is the difference between a BOD test and a COD test?

24. What is the difference between TDS and TSS? How are they measured?

25. What is an Imhoff Cone?

26. What is the difference between "hard" and "soft" water? Is it advisable to remove all the hardness from drinking water? Why?

27.   Are high concentrations of fluoride in drinking water beneficial for public health?

28.   If a sample of stream water had an unusually high chloride concentration, what might you conclude about its quality? Briefly describe the difference between chloride and chlorine residual in water.

29.   What are two effects of sulfates in water?

30.   What is the significance of nitrogen and phosphorus in water quality?

31.   What are two reasons for measuring acidity, alkalinity, and pH?

32.   Briefly describe the basic characteristics of bacteria, algae, and protozoa. List one example of a species from each group.

33.   What is the difference between an aerobe and an anaerobe?

34.   Why are indicator organisms used to evaluate the sanitary quality of water?

35.   If coliform bacteria are not detected in a water sample, what can you conclude about the possibility of recent sewage pollution? If coliforms are detected, will the water definitely cause disease among people who drink it? Explain your answer.

36.   A stream passing through a sparsely populated agricultural area was found to have a fecal coliform count of 80 per 100 mL and a fecal strep count of 100 per 100 mL. What is the most likely source of the bacterial contamination?

37.   Briefly describe the membrane filter method for determining the coliform count of a water sample. What is the basic premise or assumption underlying this method?

38.   Briefly describe the multiple-tube fermentation method for obtaining a coliform count of a water sample. What does MPN mean?

39.   Briefly discuss two different methods of collecting samples for water- (or wastewater-) quality testing.

40.   List and briefly discuss four general requirements for good sampling procedure.

PRACTICE PROBLEMS

1. Convert a concentration of 275 ppm to an equivalent value in terms of mg/L and gpg.

2. A sample of water has 4 grains per gallon of hardness as $CaCO_3$. What is the hardness in terms of mg/L? Is this level of hardness objectionable? Why?

3. A discharge from a sewage treatment plant enters a stream at a flow rate of 3 mgd. The BOD of the discharge is 50 mg/L. How many pounds of BOD are entering the stream per day?

4. A 50 lb bag of copper sulfate, $CuSO_4$, is dissolved in a lake to control algal growth. The lake volume is 30 ac–ft. If the chemical is completely dispersed throughout the lake volume, what is its concentration in mg/L?

5. How many kilograms of chlorine gas should be dissolved in 5 ML of water to result in a concentration of 2 ppm?

6. How many pounds per day of chlorine are needed to apply a chlorine concentration of 0.5 ppm in a flow of 25 mgd?

7. A 200-mL aqueous solution contains 0.005 mg of arsenic. What is the concentration of arsenic in terms of ppb?

8. A sample of water from a stream is placed into a standard 300-mL BOD bottle and is found to have a D.O. of 14.0 mg/L. After five days of incubation at 20°C, the D.O. in the bottle has dropped to 6.0 mg/L. What is the BOD_5 of the stream? What can you conclude about the quality of the stream?

9. In the preceding problem, what could you say about the BOD and the quality of the stream if after five days of incubation, D.O. was not detected in the bottle?

10. A sample of sewage is found to have a BOD_5 of 250 mg/L. Assuming the rate constant is 0.15/d, estimate the ultimate carbonaceous BOD of the wastewater.

11. A 5-mL sample of sewage is diluted to 300 mL in a standard BOD bottle. The initial D.O. in the bottle is 9.2 mg/L and after five days of incubation, the D.O. is found to be 4.7 mg/L. Determine the 5-day and ultimate BOD values for the sewage, assuming the reaction rate constant is 0.14/d.

12. A wastewater sample has an ultimate BOD of 280 mg/L. A 5-mL volume of this sample is diluted to 300 mL in a BOD bottle, and the initial D.O. is determined to be 9.0 mg/L. What would be the expected D.O. in the bottle after five days of incubation if $k = 0.1/d$?

13. The weight of an empty evaporating dish is 38.820 g. A 50-mL volume of a filtered sample is evaporated from the dish. The weight of the dish plus dried residue is found to be 38.845 g. Compute the TDS of the sample.

14. The weight of a clean glass-fiber filter is 545 mg. After filtering a 100-mL sample, the weight of the filter plus retained solids is found to be 580 mg. After ignition in a furnace at 550°C, the weight of the filter and residue is 560 mg. Compute the concentration of suspended solids in the sample and the percentage of volatile solids.

15. A 10-mL water sample was tested for fecal coliforms using the membrane filter method. A total of 22 colonies of fecal coliforms were counted on the filter after incubation at 44.5°C. What was the fecal coliform count of the sample?

16. The results of a multiple-tube fermentation test of a water sample follow. What is the MPN of the sample?

DILUTION SERIES	RESULTS AFTER INCUBATION
10 mL	3 positive
1 mL	1 positive
	2 negative
0.1 mL	2 positive
	1 negative

SELECTED REFERENCES

1. *Standard Methods for the Examination of Water and Wastewater,* American Public Health Association, American Water Works Association, Water Pollution Control Federation, 1975.

2. Sienko, M. and R. Plane, *Chemistry,* McGraw–Hill, New York, 1961.

3. Sawyer, C. N., and P. L. McCarty, *Chemistry for Sanitary Engineers,* McGraw–Hill, New York, 1967.

4. McKinney, R. E., *Microbiology for Sanitary Engineers,* McGraw–Hill, New York, 1962.

5. Pelczar, M. J. and R. J. Reid, *Microbiology,* McGraw–Hill, New York, 1965.

6. Keeton, W. T., *Biological Science,* Norton, New York, 1980.

7. *NPDES Compliance Sampling Inspection Manual,* 1979, MCD-51, U.S.E.P.A.

8. *Safe Drinking Water Act Self-Study Handbook,* The American Water Works Association.

5

WATER POLLUTION

Water has such a strong tendency to dissolve other substances that it is sometimes referred to as the "universal solvent." This is largely because of its polar molecular structure. Pure water, that is, pure H_2O, is not found under natural conditions in streams, lakes, groundwater, or the oceans. It always has something dissolved or suspended in it. Because of this, there is not any definite line of demarcation between "clean" water and "contaminated" water.

If pure water does not exist outside of a chemist's laboratory, how can we make a distinction between polluted and unpolluted water? In fact, the distinction depends on the type and concentration of impurities, as well as on the intended use of the water. We also compare the concentrations of the dissolved or suspended substances with water quality standards set for a particular use.

In general terms, water is considered to be polluted when it contains enough foreign material to render it unfit for a specific beneficial use, such as for drinking, recreation, or fish propagation. Actually, the term *pollution* usually implies that human activity is the cause of the poor water quality.

This chapter expands upon some of the topics mentioned in the preceding chapter on water quality. It focuses on many of the common types and sources of water pollutants and on their effects in streams, lakes, groundwater, and the oceans.

5.1 CLASSIFICATION OF WATER POLLUTANTS

To understand the effects of water pollution and the technology applied in its control, it is useful to classify pollutants into various groups or categories. First, a pollutant can be classified according to the nature of its origin as either a *point source* or a *dispersed source* pollutant.

A point source pollutant is one that reaches the water from a pipe, channel, or any other confined and localized source. The most common example of a point source of pollutants is a pipe that discharges sewage into a stream or river. Most of these discharges are treatment plant *effluents,* that is, treated sewage from a water pollution control facility; they still contain pollutants to some degree. But in a few isolated instances, untreated or *raw* sewage is discharged. New York City, from which millions of gallons of raw sewage are discharged each day at various points, is one of the most notorious examples of this. The city is, of course, under orders from the state and federal regulatory authorities to provide appropriate treatment.

A dispersed or nonpoint source is a broad, unconfined area from which pollutants enter a body of water. Surface runoff from agricultural areas, for example, carries silt, fertilizers, pesticides, and animal wastes into streams, but not at one particular point. These materials can enter the water all along a stream as it flows through the area. Acidic runoff from mining areas is a dispersed pollutant. Stormwater drainage systems in towns and cities are also considered to be dispersed sources of many pollutants because even though they are often conveyed into streams or lakes in drainage pipes or storm sewers, there are usually many of these discharges scattered over a large area.

The distinction between point sources and dispersed sources of pollutants is illustrated in Figure 5.1. Point source pollutants are easier to deal with than are dispersed source pollutants; those from a point source have been collected and conveyed to a single point where they can be removed from the water in a treatment plant—and the point discharges from treatment plants can easily be monitored by regulatory agencies. Under the Clean Water Act, a discharge permit is required for all point sources.

Pollutants from dispersed sources are much more difficult to control. Many people think that sewage is the primary culprit in water pollution problems, but dispersed sources cause a significant fraction of the water pollution in the United States. Perhaps the most effective way to control the dispersed sources is to set appropriate restrictions in land use. Section 208 of the Clean Water Act is addressed to the control of pollution from dispersed sources, and to the development of area-wide waste management strategies, called "208 plans."

In addition to being classified by their origin,

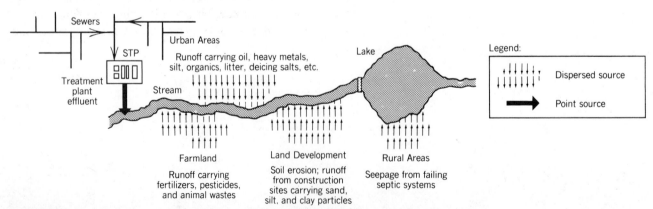

FIGURE 5.1 Dispersed source pollutants are more difficult to control than are point source pollutants, which can be collected and removed from the water.

water pollutants can be classified into groups of substances, based primarily on their environmental or health effects. For example, the following list identifies nine specific types of pollutants:

1. Pathogenic organisms
2. Oxygen-demanding substances
3. Plant nutrients
4. Toxic organics
5. Inorganic chemicals
6. Sediment
7. Radioactive substances
8. Heat
9. Oil

Domestic sewage is the primary source of the first three types of pollutants. Pathogens, or disease-causing microorganisms, are excreted in the feces of infected persons and may be carried into waters receiving sewage discharges. Sewage from communities with large populations is very likely to contain pathogens of some type.

Sewage also carries oxygen-demanding substances—the organic wastes that exert a biochemical oxygen demand as they are decomposed by microbes. This is the BOD that was discussed in some detail in the preceding chapter. BOD changes the ecological balance in a body of water by depleting the dissolved oxygen (D.O.) content. Nitrogen and phosphorus, the major plant nutrients, are in sewage too, as well as in runoff from farms and suburban lawns.

Conventional sewage treatment processes significantly reduce the amount of pathogens and BOD in sewage, but do not eliminate them completely. Certain viruses, in particular, may be somewhat resistant to the sewage disinfection process. (A virus is an extremely small pathogenic organism that can only be seen with an electron microscope.) In order to decrease the amounts of nitrogen and phosphorus in sewage, usually some form of advanced sewage treatment must be applied.

Toxic organic chemicals, primarily pesticides, may be carried into water in the surface runoff

from agricultural areas. Perhaps the most dangerous type is the family of chemicals called chlorinated hydrocarbons. Common examples are known by their trade names as chlordane, dieldrin, heptachlor, and the infamous DDT, which has been banned in the United States. They are very effective poisons against insects that damage agricultural crops. But unfortunately they can also kill fish, birds, and mammals, including humans. And they are not very biodegradable, taking more than 30 years in some cases to dissipate from the environment.

Toxic organic chemicals, of course, can also get into water directly from industrial activity. This would be from improper handling of the chemicals in the industrial plant, or, as has been more common, from improper and illegal disposal of chemical wastes. As previously mentioned, proper management of toxic and other hazardous wastes is one of the key environmental issues of the 1980s, particularly with respect to the protection of groundwater quality. Poisonous inorganic chemicals, specifically the "heavy metal" group such as lead and mercury, also usually originate from industrial activity and are considered hazardous wastes.

Oil is washed into surface waters in runoff from roads and parking lots. Accidental oil spills from large transport tankers at sea occasionally occur, causing significant environmental damage. And blowout accidents at offshore oil wells can release many thousands of tons of oil in a short period of time. Oil spills at sea may eventually move toward shore, affecting aquatic life and damaging recreation areas.

5.2 THERMAL POLLUTION

Heat is considered to be a water pollutant because of the adverse effect it can have on the oxygen levels and the aquatic life in a river or lake. Overall, the amount of water withdrawn for cooling purposes in power plants exceeds the amount of water used for any other purpose. The cooling water carries away waste heat as it passes through the condensers in the plant. (Steam is converted back to water in the condensers.) Cooling-water temperature may increase up to 15°C after it serves to condense the steam.

The discharge of warm water into a river is usually called *thermal pollution*. The warmer temperature decreases the solubility of oxygen and increases the rate of metabolism of fish. This changes the ecological balance in the river. Valuable game fish, such as trout, cannot survive in water above 25°C, and will not reproduce in water warmer than 14°C. Coarser fish, such as carp or pike, can do well in water as warm as 35°C.

Because several species of coarser fish actually prefer warmer waters, some representatives of the power industry use the term *thermal enrichment* rather than thermal pollution to refer to the warm water they return to the river. Although it is true that many fish may congregate near the outfall pipe from the power plant, a problem arises if the plant is suddenly shut down for repairs. The sudden decrease in water temperature causes a fish kill of significant proportions, leaving thousands of dead fish floating belly-up in the river or washed up along the shore.

Thermal pollution may be controlled by passing the heated water through a cooling pond or a cooling tower after it leaves the condenser. The heat is dissipated into the air, and the water can then be either discharged to the river or pumped back to the plant for reuse as cooling water. This is illustrated in Figure 5.2. There is no discharge of

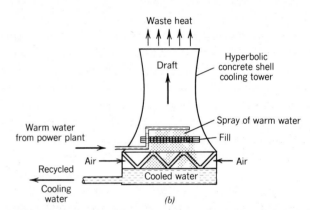

FIGURE 5.3 In a natural draft cooling tower, the waste heat is dissipated into the atmosphere. The towers are typically about 100 m in diameter and about 130 m in height. (Custodis–Cottrell, A Research–Cottrell Company)

heated water into the river, but some water will be withdrawn to make up for evaporative losses.

In locations where there is not enough room for a cooling pond, cooling towers may be built to prevent thermal pollution. These structures take up less land area than ponds do. A common type is the *natural draft hyperbolic cooling tower*, in which evaporation accounts for most of the heat transfer. A photograph of a hyperbolic tower is shown in Figure 5.3a. Hyperbolic cooling towers are usually very tall and may dominate the landscape in their vicinity.

The operation of an evaporative cooling tower is

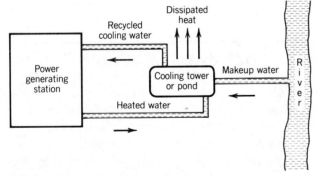

FIGURE 5.2 Thermal pollution from power plants can be eliminated by using recirculating cooling towers or ponds.

basically a simple process, as illustrated in Figure 5.3*b*. Hot water coming from the condensers is sprayed downward over vertical sheets or baffles, called fill. The water flows in thin films through the fill. Cool air enters the tower through the air inlet that encircles the base of the tower, and rises upward through the fill. Evaporative cooling takes place as the cool air passes over the thin films of hot water. A chimney effect or natural draft is maintained because of the density differential between the cool air outside and the warmer air inside the tower. The waste heat is dissipated into the atmosphere about 100 m above the base of the tower. The cooled water is collected in a basin at the floor of the tower and recycled back to the power plant condensers.

5.3 SOIL EROSION AND SEDIMENT CONTROL

The natural movement of soil particles by wind or water from one location to another is called *soil erosion*. Uncontrolled soil erosion is a significant environmental problem. Soils in agricultural regions are a precious natural resource, and the loss of these fertile soils from unwise land-use practices can be devastating. One of the most notable examples of this is the "Oklahoma Dust Bowl" of the 1930s. When a prolonged drought hit Oklahoma after many years of decreasing soil fertility, strong summer winds literally blew the dry topsoil away. In addition to environmental damage, this caused severe economic hardship and social dislocation.

Not all problems related to soil erosion are as dramatic as the Oklahoma Dust Bowl. But the cumulative effect of less extensive erosion episodes can still have adverse environmental effects, particularly with respect to water quality. Soil erosion has been identified as one of the most significant sources of water pollutants.

Soil particles suspended in water interfere with the penetration of sunlight. This in turn reduces photosynthetic activity of aquatic plants and algae,

disrupting the ecological balance of the stream. When the water velocity decreases, the suspended particles settle out and are deposited as *sediment* at the bottom of the stream or lake. Sediment smothers *benthic* or bottom-dwelling organisms and disrupts the reproductive cycle of fish and other life forms.

There are two types of water-caused soil erosion: *sheet erosion*—from land areas by raindrop impact and overland flow of storm runoff, and *stream erosion*—the removal of soils from stream beds and stream banks by the swiftly moving channelized water.

The factors that affect the rate of sheet erosion include rainfall intensity, soil texture, steepness of slope, and the amount of vegetative cover. The velocity of streamflow is one of the most important factors in stream erosion, although the type of soil is important too. The quantity of eroded material carried by some of the larger streams and rivers can be enormous. The Mississippi River, for example, transports an average of 1.5 million tons of sediment per day to the ocean. Most of this material is carried as *suspended load* in the turbulent currents, but a significant portion is also carried as *bed load,* sliding or rolling along the river bottom.

The natural vegetative cover of grass and trees provides protection against sheet erosion. Land-use activities such as agriculture and construction, which temporarily remove the natural vegetation and expose the bare soils, are the main causes of serious erosion and sediment problems. Construction projects involving major land disturbances may be the more significant of these two causes. The uncontrolled erosion of soil at major construction sites can exceed the erosion from naturally vegetated areas of the same size by a factor of 100 or more.

Construction plans and specifications should describe the location and details of specific erosion control measures to conserve the soil and prevent water pollution. In some states, a *Soil Erosion and Sediment Control Plan* is required to accompany the contract documents. The particular methods used depend primarily on the types of soil and the slope of the construction site. The Soil Conservation Service (SCS) of The Department of Agriculture sets standards for erosion and sediment control. Typical soil erosion control measures include the following:

1. *Temporary grass cover* - on exposed soils can be used to reduce wind and water erosion until permanent seeding or soil stabilization is accomplished. Application of lime and fertilization should be done on the basis of soil test data, and the proper seed mixture should be applied.

2. *Mulching* - materials, such as unrotted salt hay or woodchips, can be used for temporary cover on areas difficult to vegetate because of steep slopes, unsuitable soils, or winter operations.

3. *Diversion channels* - can be constructed across slopes to reduce open slope length, as illustrated in Figure 5.4. These channels are constructed with a ridge on the lower side of the slope, diverting water to sites where it can be disposed of safely. They may be temporary or permanent.

4. *Hay bales* - can be placed around storm-water inlets, or at the low point on the site, to intercept sediment-laden runoff and to prevent the soil from entering the storm drainage system. This is illustrated in Figure 5.5.

5. *Temporary fences* - as illustrated in Figure 5.6, can be used to reduce erosion at construction sites. They are generally placed on the perimeter of the site at the lower elevations where water runs off.

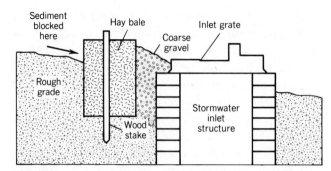

FIGURE 5.5 A typical hay-bale and gravel filter which prevents sediment from entering a drainage system and then local streams; it is usually used in the vicinity of active construction sites.

6. *Sediment basins* - or ponds can be built to intercept and retain water carrying suspended soil particles. The sediment is deposited in the pond, protecting streams or drainage systems downstream of the construction site. These ponds can be temporary or permanent earth structures and may be designed to reduce peak storm flows and flooding.

7. *Channel stabilization* - can be used to provide capacity in streams and drainage ditches for the flow of water without excessive erosion. Flow velocities should be minimized by the proper alignment and slope of the channel. Also, the channel can be protected by linings such as grass, concrete, or stone riprap.

8. *Scheduling of construction* - can be done so as to minimize exposure of bare soils prior to

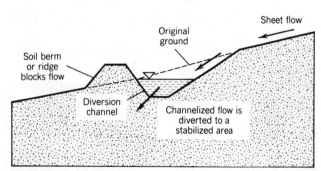

FIGURE 5.4 Diversion channels reduce the distance of overland sheetflow, thereby reducing soil erosion and sedimentation of nearby streams and lakes.

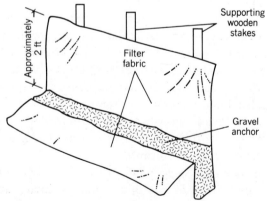

FIGURE 5.6 A temporary fence can be constructed to control erosion at a construction site.

final landscaping or paving. Drainage and soil protection facilities should be completed as early as possible.

5.4 STREAM POLLUTION

Rivers and streams are surface waters in which the entire water body continually moves downhill in natural channels, under the force of gravity. They are shallower and narrower than lakes and have a greater proportion of water exposed to land surfaces. The flowing water carries algae rapidly downstream, and tends to discourage the growth of rooted plants on the stream bed.

To a limited extent, streams and rivers have the ability to assimilate biodegradable wastes. This means that they can recover from the effects of pollution naturally, without significant or permanent environmental damage. The capacity for self-purification depends on the strength and volume of pollutants and on the stream discharge or flow rate.

It used to be said that "the solution to pollution is dilution." The effects of dilution and the constant flushing action of the flowing water are obvious factors involved in the *waste assimilative capacity* of a stream. Not as obvious, but equally important, is the effect of oxygen transfer between the air and the water. This is called *reaeration*.

The D.O. in the water is constantly replenished as atmospheric oxygen is dissolved at the water surface. Fast flowing, shallow, turbulent streams are reaerated more effectively than slow, deep, meandering streams are. This is because of the increased surface area and contact between the air and the water in the churning and well-mixed turbulent flow.

Modern-day population densities are too high for most streams and rivers to assimilate raw sewage discharges without offensive environmental conditions and public health hazards quickly developing. Some degree of treatment is required to remove enough of the BOD from the sewage so that stream dilution and reaeration can finish the job of purification. A level of treatment, called *secondary treat-*

ment, is generally sufficient for this purpose; it is the minimum level of treatment required by law in the United States. This will be discussed in more detail in the chapter on wastewater treatment.

It is important to note that not all pollutants can be assimilated in water by natural means. This is particularly true for nonbiodegradable or persistent contaminants that do not dissipate in the environment. Even the physical process of dilution is ineffective when these persistent chemicals become trapped in river sediments. Two notable examples of this problem are the accumulation of PCB (polychlorinated biphenol, a toxic industrial chemical) in the sediments of the Hudson River in New York State, and the contamination by the pesticide kepone of the sediments of the James River in Virginia. These problems are expected to persist for another 20 years, unless the sediments are removed by dredging. But dredging may even increase the pollution by "stirring up" the contaminated deposits.

DILUTION

When a point discharge of wastewater enters a flowing stream, the physical process of mixing and dilution begins immediately. But with the exception of small turbulent streams, it is unlikely that the pollutants will be thoroughly mixed in the streamflow at or near the point of discharge. Instead, a *waste plume* forms, as illustrated in Figure 5.7. The length of this gradually widening mixing zone depends on the channel geometry, the flow velocity, and the design of the discharge pipe.

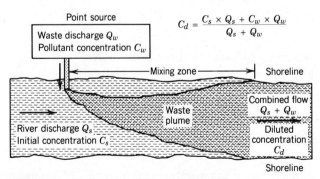

$$C_d = \frac{C_s \times Q_s + C_w \times Q_w}{Q_s + Q_w}$$

FIGURE 5.7 Dilution of pollutants from a point source, such as a sewage treatment plant, occurs within the *mixing zone* of the stream.

In water pollution control, it is often necessary to predict the BOD concentrations and D.O. levels downstream from a sewage discharge point. One of the first computations needed for this involves the effect of dilution. Assuming that the pollutant is completely mixed in the streamflow (at a point just below the end of the mixing zone), the diluted concentration of any water quality parameter can be calculated using the following equation:

$$c_d = \frac{c_s Q_s + c_w Q_w}{Q_s + Q_w} \qquad (5\text{-}1)$$

where c_d = diluted concentration or temperature
c_s = original stream concentration or temperature
c_w = waste concentration or temperature
Q_s = stream discharge
Q_w = waste discharge

EXAMPLE 5.1

The BOD_5 of an effluent from a municipal sewage treatment plant is 25 mg/L, and the effluent discharge is 4 ML/d. The receiving stream has a BOD_5 of 2 mg/L and the streamflow is 40 ML/d. Compute the combined 5-day BOD in the stream just below the mixing zone.

Solution:

Applying Equation 5-1 directly, we get

$$c_d = \frac{2 \times 40 + 25 \times 4}{40 + 4} = \frac{180}{44} = 4.1 \text{ mg/L}$$

where c_d represents the diluted BOD_5 in the combined flow.

EXAMPLE 5.2

A river has a dry-weather discharge of 100 cfs and a temperature of 25°C. Compute the maximum dis-

charge of cooling water, at 65°C, that can be discharged from a power plant into the stream. Assume the legal limit on temperature increase in the stream is 2°C.

Solution:

The maximum allowable stream temperature is

$$25 + 2 = 27°C$$

Applying Equation 5-1, we get

$$27 = \frac{25 \times 100 + 65 \times Q_w}{100 + Q_w}$$

Multiplying both sides by $(100 + Q_w)$, we get

$$2700 + 27Q_w = 2500 + 65Q_w$$

Transposing similar terms, we get

$$38Q_w = 200 \qquad \text{and} \quad Q_w = \frac{200}{38} = 5.3 \text{ cfs}$$

The discharge of warm water cannot exceed 5.3 ft³/s if the stream temperature is not to increase more than 2°C.

DISSOLVED OXYGEN PROFILE

When sewage is discharged into a stream, dissolved oxygen is utilized by microorganisms as they metabolize and decompose organic substances from the wastewater. The microbes exert a biochemical oxygen demand, or BOD, as discussed in Chapter 4. The BOD causes the dissolved oxygen level in the stream to gradually drop. This is illustrated in Figure 5.8 as curve A, called the stream *deoxygenation curve.*

While deoxygenation is occurring, oxygen from the air is dissolving into the water at the surface. The rate of oxygen transfer from the air into the water depends upon temperature as well as on the *oxygen deficit*. The oxygen deficit is the difference between the actual D.O. concentration and the saturation D.O. value. The larger the deficit, the faster the rate of oxygen transfer. This is illustrated as curve B in Figure 5.8, called the stream *reaeration curve*. Notice that the slope (rate of change) of the reaeration curve gradually increases as the deoxygenation curve falls.

At any given time, the D.O. level in the stream is a function of the combined effects of deoxygenation and reaeration. In other words, the actual D.O. is equal to the sum of the D.O. on the deoxygenation curve plus the D.O. on the reaeration curve. The graph of the combined D.O. versus time is seen as curve C, called the *dissolved oxygen sag curve*. Since the product of velocity and time equals distance [(ft/s) × s = ft], the horizontal or *x*-axis in Figure 5.8 can also be labeled "distance" for a given reach of the stream. Curve C is, in effect, a profile view of the D.O. concentrations along the length of the stream, and is also called the *dissolved oxygen profile*.

Initially, the rate of deoxygenation exceeds the rate of reaeration, so the oxygen profile begins to sag. After most of the organics are decomposed, the rate of reaeration dominates, and the oxygen profile begins to rise toward its original level. The minimum dissolved oxygen content in the stream occurs

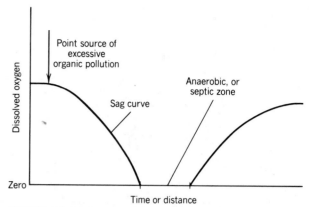

FIGURE 5.9 Under heavy loads of pollution, the D.O. level may drop to zero. This results in obnoxious odors and very unsightly conditions in the water. With additional time and distance downstream, the water will eventually be reaerated and water quality will be restored.

when the rate of reaeration equals the rate of deoxygenation. The computation of this level is of importance in water pollution studies.

In cases of extremely heavy organic pollution, or very low stream flow, the oxygen in the water may be completely depleted. The sag curve intersects the horizontal axis at D.O. = 0, resulting in anaerobic or septic conditions. This is illustrated in Figure 5.9.

ZONES OF POLLUTION

Most streams that are polluted by a point source of biodegradable organic substances can be described and evaluated in terms of four relatively distinct zones. These are illustrated in Figure 5.10. The first is the *zone of degradation* which forms below the point of waste discharge. This zone is characterized by floating solids, turbidity, and other visual evidence of pollution. The D.O. level begins to drop rapidly in the zone of degradation.

When the D.O. level drops to about 40 percent of its saturation value, the *zone of active decomposition* is considered to start. This zone is characteristic of heavily polluted water. Higher forms of aquatic life and desirable species, such as trout, either die or migrate out of the area. More tolerant

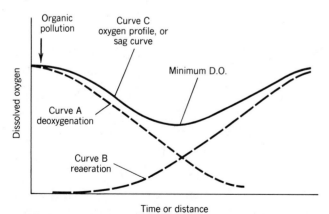

FIGURE 5.8 The oxygen sag curve shows the effect of organic pollution on the D.O. levels in a stream or river. After the organics decompose, surface reaeration will restore the original water quality. This is called *stream self-purification*.

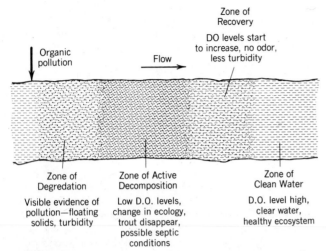

FIGURE 5.10 The zones of pollution in a stream that receives biodegradable organic contaminants.

fish species such as carp or catfish may survive. The mixture of different species is altered because of the low D.O. levels. Sludge deposits of settleable solids may form in the stream. If anaerobic conditions occur (see Figure 5.9), gas bubbles, floating sludge, and obnoxious odors may be noticeable in this zone.

After most of the organics have been decomposed by the microbes in the water, the rate of reaeration will exceed the rate of deoxygenation. When the D.O. level increases back up to 40 percent of the saturation concentration, the *zone of recovery* begins. This zone is characterized by gradually clearing water with no offensive odors; desirable aquatic species reappear. When the organic waste loading on a stream is small, or when there is considerable dilution, the zone of recovery may follow directly after the zone of degradation, with no zone of active decomposition forming at all.

Following the zone of recovery is the *zone of clean water*. This zone is characterized by clear water, high in D.O.; diverse species of aquatic organisms thrive, utilizing the stable inorganic nutrients remaining in the water. In effect, the stream has recovered its original quality through a process of natural self-purification. Of course, additional point discharges or dispersed sources of pollutants along the stream can alter this model of pollution zones

in a stream. Nevertheless, this model is of value in understanding stream pollution and in creating technical solutions to the problem.

COMPUTATION OF MINIMUM D.O.

It is important to be able to predict the minimum dissolved oxygen level in a polluted stream or river. For example, if a new sewage treatment plant is to discharge its effluent into a trout stream, it is possible that conventional (secondary) treatment levels will not remove enough BOD to prevent excessively low D.O. downstream. To determine if some form of advanced treatment is required to preserve the stream for trout spawning and survival, it is necessary to compute the minimum D.O. caused by the sewage effluent and to compare it to the allowable value for trout streams.

One of the equations used to describe and predict the behavior of a polluted stream is known as the Streeter–Phelps Equation. This equation is based on the assumption that the only two processes taking place are deoxygenation from BOD and reaeration by oxygen transfer at the surface, as was previously discussed. Two key formulas from the Streeter–Phelps model of stream pollution and oxygen sag follow. Figure 5.11 illustrates some of the variables in these equations. The minimum D.O. in the stream is the difference between the saturation D.O. level and the *critical oxygen deficit*.

$$t_c = \frac{1}{k_2 - k_1} \times \log\left[\frac{k_2}{k_1} \times \left(1 - D_i \times \frac{k_2 - k_1}{k_1 \times \mathrm{BOD_L}}\right)\right] \quad (5\text{-}2)$$

$$D_c = \frac{k_1 \times \mathrm{BOD_L}}{k_2 - k_1} \times (10^{-k_1 t_c} - 10^{-k_2 t_c}) + D_i \times (10^{-k_2 t_c}) \quad (5\text{-}3)$$

where t_c = the time it takes for the critical oxygen deficit or minimum D.O. to develop, d

D_c = the critical oxygen deficit, mg/L

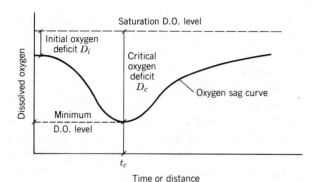

FIGURE 5.11 The critical time t_c and the critical oxygen deficit D_c can be computed using appropriate equations. The minimum D.O. is the difference between the saturation D.O. level and the computed deficit.

D_i = the initial oxygen deficit at time $t = 0$, below point of waste discharge into the stream, mg/L

BOD_L = the ultimate BOD in the stream just below the point of waste discharge, mg/L

k_1 = the deoxygenation rate constant, day^{-1}

k_2 = the reaeration rate constant, day^{-1}

The value of k_1 is generally taken to be the same as the rate constant for the BOD reaction, in Equation 4-2; it can be determined in the laboratory. The value of k_2 depends on the velocity and depth of the flow and can be determined from field studies or by an appropriate formula. The reaeration rate constant k_2 can vary from about 0.1 for a sluggish river to about 4.0 for a shallow turbulent stream. Both rate constants, k_1 and k_2, are dependent on temperature.

Equations 5-2 and 5-3 look complicated, and they are. They are presented here to illustrate the power of mathematics as a tool to model the environment and to help solve water pollution problems. But as complicated as they appear, the Streeter–Phelps equations are not quite accurate representations of the oxygen profile in a polluted stream or river. Other factors that affect the oxygen balance include photosynthesis and respiration of rooted plants and

algae, and the oxygen demand of benthic (bottom) deposits. Equations that have been developed to include these factors are even more complicated than Equations 5-2 and 5-3.

EXAMPLE 5.3

The BOD_L in a stream is 3 mg/L and the D.O. is 9.0 mg/L. Streamflow is 15 mgd. A treated sewage effluent with BOD_L = 50 mg/L is discharged into the stream at a rate of 5 mgd. The D.O. of the sewage effluent is 2 mg/L. Assuming $k_1 = 0.2$, $k_2 = 0.5$, and the saturation D.O. level is 11 mg/L, determine the minimum D.O. level in the stream. Assuming a stream velocity of 0.5 ft/s, how far downstream does the minimum D.O. occur?

Solution:

First it is necessary to compute the diluted BOD_L and D.O. using Equation 5–1, as follows:

$$BOD_L = \frac{15 \times 3 + 5 \times 50}{15 + 5} = \frac{295}{20} = 14.8 \text{ mg/L}$$

$$D.O. = \frac{15 \times 9 + 5 \times 2}{15 + 5} = \frac{145}{20} = 7.3 \text{ mg/L}$$

Now compute the initial oxygen deficit as

D_i = saturation D.O. − initial D.O. = 11.0 − 7.3 = 3.7 mg/L

Applying Equation 5-2, we get

$$t_c = \frac{1}{0.5 - 0.2} \times \log \left[\frac{0.5}{0.2} \times \left(1 - 3.7 \right. \right.$$
$$\left. \left. \times \frac{0.5 - 0.2}{0.2 \times 14.8} \right) \right]$$
$$= \frac{1}{0.3} \times \log \left[2.5 \times (1 - 0.375) \right]$$
$$= (3.33) \log 1.56$$
$$= 0.64 \text{ d}$$

It will take 0.64 days (or about 15 hours) for the minimum D.O. to occur.

Now applying Equation 5-3, we get

$$
\begin{aligned}
D_c &= \frac{0.2 \times 14.8}{0.5 - 0.2} \times (10^{-0.2 \times 0.64} - 10^{-0.5 \times 0.64}) \\
&\quad + 3.7 \times 10^{-0.5 \times 0.64} \\
&= 9.87 \times (0.745 - 0.479) + 3.7 \times 0.479 \\
&= 2.63 + 1.78 \\
&= 4.4 \text{ mg/L}
\end{aligned}
$$

The minimum D.O. in the stream is the difference between the saturation D.O. and the critical oxygen deficit, or $11.0 - 4.4 = 5.6$ mg/L. At a velocity of 0.5 ft/s, in 0.64 days the distance downstream for the minimum D.O. is 0.64 d $\times$ 24 h/d $\times$ 3600 s/h $\times$ 0.5 ft/s = 27 650 ft $\approx$ 5 miles.

5.5   LAKE POLLUTION

The pollution of natural lakes or constructed reservoirs poses problems that are different from the problems caused by pollution of streams or rivers. This is primarily because of physical characteristics. Water in a stream is constantly moving and providing a flushing action for incoming pollutants. But in lakes the water is not moving very much at all and is detained for a relatively long period of time. In some cases pollutants discharged into a lake can remain there for many years. Lakes are also significantly affected by seasonal temperature changes.

In streams organic pollutants affect the oxygen profile. In lakes water quality may be more dependent on plant nutrients than on organics from sewage. As discussed in Chapter 4, phosphorus and nitrogen are the most critical plant nutrients. When pollutants containing phosphorus and nitrogen compounds accumulate in a lake, rooted aquatic plants and free-floating algae may grow profusely.

The algae and aquatic weeds eventually die and settle to the bottom of the lake, where they are decomposed by bacteria and protozoa. This exerts an oxygen demand on the water and may deplete the D.O. in parts of the lake.

Excessive growth of algae, or *algal blooms*, form slimy mats that float on the lake surface. They are unsightly and, along with the thick growths of weeds that develop along the shore, they interfere with boating, swimming, and fishing. A lake suffering from algal blooms is not a very good recreational resource. Further, if the lake or reservoir is used for water supply purposes, the algae raise the cost of water treatment because the microscopic plant cells tend to clog the filters in the treatment facility, requiring them to be cleaned more frequently. Also, additional chemicals may be required to help control the tastes and odors imparted to the water by the algae.

Decaying plants, along with silt carried into the lake by overland runoff and feeder streams, gradually accumulate in significant amounts as sediment at the lake bottom. As the lake becomes shallower, and, as a consequence, warmer, the balance of aquatic life shifts to favoring less desirable species. For example, trout give way to perch and bass, and eventually to bullheads and carp, as the process continues.

EUTROPHICATION

Actually, the process of *nutrient enrichment* and gradual filling in of a lake, as just described, is a natural process. It is called *eutrophication,* and can be thought of as an inevitable and continual aging of the lake.

Lakes have a natural "life cycle," so to speak. Most lakes start out geologically as deep, cold, clear bodies of water. At this stage, they are called *oligotrophic* lakes. They usually have a sand or rock bottom, very few nutrients, and a scarcity of plant or fish life. Over the years, nutrients slowly accumulate and various forms of aquatic life appear. Silty sediments begin to form at the bottom as the lake passes through a *mesotrophic stage* of existence.

The *eutrophic stage* of a lake's life cycle is char-

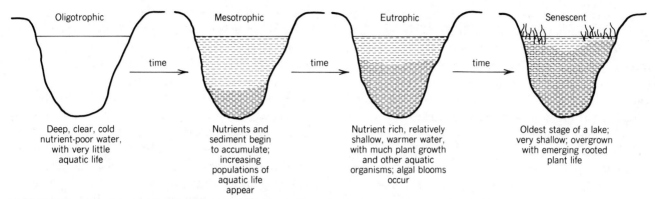

Oligotrophic — Deep, clear, cold nutrient-poor water, with very little aquatic life

time →

Mesotrophic — Nutrients and sediment begin to accumulate; increasing populations of aquatic life appear

time →

Eutrophic — Nutrient rich, relatively shallow, warmer water, with much plant growth and other aquatic organisms; algal blooms occur

time →

Senescent — Oldest stage of a lake; very shallow; overgrown with emerging rooted plant life

FIGURE 5.12 Four stages in the life of a lake. All lakes go through a natural aging process called *eutrophication*. Human activity often accelerates this process.

acterized by a relatively shallow and warmer body of water, with enough nutrients to support large populations of plants and animals. In a eutrophic lake, there are frequent algal blooms, as previously described, and at certain times of the year the water at the bottom may be devoid of dissolved oxygen. Further aging or eutrophication leads to what is called a *senescent lake,* characterized by thick deposits of organic silts and very high nutrient levels. Senescent lakes are very shallow, with much rooted emergent vegetation growing throughout the lake. Eventually, what was once a lake will become a marsh as natural geological and ecological processes continue. The aging of a lake is illustrated in Figure 5.12.

The natural process of lake eutrophication, from the oligotrophic through the senescent stages, takes many thousands of years. It is an exceedingly slow process. But many people use the term *eutrophication* synonymously with *pollution* in reference to lakes. Perhaps a more accurate characterization of the problem would be the term *cultural eutrophication.* Cultural eutrophication is the acceleration and hastening of the natural aging process because of human activity in the drainage basin or watershed of a lake.

Controlling Cultural Eutrophication

Information compiled by the Council on Environmental Quality (CEQ) indicates that as many as two thirds of the lakes in the United States are sig-

nificantly degraded as a result of eutrophication. About one third of the country's population, or about 70 million people, live within 5 miles of a lake. Sewage effluents and surface runoff carry large amounts of plant nutrients into these lakes, accelerating the eutrophication process. This can and should be controlled.

Phosphorus and nitrogen compounds are the most significant of the plant nutrients. Of these, phosphorus is generally recognized as the limiting factor and requires the greatest degree of control. It takes only a concentration of about 0.02 mg/L of inorganic phosphorus to cause algal blooms in a lake; the inorganic nitrogen concentration can be more than 10 times that level. But, on the other hand, even with very high nitrogen levels, if phosphorus concentrations are kept below 0.02 mg/L, excessive growths of algae usually do not occur.

Wastewater effluent and runoff containing phosphorus compounds can easily trigger algal blooms. More than half of the phosphorus in sewage is from phosphate-based detergents. A ban on such products is now in effect in several states. Advanced treatment of sewage can effectively remove much of the phosphorus, as well as the nitrogen, from wastewater, but this is an expensive means of control. Advanced wastewater treatment is discussed in Chapter 10. It is being applied to control nutrient enrichment of the Great Lakes, San Francisco Bay, and many other bodies of water.

In areas where most of the nutrient input is from dispersed sources, such as surface runoff from agricultural areas, advanced sewage treatment would be of little value as a control method. It has been estimated by the CEQ that more than 7 million tons of nitrogen and 0.5 million tons of phosphorus enter surface waters from agricultural areas each year in the United States. More efficient use of fertilizers, soil erosion control, and surface water diversion must be put into effect for lakes in agricultural areas.

Another way of reducing nutrient input is to divert wastewater effluents around the lake into some other body of water, such as a stream, which may be less sensitive to the nutrients. The city of Madison, Wisconsin, applies this method to protect a series of five lakes. Lake Washington in Seattle is another example of a lake protected by wastewater diversion. This method of controlling eutrophication requires construction of extensive interceptor pipeline systems.

The nuisances caused by excessive algal growth in lakes and reservoirs may be alleviated temporarily by the application of copper sulfate. The copper sulfate kills the algae, but its dose must be carefully controlled to prevent fish kills as well. In lieu of chemicals, harvesting of the algae and weeds can offer temporary relief from the problems related to eutrophication. Underwater weed cutters mounted on boats can be used to remove rooted aquatic plants and dredges can be used to remove sediments, but these are not very practical measures for very large bodies of water.

THERMAL STRATIFICATION

It was mentioned in the beginning of this section that lakes are affected by seasonal temperature changes. These effects include a layering or *thermal stratification* of the water, as well as a mixing or *seasonal overturn* of the water, because of temperature differences. Both thermal stratification and seasonal overturn can have significant impacts on pollution and the quality of the lake water. In temperate climates, the cycle of stratification and overturn occurs twice a year, whereas in warm climates where the water never freezes, the cycle occurs once.

Stratification due to temperature differences in the lake water is of most concern in the warm summer months. The lake water is warmed by the air, and the warm water forms a top layer called the *epilimnion.* Colder, and therefore denser, water remains at the lake bottom in a layer called the *hypolimnion.* A relatively thin layer of water with rapidly decreasing temperature from top to bottom, called the *thermocline,* separates the epilimnion and the hypolimnion. The thermocline acts as a physical barrier that prevents mixing of water between the top and bottom layers of the lake. This is illustrated in Figure 5.13a.

The warm water in the epilimnion is mixed by the wind and receives energy from sunlight, allowing it to support algal growths. This relatively turbid water interferes with the penetration of sunlight to greater depths. The stagnant hypolimnion waters are relatively cool and dark. Because of this, some species of fish may prefer the hypolimnion environment. But the water at the lake bottom can often be of poor quality, particularly in a mesotrophic lake. The decaying benthic sediments exert a BOD that depletes the dissolved oxygen in that zone. Sometimes anaerobic conditions may develop at the bottom of the lake.

As the air temperature decreases during the autumn months, the epilimnion waters cool, become denser, and begin to sink toward the lake bottom. Eventually, the entire lake becomes completely mixed, and the well-defined layers of the summer stratification disappear. This circulation, called the *fall overturn,* is illustrated in Figure 5.13b.

In the cold winter months, when ice covers the lake surface, a *winter stagnation* occurs. Then, in the spring, the ice melts and when the water warms to 4°C (at which temperature water is densest), it starts to sink toward the bottom. Aided by the wind, the entire lake soon becomes completely mixed again. This is called the *spring overturn.* The winter stagnation and spring overturn are illustrated in Figures 5.13c and 5.13d, respectively.

In lakes or reservoirs used for water supply, stratification and overturns can affect the quality of the water. During the fall overturn, for example, the poorer quality bottom waters in the hypolimnion become mixed throughout the volume of the

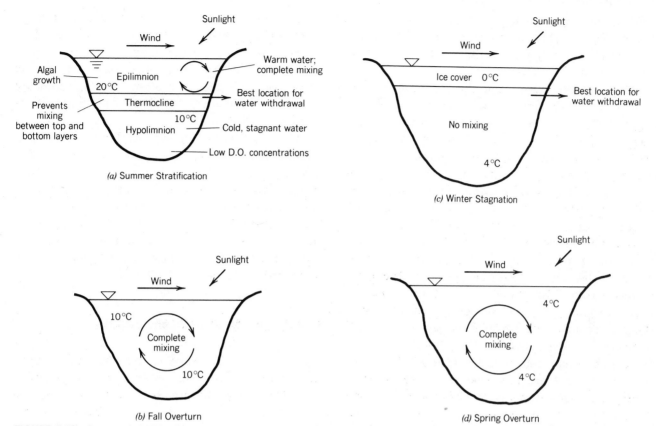

FIGURE 5.13 Seasonal stratification and overturn of a lake or reservoir have an effect on water quality.

lake. This usually intensifies taste and odor problems in the finished water, unless additional steps are taken in the treatment process. Water intake structures can be built in the lake, with inlet ports and valves at several depths. This provides flexibility in operation and permits the water of optimum quality to be taken into the treatment plant.

During the winter, the best quality water is generally withdrawn from a depth just below the ice cover. In the summer, the water in the epilimnion, as well as the water near the bottom of the hypolimnion, is of the poorest quality for reasons already discussed. At this time of the year, the best quality water is usually withdrawn from a depth just below the thermocline.

Various methods have been tried to reduce the adverse effects of stratification. When oxygen depletion and anaerobic conditions become severe in the hypolimnion, compressed air is sometimes dif-

fused through perforated pipes placed on the lake bottom, to reoxygenate the water. In some cases, mechanical mixing and destratification may be effective in improving water quality. One method is to pump cold bottom waters up to the surface. The cooler water tends to shift algal growth to less troublesome species, thus reducing taste and odor problems in the water supply.

5.6 GROUNDWATER POLLUTION

Groundwater supplies one fourth of the fresh water used for all purposes in the United States, including irrigation, industrial uses, and drinking water. About half of the U.S. population, more than 100 million people, rely on groundwater for drinking water. Despite this strong reliance on it, groundwater has for many years been one of our most neglected natural resources.

Being underground, groundwater is, of course, less visible than other environmental resources, such as streams or lakes. And being less visible, it has tended to be of less public concern. But in recent years, primarily because of well-publicized incidents of groundwater contamination, public attitude has changed. Groundwater is no longer the neglected stepchild of water resources and environmental technology.

Groundwater is usually of excellent quality. This is primarily because of the natural filtration that occurs in the layers of soil through which the water slowly moves. The distance that a pollutant can travel in the ground before being separated from the groundwater depends on the type of soil as well as on the type of pollutant. Deposits of fine sand, for example, may remove suspended solids or bacteria from the water in a short distance; whereas coarse gravel or fissured rock could allow those pollutants to travel considerable distances. And soluble pollutants are not affected at all by the filtering action of the soil, although other processes, such as adsorption, may take place.

Since the late 1970s, an increasing number of discoveries of contaminated groundwater in certain locations have been reported. The contaminants come from many different sources and include a variety of materials, most notably synthetic organic chemicals. These include a group of substances called chlorinated hydrocarbons, such as trichloroethylene and carbon tetrachloride. Many of these organic chemicals are toxic; some are suspected to be carcinogens or mutagens that pose serious public health risks at concentrations as low as 10 μg/L.

Of all types of water pollution, this is perhaps the most insidious because at low concentrations the contaminants rarely impart any noticeable taste or odor to drinking water. The water may appear to be crystal clear, but it is far from pristine pure. In some cases, the concentrations of synthetic organics found in contaminated groundwater have exceeded, by several orders of magnitude, the typical levels of those compounds found in very polluted rivers.

The slow rate of flow of groundwater in an aquifer was discussed in the chapter on hydrology. This is a most significant factor in evaluating the impact of groundwater pollution. Because it moves so slowly, typically less than 30 m/yr, a contaminated aquifer will remain contaminated for hundreds of years.

If an aquifer that supplies drinking water is polluted, it may be necessary to abandon the contaminated well(s) and drill new ones some distance away, or to seek alternative surface supplies. In some cases, it may be feasible to install special treatment units, such as aerators or activated carbon filters, to remove the contaminant, but this adds a permanent and often a big expense to the water utility budget.

SOURCES OF CONTAMINATION

Even in areas far removed from human activity, groundwater is not pure. Although it is generally free of turbidity due to natural filtration, it almost always contains dissolved minerals. This, of course, is to be expected since the water is in intimate contact with minerals in soil and rock deposits for long periods of time. As mentioned in the previous chapter, groundwater is usually harder than surface water for this reason. But for the most part, the natural contaminants found in groundwater pose no threat to public health.

The main problems with respect to serious groundwater pollution have been improper disposal of wastes and accidental spills of hazardous substances, especially from industrial activities. A brief discussion of common contamination sources follows.

Industrial Wastes

Industrial chemical wastes disposed of in surface impoundments, such as landfills or lagoons, represent a most significant source of groundwater contamination. A large fraction of the hundreds of millions of tons of industrial wastes generated each year is hazardous. Land disposal of these liquid and solid wastes is practiced because it is the least expensive way to "get rid of" the unwanted materials. But although it seems to be the most economical alternative for industry, in the long run it may be very expensive for society as a whole, with respect to the health hazards and the costs of future clean-up activities.

Most of the existing industrial impoundments in the United States, approximately 50 000 of them, do not have any bottom liners and do not meet new federal or state standards for land disposal of wastes. Contaminated liquids can leak out of these landfills and lagoons, percolate through the soil, and eventually reach a groundwater aquifer. This is illustrated in Figure 5.14. (The liquid from a solid waste landfill is called *leachate;* this is discussed in more detail in the chapter on solid and hazardous wastes.) Organic chemicals such as polychlorinated biphenol (PCB) and benzene have been found in groundwater at many industrial impoundment sites. One of the substances found most frequently is trichloroethylene (TCE), a chlorinated hydrocarbon used as a solvent and a degreaser. Heavy metals such as selenium, arsenic, and cyanide have also been found.

Industrial wastes are sometimes pumped into the ground under pressure through deep wells, in a process called *deep well injection.* This is generally an acceptable method for industrial waste disposal, but the geological conditions must be suitable. At depths over 300 m (1000 ft), ground water is often saline (high salt concentrations) and is not appro-

priate for other uses anyway. But even with deep well injection, accidental contamination of important water supply aquifers is possible, as illustrated in Figure 5.14.

Subsurface Sewage Disposal Systems

Almost one-third of the population of the United States is served by on-site subsurface sewage disposal systems. The most common of these is the septic tank and leaching field system, which is discussed in a subsequent chapter on wastewater disposal. Briefly, the septic tank traps and stores solids while the liquid effluent from the tank flows into a network of buried perforated pipes. The perforated pipes form what is called the *leaching* or *absorption field* and serve to spread out the sewage effluent over an area large enough for it to slowly seep into the soil.

It has been estimated by the Environmental Protection Agency that more than one trillion gallons of sewage enters the ground each year through on-site disposal systems. Unfortunately, not all of these systems work properly, either because of inadequate design, poor construction, or lack of maintenance. Septic disposal systems are frequently the sources of fecal bacteria and virus contamination in private wells. Also, the septic tank cleaning fluids that some homeowners use, contain organic solvents like trichloroethylene (TCE). These potential human carcinogens also pollute the groundwater in areas served by septic systems. Other contaminants that can reach the groundwater from septic effluents include detergents, nitrates, and chlorides. Pollution from septic systems is illustrated in Figure 5.15.

Municipal Landfills

Burial in the ground is one of the most common methods of disposing municipal refuse. In the past, this practice was largely uncontrolled, and the disposal sites, commonly known as "garbage dumps," were literally just that—places where municipal solid wastes were simply dumped on and into the ground. These dumps were often located in low-lying areas with high groundwater tables, or in abandoned sand and gravel pits. Leachate flowing

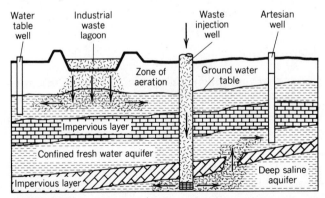

FIGURE 5.14 A diagram showing two sources of groundwater pollution from industrial waste disposal—a leaky surface impoundment or lagoon, and deep well wastewater injection. The arrows indicate the direction of flow of the pollutants. A bottom liner for the lagoon, and thorough geologic exploration of the saline aquifer, can help to prevent the pollution.

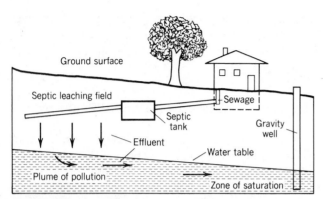

FIGURE 5.15 Groundwater can be polluted from on-site sewage disposal systems. Wells located downhill from septic absorption fields are susceptible to contamination.

through the refuse, high in BOD, chloride, nitrate, organics, heavy metals, and other contaminants, easily reaches the groundwater and enters underlying aquifers from such disposal sites. There are thousands of inactive or abandoned dumps like this throughout the United States. And the EPA estimates that only a little more than one third of existing active municipal solid waste disposal sites are in compliance with state and federal regulations.

Modern solid waste disposal technology can effectively prevent groundwater pollution. The disposal sites are properly called *landfills* rather than dumps. This is discussed in more detail in the chapter on solid wastes. Briefly, proper location of the landfill with respect to geologic conditions and the use of bottom liners are two of the ways in which groundwater quality can be protected.

Mining and Petroleum Production

Many of the active, as well as the abandoned, coal, metal, and other mines in the United States are a threat to groundwater quality. Surface water flowing in the vicinity of the mines can pick up dissolved metals and other solids, acidity, and even radioactive substances. As it infiltrates the earth, either in the open pits from strip mining, or in underground tunnels and shafts, the polluted water easily carries these contaminants into underlying groundwater aquifers. Leaching of contaminants

from tailings (residue) ponds and slag (cinder) piles is also a source of groundwater contamination from mining operations.

Many instances of groundwater pollution from petroleum production activities have been reported, particularly in the south central and southwestern United States. A basic cause of this has been the use of brine pits for disposal of the saline byproducts of drilling.

Petroleum products such as gasoline and motor oil also are groundwater pollutants. These materials can flow through the zone of aeration and reach the groundwater table. Numerous instances of local groundwater pollution due to leaky gasoline tanks at filling stations have been reported. It is estimated by the EPA that nationwide, as much as 11 million gal of gasoline leak into the ground each year. Now petroleum companies are more often replacing steel tanks with rustproof fiberglass ones to reduce this problem.

Accidental spills on the surface are also sources of oil and gasoline contamination in groundwater. Even in low concentrations, these materials cause noticeable tastes and odors in drinking water obtained from a contaminated aquifer. And gasoline contains ethylene dibromide (EDB) and benzene, which may be carcinogenic in humans.

In some cases, if the plume of petroleum pollution has not travelled too far, steps can be taken to clean up the water. The contaminated water can be pumped out of the ground, put through oil separators, and then discharged back into the ground.

Agriculture

The most significant groundwater contaminants from agricultural activities are fertilizers and pesticides. Nitrates are of particular concern in groundwater used for drinking because of the health problem called blue babies (see Chapter 4). In the farming areas of eastern Long Island in New York State, for example, many families with infants must use bottled drinking water. The soil is very sandy, and nitrates from fertilizers are easily carried through the porous soil into the groundwater, contaminating many private wells.

Urban Areas

The public works departments of cities and towns often spread salt on the roads to keep them ice-free during the winter. Eventually these salts are dissolved and carried off the pavement in sheetflow. Much of this material is carried into underlying aquifers, increasing the chloride and TDS concentrations of the groundwater.

Saltwater Intrusion

The intrusion of salty seawater into wells is a groundwater pollution problem occurring in many coastal cities and towns. Because of increasing population, urbanization, and industrialization, increasing quantities of groundwater are being used. And the amount of natural groundwater recharge is decreasing in these areas because of the construction of roads and parking lots. As a consequence, the elevation of the groundwater table is dropping.

In coastal areas, there is an interface or boundary between the fresh groundwater flowing from upland areas and the saline water from the sea. Because seawater is about 2.5 percent denser than freshwater, a pressure head of 40 ft of seawater would be equivalent to a pressure head of 41 ft of freshwater ($1/40 = 0.025 = 2.5\%$). As a result, for each foot the water table drops in elevation, the seawater boundary rises 40 ft. This is illustrated in Figure 5.16. Wells pumping water that is salty because of seawater intrusion may have to be aban-

doned as sources of drinking water. In addition to water conservation, artificial recharge of the groundwater from freshwater impoundments can be effective in halting saltwater intrusion.

PREVENTIVE MEASURES

Natural purification of chemically contaminated groundwater can take decades and perhaps centuries. And clean-up efforts are usually much too expensive to be practical. The best way, then, to control groundwater pollution is to prevent it from occurring in the first place. Recent laws related to solid and hazardous waste disposal will significantly reduce new contamination. Not only are physical barriers between the waste and the groundwater required, but monitoring wells must be installed in some cases to provide early warning of possible leakage.

Land-use management applied on the local level by towns and cities can be effective in preventing aquifer contamination. For example, zoning ordinances that prevent residential or industrial development in areas that are known groundwater recharge zones can reduce pollution problems. Strict enforcement of regulations pertaining to the siting, design, and construction of septic systems can reduce or eliminate the incidence of sewage contamination of private wells. Prudent application of pesticides and fertilizers in agricultural areas can also be effective in this regard.

At this time, there is no single federal law for groundwater protection on a national scale. Some responsibilities toward this have been assigned by Congress to the Environmental Protection Agency and the U.S. Geological Survey (USGS) of the Department of the Interior. Many of the laws passed over the years regarding water pollution and waste disposal do relate to groundwater pollution to some extent. But not all types of situations are covered and there remain gaps in the regulations regarding groundwater.

Many states have taken specific steps toward protecting groundwater and have laws in this regard. But there are many variations in state regulatory systems and little consistency in their ap-

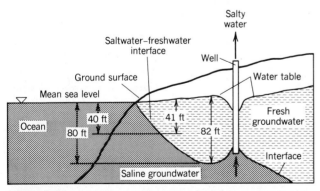

FIGURE 5.16 Saltwater intrusion into coastal area wells is a significant source of groundwater contamination.

proach to groundwater management. Possibly in the future, the EPA will provide a more uniform structure and consistent approach for protecting the nation's groundwater resources. A tentative proposal for a groundwater classification system that has been suggested, as part of a strategy to protect groundwater from further harm, includes three main groups:

1. Groundwater that serves a highly valuable human use or ecological function, warranting the most stringent levels of control.

2. Groundwater that must be protected to ensure its suitability as a drinking water source.

3. Groundwater where limited pollution would be allowed from some contaminants associated with agricultural practices, energy production, and waste disposal.

The actual controls and restrictions to enforce this classification scheme would be left to the states to develop and implement. Many states already have similar systems. Whatever form the nation's groundwater protection strategy finally takes, it is clear that this vital natural resource will receive the attention it deserves in the future.

5.7 OCEAN POLLUTION

Ocean water is naturally saline, containing about 3.5 percent dissolved solids (35 000 mg/L). This is much more than the concentration of total solids carried in raw sewage. But we do not consider the ocean to be polluted because of its natural salinity. The dissolved solids in the sea are inorganic minerals, mostly sodium chloride. The salinity, however, does make the water unsuitable for most uses. With the exception of a few isolated instances where ocean water is subjected to a desalinization process to make it useable, ocean water is not usually withdrawn as a beneficial water supply.

For these reasons, as well as because of the tre-

mendous amount of dilution it apparently provides, there is a natural tendency to consider the ocean as a convenient "sink" or receptacle for wastes of all kinds. Ultimately, all the sewage effluent discharged into streams and rivers makes its way to the ocean. In coastal areas, treated sewage effluent, and sometimes raw sewage, is discharged directly into the ocean.

Millions of tons of sewage sludge, the solids removed from wastewater in the treatment processes, are barged out to sea and dumped each year. And sediment dredged from rivers and harbors to aid navigation are dumped in the ocean. Dredging spoils are often contaminated with heavy metals or nonbiodegradable organic chemicals. The volume of dredged materials dumped in the ocean exceeds that of all other waste materials combined.

Despite the tremendous volume of the marine environment, the natural capacity of estuaries and the ocean to assimilate wastes is limited. A growing number of oceanographers, marine biologists, and other experts are concerned that, at least in certain areas, this limit has been reached or exceeded. One of the adverse effects is the destruction of marine life, particularly the phytoplankton (algae), that produces oxygen in the process of photosynthesis, and that serves as food for other organisms. In some instances, unsightly and perhaps dangerous waste materials are being washed back to shore.

Estuaries are transition zones between freshwater rivers and saline ocean waters. They are semi-enclosed bodies of water, including bays, river mouths, and salt marshes. Being adjacent to land, they are the first marine areas to receive wastes carried by river flow. Estuaries are considered to be one of the most biologically productive environments and are of critical importance to a wide variety of both terrestrial and marine organisms. Not only is pollution a threat to these vital ecosystems, but poor land-use management that allows the filling in of wetland areas for residential or commercial development also takes its toll.

Beyond the estuaries and the relatively shallow coastal ocean waters is the open ocean that comprises about 70 percent of the earth's surface. The open ocean is almost entirely dependent on the estuaries for nutrients, which are transported by currents to the deeper water, and for the support of life processes.

Diffusion of Sewage in Seawater

As previously mentioned, sewage effluent from cities and towns is discharged directly into the ocean in many coastal areas. The pipes that carry the wastewater into the ocean are called *outfalls*. These are often large-diameter conduits that may extend more than 2 km (1.2 miles) offshore.

When sewage flows out of the open end of an outfall pipe, it forms a rising column because it is warmer and less dense than seawater. This is illustrated in Figure 5.17. When it reaches the surface, the column of sewage forms a large bubble or *boil*, which moves in the direction of the surface currents. As the current carries the boil, the plume of diluted wastewater forms, similar to a plume from a smokestack. Unfortunately, the plume is sometimes carried toward shore, raising the coliform counts near recreational areas. Occasionally, beaches must be closed for swimming because of excessive coliform counts.

The effectiveness of ocean disposal of sewage effluent depends on how well the effluent is dispersed and spread out in the ocean when it exits the outfall. Sufficient dispersion of the effluent will facilitate the natural purification process, reduce bacteria concentrations, and prevent pollution at shore areas. Most outfalls are now built with *multiport*

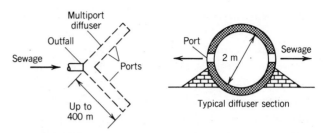

FIGURE 5.18 Multiport diffusers at the end of an ocean outfall pipe increase the mixing and dilution of sewage in the seawater, thereby preventing the formation of plumes.

diffusers at the discharge end to maximize the amount of diffusion. The diffuser distributes the effluent over a relatively large area of the ocean bottom through many circular holes or ports. This allows a much greater degree of mixing and dilution than there is from an outfall without a diffuser. Diffusers prevent the formation of sewage boils and provide greater shore protection.

The success of a diffuser in accomplishing this objective, though, depends on careful hydraulic design computations. It is necessary to achieve uniform flow distribution through the ports. A typical ocean outfall diffuser is illustrated in Figure 5.18. Dimensions are shown just to illustrate the size of these structures.

Ocean Dumping

The dumping of sewage sludge and dredging spoils is done throughout the world. In the United States, the Marine Protection, Research and Sanctuaries Act (also called the Ocean Dumping Act) was passed by Congress in 1972. This act is intended to control ocean dumping by requiring permits for various dumping activities. The Army Corps of Engineers grants permits for the dumping of dredged material, and the EPA grants permits for other materials, including sewage sludge. Theoretically, this permit system is supposed to prevent dumping that would threaten public health or welfare, and to protect the marine ecology.

For about 60 years, New York City and many of the surrounding communities in New York and New Jersey were using a sewage sludge dumping site in the Atlantic Ocean. It was located about 20

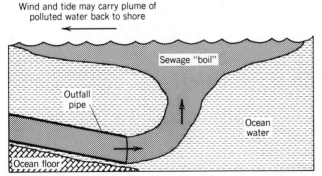

FIGURE 5.17 Sewage effluent from an ocean outfall pipe forms a rising column in ocean water because of its lower density. The column reaches the surface in a *boil*.

km (12 miles) offshore, and was only about 25 m (80 ft) deep and about 36 km^2 (14 sq mi) in area. In 1984, approximately 8 million tons of sewage sludge was dumped there. That area of the ocean had been found to have high levels of bacteria, which prevented shellfishing, as well as high levels of toxic metals and polychlorinated biphenyl (PCB), a known carcinogen. Normal marine life was virtually eliminated at the site, and ocean currents sometimes carried the sludge to beaches and parks along the shore.

In 1985, the EPA ordered that the 12-mile site be closed, and required that sewage sludge be hauled out to a more distant site, about 175 km (106 miles) off the coast. This new dumping location, just beyond the continental shelf, has an area of about 250 km^2 (100 sq mi), and is about 2500 m (8000 ft) deep. The relative locations of the old site and the new site are shown in Figure 5.19. Ocean dumping of sewage sludge at the new site will be allowed until 1990, at which time it is expected that the EPA will ban ocean dumping completely, in favor of incineration or landfill.

Oil Spills

Accidental discharges of oil can be a serious ocean pollution problem. Oil can enter ocean waters pri-

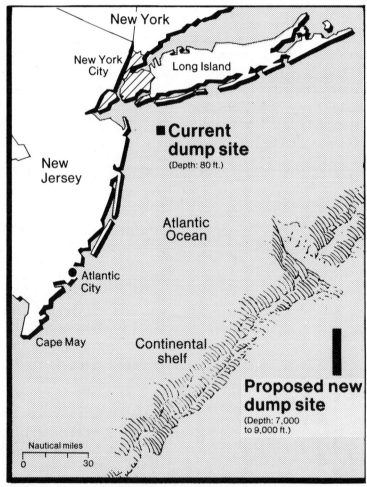

FIGURE 5.19 In the New York–New Jersey area, sewage sludge has been routinely dumped in the ocean at a site about 12 miles from the coast. A new site, more than 100 miles out to sea, has been selected for use until 1990. *(New York Times)*

marily from tanker spills or from offshore well blowouts. Worldwide, in 1979 alone, about 0.75 million tons of oil from tanker accidents was spilled into the sea. And in the same year, almost 0.5 million tons (3 million barrels) of oil was spilled from a single offshore well off the coast of Mexico, over a nine-month period. The oil and tar slick travelled 700 miles north, reaching the coast of Texas.

Oil spills on the open seas can cause the deaths of birds that live on the water surface and dive to obtain their food. Oil coats their feathers, preventing flight, and ingestion of the oil causes toxic effects. It has been estimated that the Torrey Canyon oil spill in 1967 was responsible for the deaths of about 100 000 birds. Oil slicks are often driven toward land by wind and tide action. Oil pollution on the shore harms all forms of aquatic life and interferes with bathing and recreational uses of beach areas.

In some cases, the spread of an oil slick can be controlled by employing physical barriers, and mechanical means of collecting the spilled oil have been used with varying degrees of effectiveness. Detergents used as chemical dispersants have been used to break up the oil slick, but even these detergents can be toxic and harm the marine ecosystem.

The best way to prevent environmental damage from an oil spill is to prevent it from occurring in the first place. Stricter international standards for the design and operation of oil tankers can be effective in reducing the frequency and extent of spills. Similarly, stricter requirements for safety, licensing, inspection, and monitoring of offshore drilling operations can protect the ocean environment.

5.8 WATER QUALITY STANDARDS

In the urbanized and industrialized world in which we live, it is necessary to have a legal basis for protecting water quality. It takes human effort, energy, and money to keep water clean enough for the many different uses society requires it for. Without a legal framework to allow the enforcement of *water quality standards,* environmental quality and public health would be in constant jeopardy.

Water quality standards are limits on the amount of physical, chemical, or microbiological impurities allowed in water that is intended for a particular use. These are legally enforceable by governmental agencies, and include rules and regulations for sampling, testing, and reporting procedures.

Within the past 20 years or so in the United States, Congress has passed several laws that focus on the problems of water pollution and water quality protection. They require the EPA to set minimum standards; individual states have the right to adopt the same federal standards or to establish stricter standards of their own.

There are certain disadvantages in having such rigid statutory laws, including that sometimes not enough scientific data are available to really confirm the validity of a particular standard. But the laws generally serve as a reasonable basis for pollution control and public health protection. As more research is done, the standards are revised so as to better balance the risk of contamination with the cost of cleanup.

There are three different types of standards of interest to us: *stream standards, effluent standards,* and *drinking water standards.* To put this in proper perspective, the relationship among these three kinds of water quality standards is shown schematically in Figure 5.20. Together, they serve to reinforce each other and to "cover all bases" in the ultimate goal of protecting public health and environmental quality.

Stream standards will be discussed in this section; drinking water standards will be discussed in a subsequent chapter on drinking water purification; and effluent standards will be covered in the chapter on wastewater treatment.

STREAM STANDARDS

Beginning with the *Water Quality Act* of 1965, individual states were required by federal law to classify surface waters on the basis of their "maximum

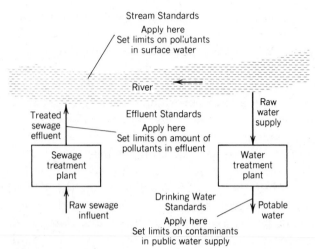

FIGURE 5.20 Three different types of water quality standards are enforced by the USEPA and state regulatory agencies to protect public health and the environment.

beneficial use." Also, the states had to establish specific criteria to limit the amount of pollutants allowed in the different classifications of surface water. These classifications and criteria are generally referred to as *stream standards,* although they have been established for lakes and coastal waters as well.

This law took into account the fact that the water quality of many streams and lakes had already deteriorated enough to prevent certain uses but to allow others. Classification on the basis of maximum beneficial use is intended to prevent even further deterioration of the water from excessive sewage discharges. Stream classifications can be upgraded by the states as progress is made in cleaning up the water, but they cannot be lowered. In this manner, existing low-quality and high-quality waters were identified, and pollution control efforts could be focussed where the maximum benefit could be obtained.

Specific classifications and criteria vary among the different states. But in general, four categories or classes of surface waters are commonly identified, as follows:

CLASSIFICATION	DESCRIPTION
A	Water suitable for primary contact recreation (i.e., swimming, etc.)
B	Water suitable for the maintenance and propagation of fish, shellfish, and wildlife, and for secondary contact recreation (i.e., boating and fishing)
C	Water suitable for public water supply after treatment and purification
D	Water suitable for agricultural or industrial use

The actual water quality criteria for the different use classifications usually include allowable limits on dissolved oxygen, coliforms, solids or turbidity, pH, and toxic wastes. For example, a minimum D.O. of 5 mg/L is typically set for the maintenance of fishlife, but a minimum D.O. of only 3 mg/L may be required for Class D waters, just to maintain aerobic conditions. Streams to be protected as trout spawning streams, however, would have minimum D.O. levels set at about 8 mg/L.

State standards for maximum coliform levels in Class A waters may vary somewhat from the EPA recommended level of 200 fecal coliforms per 100 mL. In surface water intended for public supply, the permissible level of fecal coliforms may be as high as 2000 per 100 mL, 10 times the level for bathing waters.

At first glance, this may seem to be a mistake. But it must be kept in mind that these stream standards are for the raw or untreated water. Surface water must be treated before consumption, and modern water treatment plants can easily reduce coliform levels to an average of less than 1 coliform per 100 mL, even if the water initially contained 2000 per 100 mL.

On the other hand, swimmers do not normally swallow significant quantities of water, and even at a level of 200 coliforms per 100 mL, the probability of disease transmission is very low. Moderate levels of coliforms, then, are tolerable in primary contact recreation waters. Of course, in waters intended for public drinking supplies, it would still be best to have a source with coliform levels as low as possible so as not to overburden the treatment processes.

REVIEW QUESTIONS

1. Give a brief definition of *water pollution*.

2. What is the difference between a point source and a dispersed source of pollutants? Give an example of each.

3. List nine different types or groups of water pollutants. Indicate a primary source of each type.

4. Briefly discuss the effects of thermal pollution.

5. How can thermal pollution be controlled or eliminated?

6. Why is suspended silt or clay considered to be a water pollutant? How does it get into streams and lakes?

7. Briefly describe eight soil erosion control methods.

8. What are two important factors affecting stream self-purification?

9. Briefly describe what a dissolved oxygen profile is. Make a sketch of a D.O. profile for a small slow moving stream receiving a raw sewage discharge.

10. Briefly describe the four zones of stream pollution.

11. Why is it important to be able to compute the minimum D.O. in a stream or river?

12. What is one of the basic differences between lake pollution and stream pollution?

13. Is the eutrophication of a lake a pollution problem? Why?

14. What are some methods for controlling cultural eutrophication of lakes or reservoirs?

15. Briefly explain the occurrence of thermal stratification and seasonal turnover in a lake. How does this affect the quality of the lake water?

16. Is groundwater naturally pure? Why?

17. Do you think that the contamination of an important groundwater aquifer is a serious problem? Why?

18. Briefly discuss seven sources of groundwater contamination.

19. What measures can be taken to prevent groundwater pollution?

20. Are ocean waters immune to pollution problems? Briefly discuss your answer.

21. What is an estuary? Why are they important ecosystems?

22. Briefly discuss the role of multiport diffusers in ocean disposal of sewage.

23. Briefly discuss ocean dumping and oil spills with regard to ocean pollution.

24. What is the function of stream classification standards. Briefly discuss four common classifications of streams.

PRACTICE PROBLEMS

1. An effluent from a sewage treatment plant has a TDS concentration of 500 mg/L and a flow rate of 1.5 mgd. The receiving stream has a TDS level of 100 mg/L and a discharge of 6 mgd. Compute the TDS concentration in the combined sewage and streamflow, downstream of the mixing zone. Assume the sewage is completely mixed in the stream water.

2. The BOD_5 of an effluent from a poorly operating sewage treatment plant is 100 mg/L and the discharge is 1.5 ML/d. The receiving stream has a BOD_5 of 3 mg/L. What minimum streamflow is needed for a dilution such that the combined BOD_5 of the sewage and stream water is no greater than 10 mg/L?

3. The combined BOD_L in a stream mixed with sewage effluent is 25 mg/L. The stream reaeration rate is found to be 0.4/d, and the deoxygenation constant is assumed to be 0.1/d. The initial combined D.O. in the stream is 8.0 mg/L and the D.O. saturation level is 11 mg/L. Compute the minimum D.O. level in the stream due to the sewage discharge.

4. A sewage treatment plant discharges 4 ML/d of effluent with 28 mg/L of BOD_5 into a stream. The stream discharge is 16 ML/d and its initial BOD_5 is 6 mg/L. The initial D.O. in the effluent is 2 mg/L and in the stream it is 7 mg/L. Compute the minimum D.O. level in the stream, assuming that $k_1 = 0.1/d$ and $k_2 = 0.3/d$. Assume the saturation D.O. level is 10.0 mg/L. If the velocity of streamflow is 0.1 m/s, how far downstream does the minimum D.O. occur?

SELECTED REFERENCES

1. Black, J. A., *Water Pollution Technology,* Reston, Reston, Va., 1977.

2. Andrews, W. A., et al., *Environmental Pollution,* Prentice–Hall, Englewood Cliffs, N.J., 1972.

3. Todd, D. K., *Groundwater Hydrology,* John Wiley & Sons, New York, 1980.

4. Hammer, M. J., *Water and Waste-Water Technology,* John Wiley & Sons, New York, 1977.

5. Dix, H. M., *Environmental Pollution,* John Wiley & Sons, New York, 1981.

6. *Environmental Quality—1980, The Eleventh Annual Report of the Council on Environmental Quality,* United States Environmental Protection Agency, Washington, D.C., 1981.

7. Masters, G. M., *Introduction to Environmental Science and Technology,* John Wiley & Sons, New York, 1974.

8. Rawn, A. M., et al., "Diffusers for Disposal of Sewage in Sea Water," *Journal of the Sanitary Engineering Division, American Society of Civil Engineers,* New York, Mar. 1960.

9. *Standards for Soil Erosion and Sediment Control in New Jersey,* New Jersey State Soil Conservation Service, Trenton, N.J., 1974.

DRINKING WATER PURIFICATION

Water withdrawn directly from rivers, lakes, or reservoirs is rarely clean enough for human consumption if it is not treated to purify it. Even water pumped from underground aquifers often requires some degree of treatment to render it *potable,* that is, suitable for drinking.

The nature and extent of treatment required to prepare potable water from surface or subsurface sources depends on the quality of the raw (untreated) water. Of course, better quality water needs less treatment. Generally, a source of raw water with a coliform count of up to 5000/100 mL and turbidity up to 10 units would be considered good. Water with coliform counts that frequently exceed 20 000/100 mL and turbidities that exceed 250 units would be considered a very poor source and would require expensive treatment to render it potable.

The primary objective of water purification is to remove harmful microorganisms or chemicals, thereby preventing the spread of disease and protecting public health. In addition to being safe to drink, the water must also be aesthetically pleasing. It should be crystal clear, and it should not have any objectionable color, taste, or odor. The first section of this chapter discusses the criteria and standards by which a public water supply is judged to be potable or not.

Common treatment processes that are used to prepare potable water are also discussed in this chapter. Generally, groundwater requires the least amount of treatment. It is usually free of bacteria and suspended or colloidal particles, because of the natural filtration that occurs as the water percolates through the soil. But because it is in direct contact with soil or rock, groundwater often contains dissolved minerals, such as calcium or iron.

As a minimum, most states require that public groundwater supplies be disinfected with chlorine to ensure the absence of pathogens. And if dissolved minerals are present in excessive amounts, some combination of chemical treatment, aeration, filtration, and other processes may be needed to purify the water. Some groundwater supplies have recently been found to be contaminated with very low or trace amounts of toxic organic chemicals, usually from improper land disposal of hazardous wastes. If purification using aeration, activated carbon, or other processes is not feasible, the contaminated wells may have to be abandoned.

Surface water supplies generally require more extensive treatment than groundwater supplies do. This is because most streams, rivers, and lakes are contaminated to some extent with domestic sewage and runoff. Even in areas far removed from human activity, surface water contains suspended soil par-

ticles (silt and clay) as well as organics and bacteria (from decaying vegetation and animal wastes). An aerial view of a typical modern water treatment plant required to purify river water is shown in Figure 6.1.

The most common type of treatment for surface water includes *clarification* and *disinfection.* Clarification is usually accomplished by a combination of *coagulation–flocculation, sedimentation,* and *filtration;* the most common method for disinfection in the United States is *chlorination.* A typical flow diagram that shows the sequence of the individual treatment steps, or *unit processes,* is shown in Figure 6.2. These, and other unit processes, are discussed later in the chapter.

6.1 DRINKING WATER STANDARDS

In 1974 Congress passed a law called the *Safe Drinking Water Act,* usually abbreviated *SDWA.*

FIGURE 6.1 A perspective view of a modern water treatment plant. This facility uses ozone for disinfection. (Photo courtesy Hackensack Water Company)

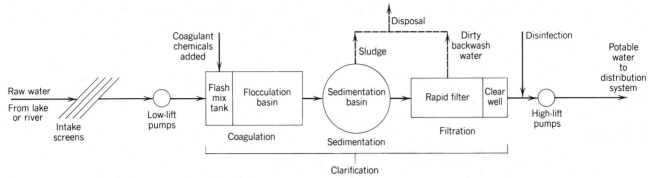

FIGURE 6.2 A flow diagram of a typical surface-water treatment plant. Screens keep fish and debris out of the plant; low pressure pumps lift the water to the flash-mix tank; coagulation, sedimentation, and filtration remove turbidity and clarify the water; disinfection destroys pathogenic organisms; high pressure pumps deliver potable water to the consumers.

Whereas the stream standards discussed in the previous chapter focus primarily on environmental quality protection, the SDWA was intended to establish standards for the water people actually drink, and thereby serve directly to protect public health. Under the SDWA, the EPA has developed water quality criteria and regulations that protect the quality of water in *public water systems.*

A public water system is one that has 15 or more service connections, or that serves 25 or more people. Two types of public water systems are *community systems,* which serve year-round residents, and *noncommunity systems,* which serve travelers or intermittent users (e.g., at motels and camping sites). The system can be owned by a private company and still be classified as a public system if it meets these requirements. All public water systems must comply with the federal SDWA regulations.

There are five different types of substances controlled by the SDWA in community water systems. These include *turbidity, bacteria, inorganic chemicals, organic chemicals,* and *radioactive substances.* In noncommunity systems, only turbidity and bacteria must be tested for on a routine basis; of the inorganics, initial sampling and testing is done only for nitrates. This is because water from a particular noncommunity system is consumed by individuals only occasionally and in small amounts.

MAXIMUM CONTAMINANT LEVELS

Based upon the results of public health studies, the EPA has set allowable concentrations of specific contaminants in drinking water, from the preceding five groups. These concentrations are called *Maximum Contaminant Levels,* or MCLs. MCLs established under the *primary regulations* are intended to protect public health by setting highest permissible levels of toxic or harmful substances. Under the *secondary regulations,* MCLs are set for substances that affect the aesthetic qualities of drinking water but have no direct effect on public health. The primary regulations are legally enforceable, and the secondary regulations are only intended as suggested guidelines.

Turbidity

Turbidity affects more than just the appearance of water. In drinking water it can be a health hazard and is therefore controlled under the primary regulations. Turbidity interferes with disinfection by shielding microorganisms and by increasing chlorine demand (chlorine is the most common disinfecting agent in the United States).

Turbidity of drinking water is analyzed daily. The MCL for turbidity is a maximum of 1 NTU, averaged over one month, or 5 NTU for a 2-day average. The 2-day limit protects against short-term surges in turbidity that would impair the bacteriological quality of the water. For treated drinking water, only the nephelometric method of turbidity measurement is allowed; nephelometers measure the amount of light scattered by turbid water, rather than the amount of light which can pass through the sample.

Bacteria

Coliform bacteria are the organisms used to gage the bacteriological quality of water, as explained in Section 4.4. Both the membrane filter method and the multiple-tube fermentation method are accepted for analysis of total coliform levels. The MCL for coliform depends on which method was used for testing, and on the number of samples tested per month; more samples are required for systems serving large populations. For example, a community with a population of 20 000 people would be required to take 24 coliform samples per month, while a population of 200 000 would need 140 samples.

Under the membrane filter method, the average monthly coliform level must not exceed an MCL of 1 coliform per 100 mL. Also, the level must not exceed 4 coliforms per 100 mL in more than 5 percent of the monthly samples. In the multiple-tube fermentation test for coliforms in drinking water, a series of five tubes with 10–mL sample portions are used. Coliforms must not be present in more than 10 percent of the sample portions tested each month.

Chlorine residual testing can give a much quicker indication of the sanitary quality of the water than either of the coliform testing methods. If a chlorine residual of 0.2 mg/L is present, it is extremely unlikely that there would be pathogenic bacteria surviving in the water. At least one residual test must be taken each day. And substitution of chlorine residual testing for coliform testing is allowed for up to 75 percent of the required bacteriological samples, at state option.

Inorganic Chemicals

Primary regulation MCLs for ten inorganic chemicals are listed in Table 6.1, as follows:

All of the above inorganic substances are part of

TABLE 6.1   MAXIMUM CONTAMINANT LEVELS: INORGANICS

Contaminant	MCL, mg/L
Arsenic	0.05
Barium	1.0
Cadmium	0.010
Chromium	0.05
Fluoride	2.0
Lead	0.05
Mercury	0.002
Nitrate	10.0
Selenium	0.01
Silver	0.05

the primary regulations because they are of public health importance. The MCL for fluoride is actually a function of local climate; a middle range value for moderate climates is shown here. Treated water is tested for inorganics on a yearly basis.

Organic Chemicals

Primary MCLs for organics regulate six pesticides, including endrin and lindane, and total trihalomethane. To illustrate the extremely small concentrations that have public health significance, the MCL for endrin is only 0.2 µg/L or 0.2 ppb. The MCL for total trihalomethane, a carcinogen that may be formed as a by-product of the chlorination process, has been set at 0.1 mg/L. Organics in drinking water are to be analyzed every three years.

As research continues on the health effects of various chemicals, the list of primary MCLs for organics will probably grow. Recently, the EPA tentatively recommended that MCLs for several industrial solvents and other synthetic organic substances be set at zero. This includes such chemicals as benzene and carbon tetrachloride, which are suspected human carcinogens. But technical as well as economic limitations will probably result in the final MCLs being somewhat higher than zero.

Radioactive Contaminants

Water can be contaminated with natural radioactive materials or with synthetic substances. The

MCL is 15 pCi/L of gross alpha activity, and 5 pCi/L of radium. Tests for radioactivity are required at least every four years. Strontium–90, one of the synthetic substances, has an MCL of 8 pCi/L.

Secondary MCLs

Several secondary standards for substances that affect the taste, odor, or appearance of drinking water are presented in Table 6.2. These secondary contaminants do not affect public health, but they are related to the general acceptability of the water to consumers. All of these substances are discussed in Sections 4.2 and 4.3 in the chapter on water quality.

Recordkeeping and Reporting

Good recordkeeping procedures are important for the proper operation of a public water system. They also provide data for future planning, public information, and legal protection. Under the requirements of the SDWA, each public water system must maintain records of water quality test results, reports, and actions taken to correct deficiencies; depending on the type of record, it may have to be kept on file for up to 10 years.

Records of MCL test data for bacteria and chemical analyses are particularly important. The records should include the time, date, and place of sampling, the name of the technician who took the sample, the type of sample, and the place, date, method, and results of analysis.

The SDWA requires that public water systems submit reports to the public as well as to the state in which the system operates. The results of routine sampling and testing must be sent to the state every month. Also, the state must be notified within 48 hours of a violation of any of the primary regulation MCLs. The requirement for routine sample reports and violation reports helps to ensure that water system deficiencies and potential health hazards will be identified and corrected. A typical monthly summary report for turbidity is shown in Figure 6.3.

Reporting to the public by public water systems is required by the SDWA to advise people of potential health hazards and to educate them about the importance of adequate financing and support for drinking water systems. The public is notified only of confirmed MCL violations, by mail, newspaper, and radio and television broadcasts. The steps leading up to public notification of MCL violations for inorganic and organic chemicals are shown in Figure 6.4.

The public notice must describe the nature of the problem and include any steps people should take to protect their health. An illustration of a public notice for violation of the nitrate MCL is shown in Figure 6.5.

TABLE 6.2   SELECTED SECONDARY DRINKING WATER MCLs

Contaminant	MCL
Chloride	250 mg/L
Color	15 color units
Copper	1 mg/L
Iron	0.3 mg/L
Manganese	0.05 mg/L
Odor	3 (threshold number)
pH	6.5 to 8.5
Sulfate	250 mg/L
TDS	500 mg/L
Zinc	5 mg/L

6.2   SEDIMENTATION

The impurities in water may be either dissolved or suspended. The easiest way to remove the suspended material is to let the force of gravity do the work. Under *quiescent* conditions, when flow velocities and turbulence are minimal, particles that are denser than water will be able to settle to the bottom of a tank. This process is called *sedimentation*, and the layer of accumulated solids at the bottom of the tank is called *sludge*. The tank may be called either a *sedimentation tank*, a *settling tank*, or a *clarifier*.

The speed at which suspended particles settle to-

FINISHED WATER TURBIDITY

_____
Name of Water Supply

_____
Location

WATER
SYSTEM
ID# _____

MONTH _____

Date	Sampled by/Loc	Routine Sample		Check Sample		Value Used in Two-Day and Monthly Average[b]	Two-Day Average of Turbidity		
		Time	Value(TUs)[a]	Time	Value(TUs)		Days	Average Value (TUs)[c]	
								Sum	Avg.
1	2	3	4	5	6	7	8	9	10
1			0.1			0.1			
2			0.3			0.3	1+2	0.4	0.2
3			0.1			0.1	2+3	0.4	0.2
4			0.2			0.2	3+4	0.3	0.15
5			0.1			0.1	4+5	0.3	0.15
6			0.8			0.8	5+6	0.9	0.45
7			0.6			0.6	6+7	1.4	0.7
8			0.3			0.3	7+8	0.9	0.45
9			0.9			0.9	8+9	1.2	0.6
10			1.1		0.9	0.9	9+10	1.8	0.9
11			1.8		1.7	1.7	10+11	2.6	1.3
12			3.7		3.6	3.6	11+12	5.3	2.65
13			5.2		5.0	5.0	12+13	8.6	4.3
14			1.3		0.9	0.9	13+14	5.9	2.95
15			0.9			0.9	14+15	1.8	0.9
16			0.7			0.7	15+16	1.6	0.8
17			0.8			0.8	16+17	1.5	0.75
18			0.8			0.8	17+18	1.6	0.8
19			0.5			0.5	18+19	1.3	0.65
20			0.2			0.2	19+20	0.7	0.35
21			0.3			0.3	20+21	0.5	0.25
22			0.3			0.3	21+22	0.6	0.3
23			0.1			0.1	22+23	0.4	0.2
24			0.5			0.5	23+24	0.6	0.3
25			0.4			0.4	24+25	0.9	0.45
26			0.5			0.5	25+26	0.9	0.45
27			0.8			0.8	26+27	1.3	0.65
28			0.7			0.7	27+28	1.5	0.75
29			1.1		0.9	0.9	28+29	1.6	0.8
30			0.9			0.9	29+30	1.8	0.9
31							30+31		

TOTAL = __24.8__

$$\text{MONTHLY AVERAGE (d)} = \frac{\text{TOTAL TUs}}{\text{\# Days in Month}} = \underline{0.83 \text{ TU}}$$

Type of Instrument used _____

(a) If value exceeds 1 TU, take check sample within one hour.

(b) Use values from original samples on days MCL was not exceeded, and check sample values for days the MCL was exceeded.

(c) To calculate: The two-day average = $\dfrac{\text{DAY 1 TU} + \text{DAY 2 TU}}{2}$

(d) If monthly average is more than 1 TU, report to state and notify the public.

_____
Person responsible for analysis

FIGURE 6.3 A typical summary report required by the SWDA (Reprinted from _Safe Drinking Water Act Self-Study Handbook,_ by permission. Copyright © 1978. The American Water Works Association.)

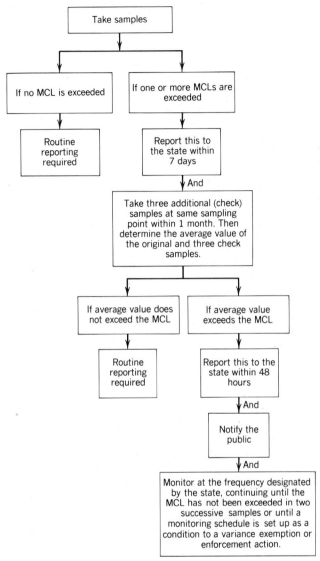

FIGURE 6.4 A diagram of the steps leading up to public notification of an MCL violation, as required by the SDWA (Reprinted from *Safe Drinking Water Act Self-Study Handbook*, by permission. Copyright © 1978, The American Water Works Association)

varies with the concentration of suspended particles and their tendency to interact with one another. In a dilute suspension where the particles are free to settle without interference, the process is called free settling or *discrete settling*. As the concentration increases, the particles tend to interact and interfere with the free movement of one another; this is sometimes called *hindered settling*. In a sedimentation tank there may be up to four different zones or types of settling that occur at different depths, and exact mathematical analysis of the process can be quite complicated. This section discusses some common factors related to discrete particle settling.

DETENTION TIME

If a volume of water was left completely undisturbed in a tank for several days or weeks, just about all of the suspended solids would have a chance to settle to the bottom. Even some bacteria, microscopic in size, would eventually settle out. But this procedure is not practical for municipal water treatment plants, because they generally handle large volumes of water on a continuous-flow basis. It is not feasible to simply shut a valve to stop the flow and let a fixed volume of water remain undisturbed in a tank for a long period of time. Too many very large tanks would be needed.

Instead, settling tanks for water (or wastewater) treatment are designed to operate as the flow slowly continues from the inlet to the outlet of the tank. But the movement of water is slow enough to allow quiescent settling for a large percentage of the suspended particles. Generally, the water remains in the tank for only a few hours before it reaches the tank outlet. The theoretical amount of time water remains in a settling tank is called the *detention time*. It can be computed as follows:

$$T_D = \frac{V}{Q} \qquad (6\text{-}1)$$

where T_D = detention time
V = volume of water in tank
Q = average flow rate (volume per unit time)

ward the bottom of a tank depends on their size as well as on their density. The larger and heavier particles will naturally settle faster than smaller or lighter particles. The forces opposing the downward force of gravity include buoyancy and friction (drag). The temperature and viscosity of the water are additional factors that affect the particle-settling rate.

The nature of the sedimentation process also

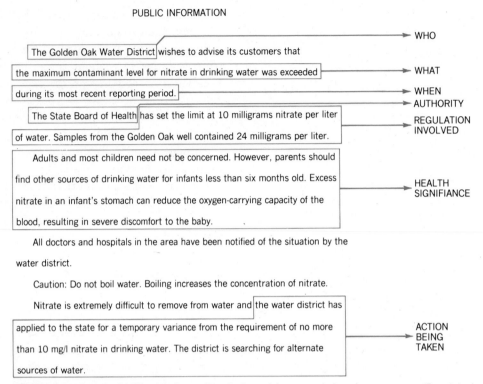

PUBLIC INFORMATION

The Golden Oak Water District wishes to advise its customers that → WHO

the maximum contaminant level for nitrate in drinking water was exceeded → WHAT

during its most recent reporting period. → WHEN
→ AUTHORITY

The State Board of Health has set the limit at 10 milligrams nitrate per liter
REGULATION
INVOLVED
of water. Samples from the Golden Oak well contained 24 milligrams per liter.

Adults and most children need not be concerned. However, parents should
find other sources of drinking water for infants less than six months old. Excess
→ HEALTH
SIGNIFIANCE
nitrate in an infant's stomach can reduce the oxygen-carrying capacity of the
blood, resulting in severe discomfort to the baby.

All doctors and hospitals in the area have been notified of the situation by the
water district.

Caution: Do not boil water. Boiling increases the concentration of nitrate.

Nitrate is extremely difficult to remove from water and the water district has
ACTION
applied to the state for a temporary variance from the requirement of no more
→ BEING
TAKEN
than 10 mg/l nitrate in drinking water. The district is searching for alternate
sources of water.

FIGURE 6.5 A typical MCL violation notice that would appear in local newspapers (Reprinted from *Safe Drinking Water Act Self-Study Handbook,* by permission. Copyright © 1978, The American Water Works Association.)

Detention time is usually expressed in terms of hours. It is important to use consistent units for V and Q, so that Equation 6-1 will be dimensionally correct. Minimum detention times of three hours are specified by most state health departments or environmental agencies to ensure sufficient settling; most of the settleable suspended solids will reach the sludge layer in this time period. The following examples illustrate the use of Equation 6-1.

EXAMPLE 6.1

A sedimentation tank has a volume capacity of 15 000 m³. If the average flow rate entering the tank is 120 ML/d, what is the detention time?

Solution:

We must first convert cubic meters to megaliters, or vice versa, for dimensional consistency. Choosing to convert volume to megaliters, we get

$$V = 15\ 000\ \text{m}^3 \times 1000\ \text{L/m}^3$$
$$= 15\ 000\ 000\ \text{L} = 15\ \text{ML}$$

Now, applying Equation 6-1, we get

$$T_D = \frac{15\ \text{ML}}{120\ \text{ML/d}} = 0.125\ \text{d}$$

and converting to hours, we get

$$T_D = 0.125\ \text{d} \times 24\ \text{h/d} = 3\ \text{h}$$

EXAMPLE 6.2

Water flowing at a rate of 6 mgd is to have a 3-h detention time in a sedimentation tank. Compute the required volume capacity of the tank, in cubic feet. If the tank has a surface area of 10 000 ft², how deep will the water be in the tank?

Solution:

Since we must solve for volume, we can first rearrange the terms of Equation 6-1 to get $V = T_D \times Q$. For dimensional consistency, we can first convert the flow rate from units of million gallons per day to gallons per hour, as follows:

$$Q = 6\ 000\ 000\ \text{gal/d} \times 1\ \text{d/24 h} = 250\ 000\ \text{gal/h},$$

and

$$V = T_D \times Q = 3\ \text{h} \times 250\ 000\ \text{gal/h}$$
$$= 750\ 000\ \text{gal}$$

Converting gallons to cubic feet, we get

$$V = 750\ 000\ \text{gal} \times \frac{1\ \text{ft}^3}{7.5\ \text{gal}} = 100\ 000\ \text{ft}^3$$

Since the volume of a tank can be expressed as the product of its depth and surface area (volume = depth × area), the depth of water can be computed as volume ÷ area or $100\ 000\ \text{ft}^3/10\ 000\ \text{ft}^2 = 10$ ft deep.

OVERFLOW RATE

Another factor or term that is of importance in the design and operation of a settling tank is the *overflow rate,* or *surface loading,* as it is also called. It can be computed as follows:

$$V_o = \frac{Q}{A_s} \qquad (6\text{-}2)$$

where V_o = overflow rate
A_s = tank surface area (top view)

In SI metric units, the overflow rate is expressed in terms of cubic meters per square meter per day ($\text{m}^3/\text{m}^2\cdot\text{d}$); flow rate is expressed in terms of cubic meters per day (m^3/d); and surface area is expressed in terms of square meters (m^2).

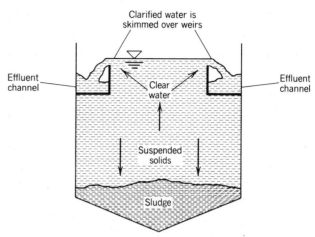

FIGURE 6.6 A schematic view of the sedimentation process.

In American units, V_o is expressed in terms of gallons per day per square foot (gpd/ft^2); Q is expressed in terms of gallons per day (gpd); and A_s is expressed in terms of square feet (ft^2). The maximum overflow rate typically allowed by state regulatory agencies is 33 $\text{m}^3/\text{m}^2\cdot\text{d}$ or 800 gpd/ft^2.

Overflow rate can be visualized as an average "up-flow" velocity of water in the settling tank. In fact, it can be easily seen that overflow rate is actually a velocity by cancelling units in the SI system: $\text{m}^3/\text{m}^2\cdot\text{d}$ = m/d (meters per day). A schematic cross-section of a tank, shown in Figure 6.6, illustrates the general flow pattern. The clarified water is skimmed from the surface as it flows over weirs into an effluent channel. All suspended particles that settle at a faster velocity than V_o reach the sludge layer at the bottom of the tank. Only a fraction of the smaller and lighter particles that settle at velocities less than V_o will be removed from the water before it leaves the tank.

EXAMPLE 6.3

What is the minimum settling velocity, in feet per hour, of suspended particles that can be completely removed in a settling tank that has an overflow rate of 700 gpd/ft^2?

Solution:

700 gal/ft^2·d × 1 ft^3/7.5 gal × 1 d/24 h = 3.9 ft/h

Only a fraction of particles that settle slower than 3.9 ft/h will be removed; the slower the settling velocity is, the smaller the percentage removed is. For example, only 20 percent of particles that settle at a velocity of 0.2 × 3.9 = 0.78 ft/h will be captured in the sludge layer.

SETTLING TANK DESIGN

By combining specified values of detention time and overflow rate, it is possible to determine the required dimensions of a settling tank. The tanks built are either rectangular or circular. The actual depth of water in the tank is called the *side water depth,* or SWD. The height of the tank wall is usually about 450 mm or 18 in. above the SWD. This is called *freeboard,* and it serves to prevent splashing of water over the tank sides. The following example illustrates how tank dimensions are determined.

EXAMPLE 6.4

A circular sedimentation tank is to have a minimum detention time of 4 h and a maximum overflow rate of 20 m^3/m^2·d. Determine the required diameter of the tank and the SWD, if the average flow rate through the tank is 6 ML/d.

Solution:

Applying Equation 6-1, we compute the required volume as

$$V = Q \times T_D = 6 \text{ ML/d} \times 1 \text{ d/24 h} \times 4 \text{ h}$$
$$= 1 \text{ ML} = 1\,000\,000 \text{ L}$$

(Note the factor 1/24 for dimensional consistency) Converting the volume to cubic meters, we get

$$V = 1\,000\,000 \text{ L} \times \frac{1 \text{ m}^3}{1000 \text{ L}} = 1000 \text{ m}^3$$

Before using Equation 6-2, we can first convert the flow rate to units of m^3/d for dimensional consistency, as follows:

$$Q = 6 \text{ ML/d} \times 10^6 \text{ L/ML} \times \frac{1 \text{ m}^3}{10^3 \text{ L}} = 6000 \text{ m}^3\text{/d}$$

Now applying Equation 6-2, we compute the surface area as

$$A_s = \frac{Q}{V_o} = \frac{6000 \text{ m}^3\text{/d}}{20 \text{ m}^3\text{/m}^2\text{·d}} = 300 \text{ m}^2$$

To determine the tank diameter, we must use the formula for the area of a circle: $A = \pi D^2/4$. From this, we get

$$D = \sqrt{4A/\pi} = \sqrt{(4 \times 300)/\pi} \approx 20 \text{ m}$$

And finally, since $V = A_s \times$ SWD, we get

$$\text{SWD} = \frac{V}{A_s} = 1000 \text{ m}^3/300 \text{ m}^2 = 3.33 \text{ m}$$

In a rectangular settling tank, the *influent* (water flowing into the tank) is directed against a baffle that distributes the water uniformly across the width of the tank and imparts a downward velocity to the flow. This is illustrated in Figure 6.7. The *effluent* (water flowing out of the tank) is skimmed from the surface over weirs placed at the opposite end of the tank. A series of redwood boards moving on a continuous chain scrapes the sludge toward a collection hopper, from where it is pumped out of the tank.

In a circular clarifier, the water usually enters at the center of the tank and flows radially outward toward an effluent weir built along the perimeter of the tank. A rotating sludge-scraper mechanism

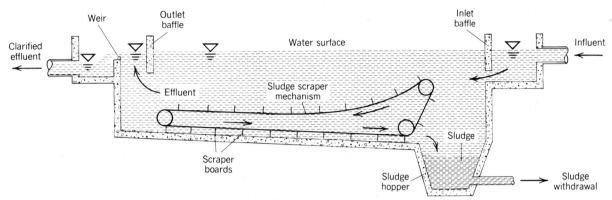

FIGURE 6.7 A simplified section view of a rectangular sedimentation tank. A slowly moving endless-chain scraper mechanism pushes sludge into a hopper for removal.

moves the sludge toward a central collection hopper. This is illustrated in Figure 6.8.

Whatever the shape of the tank, the inlets and outlets must be designed carefully to prevent currents that could resuspend the sludge. It is also necessary to avoid a condition called *short-circuiting* of the flow. In this context, *short-circuiting* refers to a condition that allows most of the water to flow through the tank in a period of time that is considerably less than the computed detention time. The effectiveness of sedimentation may be significantly reduced if short-circuiting occurs.

The effluent weirs in a settling tank are designed to operate at minimum head and velocity conditions, to reduce the chance for particles to be carried over in the effluent. The total weir length

must be long enough so that the flow rate per foot or meter of weir is less than a specified maximum value, called the *weir loading rate*. The weirs usually consist of a series of uniformly spaced V-notches in a long metal plate. The effluent flows through the notches into a channel called an *effluent launder,* which directs the flow to an outlet pipe.

The effectiveness of a settling tank in removing suspended solids depends more on the surface area than on the total volume or detention time. It can be seen from Equation 6-2 that, for a given flow rate Q, as the surface area A_s increases, the overflow rate V_o decreases. This means that the maximum particle settling velocity for complete removal also decreases with increasing surface area. There-

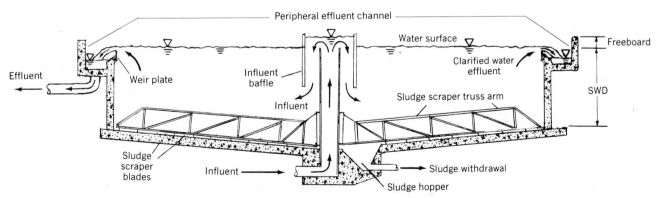

FIGURE 6.8 A simplified section view of a circular sedimentation tank. Rotating scraper blades move the settled sludge to a central draw-off hopper.

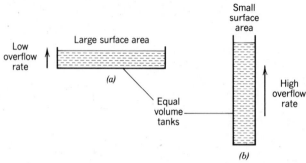

FIGURE 6.9   The shallow tank *(a)* has a lower overflow rate than tank *(b)*, because it has a larger surface area. It is more effective in removing suspended solids than the deeper tank.

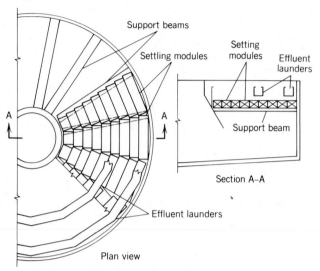

FIGURE 6.11   Tube settlers can be installed in new or in existing sedimentation tanks to improve suspended solids removal efficiency. (Figure 2-9 from *New Concepts in Water Purification*, by Culp and Culp. Copyright © 1971, by Litton Educational Pub. Co. Reprinted by permission of Van Nostrand Reinhold Company, Inc.)

fore, a shallow tank with a large surface area will be more effective than a deep tank with the same volume but smaller area. This is illustrated in Figure 6.9. Most settling tanks are not less than 3 m (10 ft) deep, however, to provide room for the sludge layer or the sludge scraper mechanism.

The concept of *shallow depth sedimentation,* as described here, is now often used when designing new settling tanks, or for increasing the capacity and efficiency of existing tanks. Prefabricated units, or modules, comprising a series of inclined and nested tubes, can be installed near the top of the tank to increase its effective surface area. This is illustrated in Figure 6.10.

As shown in Figure 6.10, the suspended particles in the water become caught in the downward flowing stream of sludge in each tube, whereas the clarified water flows upward toward the effluent weirs. A typical installation of these *tube settlers,* as they are called, is illustrated in Figure 6.11.

6.3   COAGULATION

Suspended particles cannot be completely removed from water by plain settling, even when they are given very long detention times and low overflow rates. Some of the very small turbidity-causing particles, called *colloids* (see Section 4.1), will not settle out of suspension by gravity without some help. If one rapidly mixed certain chemicals, called *coagulants,* in the water, and then slowly stirs the mixture before allowing sedimentation to occur, they will settle.

One of the properties of colloidal particles that

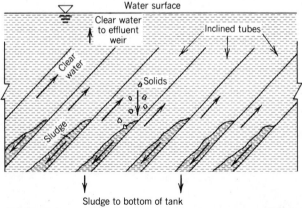

FIGURE 6.10   A series of inclined *tube settlers* increases the effective surface area of a settling tank. Settleable solids are quickly entrapped in the downward flowing sludge in each tube, whereas the clarified water flows upward.

keeps them in suspension is the small electrostatic charge they each carry. In the same way that similar poles of two bar magnets repel each other, colloidal particles push each other apart and avoid collisions. The coagulant chemical, however, neutralizes the effect of the colloidal charges. Once neutralized, the colloidal particles can collide and agglomerate (stick together), forming larger and heavier particles, called *flocs*. The coagulant also reacts with the natural alkalinity in the water, forming a "sticky" solid *precipitate* that comes out of solution and helps in the formation of the flocs by "capturing" particles.

After the initial *flash-mix* of the coagulant with the water, a gentle agitation caused by slow stirring further enhances the growth of the flocs by increasing the number of particle collisions. The slow mixing or stirring process is called *flocculation*. The combined rapid mix–slow mix process is usually referred to as *coagulation*. Most of the flocs formed during coagulation are settleable and can be removed from the water in a sedimentation tank. As illustrated in Figure 6.2, coagulation generally precedes the sedimentation process in a typical water treatment plant.

There are several different chemicals that can be used for coagulation. The most common coagulant is aluminum sulfate, $Al_2(SO_4)_3$. It is generally referred to as *alum*. Sometimes certain synthetic organic chemicals, called *polymers*, are added along with the alum to act as *coagulant aids*. These long-chain, high-molecular-weight compounds help the formation of larger, heavier floc particles.

The success of the coagulation process depends on several factors, including the chemical dosage, water temperature, pH, and alkalinity. Since the quality of a surface water supply often varies with time, it is frequently necessary to adjust the dosage. The optimum coagulant dose is usually determined in the laboratory by a procedure called the *jar test*.

In the jar test, six beakers or "jars" are filled with a sample of the raw water, and each sample is mixed with a different amount of coagulant. A stirring apparatus, illustrated in Figure 6.12, is used to provide slow mixing, thereby simulating the flocculation process.

FIGURE 6.12 A stirring apparatus for the *jar test*, which is used to determine optimum coagulat dosage. (Phipps & Bird, Inc., 8741 Landmark Road, Richmond Virginia 23288).

After the stirring paddles are stopped, the flocs are allowed to settle in the beakers. The dosage in the beaker that required the least amount of coagulant to produce a clear water with well-formed, rapidly settling floc, is used to compute the dosage for the entire water treatment plant. Additional tests can be made with the same apparatus to determine the effects of pH or alkalinity adjustments on the formation of flocs, to further optimize the process.

As previously mentioned, the first step in the coagulation process is the flash-mix or rapid-mix of water and coagulants. This involves violent agitation to spread the chemicals throughout the water and to ensure that there is a complete chemical reaction. Sometimes this is accomplished by adding the chemicals in the suction line just ahead of the centrifugal low-lift pump that brings the water into the treatment plant. The impeller of the pump provides the rapid-mix action inside the pump casing.

In most treatment plants, though, a rapidly rotating propeller is installed in a relatively small tank that provides about 1 min of detention time. These flash-mix tanks are often built immediately adjacent to the flocculation tanks in order to save on construction costs, as shown in Figure 6.13.

The size of the flocculation tank is such that it provides a detention time of up to one hour for slow stirring. Paddle-type flocculators are the most common, using redwood slats mounted horizontally on

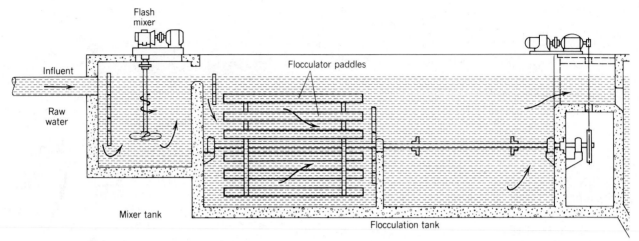

FIGURE 6.13 A section view of a flash-mix and flocculation tank used in the coagulation process. (FMC Corporation)

motor driven shafts. Rotating slowly at about one revolution per minute, the paddles provide gentle agitation that promotes the growth of flocs.

Some relatively small water treatment plants combine chemical addition, flocculation, and sedimentation in a single tank, called a *solids-contact tank* or *upflow clarifier*. A sectional view of a typical upflow clarifier is shown in Figure 6.14. Chemical addition and rapid mixing occur where the water enters the tank, in the center. The water first flows downward under a cone-shaped hood where flocculation occurs. Then it flows upward through the portion of the tank that serves for gravity settling. A sludge blanket of floc particles is formed at the bottom of the tank. Treatment units like this are particularly useful in plants where the water must also be softened by adding lime (as will be discussed in Section 6.6) and where only limited space is available.

nonsettleable floc particles. These remaining flocs can cause noticeable turbidity and may shield microorganisms from the subsequent disinfection process. In order to produce a crystal clear potable water that satisfies the SDWA requirement of 1 NTU (the MCL for turbidity), an additional treatment step following coagulation and sedimentation is needed.

This next step is a physical process called *filtration*. Filtration involves the removal of suspended particles from the water by passing it through a layer or "bed" of a porous granular material, such as sand. As the water flows through the filter bed, the suspended particles become trapped within the pore spaces of the filter material, or *filter media,* as it is called. This is shown schematically in Figure 6.15. Filtration is a very important treatment process in a surface-water purification plant. In fact, many of these facilities are called "filtration plants," even though filtration is only one step in the overall treatment sequence.

6.4 FILTRATION

Even with the help of chemical coagulation, sedimentation by gravity is not sufficient to remove all the suspended impurities from water. About 5 percent of the suspended solids may still remain as

RAPID FILTERS

The first filters built for water purification used very fine sand as the filter media. Because of the tiny size of the pore spaces in the fine sand, water takes a long time to flow through the filter bed. And when the surface becomes clogged with sus-

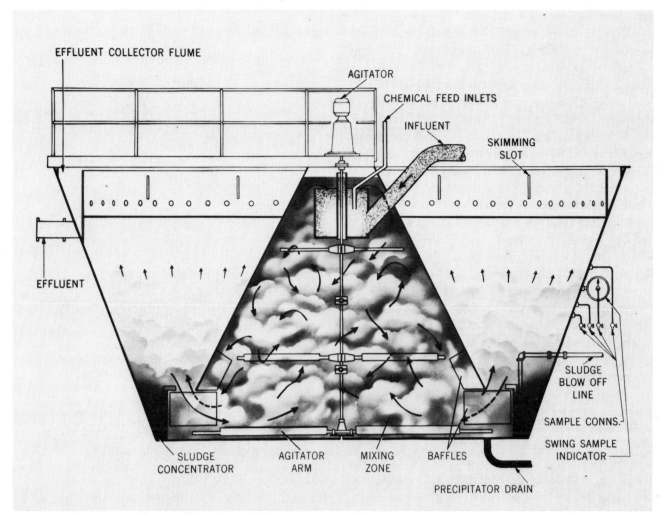

FIGURE 6.14 A suspended solids contact clarifier (Permutit Co., Inc.)

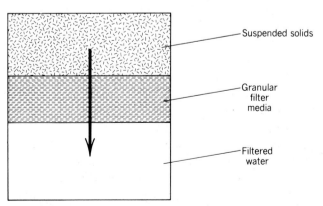

Suspended solids

Granular filter media

Filtered water

FIGURE 6.15 A schematic diagram of the filtration process.

pended particles, it becomes necessary to manually scrape the sand surface to clean the filter. These units, called *slow sand filters,* take up a considerable amount of land area because of the slow filtration rates. Slow sand filters are still used in several existing treatment plants. They are effective and relatively inexpensive to operate.

In modern water treatment plants, the *rapid filter* has largely replaced the slow sand filter. As its name implies, the water flows through the filter bed much faster (about 30 times as fast) than it flows through the slow sand filter. This naturally makes it necessary to clean the filter much more frequently. But instead of manual cleaning by scraping of the surface, rapid filters are cleaned by reversing the direction of flow through the bed. This is shown schematically in Figure 6.16.

During filtration, the water flows downward through the bed under the force of gravity. When the filter is washed, clean water is forced upward, expanding the filter bed slightly and carrying away the accumulated impurities. This process is called *backwashing*. Cleaning by a backwash operation is a key characteristic of a rapid filter.

Many rapid filters currently in operation use sand as the filter media and are called *rapid sand filters*. But the sand grains (and pore spaces) are larger than those in the older, slow sand filters. In a rapid sand filter, the effective size of the sand is about 0.5 mm and the uniformity coefficient is about 1.5 (see Section 1.4). A difficulty that rises when using only sand in the rapid filter is that after backwashing, the larger sand grains settle to the bottom first, leaving the smaller sand grains at the filter surface. This pattern of filter media gradation is shown in Figure 6.17a.

Because of this small-to-larger gradation of sand grains in the direction of flow, most of the filtering action takes place in the top layer of the bed. This results in inefficient use of the filter. The filter run time (period of time between backwashes) is reduced, and frequent backwashes are required. Also, if some of the suspended material happens to pen-

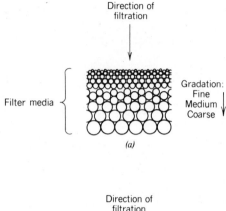

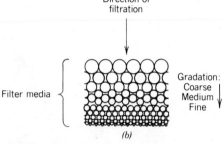

FIGURE 6.17 *(a)* Typical gradation of a rapid sand filter bed. Solids removal occurs primarily by straining action at the top of the sand bed. *(b)* Typical coarse-to-fine gradation in a "mixed-media" filter. It is preferable to the sand bed because it provides "in-depth" filtration.

etrate the upper layer of fine sand, it is then likely to pass through the entire filter bed.

A preferable size distribution of the filter material is shown in Figure 6.17b. The larger-to-smaller particle gradation allows the suspended particles to reach greater depths within the filter bed. This *in-depth filtration,* as it is called, provides more storage space for the solids, offers less resistance to flow, and allows longer filter runs. The process of filtration becomes more than just a physical straining action at the surface of the bed. The processes of flocculation and sedimentation occur within the pore spaces, and some material is adsorbed onto the surfaces of the filter media.

In order to achieve the optimum gradation for in-depth filtration, it is necessary to use two or more different filter materials. For example, if a coarse layer of anthracite coal is placed above the sand, the coal grains will always remain on top after

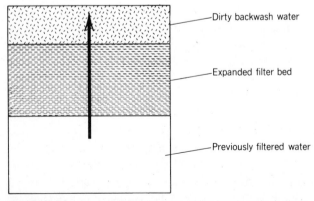

FIGURE 6.16 A schematic diagram of the backwash or cleaning cycle of a rapid filter.

backwashing occurs. This is because the coal has a much lower density than the sand does. Even though the coal grains are larger than the sand grains, they are lighter and therefore settle more slowly. The heavier sand particles settle to the filter bottom first at the end of a backwash cycle. A rapid filter that uses both coal and sand is called a *dual media filter*. In effect, the upper coal layer acts as a rough filter, removing most of the large impurities first. This allows the sand layer to remove the finer particles without getting clogged too quickly.

The coarse-to-fine gradation shown in Figure 6.17*b* is even more closely obtained by using three filter materials—coal, sand, and garnet (a very dense mineral.) After backwashing, the top layer of the filter bed is mostly coarse coal, the middle is mostly medium sand, and the bottom layer is

mostly very fine grains of garnet. This is called a *mixed media filter*. Filter material ranges in size from about 2 mm at the top to about 0.2 mm at the bottom. In recent years, dual and mixed media filters have been used to replace existing rapid sand filters in many treatment plants.

FILTER DESIGN

Rapid filters, whether of sand, dual media, or mixed media, are usually built in boxlike concrete structures, as illustrated in Figure 6.18. Multiple filter boxes or units are arranged on both sides of a central *piping gallery,* and a *clear well* used for storing filtered water is often located under the filters. Since only one unit is backwashed at a time, the

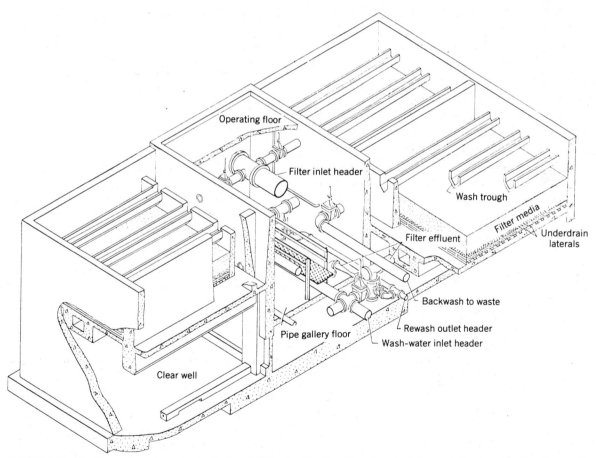

FIGURE 6.18 A perspective view of a typical rapid filter facility. The filtered water is temporarily stored in the clear well. Multiple filter "boxes" provide operational flexibility; only one filter is backwashed at a time (Permutit Co., Inc.).

filtration process can occur continuously as water flows through the treatment plant.

A typical rapid filter box is about 3 m or 10 ft deep, but the filter bed itself is only about 0.75 m or 2.5 ft deep. Located above the surface of the filter bed are *wash-water troughs,* which carry away the dirty backwash water as it flows upward through the bed and over the edge of the troughs. The filter media is generally supported on a layer of coarse gravel. Below the gravel, which only serves to support the filter bed and does not contribute to the filtering action, is a special filter bottom or *under-drain system.*

The underdrains serve to collect the filtered water, as well as to uniformly distribute the wash water across the filter bottom during the backwash cycle. They may consist of a grid of perforated pipes, leading to a common header pipe that carries the water into the clearwell. In many filters, the underdrains consist of specially manufactured porous tile blocks or steel plates with nozzles to help distribute the backwash water. A cross-section of a typical filter unit is shown in Figure 6.19.

The effectiveness of filtration and the length of a filter run depend on the *filtration rate.* Lower filtration rates generally allow longer filter runs and produce higher quality water, but they require larger filters. Filtration rate is often expressed as the flow rate of water divided by the surface area of the filter. In American units, this is usually in terms of gallons per minute per square foot (gpm/ft^2

or gal/ft^2/min). In SI metric units, it is liters per square meter per second (L/m^2·s). Rapid sand filters are usually designed to operate at an average rate of about 1.4 L/m^2·s or 2 gpm/ft^2, whereas mixed media filters can operate effectively at an average rate of about 3.5 L/m^2·s or 5 gpm/ft^2. The filtration rate is proportional to the velocity of flow through the filter bed.

EXAMPLE 6.5

A filter unit has a surface area of 50 ft^2. The flow rate through the filter is 0.25 mgd. Compute the filtration rate and the velocity at which the water flows through the filter bed.

Solution:

First convert the flow rate to gallons per minute, as follows:

$$250\,000 \text{ gal/d} \times \frac{1 \text{ day}}{24 \text{ h}} \times \frac{1 \text{ hr}}{60 \text{ min}} = \frac{174 \text{ gal}}{\text{min}}$$

Then

$$\text{Filtration rate} = \frac{174 \text{ gpm}}{50 \text{ ft}^2} = \frac{3.5 \text{ gpm}}{\text{ft}^2}$$

$$\text{Velocity} = 3.5 \frac{\text{gal}}{\text{ft}^2 \cdot \text{min}} \times \frac{1 \text{ ft}^3}{7.5 \text{ gal}} = 0.47 \text{ ft/min}$$

EXAMPLE 6.6

A square filter box is to be designed for a filtration rate of 2.8 L/m^2·s. What is the required surface area and side dimension of the unit if the flow rate is 6 ML/d?

Solution:

The flow rate in terms of liters per second is

$$6 \times 10^6 \text{ L/d} \times 1\text{d}/24\text{h} \times 1\text{h}/3600\text{s} = 69.4 \text{ L/s}$$

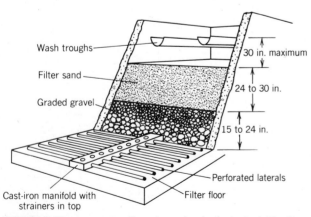

Wash troughs
Filter sand
Graded gravel
30 in. maximum
24 to 30 in.
15 to 24 in.
Perforated laterals
Filter floor
Cast-iron manifold with strainers in top

FIGURE 6.19   A cross-section view of a typical sand filter box. (National Lime Association).

Since filtration rate = flow rate/area, we get

area = flow rate/filter rate
$$= 69.4 \text{ L/s} \div 2.8 \text{ L/m}^2 \cdot \text{s} = 25 \text{ m}^2,$$

and the side dimension $= \sqrt{25 \text{ m}^2} = 5 \text{ m}$

FILTER OPERATION

Rapid filtration is usually preceded by coagulation and sedimentation. But in some cases, depending on the quality of the raw water, *direct filtration* may be used. In direct filtration, coagulant mixing and flocculation occur, but the sedimentation step is omitted. Instead, the water flows from the flocculation basin directly to the filters. This provides a saving in treatment plant area and construction cost.

A cross-sectional view of a typical rapid filter unit is shown in Figure 6.20. When filtration begins through a clean bed, the inlet valve A is fully open and the outlet valve B is throttled (i.e., only partially open). Valve B is gradually opened farther by an automatic *filter rate controller,* which operates by sensing pressure differences caused by changes in flow rate. The controller is, in effect, a venturi meter (see Section 2.3).

As solids accumulate in the filter, the resistance to flow through the bed increases and the filtration rate tends to decrease. The reduced flow is sensed by the rate controller, which causes valve B to open farther. The gradual opening of valve B compensates for the continually increasing resistance to flow in the filter bed. In this way, the rate of flow through the filter does not vary. This type of operation is called *constant rate filtration.* During constant rate filtration, the water level in the filter box remains about 1 m (3 ft) above the top of the filter bed.

Eventually, the filter bed gets clogged to the extent that valve B must be wide open to maintain the desired filtration rate. At this point it becomes necessary to clean or backwash the filter. To do this, valves A and B would be closed and valves C

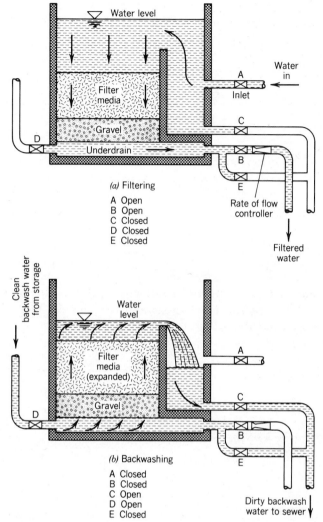

FIGURE 6.20 Schematic diagrams of a rapid filter in the *(a)* filtering cycle and *(b)* backwash cycle of operation. Valves A, B, C, D, and E control the flow. Valve E is opened briefly when filtering starts.

and D would be opened. Water from the backwash storage tank would then flow upward through the filter bed, expanding it slightly and carrying away the accumulated solids. The dirty backwash water flows into the wash water troughs and then to a drain that usually leads to a municipal sewerage system.

Rapid filters are generally backwashed at a rate of about 10 L/m^2·s (15 gpd/ft^2) for about 10 min. After the backwash flow stops, the filter material settles back in the bed and the filtration cycle begins again. But for the first 5 min of filtering, the filtered water would be wasted through valve E, to ensure that any remaining solids will not be carried into the clear well.

EXAMPLE 6.7

If the filter designed in Example 6.6 is backwashed once a day for 12 min, at a rate of 10 L/m²·s, what percentage of the total flow rate is used for cleaning the filter?

Solution:

The volume of water used for the backwash each day can be computed by multiplying the backwash rate times the filter area, and the time of backwash, as follows:

$$10 \text{ L/m}^2\cdot\text{s} \times 25 \text{ m}^2 \times 12 \text{ min} \times 60 \frac{\text{s}}{\text{min}} = 180 \text{ m}^3$$

The total daily flow is given as 6 ML = 6000 m³
The percentage of water used for backwash is therefore

$$\left(\frac{180}{6000}\right) \times 100 = 3 \text{ percent}$$

This percentage of water used for backwashing is typical for most water treatment plants.

In recent years, a mode of operation called *declining rate filtration* has been applied in some water treatment plants. In this mode of operation, rate-of-flow controllers are not used. The filtration rate is allowed to gradually decline from a maximum value at the beginning of the filter run to a minimum value when the bed is clogged. As the filter becomes clogged with accumulated solids, the water level gradually rises in the filter box. When the water level reaches a predetermined height, the filter is automatically backwashed.

Declining-rate filtration, as well as constant-rate

filtration, produces water of excellent quality. The crystal clear effluents from properly designed and operated rapid filters generally have turbidity levels less than 0.2 TU.

OTHER TYPES OF FILTERS

The gravity-flow, rapid filter is the most common type used for treating public water supplies, primarily because it is the most reliable. But there are other types of filters that are sometimes used to clarify water, including the *pressure filter* and the *diatomaceous earth filter*.

A pressure filter is very much like a conventional rapid filter, in that the water flows through a granular filter bed. But instead of being open to the atmosphere and utilizing the force of gravity, the pressure filter is enclosed in a cylindrical steel tank and the water is pumped through the bed under pressure. Because it operates under pressure, there is more of a chance that solids will get through the bed in the effluent. Since they are not as reliable as gravity filters, pressure filters are rarely used for treating public drinking water supplies. They are more commonly used for filtering water for industrial use or in swimming pools.

The diatomaceous earth filter is also used primarily for industrial or swimming pool applications. Like the pressure filter, it is less reliable than the rapid sand or mixed media filter. The filter media in this type of a filter is a thin layer of diatomaceous earth, a natural powder-like material formed from the shells of microscopic organisms called diatoms. The diatomaceous earth is supported on a cylindrical metal screen or fabric, called a septum. A typical diatomaceous earth filter is composed of many of these small septums.

6.5 DISINFECTION

The unit processes described in the previous sections—coagulation, sedimentation, and filtration, together comprise a type of treatment called *clari-*

fication. Clarification removes many microorganisms from the water along with the suspended solids. But clarification by itself is not sufficient to ensure the complete removal of pathogenic bacteria or viruses. Of course, a potable water must be more than crystal clear—it must be completely free of disease-causing microorganisms. To accomplish this, the final treatment process in a conventional water treatment plant is *disinfection,* which destroys or inactivates the pathogens.

CHLORINATION

Chlorine is the most commonly used substance for disinfection in the United States. The addition of chlorine or chlorine compounds to water is called *chlorination.* Chlorination is considered to be the single most important process for preventing the spread of waterborne disease. The effectiveness of chlorination is illustrated in Figure 1.16, which shows a steady decline in disease outbreaks as the number of chlorinated water supplies increased.

Molecular chlorine, Cl_2, is a greenish-yellow gas at ordinary room temperature and pressure. In gaseous form it is very toxic, and even in low concentrations it is a severe irritant. But when the chlorine is dissolved in low concentrations in clean water, it is not harmful. And if it is properly applied, objectional tastes and odors due to the chlorine and its byproducts are not noticeable to the average person.

Although the chlorine is effective in destroying pathogens and preventing the spread of communicable disease, there may be an indirect noninfectious health problem caused by the chlorination process. Natural waters often contain trace amounts of organic compounds, primarily from natural sources such as decaying vegetation. These substances can react with the chlorine to form compounds called *trihalomethanes (THM),* which may cause cancer in humans. Chloroform is an example of a THM compound.

The EPA has set standards that limit the maximum amount of THM compounds in drinking water. One way to prevent THM formation is to make sure the chlorine is added to the water only after clarification and the removal of most of the organics. Also, alternative methods of disinfection are available that do not use chlorine. These will be discussed later in this section.

Chlorination Chemistry

When chlorine is dissolved in pure water, it reacts with the H^+ ions and the OH^- radicals in the water. Two of the products of this reaction are *hypochlorous acid,* HOCl, and the *hypochlorite radical,* OCl^-. These are the actual disinfecting agents. If microorganisms are present in the water, the HOCl and the OCl^- penetrate the microbe cells and react with certain enzymes. This reaction disrupts the organisms' metabolism and kills them.

The hypochlorous acid is a more effective disinfectant than the hypochlorite radical because it diffuses faster through the microbe cell wall. The relative concentrations of HOCl and OCl^- depend on the pH of the water. The lower the pH is, the more HOCl there is relative to the OCl^-. In general then, the lower the pH of the water is, the more effective the chlorination–disinfection process is.

When chlorine is first added to water containing some impurities, the chlorine immediately reacts with the dissolved inorganic or organic substances and is then unavailable for disinfection. The amount of chlorine used up in this initial reaction is called the *chlorine demand* of the water. If dissolved ammonia, NH_3, is present in the water, the chlorine will react with it to form compounds called *chloramines.* Only after the chlorine demand is satisfied, and the reaction with all the dissolved ammonia is complete, is the chlorine actually available in the form of HOCl and OCl^-.

The chlorine in the form of HOCl and OCl^- is called *free available chlorine,* whereas the chloramines are referred to as *combined chlorine.* Free chlorine is often the preferred form for disinfection of drinking water. It works faster than the combined chlorine and it does not cause objectionable tastes and odors. The combined chlorine is also effective as a disinfectant, but it is slower acting and it may cause the typical "swimming-pool" odor of chlorinated water. Its advantage, though, is that it lasts longer and can serve to maintain sanitary protection throughout the water distribution system.

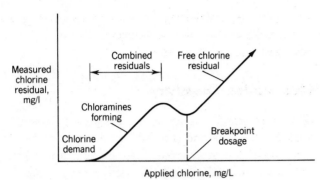

FIGURE 6.21 A *breakpoint chlorination* curve.

A process called *breakpoint chlorination* is sometimes used to ensure the presence of free chlorine in public water supplies. To do this, it is necessary to add enough chlorine to the water to satisfy the chlorine demand and to react with all the dissolved ammonia. When this occurs, it is said that the chlorine "breakpoint" has been reached. Chlorine added beyond the breakpoint will be available as a free chlorine residual, in direct proportion to the amount of chlorine added. This is illustrated in Figure 6.21. The chlorine demand and the breakpoint

dosage vary, depending on the water quality. Sometimes, chlorine dosages up to 10 mg/L are needed to obtain a free chlorine residual of 0.5 mg/L.

Chlorination Methods

Chlorine is commercially available in gaseous form, or in the form of solid and liquid compounds called *hypochlorites*. For the disinfection of relatively large volumes of water, the gaseous form of chlorine is generally the most economical. But for smaller volumes, the use of hypochlorite compounds is more common.

Gaseous chlorine is stored and shipped in pressurized steel cylinders. Under pressure, the chlorine is actually in liquid form in the cylinder; when it is released from the cylinder, it vaporizes into a gas. The cylinders may range in capacity from 45 kg (100 lb) to about 1000 kg (1 ton). Very large water (or wastewater) treatment plants may use special railroad tank cars filled with chlorine.

A device called an *all-vacuum chlorinator* is considered to provide the safest type of chlorine feed installation. It is mounted directly on the chlorine

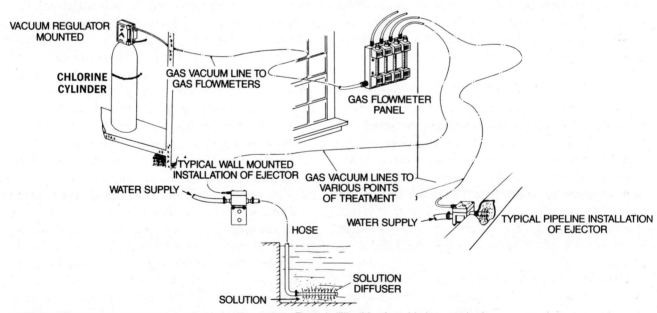

FIGURE 6.22 A typical vacuum-feed chlorination system. There is little risk of a chlorine gas leak since the chlorine feed line is at less than atmospheric pressure. (Capital Controls Company, Inc.)

cylinder. The gaseous chlorine is always under a partial vacuum in the line that carries it to the point of application; chlorine leaks cannot occur in that line. A typical vacuum chlorine feed system is shown in Figure 6.22. The vacuum is formed by water flowing through the ejector unit at high velocity. There are other types of chlorinators, some of which have the chlorine or concentrated chlorine solutions conveyed relatively long distances under pressure. These present somewhat greater risks of chlorine leaks. In any chlorine feed installation, safety factors are very important because of the toxicity of the gas.

Hypochlorites are usually applied to water in liquid form by means of small pumps, such as the one illustrated in Figure 6.23. These are *positive displacement* type pumps, which deliver a specific amount of liquid on each stroke of a piston or flexible diaphragm.

There are two types of hypochlorite compounds available for disinfection—*sodium hypochlorite* and *calcium hypochlorite*. Sodium hypochlorite is available only in liquid form and contains up to 15 percent available chlorine. It is usually diluted with water before being applied as a disinfectant. (Common laundry bleach is a 5-percent solution of sodium hypochlorite.) Calcium hypochlorite is a dry compound, available in granular or tablet form; it is readily soluble in water. Calcium hypochlorite solutions are more stable than solutions of sodium hypochlorite, which deteriorate over time.

In addition to pH, the effectiveness of chlorine and chlorine compounds in destroying bacteria depends on the chlorine concentration and the *contact time*. Contact time is the time period during which the free or combined chlorine is acting on the microorganisms. At pH values close to 7 (neutral conditions), a free chlorine residual of 0.2 mg/L with a 10-min contact time would have about the same disinfecting power as 1.5 mg/L of combined chlorine residual with a one-hour contact time.

The effectiveness of chlorination can be determined by the coliform test, or, as mentioned in Section 6.1, by a more convenient test for chlorine residual in the treated water. The method approved for chlorine residual testing, under the SDWA reg-

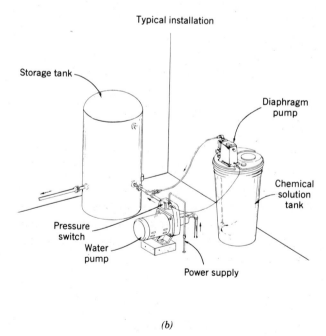

FIGURE 6.23 *(a)* A typical hypochlorinator or solution metering pump, and *(b)* a typical hypochlorinator installation. (Penwalt Corporation/Wallace & Tiernan Division)

ulations, is called the *DPD chlorine residual test.* Field test kits, such as the one illustrated in Figure 4.18, are readily available. The test procedure, which is based on a color comparison, takes only five minutes to complete.

In the DPD test, a chemical dye is added to the water sample. The dye turns red if chlorine residual is present, and the intensity of the red color is proportional to the chlorine concentration. It is assumed that the presence of a chlorine residual ensures that there are no surviving pathogenic organisms in the water. It is possible to measure either the total residual, the free residual, or the combined residual with the DPD test kit.

It is often necessary to compute the total weight or mass of chlorine used at a treatment plant, to be able to order chlorine supplies at the appropriate time. Also, it may be necessary to determine the applied chlorine dosage if the weight consumed is known. The following relationships are useful for these purposes:

$$\text{kg/d} = Q \times C \qquad (6\text{-}3a)$$

where Q = flow rate, ML/d
C = chlorine concentration, mg/L

$$\text{lb/d} = 8.34 \times Q \times C \qquad (6\text{-}3b)$$

where Q = flow rate, mgd
C = chlorine concentration, mg/L
8.34 = lb/gal of water

EXAMPLE 6.8

How many pounds per day of chlorine are required to disinfect a flow of 7.5 mgd with a chlorine dosage of 0.5 mg/L? How many 100-lb chlorine cylinders are needed per month?

Solution:

Applying Equation 6-3b, we get

$$\text{lb/d} = 8.34 \times 7.5 \times 0.5 = 31 \text{ lb/d},$$

and

31 lb/d $\times$ 30 d/month $\times$ 1 cylinder/100 lb
$$= 9.4 \text{ cylinders}$$

At least 10 chlorine cylinders would be ordered per month.

EXAMPLE 6.9

A total of 15 kg of chlorine is used in one day to disinfect a volume of 50 ML of water. What is the chlorine dosage?

Solution:

Applying Equation 6-3a, we get

$$15 \text{ kg/d} = 50 \text{ ML/d} \times C,$$

and

$$C = \frac{15}{50} = 0.3 \text{ mg/L}$$

OTHER METHODS OF DISINFECTION

Chlorination is the most widely used method for disinfection of water supplies in the United States because of its economy and its ability to maintain a protective residual. Other methods of disinfection have been receiving more attention in recent years, primarily because of the problem of THM formation and the potential effect on public health.

Ozone

Ozone, O_3, a gas at ordinary temperatures and pressures, is a very potent disinfectant that is commonly used in Europe. It is unaffected by the pH or the ammonia content of the water; it causes no taste or odor problems; and it plays no role in the formation of the harmful THM compounds. But since it is unstable and can not be stored, it must be produced on site. Also, it does not maintain a measurable residual in the water after the initial contact time. These disadvantages, plus the fact that it usually costs more than chlorine, have discouraged widespread use of ozone in the United States. A notable exception to this is the use of ozone for disinfection at the 1900 ML/d or 500 mgd Los Angeles water treatment plant (which also uses direct filtration). One of the newest facilities to use ozone is the Hackensack Water Company's 150 mgd plant in Haworth, N.J., shown in Figure 6.1. Ozone is approved by the EPA for disinfection.

Ultraviolent Radiation

Ultraviolet rays can be used to disinfect water supplies. This method involves no chemical handling; no overdoses are possible; and there are no taste or odor problems associated with its use. But its high cost of application and lack of a measurable residual make it a poor competitor with chlorine or ozone at the present time.

6.6 OTHER TREATMENT PROCESSES

Clarification by coagulation, sedimentation, and filtration serves to remove suspended impurities and turbidity from drinking water. The final step of disinfection produces potable water, free of harmful microorganisms. But other treatment processes may be required, particularly to remove some of the dissolved substances. These processes may be used in addition to clarification, or applied separately, depending on the source and quality of the raw water.

Groundwater, for example, does not ordinarily require clarification since the water is filtered naturally in the layers of soil from which it is withdrawn. Disinfection of groundwater supplies, re-quired by law for public water supply systems, is basically a precautionary step; groundwater is usually free of bacteria or other microorganisms. On the other hand, because of its contact with soil and rock, groundwater may contain high levels of dissolved minerals that must be removed. Methods to accomplish this, as well as other less common treatment processes for both surface and subsurface water, are discussed in the following paragraphs.

WATER SOFTENING

Water that contains dissolved salts of calcium and magnesium is known as "hard" water, as discussed in Section 4.3. Hardness in water interferes with the lathering action of soap, and causes deposits of scale in water heaters, pipes, and plumbing fixtures. This is basically an economic and aesthetic problem rather than a health problem. Generally, when the hardness exceeds about 500 mg/L in the raw water, it is best to remove the calcium and magnesium at a central municipal treatment plant. The process of removing these minerals is called *water softening*. The two most common methods of softening are the *lime-soda method* and the *ion exchange method*.

In the lime-soda method, two chemicals are added to the water to cause what chemists call a precipitation reaction. These chemicals are lime, $Ca(OH)_2$, and soda ash, Na_2CO_3. A reaction takes place among these chemicals and the dissolved calcium and magnesium ions in the water, causing the formation of calcium carbonate, $CaCO_3$, and magnesium hydroxide, $Mg(OH)_2$.

Since they are very insoluble in water, the calcium carbonate and magnesium hydroxide compounds precipitate out of solution as they form during the reaction. This process is then followed by sedimentation and filtration to remove the insoluble precipitates and clarify the water. Also, carbon dioxide, CO_2, may be added to the water to precipitate excess calcium and to adjust the pH which is raised by the addition of lime; this process is called *recarbonation*.

Softening by ion exchange involves passing the water through a column containing a special ion exchange material. There are several different types of ion exchange materials in use, including natural substances called *zeolites,* and synthetic resins. When water containing calcium or magnesium ions is in contact with these materials, an exchange or "trade" of ions takes place. The calcium and magnesium ions are taken up by the resin, whereas sodium ions, Na^+, are released into the water.

The ion exchange process is illustrated schematically in Figure 6.24. Eventually, the exchange capacity of the zeolite or resin is used up and the ion exchanger must be regenerated for further use. This is done by washing the exchanger with a sodium chloride, NaCl, solution. Now the sodium ions replace the calcium and magnesium ions, which are discharged to a waste disposal drain. The softening process can then begin again.

Softening by ion exchange can produce water with almost zero levels of hardness, but this is not really desirable. Very soft water may be "aggressive" or corrosive, causing damage to metal pipes and plumbing. Hardness levels of about 100 mg/L are considered optimum for drinking water.

There is also some evidence that the presence of moderate hardness levels in drinking water actually reduces the incidence of heart disease. Another factor that must be considered is that soft-

ened water from an ion exchanger contains sodium, which may be harmful to persons who already have heart disease. In such cases, the softened water may not be suitable for consumption. Finally, it should be noted that ion exchange softening does not produce a precipitate or sludge and is generally less costly than lime–soda softening. But because of the disadvantages mentioned, it is usually better adapted for treating industrial water supplies, or for use in individual home softening units.

AERATION

A physical treatment process in which air is thoroughly mixed with water is called *aeration.* Thorough contact with air and oxygen can improve water quality in a number of ways. For example, one of the common uses of aeration is for *taste and odor control.* Dissolved gases that tend to cause the taste and odor problems, such as hydrogen sulfide, are transferred from the water to the air during aeration. This application is also called *air stripping.*

Aeration is also used for the *removal of iron and manganese* from the water, particularly in groundwater supplies. The oxygen in the air reacts with the iron and manganese to form an insoluble precipitate (rust). Sedimentation and filtration are then necessary to clarify the water.

Several methods to aerate the water are available. The method selected depends primarily on the type and concentration of material to be removed from the water and on the available pressure head. Aeration using *spray nozzles* provides a large total air–water contact area, but relatively high pressures and much space is required. Spraying the water into the air can be followed by allowing the water to cascade and flow in thin sheets down several concrete or metal steps. Cascade structures require at least a 3 m (10 ft) drop.

Another common method for aeration makes use of *multiple-tray aerators.* These consist basically of a tall stack of perforated trays or slats with staggered openings. The water is applied at the top and trickles downward in thin films or sheets of flow. In some cases, a fan or blower might be used to force

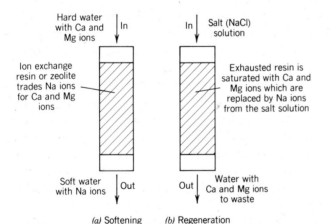

FIGURE 6.24 *(a)* An ion exchange column used for water softening; *(b)* the column may be regenerated and used again after washing with a strong salt solution.

air upwards through the stack to increase the contact with air.

For very large volumes of water, the use of *diffused-air aerators* is generally the most practical method. In this type of aerator, air is pumped by centrifugal blowers into a tank of water. The air enters the water at the tank bottom through special diffuser nozzles or porous fixtures, forming air bubbles that become thoroughly mixed with the water. *Mechanical aerators* consisting of a large propeller that churns the water at the surface are also available. (Mechanical and diffused-air aerators find frequent application in wastewater treatment, as will be discussed in a subsequent chapter.)

ACTIVATED CARBON

Activated carbon derived from coal or wood has two unique properties that make it useful for water purification. First, it is a very porous material and has an extremely high ratio of surface area to unit weight—up to 100 acres of area per pound. Second, the surface of activated carbon attracts and holds many of the impurities in water, particularly the dissolved organics. This process is called *adsorption*. (Unlike *ab*sorption, *ad*sorption is a surface phenomenon.)

Adsorption on activated carbon is an effective method for removing dissolved organic substances that cause taste and odor problems in drinking water. It is also effective in removing the organic *precursors* that react with chlorine to form harmful THM compounds after disinfection.

When the carbon surface becomes covered with impurities, it can be cleaned or reactivated by heating to a high temperature in a special furnace. The organics are driven off the carbon surface by the heat, and the carbon can be reused. But on-site reactivation, rather than complete replacement with fresh carbon, is economical only for large municipal water treatment plants.

Activated carbon is available as a very fine black powder or in granular form. The powdered carbon can be mixed with the water by a special dry-feeder device, at a point in the treatment plant that precedes the filtration process; it is removed from the water by the filters. Granular carbon is sometimes used in the filter bed itself, combining both filtration and adsorption in one treatment unit.

CORROSION CONTROL

Corrosion or rusting of metals in water supply systems can be a serious problem. Since corrosion involves a transfer of electrons, control methods are aimed at blocking the flow of electrons between the water and the metal that is susceptible to corrosion. One way to do this is to add chemicals called *complexing agents* to the water. Complexing agents such as sodium silicate or sodium phosphate are added at concentrations of about 1 mg/L. They combine chemically with the metal to form a barrier that blocks corrosion reactions. Control of the pH of the treated water is also used to prevent corrosion in the distribution system.

FLUORIDATION

As discussed in Section 4.3, fluorides are effective in preventing tooth decay, particularly in young children. Many communities intentionally fluoridate their water supplies as a public health measure. This is done by adding sodium fluoride, NaF, or sodium silicofluoride, Na_2SiF_6, to the water after filtration. It is important that the dosage of fluoride be carefully controlled; the optimum concentration is about 1 mg/L. Excessive concentrations may cause discoloration of tooth enamel, called *dental fluorosis*. Other than dental fluorosis, which only occurs if large amounts of fluoride are consumed, there are no harmful side effects of fluoridation.

REVIEW QUESTIONS

1. Which water usually requires more extensive treatment for purification—groundwater or surface water? Why?

2. Sketch a flow diagram for a typical surface-water treatment plant.

3. What is a *public water system?* What is the difference between a community system and a noncommunity system?

4. What do SDWA and MCL stand for?

5. What are the five general groups or types of contaminants that are controlled and limited under the SDWA? Briefly discuss the requirements for each group.

6. What is the difference between the SDWA primary regulations and the secondary regulations?

7. Briefly discuss the recordkeeping and reporting requirements of the SDWA.

8. Describe the meaning of *detention time* and *overflow rate,* with regard to the process of sedimentation.

9. Explain what is meant by *short circuiting* in a sedimentation tank.

10. What is meant by the term *freeboard?*

11. Is a narrow and deep settling tank more effective in removing suspended solids than a wide and shallow tank? Explain your answer.

12. What is the function of a *tube settler?*

13. Briefly describe the process of *coagulation.* How does it improve the purification of drinking water?

14. What is the purpose of a *jar test?* Describe it briefly.

15. What is an *upflow clarifier?*

16. What is the purpose of *filtration?* What is a key characteristic of a *rapid filter?*

17. What is meant by *in-depth filtration?* How can it be accomplished?

18. Briefly describe the configuration of a typical rapid filter.

19. Briefly describe the operation of a typical rapid filter.

20. What is meant by *declining rate filtration*?

21. What is meant by *direct filtration*?

22. What is considered to be the most important water treatment process with respect to preventing the spread of waterborne disease? Is there any potential harmful side effect from that process?

23. What are the disinfecting agents that kill bacteria when chlorine is added to water? What is the difference between *free chlorine* and *combined chlorine*? Compare their relative merits in disinfection.

24. What is meant by *breakpoint chlorination*?

25. Briefly describe the ways in which chlorine can be applied to water in a treatment plant.

26. What is the *DPD test*?

27. Briefly describe two methods other than chlorination that can be used to disinfect water supplies.

28. Briefly describe two methods used to soften water.

29. How can aeration improve drinking water quality? What methods are available to aerate water?

30. How can activated carbon improve drinking water quality? What are two important properties of activated carbon?

31. Why would sodium silicate be added to drinking water? Why would sodium fluoride be added?

PRACTICE PROBLEMS

1. A settling tank with a 50-ft diameter and a SWD of 9 ft treats a flow of 15 000 gpd. What is the detention time?

2. Compute the required volume of a sedimentation tank that would provide 3 h of detention time for a flow of 10 ML/d. If the tank was 10 m by 25 m in plan dimensions, how deep would the water be in the tank?

3. A clarifier operates with a surface loading of 500 gpd/ft^2. What is the slowest settling velocity of particles that will be completely removed in the tank?

4. A circular settling tank is to have a minimum detention time of 3 h and a maximum overflow rate of 800 gpd/ft^2. Determine the required basin diameter and SWD for a flow rate of 2 mgd.

5. A rectangular settling tank is to have a minimum detention time of 3.5 h and a maximum surface loading of 25 m/d. The tank length is to be twice its width. Determine the required tank dimensions, including freeboard, for a flow of 5000 m^3/d.

6. A rapid filter has plan dimensions of 10 ft by 15 ft, and it treats a flow rate of 1 mgd. Compute the filtration rate in terms of gpm/ft^2 and the velocity of flow through the filter bed.

7. Compute the required plan dimensions of a square filter box that will treat a flow of 9 ML/d at a maximum rate of 2.9 L/m^2·s.

8. If the filter designed in Problem 7 is backwashed once a day for 10 min, at a rate of 10.4 L/m^2·s, what percentage of the total flow is used for cleaning the filter?

9. How many 100-lb chlorine cylinders would be ordered per month in order to disinfect a flow of 12 mgd of water using a 0.6 mg/L dosage of chlorine? Would you recommend switching to 1-ton cylinders instead?

10. A mass of 150 kg/d of chlorine is used to disinfect a flow of 250 000 m^3/d. What is the chlorine dosage?

11. A mass of 20 kg per day of chlorine is applied to water, resulting in a chlorine concentration of 0.4 ppm. What is the flow rate of water?

SELECTED REFERENCES

1. *The Safe Drinking Water Act Self Study Handbook,* American Water Works Association, Denver, 1978.

2. *National Interim Primary Drinking Water Regulations,* Federal Register, EPA Water Programs, Vol. 40, No. 248, Dec. 24, 1975.

3. *National Secondary Drinking Water Standards,* Federal Register, EPA Water Programs, Vol. 42, No. 62, Mar. 31, 1977.

4. Culp, G. L. and R. L. Culp, *New Concepts in Water Purification,* Van Nostrand Reinhold Co., New York, 1974.

5. Steel, E. W. and T. J. McGhee, *Water Supply and Sewerage,* 5th ed., McGraw–Hill, New York, 1979.

6. Clark, J. W., et al., *Water Supply and Pollution Control,* 3rd ed., Harper & Row, New York, 1977.

7. Hammer, M. J., *Water and Wastewater Technology,* John Wiley & Sons, New York, 1977.

7

WATER DISTRIBUTION SYSTEMS

A water distribution system is an interconnected network of pipelines, storage tanks, pumps, and smaller appurtenances, including valves and flow meters. The purpose of this chapter is to describe some of the practical aspects related to the design, analysis, and operation of these systems.

The chapter begins with discussion of basic design factors, materials, and appurtenances. A section on centrifugal pumps—the "prime movers" of a distribution system—is included. Conservation and distribution reservoirs are discussed, and the analysis of pipe network hydraulics is covered. Much of this material assumes some prior knowledge of hydraulics, so it would be best to study or review Chapter 2 before starting this chapter.

7.1 DESIGN FACTORS

The design of a water distribution system begins after a study of community water requirements has been completed. A water distribution system must be able to deliver adequate quantities of water for various uses in a community. Also, sufficient pressures must be maintained throughout the system.

A survey of the service area would be required, so that maps of streets and topographical features could be prepared. On a relatively small scale map (about 1 : 24000 or 1 in. = 2000 ft), the principal elements of the system can be planned, showing the general locations of water mains, pump stations, storage tanks, and so on. On larger-scale maps (about 1 : 600 or 1 in. = 50 ft), the exact locations of the proposed facilities would be shown in detail, and existing utilities such as sewers or gas mains would be shown as well. These plan drawings would be accompanied by written specifications describing the materials and methods of construction.

REQUIRED FLOWS AND PRESSURES

It is convenient to classify water demands or water uses into four basic categories, as follows:

1. *Domestic* - water for drinking, cooking, personal hygiene, lawn sprinkling, and the like.
2. *Public* - water for fire protection and street cleaning and water used in schools or other public buildings.
3. *Commercial and Industrial* - water used by restaurants, laundries, manufacturing operations, and the like.
4. *Loss* - due to leaks in mains and house plumbing fixtures.

The total demand for water in a community varies, depending on the population, the industrial and commercial activity, the local climate, and the cost of the water. For example, in warm, dry climates, domestic use is generally a larger fraction of total consumption than it is in colder climates; lawn watering is much more common in dry climates. However, when the water bill is based on individual meter readings rather than on a flat rate, conservation is encouraged and water demand decreases.

Per Capita Demand

If the total annual water use of a community is divided by 365 days, a value of average daily water consumption is obtained. If this value is further divided by the total population served, a "per capita" value is obtained. In SI units, this is expressed in terms of liters per day per person and in American units, it is called gallons per capita per day (gpcd).

For example, if the average daily water demand is 5 megaliters per day (5 ML/d) in a system serving 10 000 people, the average per capita demand would be (5 000 000 L/d)/(10 000 people) = 500 L/d per person. Keep in mind that a figure like this includes each person's share of industrial, commercial, and public use, and leakage; it is not just individual domestic use.

Since the exact water demands of a new service area may not be known, it is common to use average per capita values from similar communities in order to design the new distribution system. New systems are generally designed to accommodate populations and water demands that are anticipated 10 to 30 years in the future. Otherwise the system would be too small soon after it was built. Table 7.1 presents projected daily water demands for the year 2000.

Variations in Water Demand

In any community, water demand will vary on a seasonal, daily, and hourly basis. For example, on a hot summer day it would not be unusual for water consumption to be as much as 200 percent of the

TABLE 7.1   ESTIMATED AVERAGE WATER REQUIREMENTS IN YEAR 2000

Type of Use	L/d per Person	gpcd	% of Total
Domestic	300	80	44
Commercial/industrial	260	70	39
Public	60	16	9
Loss	50	14	8
Total	670	180	100

average daily demand. If the average daily demand was 670 L/d, then we would have to estimate a peak daily demand to be 2 × 670 = 1340 L/d per person. Generally, the pipelines and pumps of a distribution system (as well as treatment plants and wells) must be designed to accommodate peak daily flows rather than average flows.

Water consumption also varies hourly throughout the day, according to a somewhat predictable pattern. Peak hourly demands in residential districts usually occur in the morning and evening hours, just before and after the normal work day. In commercial or industrial districts, water consumption may be uniformly high throughout the working day. Minimum flows typically occur around 4 o'clock in the morning, when almost no one is using water.

A graph illustrating typical hourly variations in water use is shown in Figure 7.1. On this graph, the peak hourly flow occurs at about 6 P.M. In extreme cases, these maximum hourly flows could be as much as 10 times the average flow, but they are usually around 3.5 times the average flow rate. As we will discuss later, these peak hourly demands are generally accommodated by water from storage tanks instead of by the pumps in the system. Otherwise, the pumps and pipes would have to be excessively large just to handle flows that occur for a relatively short time.

There can be a wide variation in average, peak daily, and peak hourly flow rates among different communities. As far as is practical, the specific water demands should be determined or estimated for each service area. Generally, big cities have higher per capita water use than small communities, and small service areas are noted for their very high peak rates.

Fire Flows

Water for fire fighting is an important part of the total demand that must be provided for in a water distribution system. Fire flows are only required once in a while, and the total amount of water used to extinguish fires in any year is small compared to all other uses. But the rate and volume of water needed in the few hours of a fire emergency can be large in a local area. Sometimes, it can be the controlling factor affecting the size of the water mains.

Municipal insurance rates depend to a large extent on the fire protection provided by the distribution system. Procedures for determining required fire flow capacity have been established by the Insurance Services Office; factors involved in this include type of building construction, occupancy, sprinkler protection, and so on. As a minimum, 30 L/s (475 gpm) of fire flow would be required for at least 2 hours. In more extreme cases, up to 760 L/s (12 000 gpm) for a 10-hour duration may be necessary. The required fire flow must be added to the peak daily demand in the system when sizing pipes and pumps.

Pressures

Water pressures in a distribution system should not drop below 140 kPa (20 psi) in order to provide for adequate operation of home plumbing fixtures and appliances, as well as for fire fighting when pumper trucks are used at fire hydrants. Maximum pressures in water mains are generally kept below 700 kPa or 100 psi, to reduce the chances for leaks or water main breaks. Pressures of about 350 kPa (50 psi) would be considered optimum. Pressure regulating valves must be installed in the distribution

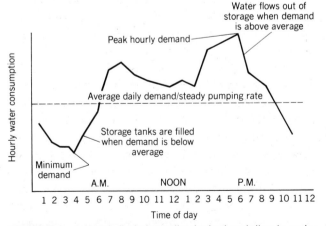

FIGURE 7.1 A graph that shows the typical variation in water demand or consumption throughout the day.

system to reduce pressures in low-lying service areas; otherwise, pressure heads in the system would be too high.

PIPELINE LAYOUT

Water mains are generally not less than 150 mm (6 in.) in diameter. They are usually located in the street right-of-way (ROW) so as to provide water to every potential customer. The so-called *gridiron* arrangement of pipes is preferred to a layout that has many dead-end branches. In the gridiron system,

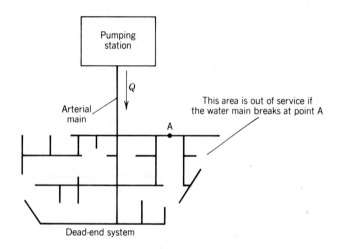

Dead-end system

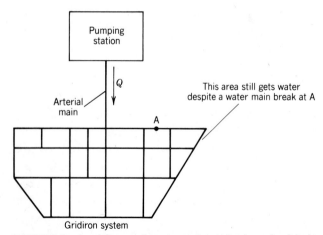

Gridiron system

FIGURE 7.2 A gridiron pattern for water mains is preferable to a dead-end type system; gridiron networks provide greater flexibility in operation and service.

water can circulate in interconnected loops. But in the dead-end system, the water may remain relatively stagnant in sections of the system, causing taste and odor problems from bacterial growth. The two types of layouts are illustrated in Figure 7.2.

In the dead-end layout, frequent flushing of the pipes at the fire hydrants is necessary to prevent consumer complaints about taste and odor. Another disadvantage of the dead-end system is that water service could be disrupted for long periods of time while repairs are made to a broken water main. But in a gridiron system, the broken section can be isolated by valves, and water can still reach consumers from the other side of the loop. Most distribution systems combine both layouts, depending on local conditions and economic factors.

Water mains may be referred to as *primary feeders* or *secondary feeders*. The primary feeders, also called *arterial mains*, carry large quantities of water from the treatment or pumping facility to areas of major water use. Secondary feeders are smaller pipes that provide a daily supply to local areas.

7.2 WATER MAINS

MATERIALS

The water mains in a distribution system must be strong and durable in order to resist applied forces and corrosion. The pipe is subjected to internal pressure from the water and to external pressure from the weight of the soil (backfill), and vehicles, above it. Another force the pipe may have to withstand is called *water hammer*. This can occur when a valve is closed too fast, for example, causing waves of high pressure to surge through the pipe. Finally, damage due to corrosion or rusting may occur internally because of the water quality, or externally because of the nature of the soil conditions. Materials commonly used in water mains to provide adequate strength and durability are discussed here.

Ductile Iron Pipe

Ductile iron is one of the most common materials used for the construction of water distribution pipelines. Because of its chemical composition, ductile iron is stronger and more elastic than gray cast iron, which was the predominant pipe material used until recently. Cast iron (CI) is also strong and durable, with many older installations still in service after 100 years or more. Ductile iron, though, is less brittle than gray cast iron; it is less vulnerable to damage during construction and it is considered to be more corrosion resistant.

Ductile iron pipe sections are available in lengths up to about 6 m (18 ft) and in diameters up to 1200 mm (48 in.). It is manufactured in several thickness classes or groups; higher-class pipe (thicker pipe walls) would be specified for deep installations or high water pressures.

Two common methods for joining individual sections of pipe together are the *push-on joint* and the *flanged joint*. In the push-on or *compression type* joint, as it is also called, a *spigot end* of one pipe is pushed into the *bell end* of an adjacent pipe; a rubber-ring gasket in the bell end is compressed when the sections are joined, creating a watertight, but flexible, connection. Flanged connections involve the bolting together of the ends of the pipe sections; they are used for above-ground installations in treatment plants or pumping stations. Both push-on and flanged joints are illustrated in Figure 7.3.

Unprotected iron pipes are subject to a process called *tuberculation*—the formation of tubercles or small projections of rust on the inside wall of the pipe. Tuberculation significantly increases the resistance to flow, reduces pipe capacity, and increases pressure losses in the system. Iron pipes are usually coated with a thin cement–mortar lining on the inside pipe wall to prevent tuberculation and to preserve the hydraulic capacity of the pipe. (The Hazen–Williams C factor may be as high as 145 for cement lined pipes—see Section 2.3.) Tarlike coatings are also applied on the outside of the pipe to prevent external corrosion.

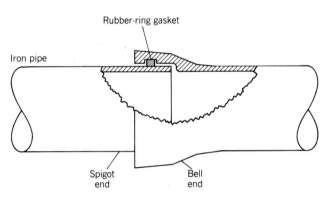

Push-on (compression) joint

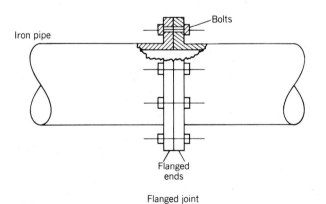

Flanged joint

FIGURE 7.3 Two common methods for joining sections of iron pipe.

Asbestos Cement (AC) Pipe

A compacted mixture of sand, cement, and asbestos fibers provides a lightweight pipe material that is smooth and corrosion resistant. Although it is not as strong as iron pipe, the absence of tuberculation and the ease of installation make AC pipe desirable in many instances.

AC pipe has long lasting hydraulic properties with high carrying capacity (C = 140). Manufactured in about 4-m (12-ft) lengths and in diameters up to about 400 mm (16 in.), it is also available in several pressure classes up to a maximum of about 1400 kPa (200 psi). The plain ends of AC pipe sections are easily joined with a coupling sleeve, as illustrated in Figure 7.4. The two rubber-ring gas-

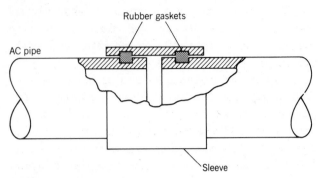

FIGURE 7.4 Asbestos–cement pipes are joined with a sleeve and two rubber-ring gaskets.

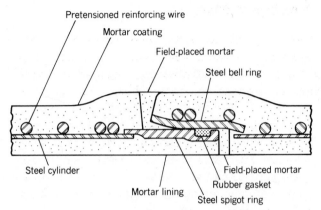

FIGURE 7.5 A section of an RCP joint (From *Water Distribution Operator Training Handbook,* by permission. Copyright 1976, The American Water Works Association)

kets in the sleeve provide a watertight, yet flexible joint.

Plastic Pipe

The use of polyvinyl chloride plastic (PVC) as a pipe material is increasing for construction of water distribution mains. These plastic pipes are strong and durable, yet they are very lightweight and are easily handled and installed. They are resistant to corrosion and they are very smooth, providing excellent hydraulic characteristics ($C = 150$). Available in diameters up to about 600 mm (24 in.), PVC pipe sections are joined using a bell-and-spigot compression type joint with a rubber-ring seal.

Other plastic materials used for service connections and domestic plumbing include polyethylene and ABS plastic. These pipes may be joined using threaded screw couplings or chemical solvent welds.

Other Pipe Materials

Reinforced concrete pipes (RCP) are made of welded steel cylinders, wrapped with steel wire and embedded in concrete. They are used primarily in long water transmission lines of large diameter. They can be precast in sections up to 5 m (16 ft) in length, and up to about 4 m (12 ft) in diameter. RCP pipes are very strong and durable, and have excellent hydraulic characteristics. Sections may be joined using a modified bell-and-spigot type connec-

tion and sealed with cement mortar, as shown in Figure 7.5.

Steel pipe is sometimes used for water transmission lines, particularly for above-ground installations. It is very strong, yet it is lighter in weight than RCP. But it must be carefully protected against corrosion; this is usually done by lining the interior and painting and wrapping the exterior. Sections of steel pipe may be joined by welding or with mechanical coupling devices as shown in Figure 7.6.

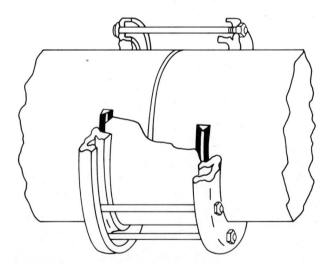

FIGURE 7.6 A mechanical coupling device used to join sections of steel pipe (from *Water Distribution Operator Training Handbook,* by permission. Copyright 1976, The American Water Works Association).

APPURTENANCES

Proper functioning of the water mains in a distribution system requires many different devices, in addition to the sections of pipe. These devices, called *appurtenances,* include hydrants, shut-off valves, throttling valves, pressure reducing valves, and other fittings.

Hydrants

The primary purpose of a hydrant is to provide convenient access to water for fire fighting and other emergencies. A hydrant also serves for flushing out water mains, for washing debris off public streets, and to provide access to the underground pipe system for pressure testing. The spacing and location of hydrants depends primarily on fire protection and insurance needs. Hydrants are also placed at dead ends and at high and low points in the pipeline.

A long valve stem inside the cast-iron barrel of the hydrant operates a shutoff valve at its base. Another valve in the pipe connecting the hydrant to the water main allows isolation of the hydrant for maintenance. The connecting pipe is usually 150 mm (6 in.) in diameter, and the hydrant has two hose connections on top. A gravel footing is provided to allow drainage from the barrel after the hydrant is used; this is particularly important in cold climates where the water can freeze and break the hydrant barrel.

Service Connections

Water from the distribution main reaches the property line of individual consumers through a service pipe, usually made of copper or plastic with a minimum diameter of 20 mm ($\frac{3}{4}$ in.).

Service connections can be made initially when the main is installed (a dry tap), or later on when the main is already in service (a wet tap). Pipe-tapping machines are available that allow wet taps to be made without affecting water service to existing users of the system. The service pipe is connected to the main by means of a special fitting called a *corporation stop.* At the user's end of the service line, there is usually a water meter and shut-off valve.

Valves

Many different types of valves are used in water distribution systems, to control the quantity and direction of flow. Many of these can be opened or closed manually by screw stems or gear train devices; large valves often are power operated using electric or hydraulic systems.

The most common function of a valve is for complete shutoff of flow. *Gate valves* are usually used for this purpose. They are placed throughout the distribution network, allowing sections of pipeline to be shut off and isolated during repairs of broken mains, pumps, or hydrants. A gate valve consists basically of a sliding disk that is moved across the path of flow by a screw-operated stem. When the valve is in the open position, the disk is enclosed in a valve cover or housing above the pipe and is completely out of the path of flow. In the closed position, the disk is lowered and tightly wedged in a valve seat, blocking the flow. A typical gate valve is illustrated in Figure 7.7a.

Gate valves are usually either in the fully open or fully closed position; they are rarely used for throttling flow by blocking it only partially. In most distribution networks, they are placed at pipe intersections and are operated manually using an extension rod to reach the operating nut on the valve; cast-iron valve boxes, which extend from the valve up to the street surface, cover and protect the underground valve. Large gate valves may be placed in underground manhole structures or valve pits for easier access and operation.

A type of valve commonly used for throttling and controlling flow rate is the *butterfly valve.* In a butterfly valve, a movable disk rotates on an axle in the path of flow. In the closed position, the disk is tightly seated against a rubber ring in its casing. In the open position, the disk is turned 90°, allowing the water to flow around and past it. The fact that the disk is always in the flow is a disadvantage of the butterfly valve, since it blocks the use of pipe cleaning tools. Because the force of flowing water

FIGURE 7.7 *(a)* A typical gate valve. (M & H Valve Company) *(b)* A typical butterfly valve. (American Darling–Valve, Division of American Cast Iron Pipe Co.)

tends to close the valve, reducing gear drives are used for manual operation, and power operators are required for the large butterfly valves. A butterfly valve is illustrated in Figure 7.7*b*.

A device called a *check valve* is used to permit flow in only one direction in a pipe; it closes automatically when the flow stops or tends to flow in the opposite direction. A common type, called a *swing-check* valve, is illustrated in Figure 7.8. The valve disk is lifted up by the force of the flowing water, and closes by gravity when the flow stops. A valve seat prevents the disk from swinging open in the opposite direction.

Check valves are usually installed in the discharge piping of a pump to prevent backflow when the pump stops. They are called *foot valves* when installed at the end of a pump suction line in a well or tank. Foot valves prevent loss of "prime" in centrifugal pumps. (The operation of centrifugal pumps is discussed in the next section.) In plumbing systems, special double-check valves may be used to prevent backflow and possible contamination of a drinking water supply when a cross-connection with another system exists.

Other types of valves that find use in water distribution systems include pressure reducing valves, air release valves, and altitude valves. Pressure reducing valves operate automatically to lower excessive hydrostatic pressure in water mains that are at a low elevation in the system. In effect, these valves form separate networks or pressure zones in a large distribution system.

Water mains generally follow the hills and valleys of the natural topography. It is not uncommon for pockets of air to develop at the high points of the main. These pockets of air reduce the flow capacity of the system and increase pressure losses.

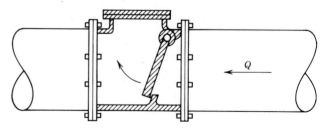

FIGURE 7.8 A swing-check valve will open to allow flow in only one direction.

Air release valves are placed in the pipeline at the peaks to automatically vent the accumulated air in the system.

Another appurtenance, called an altitude valve, is an automatic device that controls flow into an elevated water storage tank. It automatically closes when the tank is full, preventing overflow. When there is demand for water from the tank, the lower pressure in the distribution main is sensed by the valve mechanism and the valve opens to allow flow out of the tank. In effect, the water in the tank "floats" on the water in the main, and freely flows into or out of the tank, depending on pressure differentials. The hydraulics of elevated storage in water distribution systems is discussed later in this chapter.

INSTALLATION

Water mains must be installed at sufficient depths below the ground surface to provide protection against traffic loads and to prevent freezing. Generally, these depths are in the range of 1 to 2 m (3 to 6 ft). Since flow occurs under pressure instead of by gravity, the water mains can follow the general topographic shape of the ground, uphill as well as downhill.

There are several ways of placing the pipeline in an excavated trench to provide additional strength and protection. These different pipe *bedding* methods are illustrated in Figure 7.9. The material placed in the trench on top of the pipeline is called *backfill*. Compacting the backfill in layers around the pipe barrel increases the support for the pipe and can reduce the incidence of water main breaks. The type of bedding condition selected depends on the trench depth and the thickness class of the pipe.

Water mains should not be installed in the same trench with a sewer line. Generally, they should be at least 3 m (10 ft) away from a sewer, and they should be higher than sewer lines when they cross.

Pressure Testing

No matter how well a pipeline is constructed, there may be some leakage at the joints. Within certain

Laying Condition	Description
 Type 1*	Flat-bottom trench. † Loose backfill.
 Type 2	Flat-bottom trench. † Backfill lightly consolidated to centerline of pipe.
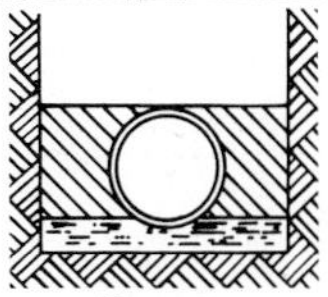 Type 3	Pipe bedded in 4-in. minimum loose soil. †† Backfill lightly consolidated to top of pipe.
 Type 4	Pipe bedded in sand, gravel or crushed stone to depth of ⅛ pipe diameter, 4-in. minimum. Backfill compacted to top of pipe.
 Type 5	Pipe bedded in compacted granular material to centerline of pipe. Compacted granular or select†† material to top of pipe.

*For 30-in. and larger pipe, consideration should be given to the use of laying conditions other than Type 1.
†"Flat-bottom" is defined as undisturbed earth.
††"Loose soil" or "select material" is defined as native soil excavated from the trench, free of rocks, foreign materials and frozen earth.

FIGURE 7.9 Standard trench bedding conditions for ductile iron pipe. (U.S. Pipe and Foundary Company)

limits, some leakage is acceptable; the construction specifications should indicate the maximum allowable rate of leakage. A common formula used for this purpose is the following:

$$Q_L = \frac{N \times D \times P^{1/2}}{C} \qquad (7\text{-}1)$$

where Q_L = allowable leakage, L/h (gal/h)
N = number of joints in length of main tested

D = pipe diameter, mm (in.)
P = test pressure, kPa (psi)
C = a constant depending on units
used: for SI metric, C = 32 600
(for American units, C = 1850)

A pressure or leakage test is conducted on a newly installed water main by filling the pipe with water and maintaining a pressure of 1000 kPa (150 psi) for one hour. If excessive leakage occurs, an amount of water greater than Q_L must be pumped into the line to maintain the pressure; repairs would be necessary before the pipeline can be accepted for use.

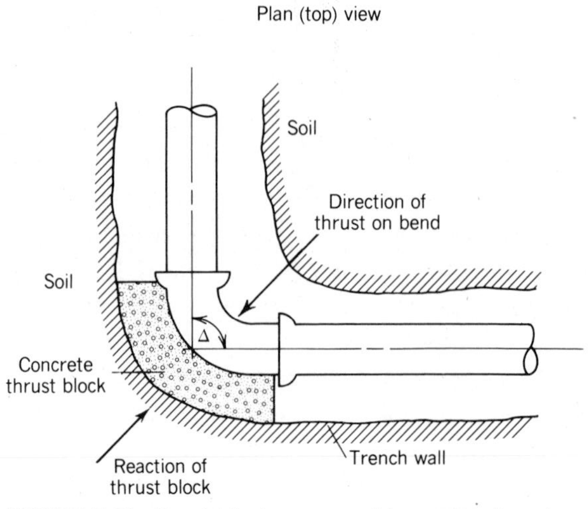

EXAMPLE 7.1

A 300-m long section of a newly installed 305-mm diameter water main is pressure tested for leakage. It was observed that during the 1-h test period, a volume of 10 L of water was pumped into the pipeline to maintain the required pressure of 1000 kPa. The pipe sections are 6 m long between joints. Has the allowable rate of leakage been exceeded?

Solution:

First compute the number of joints in 300 m of pipeline, as

$$N = \frac{300 \text{ m}}{6 \text{ m}} = 50 \text{ joints}$$

Now applying Equation 7-1, we get

$$\frac{50 \times 305 \times 1000^{1/2}}{32\ 600} = 15 \text{ L/h}$$

Since the observed leakage of 10L/h is less than the computed allowable leakage of 15 L/h, the pipe is sufficiently watertight.

Thrust Blocks

It is usually necessary to anchor the pipeline securely in the trench at dead ends and at bends, or

FIGURE 7.10 Thrust blocks are used to anchor the pipeline and to prevent movement or possible joint opening at bends.

at changes in horizontal or vertical direction. This is because of the force, or thrust, caused by the internal pressure and the kinetic energy of flow, which tends to move the pipe or fittings. Such movement could damage the joints and cause excessive leakage. One method for providing the necessary anchorage is to use concrete thrust blocks, as illustrated in Figure 7.10.

In most cases, the internal pressure causes most of the thrust and the dynamic thrust due to the flow velocity can be neglected. For a bend in a pipeline, the thrust due to static pressure can be computed from the following formula:

$$F = 2 \times P \times A \times \sin(\Delta/2) \tag{7-2}$$

where F = force or thrust, kN (lb).
P = water pressure, kPa (psi)
A = cross-sectional area of pipe, m^2 (sq. in.)
Δ = the change in direction of the pipe, degrees

EXAMPLE 7.2

An 18-in. diameter pipe carries water under a pressure of 80 psi. Compute the static thrust for a 90°

bend in the pipe, and compute the required bearing area of a concrete thrust block if the soil can support a bearing stress of 3000 lb/ft^2.

Solution:

First compute the area of the pipe section as

$$A = \frac{\pi D^2}{4} = \pi \times \frac{18^2}{4} = 254 \text{ in.}^2$$

Now use Equation 7-2 to compute the static thrust.

$$F = 2 \times 254 \times 80 \times \sin\left(\frac{90}{2}\right)$$
$$= 2 \times 254 \times 80 \times 0.707 \approx 30\,000 \text{ lb}$$

If the soil can withstand 3000 lb/ft^2, then the required area of the thrust block that is needed to spread the force out on the supporting soil is 30 000 lb ÷ 3000 lb/ft^2 = 10 ft^2.

Disinfection

A newly installed water main must be flushed clean and disinfected before being put into service. Flushing velocities of about 1 m/s (3 ft/s) are generally enough to remove dirt and debris that may accumulate in the pipe during construction. The pipe is disinfected to kill bacteria by filling it with a relatively concentrated chlorine solution for a certain period of time, as specified by local regulatory agencies. It is flushed again before being put into service.

REHABILITATION

The use of proper material and installation methods does not guarantee trouble-free operation of a water main for an unlimited period of time. Pipeline breaks and leaks occur periodically for several reasons, and emergency repairs must be made. Most water utilities have a plan of action for dealing with these emergencies and keep spare parts, tools, and equipment readily available.

Leaks that are not readily observable from wet or sunken spots in the street can be located using sounding rods for electronic amplification of the sound of the escaping water. Relatively small leaks can be repaired without shutting off the water pressure. This not only avoids inconvenience to utility customers, but also prevents contamination of the distribution system by backflow (Backflow of "dirty" water into the system can occur when the water pressure is turned off). All water is pumped out of the excavated trench. Repair clamps or sleeves are installed over the leaking section of pipe, and tightened until the leakage stops. The pipe should be flushed, hydrostatically retested, and disinfected after the repair has been made.

Cleaning

Sudden water main breaks are not the only problems that can occur in a water distribution system. Loose deposits of sediment may accumulate in the pipeline, particularly in dead-end branches. These sediments, which cause taste, odor, and color problems, can be removed by periodic flushing through hydrants.

Many water mains suffer the effects of a gradual and persistent buildup of solid deposits on the inside wall of the pipe. The longer the pipeline is in service, the worse this problem becomes. These deposits, illustrated in Figure 7.11, reduce the hydraulic capacity of the pipeline and cause high pressure losses. Pumping costs increase and there is a greater chance for regrowth of bacteria in the distribution system.

The deposits may consist of tubercles or lumps of iron oxide, if the pH of the water is low and the metal pipe is unlined. This problem is called *tuberculation*. When the pH of the water is high, the deposits consist of calcium carbonate scale. Maintaining a proper pH so that the water is neither corrosive nor scaleforming may be accomplished by adding chemicals at the treatment plant. Although this can minimize the formation of additional deposits, it does not serve to restore the lost capacity that has already occurred.

A common method for rehabilitating old water mains is to clean them using a mechanical or hydraulic scraper tool. During the cleaning operation,

FIGURE 7.11 A view of a deteriorated water main with accumulated interior deposits that reduce the pipe capacity. (Centriline Division of Raymond International Builders, Inc.).

water supply service can be provided to customers by using small-diameter temporary bypass pipes or hoses. Thick deposits can be removed using power driven mechanical cleaning devices. The cutting tool consists of rotating steel blades mounted on a series of body sections attached to a center rod.

Another type of cleaning device is a bullet-shaped resilient foam object called a "pig." It is wrapped with wearing strips in a spiral or criss-cross arrangement. Propelled by water pressure through the pipeline, the pig serves to scrape the deposits off the wall of the pipe. It is inserted into the line either through a fire hydrant or through a specially installed "launcher." Several passes of the pig may be needed to remove all the deposits. Afterwards, the pipe is flushed out and disinfected before being placed into service again.

Lining

Newly installed metal pipes are supplied with a cement–mortar lining to prevent tuberculation. But many iron water mains were installed in the past without these linings. When these mains are cleaned and the deposits removed, it is necessary to install a lining that will prevent the problem of tuberculation from recurring. One such rehabilitation method, called *sliplining,* involves placing a plastic pipe inside the cleaned pipe. The plastic pipe, of slightly smaller diameter than the original pipe, is pulled through straight sections of the transmission main.

Another method for protecting the cleaned pipe wall is to apply a cement-mortar lining to the pipe in place. For pipes less than 600 mm (24 in.) in diameter, mortar can be pumped through a hose to a lining machine that is pulled through the pipeline. The mortar is sprayed centrifugally onto the pipe wall; the thickness of the lining can be controlled by the speed at which the machine is pulled through the pipe. A lining thickness of 6 mm or less is preferred, so as not to reduce the inner diameter of the pipe excessively. For pipes over 600 mm in diameter, a worker can enter the water main with the lining equipment.

Cleaning and lining can effectively rehabilitate a water main, increasing its carrying capacity and reducing pressure drops. But the pipe must not be structurally defective if this method of rehabilitation is to be used. A deteriorated pipeline that experiences frequent breaks and leaks may have to be replaced completely with a new pipeline. Large concrete water mains, however, may be lined with pipe made of steel plate to restore structural integrity with minimum loss of capacity. It is always necessary to compare the relative economics of complete replacement versus cleaning and relining. If water demands in the area are increasing, new water mains may be needed; rehabilitation alone may not be sufficient to meet the higher demands.

7.3 CENTRIFUGAL PUMPS

A pump is a mechanical device that adds energy to water or other liquids. In most water distribution systems, pumps are needed to raise the water in elevation, and to move it through the network of water mains under pressure. One way of classifying pumps is by their application in the system. Pumps that lift the water from a river or lake and move it to a nearby treatment plant are called *low-lift pumps.* They move large quantities of water, but at relatively low discharge pressures. The pumps that discharge the treated drinking water into the transmission and distribution system are called *high-lift pumps;* they operate under relatively high heads or pressures.

Sometimes it is necessary to increase the pressure within the distribution system or to raise the water into an elevated storage tank; *booster pumps* can be used for this purpose. *Well pumps* lift water from an underground aquifer and often discharge directly into the distribution system.

Another way of classifying pumps is according to the mechanical principles on which they operate. The two basic types are *positive displacement pumps* and *centrifugal pumps.* A positive displacement pump will deliver a fixed quantity of water with each revolution of the pump rotor or piston. The water is physically pushed or displaced from the pump casing. The capacity of the pump is unaffected by changes in pressure in the system in which it operates.

Centrifugal pumps are the most common type used in water supply (as well as wastewater) systems. As we will discuss shortly, the capacity of the pump is very much a function of the pressure against which it operates in the system. A centrifugal pump adds energy to the water by accelerating it through the action of a rapidly rotating impeller. The water is thrown outward by the vanes of the impeller and passes through a spiral-shaped casing, where its velocity is gradually slowed down. As its velocity drops in the expanding spiral volute, the kinetic energy is converted to pressure head, called the discharge pressure.

Centrifugal pumps can be further classified as *radial flow* or *axial flow.* In the radial flow type, the water discharges at right angles to the direction of flow into the pump impeller; in the axial flow type, the water discharges in the same direction as the axis of the impeller. Centrifugal pumps with more than one impeller are called *turbine pumps.*

Centrifugal type pumps have several advantages over positive displacement pumps. They are simple, with only one moving part—the impeller; no internal valves are required, and there is no need for internal lubrication. Also, they operate very quietly. Disadvantages include the effect of pressure on the pump output and efficiency, and the necessity for priming the pump before it is operated. Priming involves filling the pump casing and suction line with water.

PUMP CHARACTERISTICS

The performance characteristics of a centrifugal pump define the relationships among the discharge or rate of flow, the discharge pressure head "against" which the pump operates, the power requirements, and the efficiency of operation. These characteristics depend on the diameter, speed, and shape of the impeller(s) within the pump casing. They are different for each model of pump.

Pump Head Curve

Pump manufacturers provide data in catalogs that describe the performance characteristics of their line of pumps. These data are usually presented graphically, for convenience. A typical graph showing the relationship between the rate of flow and the total pressure head for a centrifugal pump, for a fixed impeller speed and diameter, is shown in Figure 7.12. It is called a *pump head curve.*

The pressure head is plotted on the vertical axis of Figure 7.12. This is called the *Total Dynamic Head,* or TDH, and is expressed in meters (feet) of water. The TDH depends on the configuration of the system into which the pump discharges, and the flow rate; this will be explained in more detail shortly. The flow rate or discharge, Q, is plotted on the horizontal axis, and is expressed in terms of L/s (gpm).

The pump head curve clearly shows that the pump discharge rate is a function of the total pressure against which the pump works, and that a given centrifugal pump can operate over a wide

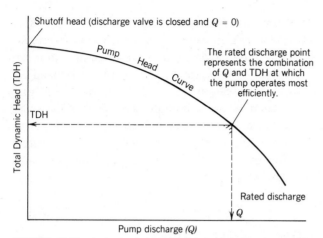

FIGURE 7.12 A typical centrifugal pump head-capacity curve. The discharge decreases as the TDH or pressure head on the pump increases.

range of Q and TDH values. *The discharge decreases as the TDH increases*. In effect, the harder the pump has to work to move the water, the less it can discharge per unit time. Specification or description of the operating condition for a centrifugal pump must always have a Q value and a corresponding TDH value.

A centrifugal pump that operates with its discharge valve completely shut will build up a maximum discharge pressure, called the *shutoff head*. At shutoff head, the discharge, of course, is zero. But the rotating impeller adds energy to the water circulating within the pump casing, and this develops pressure. When the valve is gradually opened and the water begins to flow, the pressure head will decrease, as the pump head curve shows. When the valve is completely opened, the unthrottled pump will operate at a discharge and TDH that matches that of the system in which the pump is working.

Pump head curves, or *head-discharge curves*, as they are also called, are determined by the pump manufacturer under shop test conditions. During the test, power and efficiency characteristics of the pump are also measured. Like any other mechanical device, a centrifugal pump cannot operate at 100-percent efficiency. There is always water circulating with the rotating impeller in the casing, causing energy loss. This is called *slip*, and the more slip there is in a given pump, the lower is its efficiency.

The efficiency of a centrifugal pump depends on the discharge and TDH. Obviously, at shutoff head ($Q = 0$), the efficiency is zero since the energy driving the impeller is entirely lost (as heat); no useful work is done by the completely throttled pump. As the discharge is allowed to increase, the efficiency of operation will increase to a maximum value, and then begin to decrease. The combination of Q and TDH at which the maximum efficiency is measured is called the *rated or normal discharge capacity* of the pump. The rated capacity may be indicated on the pump head curve with a mark, as shown in Figure 7.12.

Pump manufacturers often indicate the efficiency and power consumption, over a range of flows and impeller speeds and/or diameters, on the graph with the pump head curve. An example of a set of pump characteristic curves is shown in Figure 7.13.

Impeller Speed

For a given centrifugal pump, it is sometimes desirable to change the rotational speed of the impeller with a variable speed motor. The discharge of a centrifugal pump varies directly with the impeller speed, and the discharge head (pressure) varies directly with the square of the impeller speed. In other words, if the impeller speed was doubled, the discharge would double, and the discharge pressure head would increase by a factor of four. These relationships, called *affinity laws*, can be expressed as follows:

$$\frac{Q_1}{Q_2} = \frac{N_1}{N_2} \qquad (7\text{-}3)$$

$$\frac{H_1}{H_2} = \frac{N_1{}^2}{N_2{}^2} \qquad (7\text{-}4)$$

where Q_1 and Q_2 = pump discharges, L/s (gpm)
$\quad H_1$ and H_2 = pump discharge pressure heads, m (ft)
$\quad N_1$ and N_2 = pump impeller speeds, revolutions/minute (RPM)

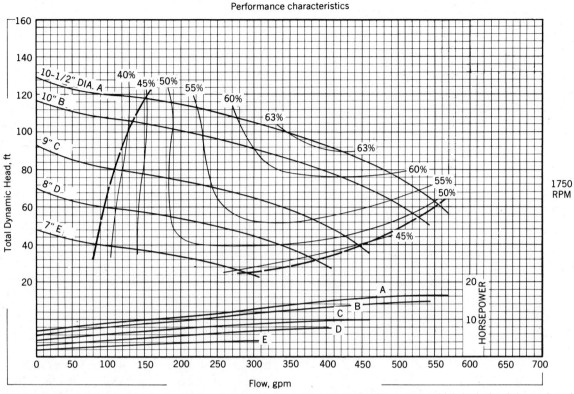

FIGURE 7.13 A typical set of manufacturer's centrigugal pump characteristic curves, which include data on head capacity, efficiency, and power requirements for various impellor sites. (FMC Corporation)

EXAMPLE 7.3

Manufacturer's data for a centrifugal pump indicate that the pump can discharge 100 L/s at a discharge pressure of 25 m, when the impeller speed is 1500 RPM. What would be the expected pump discharge and discharge pressure head if the impeller speed is increased to 2000 RPM?

Solution:

Applying Equations 7-3 and 7-4, we get

$$\frac{Q_1}{100} = \frac{2000}{1500} \quad \text{and } Q_1 = 130 \text{ L/s}$$

$$\frac{H_1}{25} = \frac{2000^2}{1500^2} \quad \text{and } H_1 = 45 \text{ m}$$

SYSTEM CHARACTERISTICS

In this discussion, the term *system* refers to the network of interconnected pipes, distribution reservoirs, valves, and other appurtenances to which the pump is connected. The piping from the water source to the inlet of the pump is called the *suction line*. One possible system configuration has the pump located below the water level on the suction side, as shown in Figure 7.14a. The vertical distance between the water level and the pump centerline is called the *static suction head*. This is a good arrangement because a suction head will always keep the pump and suction line *primed,* or in other words, filled with water and ready to operate.

When the system is arranged so that the pump is above the water level on the suction side, then the vertical distance between the water surface and the pump centerline is called *static suction lift*. The maximum height to which a water column can be

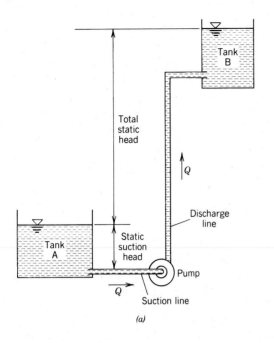

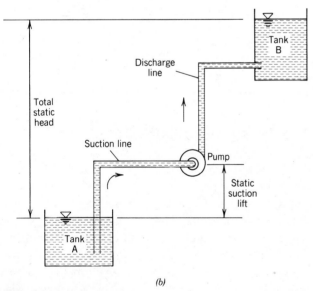

Figure 7.14 In diagram *(a)* the water in tank A is above the pump, causing a condition of *suction head*; in diagram *(b)*, the water in tank A is below the pump, causing *static suction lift*. The *total static head* is always the vertical distance between the lower and upper water surfaces, regardless of the suction line conditions.

lifted by any pump is about 10 m (33 ft), at sea level. This is equivalent to standard atmospheric pressure. But under *dynamic* conditions, that is, when the pump is operating and there is flow in the system, the suction lift is limited to a maximum

height of about 5 m (16 ft); this is because of the frictional resistance to flow in the suction line that the pump must also work against. A system configuration with static suction lift is illustrated in Figure 7.14*b*.

Under suction lift conditions, a *foot valve* is usually installed in the suction line to maintain prime on the pump. This is a check valve that allows water to flow into but not out of the pipe. Without a foot valve, water would flow out of the suction line when the pump is not operating. A centrifugal pump that loses its prime for any reason must have its casing and suction line refilled with water in order to operate. This can be done manually, or automatically for special self-priming pumps.

The vertical distance between the free water surfaces on the suction side and the discharge side of the pump is called the *total static head*. As illustrated in Figure 7.14, it represents the actual change in elevation of the water being pumped. But when the pump is operating and moving water through the system (a dynamic condition as opposed to a static or no-flow condition), the pump must also overcome the frictional resistance to flow.

System Head Curve

The amount of energy lost because of friction in the system is primarily a function of pipe diameter and flow rate. For a given diameter pipe, the greater the flow rate is, the greater the resistance to flow and friction head loss is. In addition to adding energy to lift the water and to overcome frictional resistance, the pump is adding velocity head to the system when it operates.

At a given flow rate, the sum of the total static head, the total friction head, and the velocity head is called the *total dynamic head* or TDH. (In most cases, the velocity head is small compared to the static and friction heads, and it can be neglected.) In effect, the pump "thinks" it is lifting the water a distance equal to the TDH, which is always greater than the static lift.

This *TDH* is the same term we were using in the previous discussion of pump characteristics. But the student must keep in mind that we are now looking at TDH from the perspective of the system, not the pump. A graph of the TDH as a function of

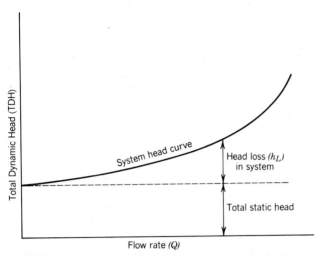

FIGURE 7.15 A *system head curve* shows the hydraulic response of a water transmission system to various flow rates. There is greater resistance to flow, and therefore a higher TDH in the system, for higher flow rates.

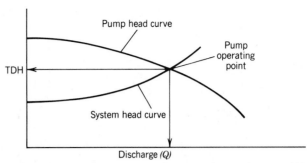

FIGURE 7.16 The intersection of the head curve for a centrifugal pump, and the system curve for the system in which it works, represents the *operating point* for the pump in that system.

flow rate is called the *system head curve*. A typical system head curve is illustrated in Figure 7.15. Note that in the system, *the TDH increases as the flow rate increases*. This is just the opposite of what happens from the pump's perspective: The TDH against which the pump can operate decreases as the flow increases.

PUMP OPERATING POINT

The pump head curve gives a picture of the hydraulic response of the pump to changes in flow rate or TDH. The system head curve gives a picture of the relationship between flow rate and TDH in the system. But when an unthrottled pump operates steadily in a given system (with a fixed total static head), there is only one point, that is, only one pair of Q and TDH values, at which the pump will operate. This point can best be determined graphically by drawing both the pump and system head curves on a single graph. The intersection of the two curves represents the *operating point* of the pump in the given system. This is illustrated in Figure 7.16.

When selecting a pump for use in a particular system, the designer must examine the system head curve together with the pump head curves

given in the manufacturer's data catalogs. The basic objective in pump selection is to find a pump that will provide the required flow rate at or close to its peak operating efficiency. In other words, the rated discharge for the pump should match the point at which it is expected to operate in the system. It is not a good practice to select an oversized pump for a system and then throttle the discharge to obtain the desired flow. The following example illustrates the procedure for determining a pump's operating point.

EXAMPLE 7.4

A centrifugal pump head curve is given in Figure 7.17. Determine the pump's operating point in a system that has a total static head of 50 m and comprises 4500 m of 305–mm diameter pipe.

Solution:

The system head curve must be plotted on the graph in Figure 7.17 in order to find the operating point. To do this, several discharge values are arbitrarily selected, and friction losses at those discharges are determined using the Hazen–Williams nomograph (Figure 2.15 in Section 2.3).

Neglecting velocity head, the TDH for each of the selected Q values is determined by adding the computed friction loss to the total static head. This is summarized in Table 7.2. The friction head loss is computed as $h_L = S \times L$, where S is the slope of

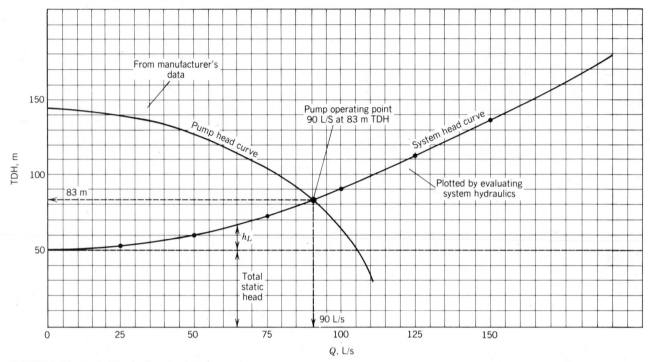

FIGURE 7.17 Illustration for Example 7.4.

the HGL and L is the pipe length of 4500 m. The TDH = 50 m + h_L at each of the selected flow rates.

Plotting the values of Q and TDH from Table 7.2, the system head curve is obtained, as shown in Figure 7.17. The intersection of the system curve with the given pump curve is the operating point for the pump in the given system. This is read from the graph to be 90 L/s at a TDH of 83 m.

TABLE 7.2 COMPUTATION OF TDH FOR EXAMPLE 7.4

Q, L/s	S, m/m	h_L, m	TDH, m
0	0	0	50
25	0.000 65	3	53
50	0.0025	11	61
75	0.005	23	73
100	0.009	41	91
125	0.014	63	113
150	0.019	86	136

PARALLEL OPERATION

Water demand and consumption vary on an hourly as well as on a daily and seasonal basis, as discussed in Section 7.1. Although the peak hourly demands are usually satisfied by water from local storage tanks, it is still necessary to provide for varying pump outputs in order to satisfy the changes in daily or seasonal demand.

One way to do this is to select a pump large enough to handle the maximum expected flow in the system, and then to reduce its discharge by throttling when the water demand is low. As previously mentioned, this is not a preferred method—the pump operating efficiency would be low since the pump would rarely be working at its rated discharge. Another method to change the pump output would be to use a variable speed motor to drive the pump. By increasing the speed of rotation of the impeller, the pump discharge can be increased. But variable speed equipment is expensive.

A third method involving *parallel operation* of two or more pumps finds the widest application. Pumps that are connected in parallel discharge into a common header or manifold pipe; the suction and

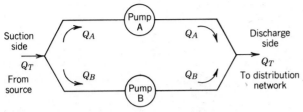

Pumps in parallel operate against
the same TDH in the system.

FIGURE 7.18 A schematic representation of pumps connected in parallel. The total flow in the system is the sum of each individual pump discharge at the same TDH, or $Q_T = Q_A + Q_B$.

discharge pressures for each of the pumps is the same. A parallel arrangement is illustrated in Figure 7.18. Parallel operation is advantageous because one or more pumps can be shut off when water demand is low, allowing the remaining pump(s) to operate at or near peak efficiency. Also, parallel connection allows maintenance work to be done on one pump without creating the need to shut down the entire pumping station.

In order to evaluate the performance of parallel pumps operating in a given system, it is first necessary to determine and plot the *combined head curve* of the pumps. This can be done simply by adding or combining the individual pump head curves horizontally. In other words, for selected values of TDH on the vertical axis, add the values of the discharge from each pump and plot the result along the horizontal axis. The next example illustrates this procedure for a system with two identical pumps in parallel. It is important to note that the combined discharge from the two pumps is not simply twice the discharge from one pump operating alone in the system.

EXAMPLE 7.5

Two identical pumps with the pump head curve shown in Figure 7.17 are connected in parallel. Sketch the combined head curve for the two pumps, and determine their operating point in the system given in Example 7.4.

Solution:

First prepare a graph with the pump head curve for a single pump, as shown in Figure 7.17. Arbitrarily select a few values of TDH, say 75 m, 100 m, and 125 m. For TDH = 75 m, the discharge of the pump is 97 L/s. For two pumps operating in parallel at the same TDH of 75 m, the combined discharge would be 97 + 97 or 194 L/s. Plot a point on the graph with coordinates Q = 194 L/s and TDH = 75 m.

If the TDH is 100 m, the flow from a single pump would be 80 L/s, and the combined flow from two pumps in parallel would be 160 L/s. Likewise, for TDH = 125 m, the combined flow is 106 L/s. Finally, at the shutoff head of 144 m, there would be no flow from either pump, and the combined Q = 0 L/s. Plotting all these points on the graph and sketching a smooth line that passes through them results in the combined pump head curve for the two pumps operating in parallel. This is shown in Figure 7.19.

The system head curve from Example 7.4 is also plotted on Figure 7.19. The intersection of the combined head curve with the system curve represents the point at which the parallel pumps would operate: From the graph, we see that they would discharge a combined flow of 130 L/s at a TDH of 115 m. Each of the two pumps will contribute 65 L/s of flow in parallel operation, whereas one pump operating alone in the system would discharge 90 L/s.

When two or more different pumps are connected in parallel, the procedure for determining the combined head curve is similar to that just described. But the combined curve will branch off from the curve for the larger pump at the higher values of TDH. Even when the smaller pump is operating, it would not contribute any discharge when the TDH in the system is above the pump shutoff head. This is illustrated in the following example.

EXAMPLE 7.6

Pump A and pump B are connected in parallel in a system that comprises 8000 ft of 8-in. diameter pipe

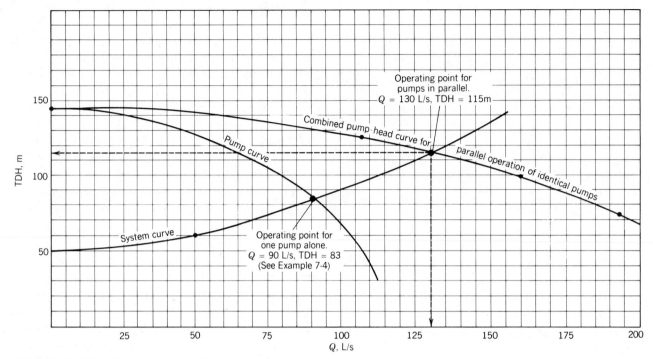

FIGURE 7.19 Illustration for Example 7.5

(with $C = 100$). The total static head of the system is 80 ft. The pump manufacturer has provided data which describe the pump head curves in tabular form, as shown in Table 7.3.

Determine the operating point of the parallel pumps in the given system. How much flow would each pump contribute?

Solution:

First plot the head discharge curves for each pump individually from the data given in Table 7.3, as shown in Figure 7.20.

Now plot the combined head curve for the two pumps operating in parallel; add the discharges for each pump at a few selected values of TDH. Note that when the TDH is between 200 and 240 ft, only pump A will be discharging into the system; pump B will be operating at a TDH equal to its shutoff head and $Q = 0$. When the TDH is less than 200 ft, both pumps can contribute flow. For example, when the TDH is 100 ft, pump A would discharge 450 gpm and pump B would discharge 335 gpm; the combined discharge at TDH = 100 ft would therefore be 785 gpm.

TABLE 7.3   PUMP OPERATING DATA FOR EXAMPLE 7.6

Discharge (gpm)	TDH, ft	
	Pump A	Pump B
0	240	200
100	225	185
200	200	155
300	165	115
400	125	60
500	65	—

Now plot the system head curve. Computations for the system TDH at four selected discharge values are summarized in Table 7.4.

As seen from Figure 7.20, the combined head curve and the system curve intersect at $Q = 530$ gpm and TDH = 160 ft, which represents the operating condition of the pumps in parallel. Of the total discharge of 530 gpm, pump A contributes about 330 gpm and pump B contributes about 200 gpm.

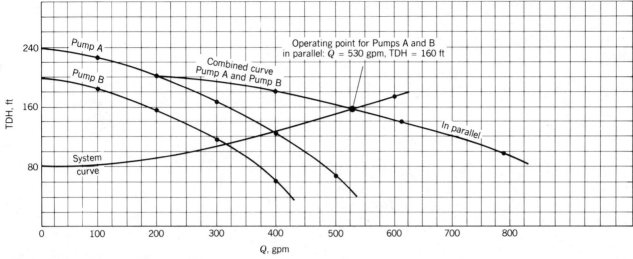

FIGURE 7-20 Illustration for Example 7.6.

TABLE 7.4 COMPUTATION OF TDH FOR EXAMPLE 7.6

Q, gpm	S, m/m	h_L, ft	TDH, ft
0	0	0	80
200	0.0014	11	91
400	0.0055	44	124
600	0.012	96	176

POWER AND EFFICIENCY

The rate at which a pump adds energy to the water is the power output of the pump, called the *water power*. Pumps are usually driven by electric motors (often with gasoline or diesel standby engines to be used in the event of a power failure). Since a pump (or any machine) can never operate at 100-percent efficiency, the water power is always less than the power delivered to the pump impeller by the motor. The efficiency of a pump can be expressed as follows:

$$\text{efficiency} = \frac{P_{\text{out}}}{P_{\text{in}}} \times 100 \qquad (7\text{-}5)$$

In SI metric units, power is expressed in terms of kilowatts (kW). Water power, P_{out}, can be computed with the following formula:

$$P_{\text{out}} = 9.8 \times Q \times \text{TDH} \qquad (7\text{-}6)$$

where Q = the pump discharge, m^3/s
TDH = total dynamic head, m

Note that 9.8 is the unit weight of water in terms of kN/m^3 and power has the units kN·m/s, which is work per unit time. In the SI system, one kN·m/s is called a kilowatt.

EXAMPLE 7.7

A pump discharges 500 L/s at a TDH of 25 m. The drive motor delivers 150 kW of power to the pump. At what efficiency is the pump operating?

Solution:

Applying Equation 7-6 to compute water power, we get

$$P_{\text{out}} = 9.8 \times 0.5 \times 25 = 123 \text{ kW}$$

From Equation 7-5, we get

$$\text{efficiency} = \frac{123}{150} \times 100 = 82 \text{ percent}$$

In American units, power is commonly expressed in terms of *horsepower (hp)*, where 1 hp = 33 000 ft·lb/min. The power input to the pump shaft is called *brake horsepower (bhp)*. Water horsepower can be computed from the following formula:

$$P_{\text{out}} = \frac{Q \times \text{TDH}}{3960} \qquad (7\text{-}7)$$

where Q is expressed in terms of gpm and TDH is expressed in terms of feet. Horsepower can be converted to kilowatts by multiplying by 0.746 (i.e., 1 hp = 0.746 kW).

EXAMPLE 7.8

A centrifugal pump with an efficiency of 65 percent discharges 1500 gpm into a system that includes 3000 ft of 10-in. diameter pipe (C = 100). The total static head is 100 ft. Compute the required brake horsepower (P_{in}).

Solution:

First determine the TDH on the pump. Enter the Hazen–Williams nomograph with Q = 1500 gpm and D = 10 in.; read S = 0.024. Then $h_L = S \times L$ = 0.024 × 3000 = 72 ft, and TDH = 100 + 72 = 172 ft. Now applying Equation 7-7, we get

$$P_{\text{out}} = \frac{(1500 \times 172)}{3960} = 65 \text{ hp}$$

From Equation 7-5, we get

$$P_{\text{in}} = \frac{P_{\text{out}}}{\text{efficiency}} \times 100 = \frac{65}{65} \times 100 = 100 \text{ hp}$$

EXAMPLE 7.9

The overall "wire to water efficiency" of a pump and motor is 50 percent. The water horsepower is 150 kW. If electric power costs $0.15 per kilowatt-hour kW·h, how much does it cost to operate the pump for 8 h?

Solution:

The electric power consumption is 150/50 × 100 = 300 kW. For 8 h of operation, the energy consumed would be

$$300 \text{ kW} \times 8 \text{ h} = 2400 \text{ kW·h}$$

The cost of operation would be

$$2400 \text{ kW·h} \times \$0.15/\text{kW·h} = \$360$$

OPERATION AND MAINTENANCE

Proper operation and maintenance (O&M) procedures must be followed to obtain satisfactory service from centrifugal pumps. Two important aspects of O&M involve keeping the pumping station clean and keeping the pump properly lubricated. Pumps located in a dirty and messy pumping station cannot be operated properly. Oil and grease cans must be covered and kept free from dirt. Dirt in lubricating oil or grease will cause excessive wear of pump bearings and shorten their life. The manufacturer's recommendations must be followed carefully for proper lubrication. It is important that the pump bearings not be over-lubricated—too much grease will cause damage.

The rotating impeller is the only moving part in the pump casing. Thrust bearings and guide bearings support the shaft that carries the impeller. A *packing gland* or seal is used where the pump shaft protrudes from the casing, to stop air from leaking in or water from leaking out. The packing gland usually consists of several graphite-impregnated asbestos rings around the shaft. These rings can be compressed to bear against the shaft; the graphite provides some lubrication. It is very important that the packing glands be properly adjusted and not over-tightened. During operation, a slight trickle of water leaking out of the casing is desirable to keep the gland cool.

A centrifugal pump must be primed or filled with water when it is started; the casing and suction piping must be free of air. When the pump is situated above the water level on the suction side, an electric or hand operated vacuum pump can be used to prime the centrifugal pump. The foot valve is designed to keep the pump and suction line filled with water when the pump is not operating, but if the valve leaks slightly, the pump will lose its prime. The ideal arrangement is to have the pump located lower than the water surface so that it will always be primed.

When the pump is started, the suction valve should be open and the discharge valve should be closed. The discharge valve is then opened slowly when the motor is up to speed. If a slight drip or flow of water from the packing gland is not observed, the packing may be too tight and cause excessive friction on the shaft; the gland should be adjusted before allowing continued operation.

Before stopping a centrifugal pump, the discharge valve must be closed slowly. If the valve is closed too fast or if the pump is stopped suddenly, a condition called *water hammer* will occur. *Water hammer* refers to a surge or pressure wave that travels throughout the pipeline. The momentary pressure caused by water hammer is often high enough to rupture the pipe or the pump casing.

Daily inspection of operating pumps is important, to check for excessive heat, noise, or gland leakage. A properly installed pump should not vibrate. Vibration is an indication of misalignment between the motor and the pump which will cause premature wear and damage of the pumping system. If vibration is noticeable, the manufacturer's specifications for proper alignment procedures and tolerances should be checked.

7.4 DISTRIBUTION STORAGE

Elevated water storage tanks and towers are familiar sights in most communities. These relatively small water storage facilities serve two basic purposes; they provide *equalizing storage* and they provide *emergency storage*. They are called distribution storage tanks or reservoirs because they are part of the localized water distribution systems. Much larger storage facilities, called *conservation reservoirs*, are generally located at a considerable distance from the distribution network and are meant to store water that can be used during long dry weather periods.

Equalizing storage refers to the volume of water in the tank that is available to satisfy the peak hourly demands for water use in the community. The hourly variation in water demand is illustrated in Figure 7.1. During the late night and early morning hours when water demand is very low, the high lift pumps move water into the distribution storage tanks. During the day, when water demand exceeds the average daily demand, water flows out of the tanks to help meet the peak hourly needs of the community.

Distribution storage tanks are often described as "floating on the line." This means that the flow into or out of the tank varies directly with the demand for water in the system, and the hydrostatic pressure in the water main is equivalent to the pressure head or elevation of water in the tank. A distribution storage tank is illustrated in Figure 7.21. Automatically controlled altitude valves can maintain the water elevations, and therefore the water main pressure, within a desired range.

The equalizing or averaging effect on flow rates provided by the stored water allows for a uniform or steady water treatment and pumping rate. When water demand is low, the extra water being pumped fills the storage tanks, and when demand exceeds the pumping rate, the tanks are emptied to make up the difference. This has the advantage of reducing the required sizes and capacities of the pipes, pumps, and treatment facilities, resulting in reduced construction and operating costs.

Generally, the volume of water needed to balance or equalize the peak hourly flows is about 20 percent of the average daily water demand in the service area. For example, if a community has an average water demand of 1 ML/d, then at least 0.2 ML or 200 000 L of storage volume should be provided for equalizing purposes. In communities

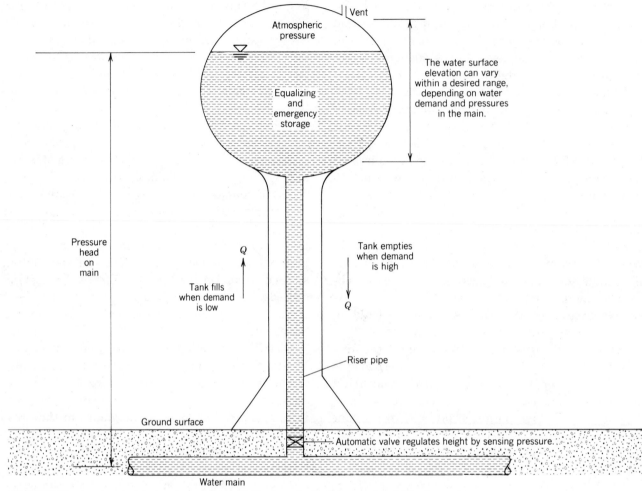

FIGURE 7.21   An elevated water storage tank will "float on the line" in a water distribution system.

where adequate records of water use and demand are available, the summation hydrograph method (see Section 3.6) can be used for a more accurate determination of the required equalizing volume in a new tank.

Distribution storage tanks are not constructed for the sole purpose of providing equalizing storage. As previously mentioned, emergency storage volume is also provided by these tanks. This provides additional water for fire fighting needs, power blackouts, or pump station failure.

The amount of water required for fire control varies depending on the type of service area and the capacity of the water pumping station. For example, if storage for a fire flow of 30 L/s for a 2-h duration was needed, the required volume would be 30 L/s × 3600 s/h × 2 h = 216 000 L. This would be added to the volume needed for equalizing purposes, to determine the total tank volume. If 200 000 L were needed for equalizing storage and 216 000 L were needed for fire control, the minimum tank volume would have to be 416 000 L or 416 m³.

If an emergency power generator or stand-by diesel engine were not available at the high lift water pumping station, it could be necessary to provide additional storage so that both domestic water demand and emergency fire fighting needs could be met even during a temporary power failure. In order to accommodate all these distribution needs,

FIGURE 7.22 A ground-level distribution storage reservoir. (Natgun Corporation)

some states simply require that the storage volume be equal to one day's average water demand, unless it can be demonstrated with available data that a smaller volume would suffice.

Types of Distribution Reservoirs

As previously discussed, distribution reservoirs store water for relatively short periods of time (one day or less) and are small enough to be covered to prevent contamination and to reduce evaporation. In areas with flat topography, the storage tanks are elevated above the ground on towers to provide adequate pressures in the water mains. These elevated tanks are usually constructed of steel.

In hilly areas, distribution tanks can be built at ground-level on hilltops higher than the service area, and still "float" on the line while maintaining adequate pressures in the system. They may be constructed of either steel or reinforced concrete. A typical ground-level distribution reservoir is illustrated in Figure 7.22.

When the height of the storage tank is greater than its diameter, the structure is called a *standpipe,* as illustrated in Figure 7.23. Standpipes provide more storage capacity than elevated tanks.

But the storage capacity that is useful for equalizing purposes is only that volume above the elevation required for minimum pressure in the water main. The water below that elevation can be used for fire protection with pumper trucks, or during

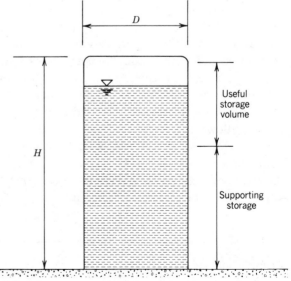

FIGURE 7.23 A water storage tank is called a *standpipe* when its height is greater than its diameter ($H > D$).

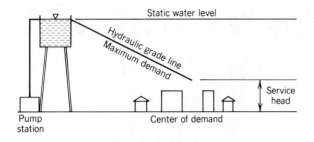

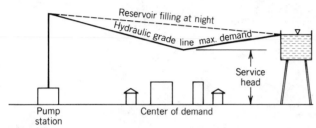

FIGURE 7.24   Location of distribution storage tanks opposite the source (B) is preferable to location at the source (A). (From *Water Distribution Operator Training Handbok,* by permission. Copyright 1976, The American Water Works Association.)

other emergency conditions. Standpipes are constructed of steel, with thicker walls at the bottom to withstand the hydrostatic pressure.

Location of Storage Tanks

Using several small storage tanks near the major centers of water withdrawal is preferable to using one large tank near the pumping station. Also, it is best to locate the tanks on the opposite side of the demand center from the pumping station. This allows for more uniform pressures throughout the distribution network, as illustrated in Figure 7.24. It also allows the use of smaller diameter mains and pumps than would otherwise be needed. The following example illustrates the hydraulic computations involved and the effect of storage on the required discharge pressure at the pump station.

EXAMPLE 7.10

A peak hourly flow rate of 100 L/s is required at point A, as illustrated in Figure 7.25. Pressure at that point is not to drop below 150 kPa. Determine (*a*) the required pressure at the pump station if the demand is satisfied without any distribution storage tank, and (*b*) the required pressure head at the pumps if a storage tank is used to help meet peak demand by floating on the line at elevation 20 m, as shown.

Solution:

(*a*) Compute the drop in pressure head from the pump station to the point of withdrawal, as follows:

For $Q = 100$ L/s and $D = 250$ mm, read $S = 0.024$ on the Hazen–Williams nomograph (Figure 2.15). Compute $h_L = S \times L = 0.024 \times 2000 = 48$ m.

At point A, the pressure is not to drop below 150 kPa. Using Equation 2-3a, we see that 150 kPa is equivalent to 0.1×150 or 15 m of pressure head. Since flow occurs in the direction of sloping HGL, we must add the 48 m of headloss to the 15-m minimum pressure head requirement at point A and sketch the HGL as shown in Figure 7.24. The pressure head is then $48 + 15 = 63$ m, and using Equation 2-2a, we find that the required pressure at the pump station is $P = 9.8 \times 63 \approx 620$ kPa.

(*b*) In this part of the problem, the storage tank is contributing flow to the withdrawal point. We can see that the headloss in the 500 m of pipeline is $20 - 15 = 5$ m. The slope of the HGL is then $5/500 = 0.01$, and from the Hazen–Williams nomograph, with $D = 200$ mm and $S = 0.01$, we read $Q = 32$ L/s.

Since the total withdrawal is still 100 L/s, the flow from the pump station must be $100 - 32 = 68$ L/s. From the Hazen–Williams nomograph, with $Q = 68$ L/s and $D = 250$ mm, we read $S = 0.012$. The headloss between the pump station and point A is then $h_L = 0.012 \times 2000 = 24$ m, and the required pressure head at the pump station is then $15 + 24 = 39$ m; the required pressure is $P = 9.8 \times 39 \approx 380$ kPa.

It can be seen in this problem that the required pump capacity decreases from 100 L/s at a discharge pressure of 620 kPa to a capacity of 68 L/s at a pressure of 380 kPa, when peak hourly demand is partially satisfied by water from a distribution storage tank.

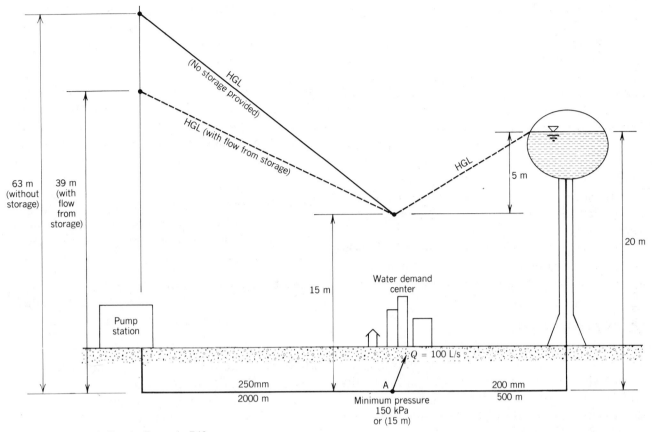

FIGURE 7.25 Illustration for Example 7.10.

Maintenance

Distribution storage reservoirs should be inspected frequently. Cracks and leaks must be repaired, and air vents should be kept cleár. Blocked air vents can cause excessive pressures or vacuums to develop in the tank, which can result in structural damage.

Storage tanks are occasionally drained, cleaned, and disinfected. Accumulated silt should be removed. Sometimes the inside of a steel tank will have to be painted with an approved bituminous or vinyl coating. All traces of rust must first be removed. Disinfection can be accomplished by spraying the tank walls with a 500 mg/L chlorine solution.

Corrosion of steel tanks can be a major problem. About 10 kg (22 lb) of steel per year can be lost because of corrosion. In addition to protecting the steel with high-quality bituminous coatings, a

method called *cathodic protection* is sometimes used. Corrosion involves a flow of electric current through the water, in the form of positively charged metal ions. In cathodic protection, a voltage is maintained in the tank that tends to reverse the direction of the current. This applied voltage keeps the metal from ionizing and thus prevents corrosion. A cathodic protection system must be custom-designed by specialists for each tank.

7.5 FLOW IN PIPE NETWORKS

Water distribution systems, particularly those serving densely populated cities, consist of a complex network of interconnected pipes and appurtenances. The hydraulic conditions in these systems must be analyzed to determine flow capacities and pressures in the main and secondary feeders and at points of significant water withdrawal.

It is convenient to first simplify or "skeletonize" the system by replacing the many smaller water mains with "equivalent pipes," in order to reduce the number of loops and interconnections. This is done only conceptually, on paper; the existing pipes are not actually replaced in the streets. After the distribution system has been skeletonized, network analysis methods can be applied to study the hydraulics of the system. Studies of this type are done to plan for possible expansion or upgrading of the water supply network.

This section presents the basic hydraulic techniques for determining hydraulically equivalent pipes and for analyzing the pipe network.

EQUIVALENT PIPES

An *equivalent pipe* is one that has the same hydraulic characteristics as the pipes it theoretically replaces. In other words, for any given flow rate, the pressure drop through the equivalent pipe would be the same as the total pressure drop through the original pipes. And for any given pressure drop, the flow rate in the equivalent pipe would be the same as the total discharge through the original pipes.

Equivalent pipes can be determined to replace pipes in series (connected end to end) or pipes in parallel (forming loops). For any given series or parallel system, there is no limit to the combinations of theoretical pipe diameters and lengths for the equivalent pipe. Usually, either the diameter (or length) is first specified, and the required length (or diameter) for hydraulic equivalence is then determined.

Pipes in Series

The following five steps summarize the method for determining an equivalent pipe to replace several different pipes connected end to end:

Step 1: Assume any flow rate Q. For our purposes, the flow rate should be selected within the range of flows on the Hazen–Williams nomograph of Figure 2.15, but theoretically, any flow rate can be selected.

Step 2: Using the nomograph, line up flow rate Q and diameter D for each section of the original series pipeline; read slope S and compute headloss $h_L = S \times L$, for each section, where L is the length of the section.

Step 3: Add up the headlosses for all sections in the series to determine a total headloss H_L for the assumed discharge Q in the pipeline.

Step 4: Convert the total headloss H_L to an overall hydraulic gradient $S' = H_L/L'$, where L' is the specified total length of the equivalent pipe.

Step 5: Enter the nomograph again with the assumed value of Q and the computed value S'; read the required value for the diameter D of the equivalent pipe.

EXAMPLE 7.11

A series pipeline is shown in Figure 7.26. It consists of 1500 feet of 8-in. diameter pipe from point A to point B, and 2500 ft of 12-in. diameter pipe from point B to point C. Determine the equivalent diameter of a single 4000-foot-long pipeline from A to C that could theoretically replace the pipes AB and BC.

Solution:

Step 1: Assume a flow rate Q = 450 gpm

Step 2: For section AB, line up Q = 450 gpm and D = 8 in. on the Hazen–Williams nomograph.
Read S = 0.0064
Compute $h_L = S \times L = 0.0064 \times 1500$ ft = 9.6 ft

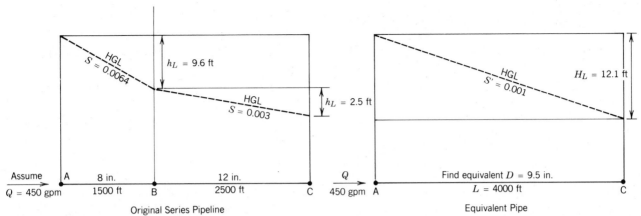

FIGURE 7.26 Illustration for Example 7.11.

Similarly for section BC, line up $Q = 450$ gpm and $D = 12$ in., then read $S = 0.001$ and compute

$$h_L = 0.001 \times 2500 = 2.5 \text{ ft}$$

Step 3: Compute the total headloss from A to C as

$$H_L = 9.6 + 2.5 = 12.1 \text{ ft}$$

Step 4: Compute the overall hydraulic gradient from A to C as

$$S' = \frac{12.1}{4000} = 0.003$$

Step 5: Now line up $Q = 450$ gpm and $S' = 0.003$ on the nomograph. Read $D = 9.5$ in.

An equivalent pipe for this problem, then, is one that would have a diameter of 9.5 in. and a length of 4000 ft.

Notice that the equivalent diameter of 9.5 in. in the preceding example is not simply a weighted average of the original two diameters from A to B and

B to C. It is the diameter that for a 4000-ft length of pipe would have the same hydraulic characteristics as the original 8-in. and 12-in. pipes. The original pipeline could also be replaced by an equivalent 3000-ft length of an 8.9-in. pipe, a 2000-ft length of an 8.1-in. pipe, and so on; there are an unlimited number of equivalent pipes for any given system.

For Example 7.11, if a flow rate other than $Q = 450$ gpm were assumed in Step 1, the same final answer of 9.5 in. would have been obtained for the equivalent diameter. Try the same problem with an assumed $Q = 1000$ gpm, or any other flow rate you choose, to verify this.

Pipes in Parallel

The procedure for determining an equivalent pipe to replace a loop configuration of pipes differs somewhat from the preceding procedure for pipes in series. The following seven steps outline the method for a parallel or looped pipe system (see Figure 7.27):

Step 1: Assume a total headloss H_L across the loop from the first pipe junction at A to the second pipe junction at B. This assumed headloss must be the same for both the top branch and the bottom branch of the loop, since the pressures at the ends of each pipe are the same at the junctions.

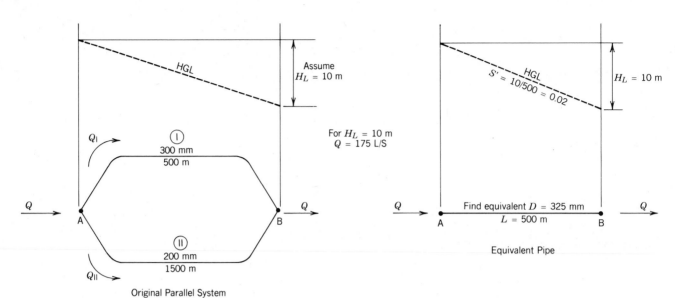

FIGURE 7.27   Illustration for Examples 7.12 and 7.13.

Step 2:   For branch AIB, compute $S = H_L/L$, where L is the length of AIB.

Step 3:   Enter the Hazen–Williams nomograph with D and S for branch AIB; determine Q_I in that branch.

Step 4:   Repeat steps 2 and 3 for branch AIIB to determine Q_II.

Step 5:   Compute the total flow rate Q entering junction A, as $Q = Q_\text{I} + Q_\text{II}$.

Step 6:   Determine an overall hydraulic gradient S' as follows: $S' = $ assumed H_L/L', where L' is the specified length of the equivalent pipe.

Step 7:   Enter the Hazen–Williams nomograph with Q and S' to determine the equivalent diameter D.

EXAMPLE 7.12

Two pipelines are connected in parallel from junction A to junction B, as shown in Figure 7.27. Branch AIB consists of 500 m of 300-mm diameter pipe, and branch AIIB consists of 1500 m of 200-mm pipe. Determine the equivalent diameter of a single 500-m-long pipeline from A to B that could replace the given loop.

Solution:

Step 1:   Assume $H_L = 10$ m

Step 2:   For branch AIB, $S = 10/500 = 0.02$

Step 3:   From the Hazen–Williams nomograph, with $D = 300$ mm and $S = 0.02$, read $Q_\text{I} = 143$ L/s

Step 4:   For branch AIIB, $S = 10/1500 = 0.0067$. From the nomograph, with $D = 200$ mm and $S = 0.0067$, read $Q_\text{II} = 27$ L/s

Step 5:   The total flow into junction A is $Q = Q_\text{I} + Q_\text{II} = 143 + 27 = 170$ L/s

Step 6:   The overall hydraulic gradient $S' = 10/500 = 0.02$

Step 7:   From the nomograph, with $Q = 170$ L/s and $S' = 0.02$, read an equivalent diameter of 320 mm

We have found that a 500-m-long, 320-mm diameter pipeline is hydraulically equivalent to the given loop from A to B. An unlimited number of other equivalent pipes with different combinations of length and diameter can be found. Also, if a headloss other than 10 m was assumed in Step 1, we would have arrived at the same final answer of 320 mm for a length of 500 m. Try the same problem with an assumed $H_L = 20$ m to verify this.

PIPE NETWORK ANALYSIS

After a complex distribution network has been skeletonized as far as practical using equivalent pipes, the system can be analyzed to determine pressures at critical points under assumed flow conditions. Flow rates and headlosses in the skeletonized system can be determined using a method called *Hardy Cross Network Analysis.*

The Hardy Cross method is a controlled "trial-and-error" procedure; corrections are applied to assumed flow rates in a manner that leads quickly to a *hydraulically balanced system.* A balanced system is one in which the computed flows and headlosses "match-up" at the pipe junctions of the distribution network. In other words, no matter which path or branch is followed to get to a specific junction, the computed pressure would be the same. And the total flow into the junction would equal the total flow out of that junction.

The corrections applied to the assumed flows are determined from the following formula, which is derived from the Hazen–Williams Equation:

$$\Delta Q = -\frac{\Sigma h_L}{1.85 \times \Sigma\, h_L/Q} \qquad (7\text{-}8)$$

where ΔQ (pronounced "delta Q") = the flow correction
Σh_L (pronounced "sigma h_L") = the sum of headlosses
$\Sigma h_L/Q$ = sum of h_L/Q ratios for each pipeline in a loop

A sign convention is used in the Hardy Cross procedure to indicate the direction of flow in a loop: flows in a clockwise direction (↻) are considered to be positive (+), and flows in a counterclockwise direction (↺) are considered to be negative (−). Headlosses caused by clockwise flows are also considered to be positive, and headlosses from counterclockwise flows are negative.

In a hydraulically balanced system, the algebraic sum of the headlosses around a loop (Σh_L) will add up to zero. For example, in the parallel pipe system shown in Figure 7.27, the flow rate of 143 L/s in AIB is positive (clockwise), and the 10-m headloss in that pipe is also positive (+10 m). The flow rate of 27 L/s in AIIB is negative (counterclockwise), and the 10-m headloss in that pipe is also negative (−10 m). In this simple loop, $\Sigma h_L = (+10) + (-10) = 0$, as it should be for a balanced system.

EXAMPLE 7.13

Consider again the single loop of parallel pipes shown in Figure 7.27. Instead of replacing them with an equivalent pipe, the problem here is as follows: If a flow of 400 L/s enters the loop at junction A, what will be the flow rates Q_I in branch AIB and Q_{II} in branch AIIB?

Solution:

From the principle of continuity of flow, we know that the flow entering a junction must equal the total discharge from the junction, or $Q_I + Q_{II} = 400$ L/s. If branches AIB and AIIB were identical in all respects, then the flow of 400 L/s would be evenly split between the two. But this is not the case here; AIIB is longer and narrower than AIB, and it will therefore offer more resistance to flow. It is reasonable to assume that the flow rate in AIIB will be less than that in AIB.

Let us start by assuming that the flow rate $Q_I = 300$ L/s and $Q_{II} = -100$ L/s. (The minus sign for Q_{II} indicates a counterclockwise direction of flow.) The magnitudes of the flows must still add up to 400 (i.e., $300 + 100 = 400$).

If our assumption for flow rate is correct, then the headlosses in AIB and AIIB should be equal in magnitude (but opposite in sign). Using the Hazen–Williams nomograph to check this, we find:

Pipe AIB: For $Q = 300$ L/s and $D = 300$ mm, $S = 0.075$ and $h_L = S \times L = 0.075 \times 500 = 37.5$ m

Pipe AIIB: For $Q = -100$ L/s and $D = 200$ mm, $S = -0.07$ and $h_L = S \times L = -0.07 \times 1500 = -105$ m

Since $\Sigma h_L = 37.5 + (-105) = -67.5$ m, instead of zero, the loop is not balanced and our assumed flows are incorrect. Instead of simply guessing at new flows to try, we will use the Hardy Cross correction formula, Equation 7-8, as follows:

$$\Delta Q = -\frac{-67.5}{1.85 \times (37.5/300 + 105/100)}$$
$$= -\frac{-67.5}{1.85 \times 1.175} = 31 \text{ L/s}$$

Adding the flow correction $\Delta Q = 31$ L/s to the assumed flows, we get new flow rates as follows:

Pipe AIB: $Q_I = 300 + 31 = 331$ L/s
Pipe AIIB: $Q_{II} = -100 + 31 = -69$ L/s
Check continuity: $331 + 69 = 400$ OK

Note the importance of using the algebraic signs properly.

Now we will use the nomograph again to see if the loop is hydraulically balanced with the new flows:

Pipe AIB: For $Q_I = 331$ L/s and $D = 300$ mm, $S = 0.085$ and $h_L = S \times L = 0.085 \times 500 = 42.5$ m

Pipe AIIB: For $Q_{II} = -69$ L/s and $D = 200$ mm, $S = -0.037$ and $h_L = S \times L = -0.037 \times 1500 = -55.5$ m

Now $\Sigma h_L = 42.5 + (-55.5) = -13$ m. This is still not zero, so we can make another flow correction as follows:

$$\Delta Q = -\frac{-13}{1.85 \times (42.5/331 + 55.5/69)}$$
$$= -\frac{-13}{1.726} = 7.53 \text{ L/s}$$

For practical purposes, we can round up the 7.53 to a $\Delta Q = 8$ L/s, and for corrected flows, we get the following:

Pipe AIB: $Q_I = 331 + 8 = 339$ L/s
Pipe AIIB: $Q_{II} = -69 + 8 = -61$ L/s
Check continuity: $339 + 61 = 400$ L/s OK

Now checking headlosses again to see if the loop is balanced,

Pipe AIB: For $Q_I = 339$ L/s and $D = 300$ mm, $S = 0.9$ and $h_L = S \times L = 0.9 \times 500 = 45$ m

Pipe AIIB: For $Q_{II} = -61$ L/s and $D = 200$ mm, $S = -0.03$ and $h_L = S \times L = -0.03 \times 1500 = -45$ m

Since $\Sigma h_L = 45 + (-45) = 0$, the loop is balanced and the final answers for the flow rates are $Q_I = 339$ L/s, and $Q_{II} = 61$ L/s. (For practical purposes, we should round off the answers to 340 L/s and 60 L/s.) The pressure drop from A to B would be $9.8 \times 45 \approx 440$ kPa.

Complex Networks

In practice, large and complex pipe networks are usually analyzed using digital electronic computers. But the beginning student must have a clear understanding of the basic principles involved. The best way to obtain this understanding is to solve some problems "manually" with a nomograph and a hand-held calculator. For networks with more than one loop, having several points of water inflow or outflow, it is convenient to set up a table to organize and keep track of the Hardy Cross computations. This will be illustrated in Example 7.14; first, the general procedure is outlined as follows:

Step 1: For each pipe in the network, assume a flow rate and flow direction. The only restriction on this initial assumption for flows is that the total flow going into a pipe junction must equal the total flow going out of that junction.

Step 2: Working with only one loop at a time, use the Hazen–Williams nomograph to determine S and h_L for each pipe in the loop. Also, compute h_L/Q for each pipe.

Step 3: Compute Σh_L and $\Sigma h_L/Q$, using the appropriate signs ($+$ or $-$). The h_L/Q terms are always positive.

Step 4: Compute ΔQ, the flow correction for the loop, using the Hardy Cross formula (Equation 7-8). Add ΔQ to the flow in each pipe of that loop, taking care to use the appropriate sign. Do not change the flows in the other loop(s) with this ΔQ.

Step 5: Repeat steps 2 through 4 for an adjacent loop in the network. Note that at least one of the pipes in the first loop is also part of the adjacent loop. Use the previously corrected flow(s) in the common pipe(s), but keep in mind that the algebraic signs of the flows and headlosses in the common pipe(s) change, depending on which loop you are looking at.

Step 6: Alternately repeat steps 2 through 4 for each loop in the network as many times as needed to arrive at a reasonably balanced system. Generally, a difference of less than 10 percent between the positive and negative headlosses in a loop is acceptable, in lieu of $\Sigma h_L = 0$.

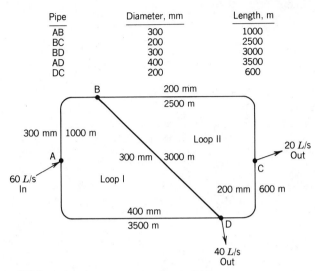

FIGURE 7.28 Illustration for Example 7.14.

drawal points, at C and D, discharge 20 L/s and 40 L/s, respectively. Determine the flow rates in all the pipes of the network.

Solution:

The computations outlined in steps 1 through 6 of the Hardy Cross procedure are summarized in Table 7.5.

A running tabulation of corrected flows is shown in Figure 7.29. Notice that the corrected flow in BD ($Q + \Delta Q = 4$ L/s) from loop I, first try is used with opposite sign in the computations for loop II/first try. This is because pipe BD is common to the two loops; while the flow is clockwise in loop I, it is counterclockwise in loop II. This reversal in sign for flows in BD occurs repeatedly, as indicated in the rest of the computations. The final flow rates in the balanced network are shown in Figure 7.30.

EXAMPLE 7.14

A water distribution system has been skeletonized and reduced to the two-loop network shown in Figure 7.28. A flow rate of 60 L/s is pumped into the network at point A, and two major water with-

TABLE 7.5 HARDY CROSS ANALYSIS FOR THE SYSTEM IN EXAMPLE 7.14

Loop	Pipe	D, mm	L, m	Q, $^L/_s$	S, $^m/_m$	h_L, m	h_L/Q	$Q + \Delta Q$
I	AB	300	1000	35	0.0015	1.5	0.043	29
(first	BD	300	3000	10	0.000 15	0.45	0.045	4
try)	AD	400	3500	−25	−0.0002	−0.7	0.028	−31
						1.25	0.116	

$$\text{I: } \Delta Q_1 = -\frac{1.25}{1.85 \times 0.116} \approx -6 \; ^L/_s$$

Loop	Pipe	D, mm	L, m	Q, $^L/_s$	S, $^m/_m$	h_L, m	h_L/Q	$Q + \Delta Q$
II	BC	200	2500	25	0.0055	13.75	0.55	12
(first	CD	200	600	5	0.0003	0.18	0.036	−8
try)	BD	300	3000	−4	neglect	neglect	neglect	−17
						13.9	0.586	

$$\text{II: } \Delta Q_1 = -\frac{13.9}{1.85 \times 0.586} \approx -13 \; ^L/_s$$

Loop	Pipe	D, mm	L, m	Q, $^L/_s$	S, $^m/_m$	h_L, m	h_L/Q	$Q + \Delta Q$
I	AB	300	1000	29	0.001	1.00	0.034	24
(second	BD	300	3000	17	0.0004	1.20	0.071	12
try)	AD	400	3500	−31	−0.0003	−1.05	0.034	−36
						1.15	0.139	

$$\text{I: } \Delta Q_2 = -\frac{1.15}{1.85 \times 0.139} \approx -5 \; ^L/_s$$

Loop	Pipe	D, mm	L, m	Q, $^L/_s$	S, $^m/_m$	h_L, m	h_L/Q	$Q + \Delta Q$
II	BC	200	2500	12	0.0015	3.75	0.313	8
(second	CD	200	600	−8	−0.000 75	−0.45	0.056	−12
try)	BD	300	3000	−12	−0.0002	−0.60	0.050	−16
						2.7	0.419	

$$\text{II: } \Delta Q_2 = -\frac{2.7}{1.85 \times 0.419} \approx -4 \; ^L/_s$$

Loop	Pipe	D, mm	L, m	Q, $^L/_s$	S, $^m/_m$	h_L, m	h_L/Q	$Q + \Delta Q$
I	AB	300	1000	24	0.0007	0.7	0.029	23
(third	BD	300	3000	16	0.000 34	1.02	0.064	15
try)	AD	400	3500	−36	−0.0004	−1.4	0.039	−37
						0.32	0.132	

$$\text{I: } \Delta Q_3 = -\frac{0.32}{1.85 \times 0.132} \approx -1 \; ^L/_s$$

Loop	Pipe	D, mm	L, m	Q, $^L/_s$	S, $^m/_m$	h_L, m	h_L/Q	$Q + \Delta Q$
II	BC	200	2500	8	0.0007	1.75	0.219	
(third	CD	200	600	−12	−0.0015	−0.9	0.075	
try)	BD	300	3000	−15	−0.0003	−0.9	0.06	
						−0.05	0.354	

$$\text{II: } \Delta Q_3 = -\frac{-0.05}{1.85 \times 0.354} = 0.07 \text{ (negligible)}$$

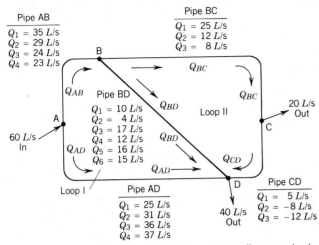

Pipe AB

$Q_1 = 35\ L/s$
$Q_2 = 29\ L/s$
$Q_3 = 24\ L/s$
$Q_4 = 23\ L/s$

Pipe BC

$Q_1 = 25\ L/s$
$Q_2 = 12\ L/s$
$Q_3 = 8\ L/s$

Pipe BD

$Q_1 = 10\ L/s$
$Q_2 = 4\ L/s$
$Q_3 = 17\ L/s$
$Q_4 = 12\ L/s$
$Q_5 = 16\ L/s$
$Q_6 = 15\ L/s$

Pipe AD

$Q_1 = 25\ L/s$
$Q_2 = 31\ L/s$
$Q_3 = 36\ L/s$
$Q_4 = 37\ L/s$

Pipe CD

$Q_1 = 5\ L/s$
$Q_2 = -8\ L/s$
$Q_3 = -12\ L/s$

FIGURE 7.29 A summary of initial flow assumptions, and subsequent corrected flows, for Example 7.14.

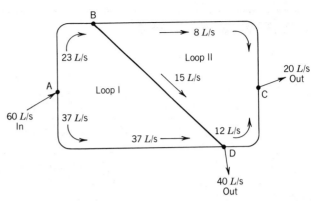

FIGURE 7.30 Balanced flow rates for Example 7.14. Note that continuity of flow applies at all pipe junctions ($Q_{IN} = Q_{OUT}$). At point A, 60 = 23 + 37; at point B, 23 = 15 + 8; at point C, 8 + 12 = 20; at point D, 15 + 37 = 40 + 12 = 52.

REVIEW QUESTIONS

1. List four basic categories of water use. What is the effect of local climate and the use of individual water meters on water demand?

2. What does *gpcd* stand for? Explain briefly.

3. Briefly discuss variations in water demand over time. Sketch a graph that would illustrate hourly variations over a 24-hour period.

4. What is the range of working pressures in a typical water distribution main?

5. What is the minimum size of a water main in a public water system?

6. Why is a gridiron pipe layout preferable to a system with many dead ends?

7. List five materials commonly used in the manufacture of water distribution pipes. Briefly discuss each type.

8. List three common types of valves used in water distribution systems. Briefly explain the use and operation of each type of valve.

9. Are water mains completely watertight?

10. What is the purpose of a thrust block?

11. Briefly discuss methods for rehabilitating a water main that has reduced flow capacity due to internal deposits on the pipe wall.

12. What is a low-lift pump? What is a high-lift pump?

13. Make a sketch showing the difference between static lift and static suction head for a pump. Also show total static head.

14. What happens when a centrifugal pump loses its prime?

15. What is the maximum practical height to which a column of water can be lifted by a pump under suction lift conditions?

16. What is the effect of increasing the impeller speed of a centrifugal pump on discharge and on pressure head?

17. What is TDH? Sketch a graph showing the relationship between TDH and the discharge of a centrifugal pump.

18.   What is meant by the term *shutoff* head?

19.   How does TDH vary with discharge in a water distribution system? Sketch a graph showing this relationship on the same paper you used for Problem 17. What does the intersection between the two curves represent?

20.   If two identical pumps operate in parallel in a water distribution system, would the total discharge be twice the discharge from one pump alone? Why?

21.   What is meant by the term *water power* of a pump? Is it greater than, the same as, or less than the power used to operate the pump?

22.   What is meant by the term *equalizing storage?* What benefits does equalizing storage provide in a water distribution system?

23.   What are the basic functions of a distribution storage tank other than to provide equalizing storage?

24.   Why are distribution storage tanks often elevated above the ground by a tower? What does it mean to say that the tank floats on the line?

25.   Is it preferable to have a single large water storage tank located near the pumping station, or to have several smaller tanks located near the major centers of water demand? Explain briefly.

26.   What is the difference between a standpipe and a reservoir?

27.   What is an equivalent pipe? What is the purpose of determining equivalent pipes in water distribution systems?

28.   What is the first step in determining an equivalent pipe for pipes in series? For pipes in parallel?

29.   What is a hydraulically balanced pipe network? What is the sum of the headlosses around a balanced pipe loop?

30.   In the Hardy Cross analysis of a pipe network, what is the only restriction on the initial assumption of flow rates in each pipe of the system?

PRACTICE PROBLEMS

1.   A community has an estimated population of 52 500 people for the year 2000. What would be the total expected daily water demand in that year? (Refer to Table 7.1.)

2. Referring to Problem 1, compute the peak hourly flow in liters per second for the day of maximum use. Assume that the peak daily demand is 200 percent of the average daily demand, and the peak hourly flow is 400 percent of the daily demand.

3. A 240-m-long section of 600-mm water main is pressure tested for 1 h, in which time a volume of 100 L of water was pumped into the pipeline to maintain the test pressure of 1000 kPa. Pipe sections are 6 m long. By how much is the allowable leakage exceeded?

4. Compute the static thrust on a 90° bend in a 305-mm pipe that carries water at a pressure of 600 kPa. If the soil has a bearing capacity of 100 kN/m^2, what area of thrust block would be needed to anchor the bend?

5. Manufacturer's data for a centrifugal pump indicate that the pump can discharge 500 gpm at a discharge pressure of 100 ft when the impeller speed is 2000 rpm. What would be the expected pump discharge and pressure head if the impeller speed is increased to 2500 rpm?

6. A centrifugal pump with the following characteristics is installed in a system to raise water from one reservoir to another. The water surface elevation in the first reservoir is 500 ft and in the second reservoir it is 650 ft above mean sea level. The pipeline connecting the reservoirs includes 2 miles of 10-in.-diameter pipe. Determine the operating point of the pump in this system.

Q, gpm	TDH, ft
0	210
200	205
400	190
600	165
800	125
1000	70

7. If two identical pumps with the characteristics tabulated in Problem 6 were operating in parallel in that system, what would be the operating point of the pumps?

8. Pump A and pump B are connected in parallel in a system that includes 10 km of 600-mm pipe. The following data describe the head–discharge relationship for each pump:

Q, L/s	TDH, m	
	PUMP A	PUMP B
0	28	34
50	27	33
100	24	31
150	20	26
200	12	19
250	—	10

The total static head in the system is 16 m. For the pumps operating in parallel, what is the operating point? Assuming the discharge line is throttled by a butterfly valve to a point that increases the system TDH to 30 m, describe the operating condition of the pumps.

9. For the pump described in Problem 6, compute the water horsepower. Assuming the pump efficiency is 75 percent, what horsepower must the motor deliver to drive the pump?

10. For Problem 8, the total power input to the pumps is 75 kW. At what efficiency are the pumps operating?

11. A town has a population of 32 000 people and an average per capita water demand of 500 L/d. Assuming that the need for equalizing storage is 20 percent of the average daily demand, and that storage for a fire flow of 60 L/s for a 4-h duration is required, compute the required volume of a distribution storage tank for the town.

12. Referring to Figure 7.25, determine the rate at which water is being withdrawn from the system at point A when the pump discharge pressure is 350 kPa and the pressure at point A is 250 kPa. The storage tank floats on the line with water at a height of 20 m, as shown. Sketch the HGL for the system. Determine at what rate the storage tank is being filled (or emptied) under these conditions.

13. Determine the theoretical diameter of a single 2000-m-long pipeline from A to D that would be equivalent to the series pipeline shown in Figure 7.31.

14. Determine the theoretical diameter of a single 1000-m-long pipeline from A to C that would be equivalent to the looped pipeline shown in Figure 7.32.

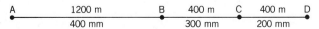

FIGURE 7.31 Illustration for Problem 13.

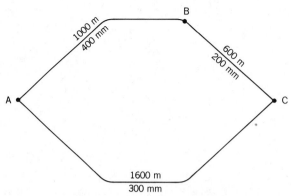

FIGURE 7.32 Illustration for Problems 14 and 16.

15. Determine the theoretical diameter of a single 1-mile-long pipeline from A to C that would be equivalent to the pipe system shown in Figure 7.33.

16. A flow of 500 L/s enters junction A of the loop shown in Figure 7.32. Determine the flow rate in each pipe if 200 L/s is withdrawn at junction B and 300 L/s is withdrawn at junction C.

17. Under the conditions stated in Problem 16, if the pressure at junction A was 750 kPa, what would be the pressure at junction C, assuming A and C were at the same elevation?

18. Consider the two-loop pipe network shown in Figure 7.28. If the flow into junction A is 100 L/s and the flows out of junctions B and C are 60 L/s and 40 L/s, respectively, compute the flows in each pipe of the network.

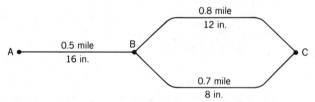

FIGURE 7.33 Illustration for Problem 15.

SELECTED REFERENCES

1. Babbitt, H. E., J. J. Doland, and J. L. Cleasby, *Water Supply Engineering,* 6th ed., McGraw–Hill, New York, 1967.

2. Hammer, M. J., *Water and Wastewater Technology,* John Wiley & Sons, New York, 1977.

3. Hardenbergh, W. A. and E. B. Rodie, *Water Supply and Waste Disposal,* International Textbook Company, Scranton, Pennsylvania, 1960.

4. *Manual of Water Utility Operations,* Texas Water Utilities Association, Lancaster, Pennsylvania, 1979.

5. Steel, E. W. and T. J. McGhee, *Water Supply and Sewerage,* 5th ed., McGraw–Hill, New York, 1979.

6. *Water Distribution Operator Training Handbook,* American Water Works Association, Denver, 1976.

SANITARY SEWER SYSTEMS

A sewage collection system consists of a network of pipes, pumping stations, channels, and appurtenances that convey wastewater to a point of treatment, storage, or disposal. The wastewater may be *sanitary sewage, industrial sewage, storm sewage,* or a mixture of the three.

Sanitary sewage, also called domestic sewage, carries human wastes and washwater from homes, public buildings, or commercial and industrial establishments. Industrial sewage is the used water from manufacturing processes, usually carrying a variety of chemical compounds. Storm sewage, or stormwater, is the surface runoff caused by rainfall; it carries organics, suspended and dissolved solids, and other substances picked up as it travels over the ground.

Pipelines that carry a mixture of these three types of liquid wastes are called *combined sewers*. Combined sewers were commonly built in older cities and towns. Most of these systems, some of which are more than 100 years old, are still in use today.

Combined sewers typically consist of large diameter pipes or tunnels. This is because the volumes of stormwater that must be carried away during wet weather periods are so large. The volume of sanitary sewage is very small when compared to stormwater flow. To keep the sanitary sewage flowing swiftly in the large diameter conduits, a combined sewer may have the shape shown in Figure 8.1a. During dry weather, the sanitary sewage flows in the smaller channel at the bottom.

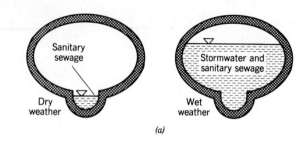

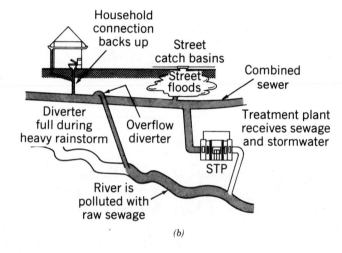

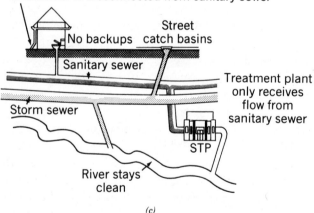

FIGURE 8.1 (a) Cross-section of a *combined sewer;* combined sewers, which were built in many older cities, carry both storm and sanitary sewage. A smaller channel at the bottom serves to carry the sanitary sewage at self-cleansing velocities during dry weather periods. During rainstorms, the storm runoff and sanitary sewage become mixed together. (b) During very heavy rainstorms, most of the *combined sanitary* and storm sewage must be diverted directly into the receiving stream because it is too much for the treatment plant to handle; this causes water pollution. (c) Separate sewers are used in new construction. (From *Waterworld News,* The American Water Works Association).

Today combined sewers, as in Figure 8.1*b,* are no longer built. Instead, separate sewer systems are constructed. Stormwater is carried in separate *storm sewers* to a point of storage or disposal; sanitary sewage and pretreated industrial wastewater is carried in separate *sanitary sewers,* usually to a municipal wastewater treatment plant. This is shown in Figure 8.1*c.*

The basic reason for building separate systems is that conventional wastewater treatment plants do not have the capacity to handle the huge volumes of stormwater that develop when it rains. Consequently, stormwater mixed with sanitary sewage must be bypassed around the treatment plant directly into receiving waters, causing water pollution. This is called a *combined sewer overflow,* or CSO. In some large cities, CSO may be directed to a large storage basin and disinfected before flowing into the receiving waters. Separate sewers are designed to prevent the problem of CSO, by conveying only sanitary sewage to the sewage treatment plant (STP).

In most sewer systems, the wastewater flows downhill, by gravity, in partially filled pipes that are not under pressure. Sometimes, sewage must be conveyed under pressure in pipelines called *force mains,* to a treatment plant or to a point where it again can flow downhill under the force of gravity. In some cases, the entire sanitary sewer system for a localized area may consist of relatively small diameter pipes under pressure; however, these *pressure sewer systems,* as they are called, are relatively uncommon.

This chapter focuses on sanitary sewer systems and includes design factors, materials, appurtenances, construction, infiltration–inflow, and rehabilitation. Stormwater drainage and management is covered in the next chapter.

8.1 SANITARY SEWER DESIGN

Sewers that collect wastewater directly from a series of homes or buildings are called *lateral sewers.* Except for the individual house connections, these are the smallest diameter sewers in the system. Laterals carry the sewage by gravity flow into larger *submains* or *collector sewers,* which in turn tie into an even larger main sewer, called a *trunk line* or *interceptor.* The interceptor carries the sewage to the treatment plant, where most of the pollutants are removed before the sewage flows into the receiving waters. It is generally located in the lowest part of the service area or drainage basin and may be built parallel to a valley floor or river bed.

MATERIALS AND APPURTENANCES

The pipes in a sanitary sewer system must be strong and durable to resist the abrasive and corrosive properties of the wastewater. They also must be able to withstand stresses caused by the soil backfill material, which is placed into the excavated trench to cover the pipe, and the effect of vehicles passing above the pipeline.

The joints between sewer pipe sections should be flexible but tight enough to prevent excessive leakage, either of sewage out of the pipe or groundwater into the pipe. In addition to the use of appropriate materials to meet all of these requirements, a variety of appurtenances are necessary for the proper operation of a sewer system. These include manholes, inverted siphons, lift stations, and flow meters. Flow meters are discussed in Section 2.4 and lift stations are discussed in Section 8.2.

Vitrified Clay Pipe (VCP)

Vitrified clay is a good sewer material because it is very resistant to corrosion or deterioration from acids and other chemicals. It also resists erosion and scour from abrasive materials carried in the flow; but it is brittle and can break easily. Careful handling during construction is important, and proper placement in the trench, called *bedding,* will provide support to resist external loads. Vitrified clay pipe is limited in size; it is available in diameters up to about 1 m (3 ft), and in lengths up to about 2 m (6 ft). Bell-and-spigot O-ring compression joints are generally used to connect the pipe sections.

Reinforced Concrete Pipe (RCP)

Precast sections of reinforced concrete pipe are suitable for larger sewer systems. RCP is available in diameters up to about 4 m (12 ft) and in lengths up to about 8 m (25 ft). It is a strong pipe material, but the concrete is susceptible to deterioration in the presence of hydrogen sulfide gas. This problem, called *crown corrosion,* is discussed in more detail later on in this section. A variety of protective linings applied on the inside pipe wall can prevent crown corrosion. Joints are generally bell-and-spigot with O-ring rubber gaskets.

Plastic Pipe

The use of plastic pipe for gravity sewer lines is increasing because of its lightness and ease of handling during construction. It is corrosion resistant, and the smooth plastic pipe wall provides good hydraulic characteristics. One type of plastic commonly used is polyvinyl chloride (PVC); PVC pipe is available in diameters up to 750 mm and lengths up to 6 m. Bell-and-spigot ends can be joined using either an O-ring gasket or a chemical weld joint.

Other Materials

Asbestos cement (AC), ductile iron, and steel sometimes are used in sanitary sewer systems. AC pipe is light and easy to handle during construction, but it is subject to deterioration when acids or hydrogen sulfide is present. Iron and steel pipes are generally used for force mains or in pumping stations, where the sewage is under pressure, for unusual external loading conditions, or when the sewer line is installed very close to or above a water main.

Manholes

Manholes are structures that provide access to the sewer pipeline for cleaning, repair, sampling, and flow measurement. They are generally circular in cross-section, with a diameter of at least 1.25 m (4 ft) so that a worker can move inside it without too much difficulty.

A typical manhole for a lateral sewer is shown

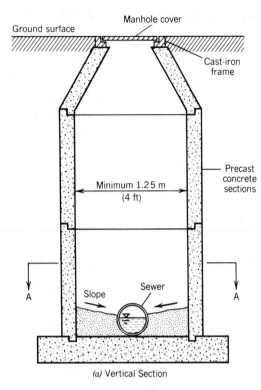

(a) Vertical Section

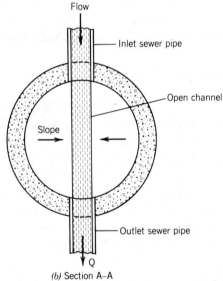

(b) Section A–A

FIGURE 8.2 Section views of a lateral sewer manhole, which provides access for pipeline inspection, cleaning, repair, and flow sampling or measurement.

in Figure 8.2. The bottom is made of concrete, with a semicircular channel connecting the inlet and outlet sewer pipes. It is common practice to use precast concrete pipe sections to build the manhole up to required grade, although brick, concrete block, or

plain concrete may also be used in its construction. A cast iron frame and cover are provided to carry traffic loads and keep out surface water.

Manholes are located over the pipe centerline under the following circumstances:

1. When there is a change in pipeline diameter.
2. When there is a change in pipeline slope.
3. When there is a change in pipeline direction.
4. At all pipe intersections.
5. At the uppermost end of each lateral.
6. At intervals not exceeding about 150 m (400 ft).

Sometimes when a lateral sewer joins a deeper submain sewer, the use of a *drop-manhole* will reduce the amount of excavation needed by allowing the lateral to maintain a shallow slope. This is illustrated in Figure 8.3. The wastewater drops into the lower sewer through the vertical pipe at the manhole. In this way, a worker is protected from the incoming sewage flow.

Manholes should be built so as to cause minimum head loss and interference with the hydraulics of the sewer line. One way to maintain a relatively smooth flow transition through the manhole, when a small sewer joins one of a larger diameter, is to match the pipe crown elevations at the manhole. This is illustrated in Figure 8.4.

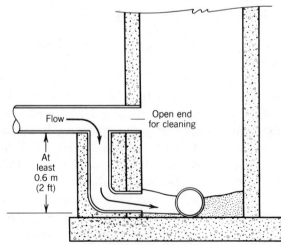

FIGURE 8.3 A typical drop-manhole structure.

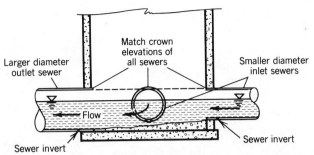

FIGURE 8.4 When intersecting pipes at a manhole are of different diameters, the pipe crown elevations may be kept the same to allow for a smooth flow transition.

Sewer Crossings

A sewer line can be built across a stream, highway cut, or other obstruction, either below ground or above ground. When built below ground, the section of sewer that goes under the stream or road is called an *inverted siphon,* or, more accurately, a *depressed sewer.* A sketch of a depressed sewer is shown in Figure 8.5.

It can be seen that this section of sewer is under the hydraulic grade line; it flows full and under pressure. To maintain flow velocities high enough to prevent solids from settling out in the pipe, two or three different sizes of parallel pipes are used to carry the minimum, average, or peak flows. Since the siphon is under pressure, ductile iron pipes or a concrete encasement is provided to prevent leakage.

In some cases, it may be more economical to cross a stream or other topographic low area with the sewer above ground. For example, a pipeline may be suspended on an existing bridge. Occasionally, a shallow depression can also be crossed by supporting the pipe on concrete piers.

SEWER LOCATION AND LAYOUT

Before the detailed design of a sewer system can begin, the area to be sewered must be surveyed to obtain topographic data. A map must be prepared so that a preliminary layout can be made. Sewers are generally located in the public right-of-way (ROW), near the center of the street to conveniently serve houses on both sides. Sometimes they are located in alleys or in easements across private property.

The designer sketches the location of the sewer lines and manholes in the streets or easements on the map, with the objective of obtaining gravity (downhill) flow as much as possible. Force mains, and the required pumping stations, are avoided because of the added expense and potential operating problems. Of course the slope of the ground governs the degree to which gravity or open channel flow conditions can be maintained. A typical sewer system layout is shown in Figure 8.6.

From the field survey data, detailed maps of each street are prepared at a relatively large scale, showing the locations of the houses to be served by the sewers as well as the locations of other utilities such as storm sewers and water and gas mains. In urban areas, the congestion of other underground utilities often complicates the design and construction of a new sewer line.

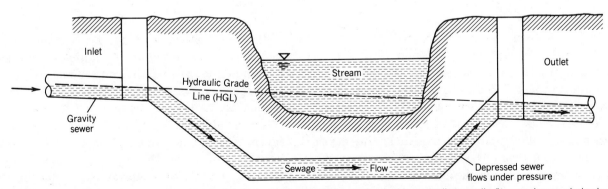

FIGURE 8.5 A *depressed sewer,* or *inverted siphon,* as it is called, is constructed to carry the gravity flow under an obstacle along the pipeline route.

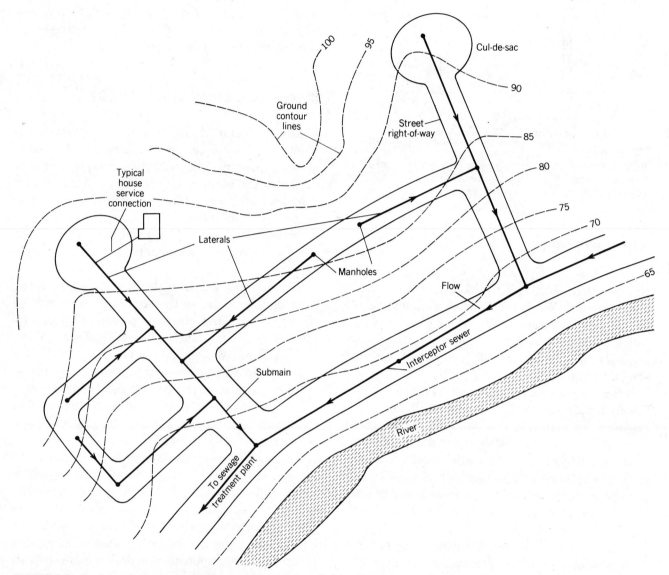

FIGURE 8.6 Sanitary sewers are located so as to provide gravity (downhill) flow as much as possible. It is best to avoid the need for pumping stations.

Profile views of the streets and utilities are generally drawn directly below the plan views. These plan and profile drawings serve as working drawings during the detailed design phase. The final design of the system is then added to the drawings, which serve as part of the construction contract documents. Data from soil borings, including depth to bedrock and groundwater, are also shown on the profiles and sometimes the boring logs or records are included with the contract documents.

The construction drawings for a sewer system must show the *location, depth, diameter,* and *slope* of the pipeline, so that the construction contractor can readily excavate the trench and place the pipe sections at the proper position and grade. A typical sewer plan and profile drawing is shown in Figure 8.7. In addition to showing the sewer diameter and slope, the drawings show manhole locations, pipe-invert elevations, and existing utilities.

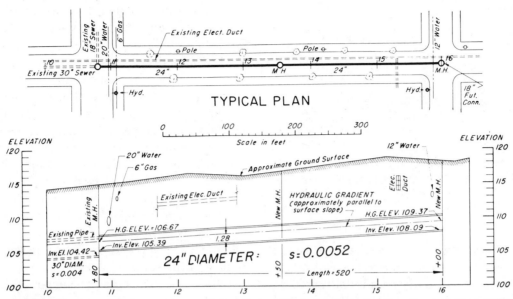

FIGURE 8.7 A typical sewer plan and profile will show the location, depth, diameter, and slope of the pipeline, as well as the location of other utility lines. (National Clay Pipe Institute)

QUANTITY OF SEWAGE

Before the diameter and slope of a sewer can be established, it is necessary to have an estimate of the quantity of wastewater it will carry. Actually, the quantity of sewage carried by a sewer is not constant with time. The flow rate varies throughout the day in a pattern similar to the variation in water demand, which was illustrated in Figure 7.1.

Sewers must be designed to carry the peak or maximum flow rates, not just the average flow rates. They also must maintain self-cleansing flow velocities, as will be discussed shortly. State environmental agencies or local health departments generally have specific requirements for sewer design flows. For example, a typical requirement for lateral and submain sewers is that they be designed to have a full-flow capacity that is 4 times the average flow rate. Main or trunk sewers must have a capacity of 2.5 times the average flow.

The flow rate peaks are less pronounced in the larger diameter sewers because it is less likely that all the sources of wastewater in a large service area will be contributing sewage into the system at the same time. In effect this smooths out the daily flow pattern; by the time the sewage reaches the treatment plant, there is considerably less variation among peak, average, and minimum flows, as there is in the laterals.

It is very important to have a good estimate of flow rates before designing the sewer system. If the quantity of sewage flow is underestimated, the selected pipe diameters and slopes may be too small to carry the peak flow rates. This is called underdesign; it can result in the *surcharging* of part of the sewer system.

A sewer system is surcharged when the wastewater is backed up into the manholes and the hydraulic grade line is above the pipe crown. This is illustrated in Figure 8.8. In severe cases of surcharging, the sewage may overflow out of the manholes and into the street, and it also can back up into basements of houses connected to the line. Surcharging may occur in a sewer system because of poor design, or because of excessive infiltration or inflow, as will be discussed later in this chapter.

Underdesign and surcharging is one problem related to erroneous sewage flow estimates. Another problem may occur if the amount of sewage is overestimated, resulting in "overdesign" of the system. Not only does construction of an overdesigned system become unnecessarily expensive, but it also

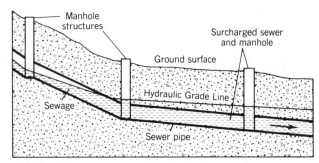

FIGURE 8.8. When the sewer pipe diameter and slope are not adequate to carry the peak wastewater flows, the system will temporarily become *surcharged.*

tends to attract extra land development sooner than expected. This premature development is usually undesirable, particularly because other municipal services and utilities may not have sufficient capacity to support it. In an EIS (see appendix), land development following overdesign of sewers is considered a "secondary" environmental effect of sewer construction.

Sewage flow rates depend on several factors, including population density, per capita water consumption, and commercial and industrial activity. The amount of groundwater seeping into the pipeline, called *infiltration,* must also be taken into consideration. Infiltration (and inflow) are discussed in more detail in Section 8.4.

Population densities are expressed as the number of people per hectare or acre. Sewer pipelines are generally designed to carry peak flows from the anticipated saturation (maximum possible) population for the local area to be served. These population estimates are usually available from county or state planning agencies.

Roughly 75 percent of the per capita water use (see Section 7.1) actually becomes sewage. This is because some water is "lost" in lawn watering, car washing, and other uses that keep it out of the sewers. But in some cases, the quantity of infiltration and inflow into the system can more than make up for these losses. For practical purposes, then, it is often assumed that the average sewage discharge is about the same as the average water use in the community. On a nationwide basis, the average sewage discharge is presently about 400 L/d per person (roughly 100 gpcd). Of course, if more accu-

rate data are available, such as from sewage flow studies of similar neighborhoods in the area, they should be used to estimate the flow quantities.

On the map or maps showing the sewer layout, the total areas that would contribute sewage to segments of the system are outlined by the designer. A segment, for design purposes, is typically a length of the sewer line between manholes, called a *sewer reach.* A reach of a sewer line is constant in slope and diameter. The boundary of the tributary area is similar in concept to the drainage divide line discussed in Section 3.4, but it does not follow the same set of rules regarding contour lines and direction of flow. This is because the flow is confined in, and must follow the direction of, the service connections and the sewer line itself.

After the tributary boundaries have been sketched for every reach of the system, the areas can be determined. For square or rectangular shapes, computation of the areas using scaled map dimensions is easily done. For irregularly shaped areas, a planimeter is used to trace the boundary and determine the enclosed area. A high degree of precision in defining the boundary is usually not warranted because of the uncertainties in the assumed population densities and the per capita flow rates.

For a given sewer reach, the design flow is taken as the product of three factors—the tributary area, the population density for that area, and the assumed peak per capita sewage flow rate. This is illustrated in Example 8.1.

EXAMPLE 8.1

A reach of a submain sewer is to be designed to receive flow from 100 ha of a community where the population density is estimated to average 25 people/ha. The average per capita sewage flow is estimated to be 400 L/d. Compute the design flow for the reach in liters per second.

Solution:

First determine the peak per capita flow rate, assuming that the submain must be designed to carry a flow rate four times the average.

peak flow = 4 × 400 L/d/person

$$= 1600 \text{ L/d per person}$$

The design flow rate for the reach is then computed as

100 ha × 25 persons/ha × 1600 L/d/person

$$= 4\,000\,000 \text{ L/d or 4ML/d}$$

and

$$4 \times 10^6 \text{ L/d} \times \frac{1 \text{ d}}{24 \text{ h}} \times \frac{1 \text{ h}}{3600 \text{ s}} \approx 46 \text{ L/s}$$

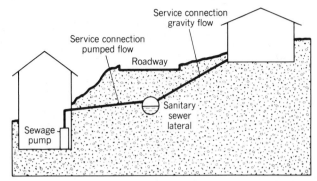

FIGURE 8.9 Sewers cannot always be placed deep enough in the ground to allow gravity flow from all service connections. Individual sewage pumps or ejectors may have to be used by some homeowners.

Design generally starts at the uppermost reach in the system and proceeds downstream, accumulating tributary areas and populations along the way. This is continued until a junction with an incoming branch sewer occurs, at which point the uppermost reach, and then the rest of the branch, is designed. The reach just downstream of a junction must, of course, be designed to accommodate the total population and sewage flows from the branches above it.

DEPTH AND VELOCITY LIMITATIONS

As far as is practical, sewer pipelines are placed deep enough in the ground to be able to receive wastewater flow by gravity from the houses and buildings to be served. But very deep basements, or buildings located below street level because of steep topography, may require individual pumping units to lift the sewage up into the public sewer. This condition is illustrated in Figure 8.9. Generally, a minimum cover of about 2 m (6 ft) above the crown of the sewer pipe is maintained, but design practice may vary in this regard. Of course greater depths require more excavation and increase the cost of construction. Generally, sewer depths do not exceed about 6 m (20 ft).

It is common practice to select the size and slope of a sanitary sewer so that the full-flow velocity would not be less than about 0.6 m/s (2 ft/s). This is called the *minimum self-cleansing velocity* because it is the velocity that will keep sewage solids suspended in the flow. Lower velocities tend to allow solids to settle out in the pipeline, blocking the flow and reducing the capacity of the sewer line.

The upper limit of flow velocity is usually about 3 m/s (10 ft/s), to prevent excessive abrasion and wear of the pipe wall from sand and grit carried in the sewage. For a given diameter or slope of sewer, Manning's Formula or nomograph can be used to check that the full-flow velocity is within the allowable range. For example, the minimum slope for a 300-mm (12-in.) pipe flowing full is 0.2 percent, in order to maintain the self-cleansing velocity.

Crown Corrosion

When flow velocities in a sanitary sewer drop below the minimum self-cleansing velocity, sludge deposits form in the pipeline. As the sludge decomposes, the dissolved oxygen in the sewage is depleted, resulting in septic or anaerobic conditions. Even the remaining suspended solids have time to undergo decomposition in the pipeline, because of the long travel time, before reaching the treatment plant.

Sanitary wastewater contains sulfate, SO_4^{2-}, compounds, which are converted to hydrogen sulfide gas, H_2S, by anaerobic bacteria. This results in the characteristic rotten-egg odor of "stale" sewage. But a serious structural problem can also develop in concrete pipelines. In gravity sewers, which have air and moisture in the pipeline above the flowing

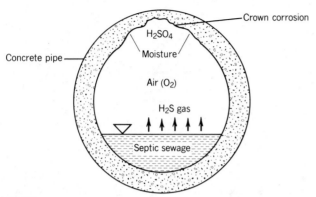

FIGURE 8.10 Crown corrosion in unlined concrete sewers can eventually lead to structural collapse of the pipeline.

wastewater, aerobic bacteria attached to the pipe crown oxidize the H_2S, converting it to sulfuric acid, H_2SO_4. This is illustrated in Figure 8.10. The sulfuric acid reacts with and weakens the concrete, causing the problem called *crown corrosion*. In extreme cases, concrete sewer lines have actually collapsed from crown corrosion, necessitating immediate and expensive repairs. Proper hydraulic design of the sewer system, to maintain self-cleansing flow velocities, can help to avoid this problem.

SIZE, SLOPE, AND INVERT ELEVATIONS

In order to minimize the amount of excavation, the slope or grade of the sewer should follow the slope of the ground as much as possible. A typical design procedure starts with an examination of the street profile to determine the average ground slope for a reach between two manholes. Manning's nomograph is used to determine the smallest standard pipe diameter that would carry the design flow for that reach at the same slope as the ground surface. For that diameter and slope, the velocity is then checked to make sure it is within the acceptable limits.

Appropriate adjustments are then made either in the pipe slope or diameter, as needed. For example, if the velocity is too low or if a very large diameter pipe is required, the designer might opt to increase the sewer slope. A steeper slope would increase the

velocity of flow and reduce the required pipe diameter. But the designer must keep in mind that steeper slopes make it necessary to place the pipeline deeper in the ground, increasing excavation costs.

Lateral sewers are generally not less than 200 mm (8 in.) in diameter, no matter how small the design flow; this minimum size reduces the occurrence of pipe blockages and facilitates maintenance activities. But the low flow rates contributed at the upper reaches of the system lead to low flow velocities in these "oversized" laterals, and they need to be routinely flushed and cleaned to keep them clear of sludge deposits and blockages.

Once the pipe diameter and slope have been established, the *invert elevations* of the pipe can be determined and the proposed sewer can be drawn on the profile. (The invert is the bottom inside wall surface of the pipe.) Usually a minimum depth of earth cover above the pipe crown is specified. The invert elevation is computed by subtracting the cover and the pipe diameter from the ground elevation at the manhole, or by matching crown elevations (see Figure 8.4). This design procedure for a sewer reach is illustrated in Example 8.2.

EXAMPLE 8.2

A 120-m reach of sewer is to be designed with a flow capacity of 100 L/s. The street elevation at the upper manhole is 90.00 m and at the lower manhole it is 87.60 m, as shown in Figure 8.11. Determine an appropriate pipe diameter and slope for this reach, and establish the pipe invert elevations at the upper and lower manholes. Assume a minimum earth cover of 2 m above the crown of the pipe.

Solution:

The ground elevation drops $90.00 - 87.60 = 2.4$ m.

The ground slope is the change in elevation divided by the horizontal distance, or $S = 2.4/120 = 0.020$. Now enter Manning's nomograph (Figure 2.21) with $S = 0.02$ and $Q = 100$ L/s or 0.1 m^3/s.

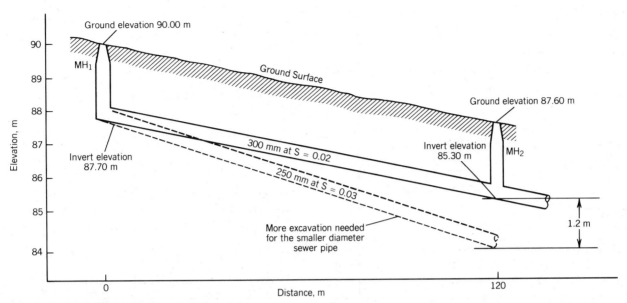

FIGURE 8.11 Illustration for Example 8.2.

The straight line connecting S and Q crosses the diameter axis at 26 cm or 260 mm.

It is necessary to choose a standard pipe size—one that is readily available from pipe manufacturers. A 250-mm pipe could be selected, but its slope would have to be steeper, about 0.03, in order for it to have a capacity of 100 L/s, according to the nomograph solution of Manning's Formula. At a slope of 0.03, the sewer reach would drop 0.03 × 120 m = 3.6 m in elevation. Starting with 2 m of cover at the upper end, it then would have 3.2 m of cover at the lower end. In other words, the extra cover is 3.6 m − 2.4 m = 1.2 m.

It may be preferable to select the larger 300-mm-diameter pipe and install it at a slope of 0.02, parallel to the ground surface. This would keep the depth of cover constant at 2 m, and involve less excavation than the 250-mm pipe would require. It is important to note that the savings in excavation would not just be for that one reach. If the 250-mm pipe is used, the extra depth of 1.2 m would be carried throughout the rest of the sewer line downstream of that reach. This can be seen in the sewer profile shown in Figure 8.11.

It also should be noted that the full-flow capacity of the 300-mm pipe placed on a 0.02 slope is more than the required 100 L/s; from Manning's nomo-

graph, it is seen that the actual full-flow capacity would be 0.135 m³/s or 135 L/s. Partial flow conditions can be evaluated using Figure 2.22. If q = 100 L/s, then the ratio q/Q = 100/135 = 0.74, and we read d/D = 0.63 from the partial flow diagram. From this we get d = 0.63 × D = 0.63 × 300 mm = 190 mm. The full-flow velocity V = 1.95 m/s, from the nomograph. For d/D = 0.63, we read v/V = 1.08 from the partial flow diagram. From this we get v = 1.08 × V = 1.08 × 1.95 m/s × 2.1 m/s.

Let us select the 300-mm-diameter pipe and a slope of 0.02 for this reach, and compute the invert elevations. The upper invert elevation is computed as follows:

upper invert elevation
    = ground elevation − cover − pipe diameter
    = 90.00 m − 2.00 m − 0.300 m
    = 87.70 m

The fall or drop in elevation of the sewer over the length of the reach is the product of the slope and the distance, as follows:

fall of sewer = 0.020 × 120 m = 2.40 m

and therefore

lower invert elevation

$$= \text{upper invert elevation} - \text{fall of sewer}$$
$$= 87.70 - 2.40$$
$$= 85.30 \text{ m}$$

This is illustrated in Figure 8.11.

EXAMPLE 8.3

Design the two sewer reaches shown in Figure 8.12a. The design flow for reach 1 is 1 mgd, and for reach 2 it is 2 mgd. The ground elevation at the first manhole, MH1, is 1100 ft; at MH2 it is 1093 ft; and at MH3 it is 1090 ft. Use a minimum cover of 8 ft.

Solution:

For reach 1, the ground slope is computed as

$$\frac{(1100 - 1093)}{300} = \frac{7}{300} = 0.023$$

To use Manning's nomograph in Figure 2.21, we must first convert the flow rate from units of million gallons per day to gallons per minute, as follows:

$$Q = 1\,000\,000 \, \frac{\text{gal}}{\text{d}} \times \frac{1 \text{ d}}{24 \text{ h}} \times \frac{1 \text{ h}}{60 \text{ min}} = 690 \, \frac{\text{gal}}{\text{min}}$$

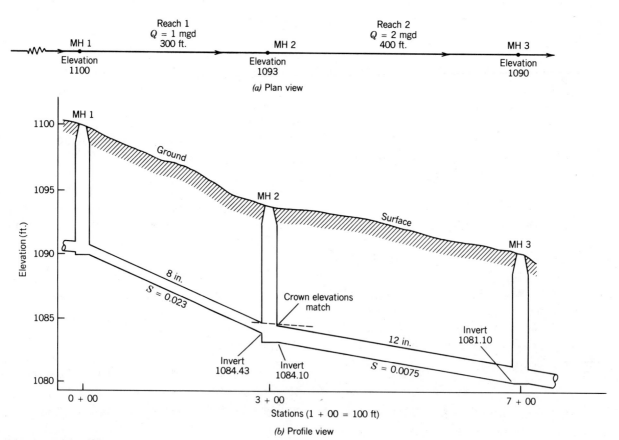

FIGURE 8.12   Illustration for Example 8.3.

Now entering the nomograph with $Q = 690$ gpm and $S = 0.023$, we select an 8-in.-diameter pipe. The actual full-flow capacity of an 8-in.-diameter pipe at that slope is about 840 gpm, and the velocity is about 5.3 ft/s. Invert elevations are computed as follows:

$$\text{upper invert elevation} = 1100 \text{ ft} - 8 \text{ ft} - \frac{8}{12} \text{ ft}$$
$$= 1091.33 \text{ ft}$$
$$\text{fall of sewer} = 0.023 \times 300 \text{ ft} = 6.90 \text{ ft}$$
$$\text{lower invert elevation} = 1091.33 - 6.90 = 1084.43$$

For reach 2, the ground slope is computed as

$$\frac{(1093 - 1090)}{400} = \frac{3}{400} = 0.0075$$

From Manning's nomograph with $S = 0.0075$ and $Q = 1380$ gpm, we would select a 12-in.-diameter pipe for this reach. The velocity would be 3.9 ft/s. To establish the upper invert elevation of the 12-in. pipe, we would match its crown elevation with that of the 8-in. pipe in reach 1, as follows:

The crown elevation of the 8-in. pipe is simply the sum of its invert elevation and the pipe diameter, or

$$1084.43 \text{ ft} + \frac{8}{12} \text{ ft} = 1085.10 \text{ ft}$$

The upper invert elevation for the 12-in. (1.00-ft) diameter pipe is therefore

$$1085.10 - 1.00 \text{ ft} = 1084.10$$

The fall of sewer is 0.0075×400 ft = 3.0 ft, and

$$\text{lower invert elevation} = 1084.10 - 3.0 = 1081.10$$

This is illustrated in Figure 8.12*b*.

8.2 SEWAGE LIFT STATIONS

Sanitary sewers are usually built to follow the general slope of the land so that the wastewater will flow downhill by gravity. In some cases, however, it becomes necessary to pump the sewage up from a low point to a higher elevation, either to reach a treatment plant or to reach another gravity sewer.

Special "nonclog" centrifugal pumps are available for pumping raw sewage. The impellers and casings of these pumps are designed to allow sewage solids, including rags and small sticks, to pass through without causing any damage or blockages. The structure in which these pumps operate is called a *lift station*.

A lift station may be used, for example, in an area where, because of flat topography, the sewer line would eventually become excessively deep, even at the minimum slope needed for self-cleansing velocity. This is illustrated in Figure 8.13.

Sewer construction costs increase with increasing trench excavation depths. When the slope of the pipeline carries it too deep, it may be more economical to install and operate a lift station in order to raise the hydraulic grade line. Another situation in which a lift station is necessary is when a new residential or industrial development is built in a topographically low area that is close to an existing municipal sewer system. A lift station would be needed to pump the sewage from the new development up into an existing submain or trunk sewer.

There are two basic types of lift stations—*dry well* installations and *wet well* installations. A wet well installation has only one chamber to receive and hold the sewage until it is pumped. One such arrangement involves the use of specially designed submersible pumps and motors, as shown in Figure 8.13. It is also possible to use a submerged pump, powered by a vertical shaft connected to a motor that is located above the wet well. In both cases, the pumps must be raised out of the wet well for maintenance.

In yet another arrangement, both the motor and the pump may be located above the wet well in a protective enclosure, but then suction lift and pro-

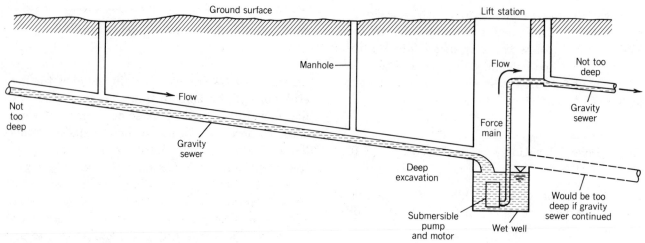

FIGURE 8.13 Lift stations are used to raise the hydraulic grade line in the sewer system.

vision for priming the pumps are necessary. A typical prefabricated "package" lift station of this type is shown in Figure 8.14. The displacement switches serve to automatically shut the pumps off and on by sensing the water level in the wet well.

A dry well installation has two separate chambers, one to receive the wastewater and another to house the pumps and controls. The protective dry chamber allows easy access to the pumps and controls for inspection and maintenance. An advantage of the dry well installation is that the pumps may be located below the sewage level. This provides suction head instead of suction lift, eliminating the need to prime the pumps. A dry well installation is sketched in Figure 8.15. The back pressure sensed by the air bubbler tube serves to indicate the sewage level and start or stop the pumps automatically.

The hourly variation in sewage flow rates must be known or estimated in order to design a lift station or to select and specify a suitable prefabricated "package" unit. Even the smallest lift station should have at least two pumps, so that one could operate while the other was removed from service for maintenance or repair. Use of these two pumps could be automatically alternated so that one would not sit idle for very long periods of time.

The volume of the wet well should be large enough so that the pumps do not cycle on and off too frequently; yet it should be small enough so that the sewage does not remain stagnant for too

long. If the wet well is too small, the pumps would run continuously or cycle on and off so frequently that they would soon be worn out. If the wet well is too large, the long detention time of sewage before

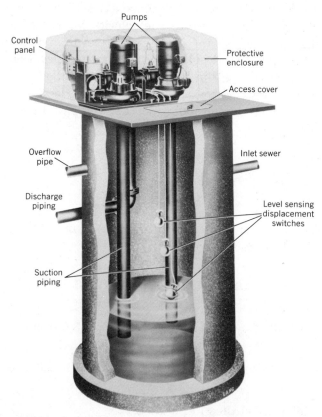

FIGURE 8.14 A wet-well mounted pump station. (Smith and Loveless, Inc.).

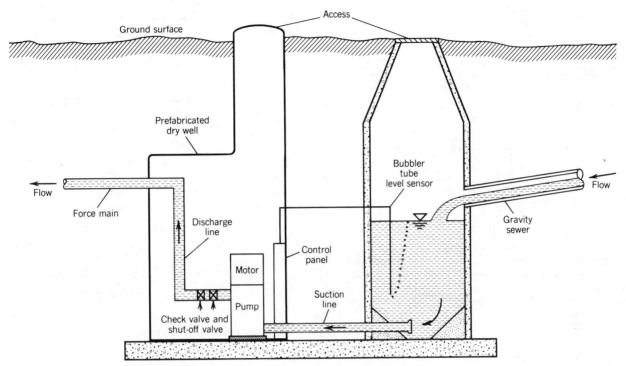

FIGURE 8.15 A dry-well type lift station.

pumping started would lead to anaerobic decomposition and septic conditions in the lift station. The designer must have reliable flow rate data to determine an appropriate wet well volume.

8.3 SEWER CONSTRUCTION

After the engineering plan drawings and specifications for a sewer system are completed, they must be approved by the local regulatory agencies before construction can begin. The most common type of sewer construction, which will be discussed here, uses open trenches and prefabricated circular pipe sections. Larger sewer systems or unusual construction situations sometimes require tunneling, jacking of pipe through the soil, or cast-in-place concrete sewers.

In order to ensure that there is quality construction and a well-built system, careful and continuous surveillance of the project by trained inspectors is required. The basic responsibility of an inspector during sewer construction is to ensure that there is compliance with the project plans and specifications. Some specific tasks of the inspector include:

1. Making sure each pipe section is uncracked and fully usable.
2. Checking for proper placement (bedding) of the pipe sections in the open trench.
3. Checking for proper joining of pipe sections.
4. Checking for proper alignment (direction and slope) of the pipeline.
5. Making sure the pipe is covered (backfilled) properly with clean fill material.
6. Determining the need for trench dewatering.

STRUCTURAL REQUIREMENTS

Sewers must be able to support the load caused by the backfill soil above them, called the *dead load,* and the force due to vehicular traffic, called *live load.* This is illustrated in Figure 8.16. The depth

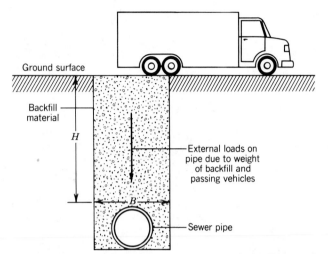

FIGURE 8.16 A buried pipeline must be able to resist external forces without excessive deflection or cracking. The load on the pipe due to backfill depends on the type of soil, the depth of cover over the pipe crown *(H)*, and on the width of the trench *(B)*.

TABLE 8.1 SELECTED MINIMUM VCP CRUSHING STRENGTHS

Nominal size	Standard Strength		Extra Strength	
mm (in.)	kN/m	lb/ft	kN/m	lb/ft
200 (8)	20.4	1400	32.0	2200
250 (10)	23.2	1600	35.0	2400
300 (12)	26.3	1800	37.9	2600
380 (15)	29.2	2000	42.3	2900
460 (18)	32.0	2200	48.1	3300

$$\text{load factor} = \frac{\text{field supporting strength}}{\text{crushing strength}} \quad (8\text{-}1)$$

Another way of expressing this is:

field supporting strength
$$= \text{load factor} \times \text{crushing strength}$$

of cover, width of trench, and type of backfill material are the key factors that affect the dead load. The key factors that affect the load carrying capacity of the pipe are the *pipe crushing strength* and the type or class of *pipe bedding.*

The crushing strength of a pipe is determined by a standard laboratory procedure, and it is specified in terms of load or force per unit length. The procedure is called a "three-edge bearing test," the load is applied to the test sections only along three "edges" of the pipe barrel: one on the top and two on the bottom. Minimum required crushing strengths for various pipe materials and sizes are published by the American Society for Testing Materials (ASTM), and pipe manufacturers must meet these requirements. For example, typical values for the crushing strength of vitrified clay pipe (VCP) are presented in Table 8.1.

Bedding refers to the way in which the pipe is placed on the bottom of the trench. Proper bedding always increases the actual supporting strength of the installed pipe above the reported crushing strength value, by distributing the load over the pipe circumference. The ratio of the actual field supporting strength to the crushing strength is called the *load factor.*

Four types or classes of bedding are illustrated in Figure 8.17. Class D bedding is the weakest and least desirable type, and is not recommended for sewer construction in most circumstances. The bottom of the trench is left flat and the barrel of the pipe is not fully supported because of the protruding bell-ends; backfill is placed loosely over the pipe without proper compaction.

Class C, called *ordinary bedding,* has compacted granular material placed under the pipe and extending partially up the pipe barrel. This provides good support, with a load factor of 1.5. In other words, the field supporting strength is one-and-a-half times greater than the crushing strength.

Class B, or *first-class bedding,* has the compacted granular material extending halfway up the pipe barrel, and the backfill is carefully compacted over the top of the pipe; the load factor is 1.9. In Class A bedding, the pipe barrel is cradled in concrete and the backfill is carefully compacted, providing a load factor of 2.8.

In addition to the load factor provided by the pipe bedding, a *safety factor* is applied in the com-

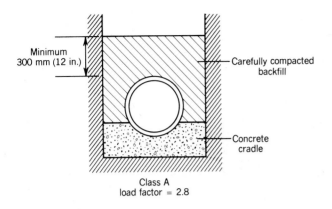

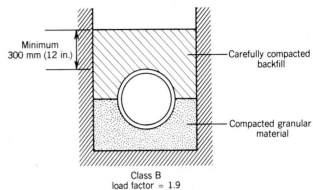

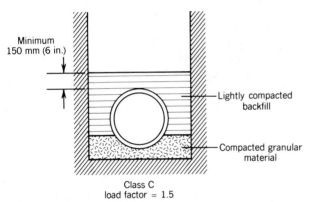

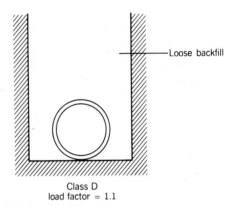

FIGURE 8.17 Different types of pipe bedding conditions affect the safe supporting strength of the pipe.

putations to arrive at the *safe supporting strength* of the pipeline, as follows:

safe supporting strength

$$= \frac{\text{field supporting strength}}{\text{safety factor}} \quad (8\text{-}2)$$

and, substituting from Equation 8-1, we get

safe supporting strength

$$= \frac{\text{load factor} \times \text{crushing strength}}{\text{safety factor}} \quad (8\text{-}3)$$

A safety factor of 1.5 is commonly used for clay or unreinforced concrete sewers, to compensate for the possibility of use of poor quality materials or for faulty construction.

Marston's Formula

In order to select an appropriate bedding condition for a pipeline, the total live load and dead load on the pipe must first be determined. The class of bedding is selected so that the safe supporting strength is equal to or greater than the computed total load on the pipe.

For pipes in shallow trenches, such as storm sewers and some water mains, the vehicular traffic load may be a significant part of the total load; tables are available to help the designer estimate these live loads. But for pipes in relatively deep trenches, like sanitary sewers, the live traffic loads are often insignificant when compared to the dead load due to backfill. For our purposes, we will consider only the dead load on the pipe.

An equation commonly used to estimate the dead

load due to backfill is known as Marston's Formula, and is expressed as follows:

$$W = CwB^2 \qquad (8\text{-}4)$$

where W = dead load due to backfill, kN/m (lb/ft)

C = a dimensionless coefficient

w = unit weight of the backfill soil, kN/m^3 (lb/ft^3)

B = trench width at the pipe crown, m (ft)

The value of the coefficient C depends upon the depth of cover, the trench width, and the type of backfill material. A chart for determining values of C is presented in Figure 8.18. The horizontal axis represents the ratio of cover H to trench width B.

Typical values of unit weights for a few selected soil types are presented in Table 8.2. The following simplified examples will illustrate the use of Mar-

ston's Formula and the analysis of pipe bedding conditions.

EXAMPLE 8.4

A 300-mm-diameter pipe is placed in a 3-m-deep rectangular trench that is 0.60 m wide. The trench is backfilled with clay that has a unit weight of 18.8 kN/m^3. Compute the dead load due to backfill that the pipe must support.

Solution:

The cover H is equal to the total trench depth minus the pipe diameter, or $H = 3$ m $- 0.3$ m $= 2.7$ m.

The ratio of cover to width is $H/B = 2.7/0.6 = 4.5$. From Figure 8.18, we read $C = 2.6$ for clay soil. Now applying Marston's Formula (Equation 8-4), we get

$$W = CwB^2 = 2.6 \times 18.8 \times 0.60^2 \approx 18 \text{ kN/m}$$

EXAMPLE 8.5

If the pipe in the previous example were standard strength VCP, what class of bedding should be

TABLE 8.2   TYPICAL VALUES OF SOIL UNIT WEIGHTS

Soil Type	kN/m^3	lb/ft^3
Sand and gravel	17.2	110
Clay	18.8	120
Saturated clay	20.4	130

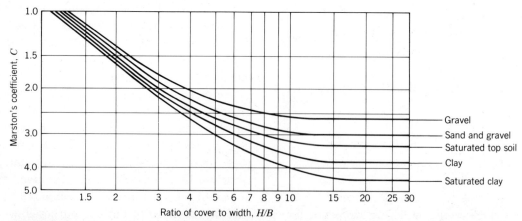

FIGURE 8.18 Values of the C coefficient used in Marston's Formula. (Adapted from *Design and Construction of Sanitary and Storm Sewers,* with permission of the American Society of Civil Engineers, Water Pollution Control Federation)

specified for construction, using a safety factor of 1.5?

Solution:

From Table 8.1 we see that the crushing strength of a standard strength 300-mm VCP is 26.3 kN/m. From the solution to Example 8.4, we know that the safe supporting strength of the pipeline must be equal to (or greater than) the dead load of 18 kN/m. Applying Equation 8-3, we get

safe supporting strength
$$= \frac{\text{load factor} \times \text{crushing strength}}{\text{safety factor}}$$
$$18 \text{ kN/m} = \frac{\text{load factor} \times 26.3 \text{ kN/m}}{1.5}$$

and from this

$$\text{load factor} = \frac{1.5 \times 18}{26.3} = 1.0$$

According to these computations, Class D bedding, with a load factor of 1.1, would be adequate. But use of Class D bedding is generally not considered good construction practice, and preferably Class C bedding would be specified instead.

EXAMPLE 8.6

An 8-in.-diameter VCP sewer is to be placed in a 2-ft-wide trench with 11 ft of cover. Backfill is saturated clay. Determine the required bedding condition for standard strength as well as for extra strength pipe, using a safety factor of 1.5.

Solution:

First compute the cover to width ratio and determine C.

$$\frac{H}{B} = \frac{11}{2} = 5.5$$

and from Figure 8.18, $C = 3.2$. From Table 8.2 we read $w = 130$ lb/ft^3.

Now applying Marston's Formula, we get $W = 3.2 \times 130 \times 2.0^2$, and the dead load $W = 1700$ lb/ft. (rounded to 2 significant figures). The safe supporting strength must be equal to or greater than this load.

For standard strength 8-in. VCP, the crushing strength is equal to 1400 lb/ft. Applying Equation 8-3 we get

$$1700 = \frac{\text{load factor} \times 1400}{1.5}$$

and

$$\text{load factor} = \frac{1.5 \times 1700}{1400} = 1.8$$

Referring to Figure 8.17, we see that Class B bedding, which has a load factor of 1.9, is needed for the standard strength VCP in this problem; Class C does not have a high enough load factor. For extra strength VCP, the crushing strength is 2200 lb/ft, and recomputing the required load factor, we get

$$\text{load factor} = \frac{1.5 \times 1700}{2200} = 1.2$$

Therefore, if extra strength pipe rather than standard strength pipe is used, Class C bedding with a load factor of 1.5 would be sufficient. The additional cost for the stronger pipe may be justified because it could offset the time, material, and money that would be spent for Class B bedding.

FIELD LAYOUT AND INSTALLATION

A basic consideration in sewer construction is the accurate field layout of the *line and grade* of the pipeline, as shown in the plan drawings. The line, or horizontal alignment, defines the location and direction of the pipeline within the right-of-way. The grade, or pipe slope, must also be accurately established to provide the required hydraulic capacity of the system.

In the construction survey, the location of the sewer trench is usually established and laid out as an *offset line,* which runs parallel to the proposed sewer centerline. The offset line is marked by wooden stakes driven into the ground at uniform intervals, usually about 15 m (50 ft) long. The offset line is far enough away from the pipe centerline so that it will not be disturbed during construction, yet it is close enough so that transfer of measurements to the excavated trench can readily be done by the builder. The offset stakes may be set so that their tops are a specific height above the required trench bottom. They can then be used periodically to check the depth of the trench during excavation.

There are two methods used to set the pipe sections properly in the open trench—*batter boards* and *lasers.* The batter board method is the older of the two, but the application of low-power laser instruments in surveying and construction activities is becoming more popular. Placing the pipe using lasers is a more accurate, quicker, and less labor intensive method than using the traditional batter board method. Nevertheless, batter boards are still used in sewer construction.

Batter Boards

After the trench has been excavated, batter boards are set across the trench at uniform intervals, as shown in Figure 8.19. The tops of the boards are usually set at some even height above the required sewer invert elevation. The sewer centerline is marked on the boards by extending a line of sight with a transit or theodolite; a string is stretched from board to board along this line.

The centerline is transferred down into the trench with a plumb-bob line held against the top centerline string. Invert elevations are set and checked using a vertical rod marked off in even increments. The lower end of the rod is placed on the pipe invert and the batter board string is checked to see if it matches with the proper elevation mark on the rod. If it does not match with the required

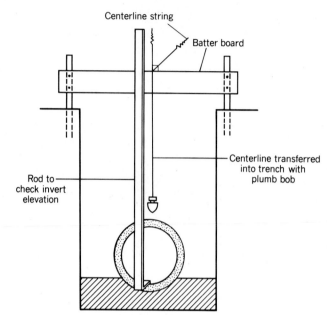

FIGURE 8.19 Batter board method of construction for a sewer line.

mark, an appropriate adjustment of the pipe invert elevation is made by a worker in the trench.

Lasers

A laser is an instrument that can project an intense but narrow beam of light for a long distance. This pencil-thin light beam can be aimed through a pipe and be seen on a target placed in the other end of the pipe, as shown in Figure 8.20.

The laser can be securely mounted in the manhole, and the slope of the light beam can be accurately set to match the required slope of the pipe. A transit mounted above the manhole is used to establish pipe alignment from field reference points, and to transfer the alignment down to the laser instrument. Lasers can maintain accuracies in the pipe slope of 0.01 percent over a distance up to 300 m (1000 ft). In other words, the invert elevations can be set accurately to within 30 mm in a 1000-m length of pipeline (0.1 ft in a 1000-ft-long pipeline).

Excavation

The most common type of equipment used for digging a sewer trench is the backhoe, although this

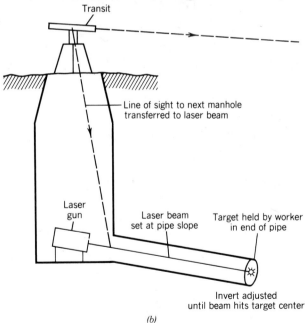

FIGURE 8.20 (a) In modern sewer construction, laser beams are used to establish the specified slope of the pipe (Spectra–Physics, Construction and Agricultural Division, Dayton, Ohio) (b) The laser beam is set in the proper horizontal direction with a transit.

may vary depending on the depth and type of material (soil or rock) being excavated. The trench width should be kept as narrow as possible while allowing enough room for a worker. Keeping the trench narrow not only reduces excavation costs, but it also reduces the backfill load on the pipe. Generally, a minimum working room allowance of at least 300 mm (1 ft) from each side of the pipe is required.

For safety, the sides of trenches more than 1.5 m (5 ft) deep should be supported with sheeting and bracing to prevent collapse and to protect workers. Sheeting is wood planking or other material that is in contact with the trench sides; the braces extend across the trench from one side to the other.

If the trench is flooded with groundwater that seeps in through the bottom and sides, dewatering with a pump or well-point system may be necessary. The effect of a well-point system on lowering the groundwater table in the vicinity of the trench is illustrated in Figure 8.21.

Placement and Backfill

The importance of proper bedding design was discussed in the previous section. Improper bedding can significantly reduce the supporting strength of the pipe, causing pipe deflection, cracking, and excessive infiltration. Of prime concern in bedding is that the entire length of pipe barrel be uniformly supported by the soil or gravel bed under the pipe. The pipe sections should be handled carefully and placed with minimum disturbance of the supporting material on the trench bottom. Care must be taken in joining pipe sections so that excessive infiltration through the joints will not be a problem.

Backfilling of the sewer trench should be done immediately after the pipe has been placed at the proper grade. The backfill must not contain boulders or large cobbles, frozen material, tree stumps, or other debris. It should be placed and compacted in uniform layers about 150 mm (6 in.) deep, to a height of about 300 mm (1 ft) above the top of the pipe; after that, backfilling can usually proceed more rapidly.

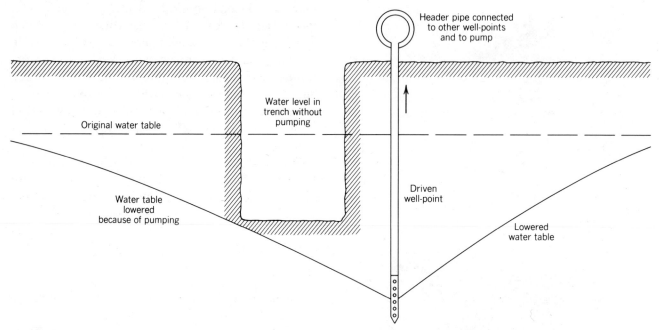

FIGURE 8.21 A series of well-points attached to a main header pipe and a pump can be used to dewater a trench that contains groundwater.

8.4   INFILTRATION AND INFLOW

When a sewer line lies below the groundwater table, the groundwater will seep into the sewer through poorly constructed pipe joints, cracked pipe sections, and leaky manhole structures. This flow of groundwater into the system is called *infiltration*.

Control of infiltration depends primarily on the quality of construction of the pipeline. But even in well-constructed sewer systems, there will be some infiltration; building a completely watertight gravity flow pipeline is not feasible nor economical. Generally, up to 45 L/d per mm of pipe diameter per km of pipe length is allowed in the specifications for a sewer construction project. This is roughly equivalent to 500 gal/d per inch of pipe diameter per mile of pipe length.

EXAMPLE 8.7

What is the total allowable rate of infiltration in a 1500-m-long, 200-mm-diameter lateral sewer if 45

L/d/mm/km is acceptable? How does it compare to the minimum pipe capacity?

Solution:

The allowable infiltration is computed as follows:

$$45 \text{ L/d} \times 200 \text{ mm} \times 1.5 \text{ km} = 13\,500 \text{ L/d}$$

The minimum capacity of a 200-mm pipe is the discharge that would have a self-cleansing velocity of 0.6 m/s. Using Manning's nomograph in Figure 2.21, we read a discharge of 0.019 m³/s or 19 L/s. Converting to L/d, we get

$$19 \text{ L/s} \times 3600 \text{ s/h} \times 24 \text{ h/d} \approx 1\,600\,000 \text{ L/d}$$

The ratio of infiltration to pipe capacity is therefore

$$\frac{13\,500}{1\,600\,000} = 0.008 \text{ or } 0.8 \text{ percent}$$

This amount of infiltration is negligible compared to the pipe capacity.

Surface water that enters a sewer system through poorly sealed manhole covers or that comes from intentional connections to roof drains and basement sump pumps on private property is called *inflow*. Most communities have sewer-use ordinances that prohibit roof or cellar drain connections to the sanitary sewer, in an attempt to control the inflow problem.

It is very important to prevent excessive infiltration or inflow in a sanitary sewer for two basic reasons:

1. Too much infiltration and inflow can cause surcharging of the sewer system during wet weather periods.
2. Sewage treatment plants are not designed to handle the extra volume of water from infiltration and inflow. During wet weather, the rapid increase of flow through the treatment plant causes hydraulic overloading and failure of the treatment process.

Newly constructed sewer lines must be tested to ensure compliance with infiltration specifications before they are put into service. Older systems may also be tested to determine the extent of leakage. The objective is to ensure that tax dollars will not be spent on expensive new treatment plants, or treatment plant additions, that would have operating problems due to hydraulic overloads. This process of examining the integrity of older systems is called an *I/I survey*.

There are several methods of infiltration and inflow testing, including *direct measurement, exfiltration testing, smoke testing,* and *low pressure air testing*.

Direct measurement for infiltration in a new sewer is done when the pipeline lies below the groundwater table at the time of the test, and before service connections are made. A V-notch weir can be placed in the end of a known length of the sewer line to gage the flow; any flow in the line must be infiltration. Direct measurement can also be done as part of an I/I survey of an existing sewer system. But in existing systems, the flow measurements must be made at about 3 o'clock in the morning, when little or no sanitary flow would be expected.

If a new sewer must be tested for watertightness when it lies above the water table, as is often the case, an exfiltration test is conducted in lieu of direct measurement. This is, in effect, just the opposite of measuring infiltration; it measures the flow going out of the pipeline instead of the flow coming in. After plugging the pipes as shown in Figure 8.22, the reach of sewer being tested is filled with water. The reach is kept flooded for a few hours so that entrapped air can be removed and the pipeline material becomes saturated with water.

If water is leaking out of the sewer line, the water level in the manhole will be observed to drop in

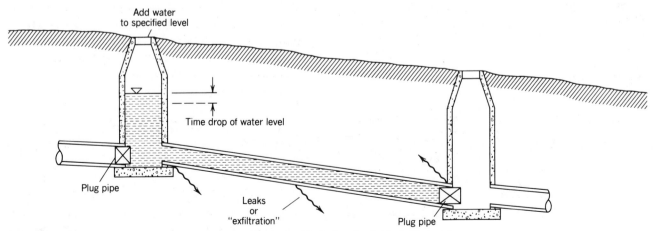

FIGURE 8.22 An *exfiltration test* is one of the methods used to judge the watertightness of a new sanitary sewer.

elevation. By measuring the amount of drop in the water surface over a known interval of time, the rate of exfiltration can be computed. Obviously, if the water can leak out of the pipe, it is an indication of a potential infiltration problem during wet weather and high groundwater table conditions.

EXAMPLE 8.8

An exfiltration test is conducted on a 400-ft-long, 12-in.-diameter sewer reach. The water level in a 4-ft-diameter manhole is observed to drop 2 in. in 1 h. Compute the rate of exfiltration in terms of gallons per day per inch per mile of pipe.

Solution:

The volume of water that leaked out of the system is equal to the product of the manhole cross-sectional area and the drop in water elevation. This is computed as follows:

$$\text{area} = \pi D^2/4 = \pi \times 4^2/4 = 12.6 \text{ ft}^2$$
$$\text{drop} = 2 \text{ in.} \times 1 \text{ ft}/12 \text{ in.} = 0.167 \text{ ft}$$
$$\text{volume} = 12.6 \text{ ft}^2 \times 0.167 \text{ ft} = 2.1 \text{ ft}^3$$

Converting the volume to gallons, we get

$$\text{volume} = 2.1 \text{ ft}^3 \times 7.5 \text{ gal/ft}^3 \approx 16 \text{ gal}$$
$$\text{sewer length} = 400 \text{ ft} \times 1 \text{ mile}/5280 \text{ ft}$$
$$= 0.076 \text{ mile}$$

Since the time interval for the drop was 1 h, we get

$$\text{leakage rate} = 16 \text{ gal/h} \times 24 \text{ h/d} = 384 \text{ gal/d}$$

and

$$\text{exfiltration} = 384 \text{ gal/d} \div 12 \text{ in.} \div 0.076 \text{ mile}$$
$$\approx 420 \text{ gal/d/in./mile}$$

The tightness of construction of a sewer system can also be evaluated by using a low-pressure air testing procedure; the pressure used is about 28 kPa or 4 psi. Specifications for allowable air loss rates, in terms of pressure drop, are used in lieu of allowable infiltration or exfiltration rates. In the air pressure test, the sewer pipes are tightly plugged with special "air-loc" balls or other devices that allow the introduction of pressurized air into the sewer reach. The time it takes for the air pressure to drop 7 kPa (1 psi) is determined. This time is compared to a specified allowable time, which is a function of the pipe length and diameter. If the pressure drops in less than this specified allowable time, the sewer line has failed the test and repairs must be made.

Smoke testing can be used effectively to locate sewer leaks or illegal sources of inflow. In the smoke test, a sewer reach is isolated (but flow is not completely blocked), and smoke from a nontoxic smoke bomb is forced through by a blower. The smoke will appear above ground wherever there is a hole or crack in the pipe. It also will appear rising from the roof drain on a home where there is an illegal connection to the sewer line.

8.5 SEWER REHABILITATION

Sewer systems are part of urban *infrastructure,* the physical facilities and public works that allow our communities to function efficiently. Unlike transportation infrastructure, particularly highways and bridges which are quite visible to the public, sewers receive almost no attention once they are put in the ground in many communities. The old saying, "Out of sight, out of mind," seems to apply.

There is often little thought given to the need for maintenance of these wastewater collection systems, and only a bare minimum of funds is allocated for this purpose. As a result, many sewer systems in the United States are in a state of disrepair. They are leaky and carry large volumes of infiltration water; they frequently become blocked and surcharge as a result; and occasionally they even collapse.

The cost of excavating and replacing a section of a poorly functioning sewer system is very high. It is usually more economical to apply one of several available methods to repair and rehabilitate the system internally, without having to excavate. It is even more economical for municipalities and sewer agencies to have continuing sewer maintenance programs designed to prevent unnecessary deterioration of the system in the first place.

PIPELINE REPAIR

There are several methods of restoring flow capacity, or sealing of leaks in sewer lines, that do not require excavation. To a certain extent, the structural integrity of a pipe weakened by crown corrosion can also be strengthened, as long as a complete collapse has not already occurred. The particular method selected depends on the nature of the problem, the need to maintain flow during the repair, possible traffic disruption, safety, and cost.

Cleaning and Inspection

Sewer lines must be cleaned before a visual inspection and the selected rehabilitation efforts can be made. Flushing the line, using a fire hose attached to a hydrant and discharging the flow into a manhole, is a common cleaning method. But this must be done with caution to prevent backups into nearby houses connected to the system.

Another cleaning method, for sewers without serious blockages, makes use of a soft rubber ball that is inflated to match the diameter of the pipe and then pulled by rope through a reach between manholes. Sometimes, power rodding machines can be used for mechanical cleaning of a line with blockages. In sewers with accumulations of grit because of relatively low flow velocities, power winches can be used to pull a bucket through the line to scrape up the deposits. Whichever method of cleaning is used, collected sediment and debris should be removed and disposed of properly.

After cleaning, inspections can be made with a

flashlight during low flow periods. A much more thorough inspection can be accomplished, however, using closed-circuit television systems. TV inspection allows accurate location of leaks, root intrusion, and structural problems. The camera can be pulled through a sewer reach on a special mounting, and a photographic or videotape record of the inspection can be made.

Grouting

Sealing leaks with chemical grout in structurally sound sewer lines is a common method for rehabilitation. The grout can be applied internally to joints, holes, or cracks using special tools and techniques, without the need for excavation. The gel type or foam type grouts that are used solidify after being forced into the joints or cracks under pressure. In small or medium sewer lines, a "sealing packer" with inflatable rubber sleeves can be pulled through a reach of the system, as illustrated in Figure 8.23. Closed-circuit TV is used to position the packer over the joint or crack to be repaired.

In large-diameter sewer lines, workers must enter the line to place a sealing ring manually over the defective joint, as illustrated in Figure 8.24. The grout is pumped through a hand-held probe. The air in the sewer must be tested for carbon monoxide, hydrogen sulfide, and explosive gases before the workers enter, and appropriate safety precautions should be taken.

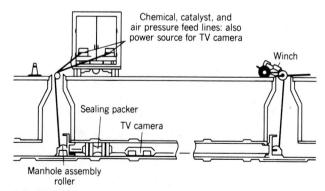

FIGURE 8.23 In small sewers, a sealing packer can be pulled through the line for repair. (From *Existing Sewer Evaluation and Rehabilitation*, with permission of American Society of Civil Engineers, Water Pollution Control Federation, 1983.)

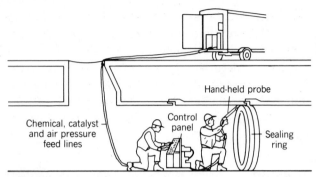

FIGURE 8.24 In large sewers, workers must enter the line for repairs. (From *Existing Sewer Evaluation and Rehabilitation,* with permission of American Society of Civil Engineers, Water Pollution Control Federation, 1983.)

Linings

Large sewers with structural damage caused by crown corrosion can be reinforced by applying an internal lining of *gunite,* which is a mixture of fine sand, cement, and water. Gunite is applied pneumatically by spraying; it adheres to the vertical and overhead surfaces of the pipe. Long lengths of concrete sewers can be effectively renewed with a gunite lining.

A method called *sliplining* can be used to rehabilitate extensively cracked pipelines. This involves pulling a flexible plastic liner pipe into the existing pipe and then reconnecting individual service connections to the liner. This is illustrated in Figure 8.25. Sometimes the narrow annular space between the liner and the old pipe is filled with grout to prevent movement. Multiple excavations are required to reconnect each service line to the new liner.

A relatively new method is now available, though, called *inversion lining.* Inversion lining avoids the need for service line excavation. A flexible liner that expands to fit the pipe geometry is thermally hardened and cured. A special cutting device can be used with a closed-circuit TV camera to locate and reopen the service connections. The liner is installed through a tube placed in a manhole, using water or air pressure to push it through the pipe. When the liner is in place, the water or air is heated to begin the curing and hardening process.

MANHOLES AND SERVICE CONNECTIONS

Manholes often need repair to eliminate surface water inflow or groundwater infiltration. Inflow can enter the manhole through holes in the manhole cover, through spaces between the cover and the frame, and under the frame itself if it is poorly sealed. Self-sealing frames and covers are available, but the seals are often damaged by heavy traffic, road work, or snow plowing. Frames may be resealed in place by applying hydraulic cement and a waterproof epoxy coating.

Sometimes an effective way to reduce inflow is to raise the frame and cover by adding a manhole adjusting or extension ring and by coating the exposed portion with cement or asphalt. Another method is to install a special insert between the frame and the cover. The insert, illustrated in Figure 8.26, prevents water and grit from entering the manhole, but allows gas to escape through relief vents.

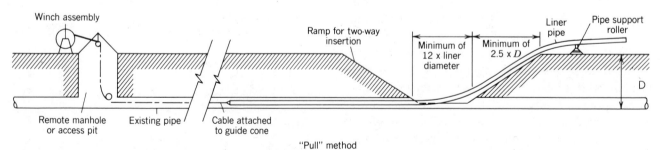

"Pull" method

FIGURE 8.25 Sliplining is one method used to repair cracked sewers. (From *Existing Sewer Evaluation and Rehabilitation,* with permission of American Society of Civil Engineers, Water Pollution Control Federation, 1983.)

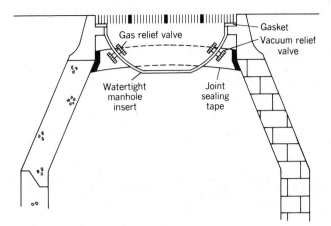

FIGURE 8.26 An insert device installed under the manhole cover prevents surface water from entering the system. (PRECO)

Concrete manholes are also subject to sulfuric acid corrosion and therefore sometimes need structural rehabilitation. Severe structural deterioration is usually solved by excavating and replacing the manhole and applying corrective measures to eliminate the cause of deterioration. Structural repair of less severely damaged manholes first involves removal of the deteriorated materials, using water or sand blasting, or mechanical tools. The remaining material is then stabilized using special chemical preparations; high strength patching mortar is used to fill in surface irregularities; and a lining or coating is then applied.

One of the most common manhole maintenance problems is infiltration of groundwater through the sidewall and base and around pipe entrances.

Chemical grouting can be used effectively to solve this problem. It is less costly than lining or coating methods and it does not require prior surface restoration. The cracks and openings are sealed by pressure injection of gel or foam grouting materials.

Service or house connections are small-diameter pipelines, 100 mm (4 in.) that connect the lateral sewer line in the street to the buildings it serves. Also called building sewers or service laterals they can be as long as 30 m (100 ft). The section of the service connection between the sewer line and the property line is installed and maintained by the local public works or sewer department; the section between the property line and the building's drainage system is installed under local plumbing or building code regulations and must be maintained by the property owner.

Defects in service connections, including cracked or open-jointed pipes, can be the cause of a significant portion of the infiltration problem in a sewerage system. This is because the total length of service connections in a community is often greater than the length of sewer mains. In view of this often neglected fact, repair of service connections is an important aspect of a sewer system rehabilitation effort. There are several methods available for applying chemical grout as a repair material, and the inversion lining method can be used. Efforts also must be made to eliminate illegal hookups of roof drains and basement sump pumps to service connections on private property, to reduce the inflow problem.

REVIEW QUESTIONS

1. What is the difference between a *combined* sewer system and a *separate* sewer system? Which is preferable? Why?

2. What is the difference between a *lateral* and a *submain* sewer? What is an interceptor sewer?

3. List five different materials used in sewer-line construction. Briefly describe the characteristics of three of these.

4. List four purposes of a sewer manhole.

5. List six factors that determine where a manhole is located in a sewer system.

6. What is a *drop-manhole*? What is an *inverted siphon*?

7. Describe the general procedure for sanitary sewer design.

8. What key information must be shown on the plan drawings for sanitary sewer construction?

9. Why are sewer pipelines designed to accommodate peak hourly flows? Why is the peak flow factor in a trunk sewer less than that in a lateral?

10. What happens when a sewer system is *surcharged*? Under what circumstances might this occur?

11. Is it good practice to overdesign a sewer pipeline? Why?

12. Describe in general terms how the design flow for a sewer reach is determined.

13. What are typical limitations on the pipeline depth and wastewater flow velocity for a sanitary sewer system?

14. Describe the problem called *crown corrosion* of sewers.

15. Describe two types of sewage lift stations. Under what circumstances might a lift station be needed?

16. List six important factors in sewer construction inspection.

17. What are three key factors that affect the dead load on a buried pipeline?

18. List two key factors that affect the external load carrying capacity of a pipeline.

19. Describe two different methods for setting line and grade of a sewer pipeline.

20. What is the difference between *infiltration* and *inflow* in a sanitary sewer system?

21. Why is it important to limit the amount of infiltration and inflow into a sewer line? Briefly indicate how each may be controlled.

22. Describe three methods for infiltration testing.

23. Briefly discuss methods used to rehabilitate sanitary sewer systems.

PRACTICE PROBLEMS

1. A collector sewer is to be designed to receive flow from 250 ac of a community where the population density is estimated to be 12 persons/ac. The average per capita sewage flow is taken to be 100 gpcd. What is the required design flow for the collector, in gallons per minute?

2. A trunk sewer is to be designed to receive flow from a 1-km² area of a community where the population density is 50 persons/ha. The average per capita sewage flow is taken to be 400 L/d. What is the design flow for the trunk sewer, in liters per second?

3. A 100-m reach of sewer is to have a minimum capacity of 200 L/s. The street elevation at the upper manhole is 305.55 m and at the lower manhole it is 303.05 m. Determine an appropriate pipe diameter and slope for this reach, and establish the pipe invert elevations at the upper and lower manholes. Assume that a minimum earth cover of 2 m above the pipe crown is required.

4. Design the two sewer reaches shown in Figure 8.27. The design flow for reach 1 is 40 L/s and for reach 2 it is 80 L/s. The ground elevation at MH1 is

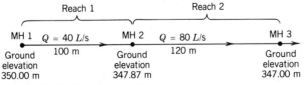

FIGURE 8.27 Illustration for Problem 4.

350.00 m; at MH2 it is 347.87 m; and at MH3 it is 347.00 m. Use a minimum cover of 2.5 m. Sketch a profile of the street and sewer, using a horizontal scale of 1:1000 and a vertical scale of 1:100.

5. A 12-in.-diameter pipe is placed in a 10-ft-deep, 2.5-ft-wide trench that is backfilled with saturated clay. Compute the dead load on the pipe.

6. A 200-mm-diameter pipe is placed in a 3-m-deep, 0.9-m-wide trench and backfilled with sand. Using a safety factor of 1.5, select an appropriate class of bedding for the pipe if it is standard strength VCP.

7. What would be a suitable type of bedding for the pipe in Problem 6 if it were extra-strength VCP instead of standard strength?

8. What is the total allowable rate of infiltration in a 750-m-long, 600-mm-diameter sewer, if 45 L/d/mm/km is allowed?

9. An exfiltration test is conducted on a 350-ft-long, 18-in.-diameter sewer reach. The water level in a 4-ft-diameter manhole is observed to drop 5.5 in. in 1 h. Compute the rate of exfiltration.

REFERENCES

1. *Design and Construction of Sanitary and Storm Sewers,* American Society of Civil Engineers, Water Pollution Control Federation, New York, 1969.

2. *Existing Sewer Evaluation & Rehabilitation,* American Society of Civil Engineers, Water Pollution Control Federation, New York, 1983.

3. Steel, E. W. and T. J. McGhee, *Water Supply and Sewerage,* 5th ed., McGraw–Hill, New York, 1979.

4. Hardenbergh, W. A. and E. B. Rodie, *Water Supply and Waste Disposal,* International Textbook Company, Scranton, Pennsylvania, 1960.

5. Parker, H. W., *Wastewater Systems Engineering,* Prentice–Hall, Englewood Cliffs, New Jersey, 1975.

6. Hammer, M. J., *Water and Wastewater Technology,* John Wiley & Sons, New York, 1977.

STORMWATER CONTROL

Uncontrolled stormwater and surface runoff can cause significant environmental damage. Flooding is one obvious example of a stormwater problem, with the accompanying loss of property, and, sometimes, human life. Even when the amount of stormwater runoff is not enough to be characterized as a flood, water pollution problems can be severe, particularly from soil erosion and sedimentation (see Section 5.3). Storm runoff is a major nonpoint source of water pollutants, including fertilizers, pesticides, oil, organics, and other substances.

Land development and urbanization increase the frequency and the severity of these problems. Technical personnel involved in the design and construction of new residential, commercial, or industrial land-use facilities must take steps to reduce the harmful environmental impacts of stormwater runoff.

Until the mid 1970s, the basic approach toward stormwater control was to collect it in underground pipes and to dispose of it as soon as possible at some convenient downstream location. But this is precisely what led to many of the flooding and pollution problems seen today, particularly those in rapidly developing suburban areas. This approach is no longer acceptable in most developing communities.

It is still necessary to provide storm drains to remove excess water from streets and parking lots, in order to prevent inconvenience or flood damage in localized areas. But modern drainage practice recognizes the need to further control or restrain the flow of the collected stormwater, in order to prevent flooding and pollution problems in the lower part of the drainage basin. In fact, stormwater is increasingly being viewed as a natural resource for use in a beneficial manner, rather than as a waste material to be disposed of quickly.

The use of on-site storage or detention basins is becoming a common method for controlling stormwater. In some cases, the stored stormwater can percolate to the groundwater and help to recharge an important aquifer; this is a beneficial use of the water. Even if the stormwater is not used to recharge an aquifer, the fact that it is held temporarily in storage and released slowly from the detention basin protects the environment; some water pollutants are removed, and downstream runoff flow rates are reduced.

This chapter focuses on some of the common ways stormwater control is accomplished. It will build upon some of the topics covered in Chapter 3, "Hydrology," particularly those sections dealing with rainfall and surface water. Also, a knowledge of basic hydraulics will be necessary to study and understand stormwater control technology. It may be necessary to review the material in Chapter 2, "Hydraulics," particularly the section on gravity or open channel flow, as well as the pertinent material in Chapter 3.

9.1 ESTIMATING STORM RUNOFF

The design or analysis of stormwater control facilities begins with an estimate of the rate and volume of surface runoff to be controlled. Usually, the most important figure to be estimated is the peak flow rate from a storm of a specified frequency and duration. In most cases, it is necessary for the designer to rely on local rainfall data and to use an acceptable formula that relates rainfall intensity and duration to the volume or rate of surface runoff.

RATIONAL METHOD

The most popular formula used to correlate rainfall with runoff in relatively small drainage basins is called the *Rational Formula*. The Rational Formula expresses the relationship between *peak runoff* and rainfall, as follows:

$$Q = C \times i \times A \qquad (9\text{-}1)$$

where Q = peak or maximum rate of runoff
C = a dimensionless runoff coefficient
i = rainfall intensity
A = drainage basin area

It is important to note that Q in the Rational Formula represents only the maximum discharge caused by a particular storm; it can be visualized as the peak of the storm hydrograph illustrated in Figure 9.1.

In SI metric units, Q is computed in terms of cubic meters per hour (m³/h) and then converted to cubic meters per second or liters per second; in American units, Q is computed directly in terms of cubic feet per second (ft³/s or cfs). It will be important to use appropriate units for rainfall intensity i and area A in each case.

American Units

In this system of units, rainfall intensity is expressed in terms of inches per hour, and area is expressed in terms of acres. No conversions are nec-

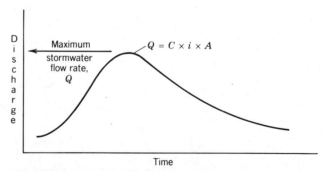

FIGURE 9.1 The *Rational Formula* is used to estimate the peak or maximum rate of surface runoff due to a particular storm in a specific drainage basin.

essary. The resulting dimensions for Q become inch-acre/hour, but 1 in.-ac/h is so close to 1 ft^3/s that the difference is neglected for practical purposes. (Try the conversion yourself to see how close it is.)

EXAMPLE 9.1

Estimate the peak rate of runoff on a 5-ac watershed from a storm with rainfall intensity of $i = 3$ in./h. Assume that the dimensionless runoff coefficient $C = 0.4$. (The meaning of this coefficient will be explained shortly.)

Solution:

Applying the Rational Formula (Equation 9-1), we get

$$Q = C \times i \times A = 0.4 \times 3 \times 5 = 6 \text{ ft}^3/\text{s}$$

Remember, there is no need to convert the units for i or A; keep them as inches per hour and acres respectively, and the value for Q is in terms of cubic feet per second.

SI Metric Units

In this system of units, rainfall intensity is usually expressed in terms of millimeters per hour (mm/h). To obtain a correct value for Q, it is necessary to convert millimeters per hour to meters per hour (m/h), and to express the area A in terms of square meters (m^2). Drainage areas in SI units are usually expressed in terms of hectares (ha), or square kilometers (km^2). The following conversions will be used: 1 ha = 10 000 m^2, and 1 km^2 = 1 $\times$ 10^6 m^2.

EXAMPLE 9.2

Estimate the peak rate of runoff from a 2-ha drainage basin that has a runoff coefficient of 0.4 for a rainfall intensity of 75 mm/h.

Solution:

Converting the intensity from millimeters per hour to meters per hour is simply a matter of dividing by 1000, since 1 m = 1000 mm; therefore $i = 0.075$ m/h. The area $A = 2$ ha $\times$ 10 000 m^2/ha = 20 000 m^2. Applying the Rational Formula, we get

$$Q = 0.4 \times 0.075 \text{ m}^2/\text{h} \times 20\,000 \text{ m}^2 = 600 \text{ m}^3/\text{h}$$

and

$$Q = 600 \text{ m}^3/\text{h} \times 1 \text{ h}/3600 \text{ s} \approx 0.17 \text{ m}^3/\text{s or } 170 \text{ L/s}$$

Area

The watershed area A can be easily measured from a topographic map, as described in Section 3.4. Of all the terms in the Rational Formula, area is the only one that can be determined with some degree of precision. The Rational Formula has been applied to drainage areas ranging in size from a fraction of an acre to about 5 square miles, but for areas exceeding about 250 ac (100 ha), it is considered best to apply some other more sophisticated hydrologic method to estimate runoff.

Runoff Coefficient

The coefficient C is a dimensionless number that represents the fraction of rainfall that appears as surface runoff in the drainage basin. If all the rainfall became runoff, as it would on a completely impervious surface, the value of C would be 1.0, its maximum possible value. But as discussed in Section 3.2, some of the rain water is intercepted, evaporates, or infiltrates the ground surface. Therefore, the value of C is always a decimal less than 1.0.

Factors that affect the value of the runoff coefficient include the type of land use in the drainage basin, the ground cover or vegetation, and the ground slope. Typical values of the coefficient C are presented in Table 9.1. Each type of surface or land use has a wide range of possible C values; the selection of a runoff coefficient for a drainage basin

TABLE 9–1 TYPICAL RUNOFF COEFFICIENTS

Type of Surface or Land Use	Runoff Coefficient C
Woodland areas	0.01 to 0.20
Grassland or lawns	0.10 to 0.25
Pavements and roofs	0.70 to 0.95
Suburban residential areas	0.25 to 0.40
Apartment housing areas	0.50 to 0.70
Industrial areas	0.60 to 0.90
Business areas	0.70 to 0.95

requires good technical judgment and careful evaluation of the physical characteristics of the basin.

Composite Runoff Coefficient

When several different surface types or land uses comprise the watershed, a composite or weighted average value of the runoff coefficient can be computed using the following formula:

$$C = \frac{1}{A_T} \times (C_1 \times A_1 + C_2 \times A_2 + C_3 \times A_3 + \ldots) \quad (9\text{-}2)$$

where C = composite runoff coefficient
A_T = total basin area $(A_1 + A_2 + A_3 + \ldots)$
A_1, A_2, etc. = areas of different type in the basin
C_1, C_2, etc. = respective runoff coefficients for A_1, A_2, etc.

EXAMPLE 9.3

From an air photo of a 15-ha watershed, it is determined that 6.5 ha is a flat densely wooded area, 6.0 ha is lawn, and 2.5 ha is paved roadway and parking area. Compute a composite runoff coefficient for the total watershed area.

Solution:

Refer to Table 9.1. For the wooded area, use $C_1 = 0.01$, since it is flat and densely wooded. For the

lawn, use $C_2 = 0.20$, the middle of the range, and for the paved areas, use $C_3 = 0.95$, a conservative judgment. Applying Equation 9-2, we get

$$C = \frac{1}{15} \times (6.5 \times .01 + 6.0 \times 0.20 + 2.5 \times 0.95) = 0.24$$

If we had more detailed information about the characteristics of the land, other values of C might have been used. For example, if the lawn area was very steep, then we might have selected $C = 0.25$ instead of $C = 0.20$. Using the upper and lower limits of each range of C for this problem, the composite C could fall within the range of 0.16 and 0.35. (Try the computation yourself, to check it out.)

Rainfall Intensity

The value of rainfall intensity i, used in the Rational Formula, depends on the storm recurrence interval and the storm duration. It is determined from local rainfall intensity–duration–frequency relationships, as discussed and illustrated in Section 3.3.

The first step in establishing the value of rainfall intensity to use in the Rational Formula is to *select a storm frequency or recurrence interval.* More often than not, the recurrence interval is specified by local township, county, or state agencies for various types of projects. Longer recurrence intervals reduce the possibility of the selected storm intensity being equalled or exceeded in any given year.

Accordingly, a local township engineering department may specify a 5-year return period for the design of storm drains in a middle-class residential neighborhood. This would indicate a willingness to accept some occasional street flooding, in order to have a reasonably economical system. On the other hand, a high-value commercial district may have storm sewers designed for a 15- or 25-year storm, reducing the chance of flooding, but increasing the cost of construction.

The spillway structure of a dam must be designed to accommodate stormwater flows from at least the 100-year frequency storm, in order to minimize the chance for failure of the dam and loss of life or property downstream. In general, the risks of failure must be balanced with the cost of construction for all stormwater control projects.

The second step in determining the value of rainfall intensity to use is to *select an appropriate storm duration*. The storm duration used in the Rational Method is called the *time of concentration* of the drainage basin, abbreviated T_c. It is defined as the time it takes a drop of water to flow from the hydraulically most distant point of the basin to the outlet of the basin.

If the storm lasts at least as long as the time of concentration, then the entire watershed will be contributing flow to the outlet; only under this condition is there enough time for the maximum peak flow to develop. For shorter-duration storms, not all of the basin area contributes flow simultaneously; for storms longer than T_c, the rainfall intensity, and therefore the peak flow, would be less than the maximum computed by using the time of concentration of the basin as the storm duration.

Time of Concentration

There are two parts or components of the time of concentration. These are the *overland flow time* and the *channel flow time*, as illustrated in Figure 9.2. The overland flow time is the time it takes the surface sheetflow to reach the beginning of the stream or storm drain inlet at the upper end of the basin.

Several charts or nomographs are available that correlate overland flow time with ground slope and runoff coefficient or land use. One such chart is illustrated in Figure 9.3. As illustrated, the chart is entered on the left with the known overland travel distance, and a horizontal then vertical path is traced to the known values of slope and runoff coefficient curves, respectively; finally, the overland travel time is read from the vertical axis on the right side of the chart.

If the velocity of flow in the stream or channel is known or estimated, the channel flow time can be computed from the simple relationship: *channel*

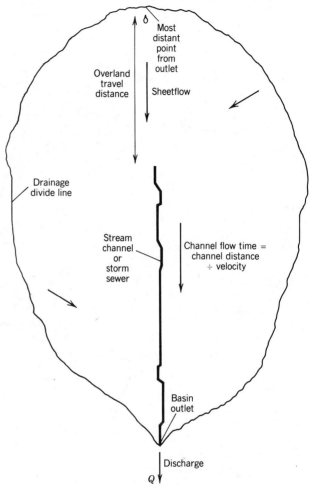

FIGURE 9.2 The *time of concentration* for a drainage basin includes overland flow time and channel flow time.

flow time = flow distance ÷ flow velocity. If the channel geometry is known, Manning's Formula can be used to estimate the velocity of flow. Otherwise a nomograph, such as the one illustrated in Figure 9.4, may be used.

EXAMPLE 9.4

A 0.25-km² watershed has a composite runoff coefficient $C = 0.25$. The overland flow distance to the beginning of the stream that drains the watershed is 150 m and the slope is 7 percent. The stream is 600 m long and drops 30 m in elevation by the time

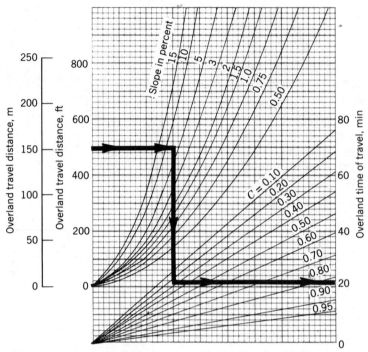

FIGURE 9.3 A chart like this may be used to estimate the overland flow time when the average travel distance, slope, and runoff coefficient are known. (Federal Aviation Administration)

it reaches the watershed outlet. Estimate the peak rate of runoff for the 5-year storm and the 100-year storm.

Solution:

The drainage area $A = 0.25 \text{ km}^2 \times 10^6 \text{ m}^2/\text{km}^2 = 250\,000 \text{ m}^2$. Enter Figure 9.3 with 150 m or 490 ft for overland travel distance, move horizontally to the right to the slope curves and estimate where the 7-percent curve would be (about midway between the 5- and 10-percent curves), then move down vertically to the runoff coefficient curves and estimate the position for $C = 0.25$. Finally, move horizontally to the right and read 22 min for the time of overland travel.

Use Figure 9.4 to estimate channel flow time; line up the drop in elevation of 30 m (100 ft) with the length of channel, 600 m (2000 ft), and read a channel flow time of about 8 min.

The time of concentration for this watershed is the sum of the overland flow time and the channel flow time, or

$$T_C = 22 \text{ min} + 8 \text{ min} = 30 \text{ min}$$

This value of T_c will now be used as the storm duration to determine the peak flow at the basin outlet. From Figure 3.5, a rainfall intensity for a 5-year, 30-min-duration storm is 70 mm/h, and for the 100-year, 30-min-duration storm, the intensity is 150 mm/h. Applying the Rational Formula, we get

5-year storm: $Q = 0.25 \times 0.070 \text{ m/h} \times$
$250\,000 \text{ m}^2 \approx 4400 \text{ m}^3/\text{h} \approx 1200 \text{ L/s}$
100-year storm: $Q = 0.25 \times 0.150 \text{ m/h} \times$
$250\,000 \text{ m}^2 \approx 9400 \text{ m}^3/\text{h} \approx 2600 \text{ L/s}$

SCS METHOD

The Soil Conservation Service (SCS) of the U.S. Department of Agriculture has developed methods for

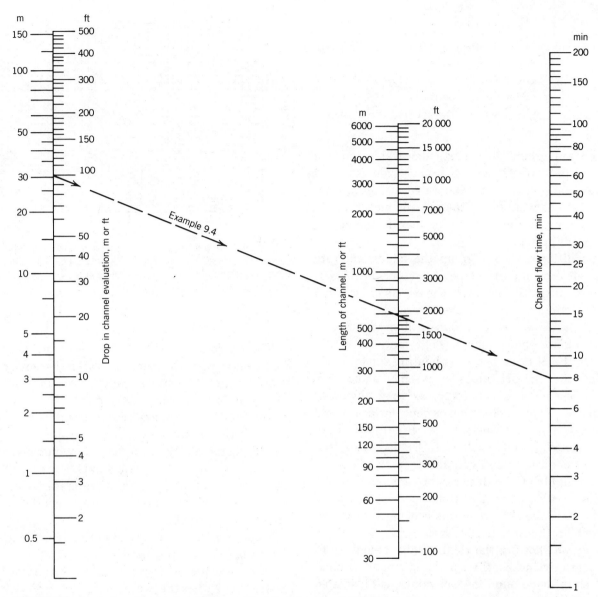

FIGURE 9.4 A rough approximation of channel flow time may be obtained with this nomograph, using a straight edge. For example, the time of travel in a channel that drops 4 m in elevation over a distance of 500 m is about 15 min. A more accurate estimate can be obtained using Manning's Formula.

estimating the volume and rate of storm runoff that can be applied to developing suburban areas. The SCS methodology is finding widespread application in recent years, and in some areas of the United States its use is required by regulatory agencies for project approval.

A basic distinction between the SCS and Rational methods is in the emphasis that the SCS places on the correlation between the type of soil cover in the watershed and the runoff. In this section, a brief overview of the *SCS Graphical Method* is presented, for relatively small watersheds. Only American units are used since the SCS has not yet converted their manuals to SI metric units.

Four *hydrologic soil groups* are defined in the SCS method, as follows:

Soil Group	Description
A	High infiltration rate/low runoff potential
B	Moderate infiltration rate
C	Slow infiltration rate
D	Very slow infiltration/high runoff potential

The soil groups in a particular watershed can usually be determined from countywide SCS soil survey maps, of the type illustrated in Figure 1.14. Data from field studies of the site and measured infiltration rates can also help to identify the appropriate soil group.

In addition to the soil group, the volume and rate of runoff depend on the type of land-use in the watershed. In the SCS method, the effects of both soil group and land use are characterized in a term called the *runoff curve number,* abbreviated CN. Table 9.2 presents a summary of typical CN values used in the SCS method of estimating runoff.

Note that the CN values for pavements and roofs are independent of soil type, as they obviously should be. A completely impervious surface would have a CN = 100, and all the rainfall would become runoff. As the value of CN decreases, the amount of direct runoff will also decrease. Composite or weighted CN values can be computed for watersheds comprising more than one type of soil or land use, in the same way this is done for the composite C in the Rational Method.

A graph showing the relationships among rainfall depth, in inches, the amount of runoff, also expressed in inches, and the CN values, is illustrated in Figure 9.5.

TABLE 9.2 TYPICAL SCS RUNOFF CURVE NUMBERS

	CN Value for Hydrologic Soil Group			
Land-use Description	A	B	C	D
Meadow	30	58	71	78
Forest	25	55	70	77
Grass lawns	39	61	74	80
Commercial–business	89	92	94	95
Residential	54	70	80	85
Pavement–roofs	98	98	98	98

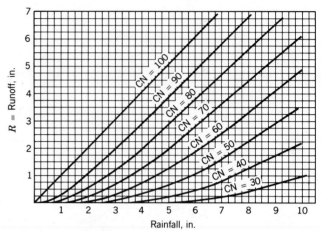

FIGURE 9.5 A selection of SCS rainfall–runoff relationships for several CN values. (Adapted from "A Method for Estimating Volume and Rate of Runoff in Small Watersheds," with permission of the Soil Conservation Service/USDA.)

The chart in Figure 9.5 is entered on the horizontal axis with the depth of rainfall from an *N*-year storm of 24 hours' duration. First moving vertically up to the curve matching the CN for the watershed (or the estimated curve position if it falls between the values shown on the graph), and then moving horizontally to the left, the volume of runoff can be read on the vertical axis. For example, a 24-hour rainfall of 6 in., on a watershed with CN = 70, would produce 2.8 in. of direct runoff over the area of the watershed.

To determine the peak rate of runoff, the graph shown in Figure 9.6 is used.

The horizontal axis represents the time of concentration T_c for the watershed; this was defined previously in the discussion for the Rational Method. The vertical axis has the units of *cubic feet per second per square mile of watershed, per inch of runoff (csm/in.).* This value is then applied in the following equation to compute the peak rate of runoff:

$$Q = q \times A \times R \qquad (9\text{-}3)$$

where Q = peak rate of runoff, ft^3/s
 q = unit peak discharge, csm/in. (from Figure 9.6)
 A = drainage area, square miles
 R = direct runoff, in. (from Figure 9.5)

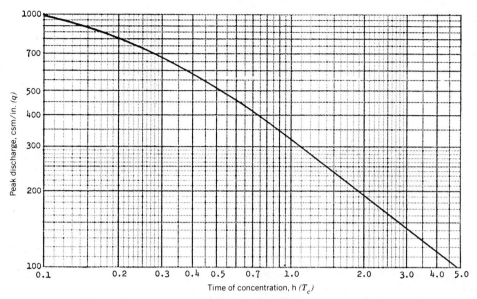

FIGURE 9.6 Unit peak discharge in csm/in. of runoff versus time of concentration, for a 24 hour storm duration. (From "Urban Hydrology for Small Watersheds," with permission from the Soil Conservation Service, USDA.)

This SCS method for estimating runoff is generally applicable for drainage areas up to about 2000 ac or 3 miles^2. It is considered to be more conservative than the Rational Method and usually results in higher estimates of runoff. The following example illustrates the SCS graphical procedure to determine peak discharge.

EXAMPLE 9.5

A 1000-ac watershed has a composite CN = 70 and a time of concentration T_c = 66 min. From rainfall data, it is determined that the 100-year, 24-hour storm would cause a total of 7.5 in. of rainfall in the watershed. Compute the peak rate of stormwater runoff from the watershed by the SCS method.

Solution:

From Figure 9.5, we find that R = 4.0 in.
The time of concentration T_c = 66 min × 1 h/60 min = 1.1 h
From Figure 9.6, we find that q = 300 csm/in.

Applying the conversion 1 mile^2 = 640 ac, we get watershed area A

$$= 1000 \text{ ac} \times 1 \text{ mile}^2/640 \text{ ac} = 1.56 \text{ mile}^2$$

Now applying Equation 9-3, we get

$$Q = q \times A \times R = 300 \times 1.56 \times 4.0 \approx 1900 \text{ ft}^3/\text{s}$$

EFFECTS OF LAND DEVELOPMENT

The construction of homes, factories, roads, and other facilities for a growing community changes the runoff patterns of the watershed in which it takes place. Unless appropriate steps are taken, land development can increase the frequency and severity of flooding and water pollution from stormwater runoff.

As land is developed and urbanization takes place, much of what was originally woodland, meadow, or farmland is covered with relatively impervious surfaces—paved roads, driveways, parking lots, and buildings. This has two direct effects on the local hydrology:

1. The amount of infiltration decreases, and therefore, the volume of direct surface runoff increases.

2. The time of concentration of the watershed decreases, since the runoff flows faster to the outlet over the modified land surfaces and in the drainage channels.

Since the time of concentration is shorter, a shorter duration storm will be able to cause the peak flow to develop, and the rainfall intensity is greater for a shorter duration storm. The net result, then, of land development and urbanization is a definite increase in both the peak rate and total volume of runoff from a given storm. This effect is depicted in Figure 9.7.

On a short-term basis, soil erosion and stream siltation problems may be particularly severe during the period of land development and construction. In the long run, the chances for severe flooding increase as a result of land development. And urban runoff can carry oil, lawn fertilizers, organics, and other pollutants into the receiving waters.

Modern land planning practices, however, generally require the implementation of methods to prevent any significant increase in the volume and rate of runoff, due to urbanization. Some of the technical solutions to this problem are discussed in Section 9.3. The following example illustrates the effects of land development on runoff.

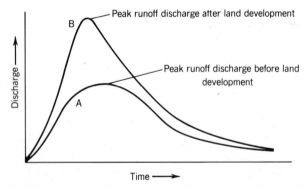

FIGURE 9.7 Curve A shows a storm hydrograph for the original or predevelopment conditions, and curve B shows a post-development hydrograph for the same area. Land development or urbanization causes the volume and rate of stormwater runoff to increase.

EXAMPLE 9.6

The 0.25-km^2 watershed described in Example 9.4 is now to be developed as a residential and business area. The resulting composite runoff coefficient C is estimated to be 0.60. Part of the stream is to be carried in a pipeline, and it is estimated that the new channel flow time after development will be only 5 min. For both the 5-year and the 100-year storms, determine the estimated peak flow rates under developed conditions, using the Rational Method. Compare the increases in peak discharges to those during the predevelopment conditions of Example 9.4.

Solution:

Entering Figure 9.3 with overland flow distance = 150 m, slope = 7 percent, and C = 0.60, we read overland travel time = 12 min. Adding the channel flow time of 5 min, we get a new time of concentration T_c = 17 min for the developed watershed.

Now, from Figure 3.5, we read rainfall intensity i = 90 mm/h for the 5-year storm, and i = 190 mm/h for the 100-year storm. Applying the Rational Formula, we get

5-year storm: $Q = 0.60 \times 0.090$ m/h
$$\times\ 250\ 000\ \text{m}^2 \approx 14\ 000\ \text{m}^3/\text{h}$$

Under predevelopment conditions (Example 9.4), the peak flow was computed to be 4400 m^3/h. Therefore, land development is expected to increase the peak stormwater discharge by a factor of 14 000/4400 = 3.2, for the 5-year storm.

100-year storm: $Q = 0.60 \times 0.190$ m/h $\times$
$$250\ 000\ \text{m}^2 = 29\ 000\ \text{m}^3/\text{h}$$

Land development will increase the peak discharge by a factor of 29 000/9400 = 3.1, for the 100-year storm.

9.2 STORM SEWER SYSTEMS

Storm sewers serve to convey surface runoff to a point of storage or disposal. In separate sewer systems, sanitary or industrial wastewater is excluded from the storm sewers. There are several basic differences in design between storm sewers and sanitary sewers that should be noted.

First, it can be said that storm sewers are actually designed to "fail" with a predictable recurrence interval. In other words, a storm drainage system will periodically surcharge and overflow, causing local flooding, with a known frequency. This is because the designer must first select a storm return period in order to estimate peak discharge and then size the pipeline. This limits the capacity of the system to storm intensities equal to or less than the one selected for design.

Of course the probability or frequency of failure can be reduced by selecting a large storm recurrence interval. But the chance for surcharge of a storm sewer can never be completely eliminated. Sanitary sewers, on the other hand, are designed to carry a peak flow rate of sanitary wastewater from a "saturation population," without surcharge or overflow. When sanitary sewers do surcharge, it is usually from excessive infiltration or inflow (i.e., poor construction or maintenance).

There are other differences between storm servers and sanitary sewers. Storm sewers are usually much bigger in diameter than the separate sanitary sewers serving the same area, as illustrated in Figure 9.8. Although they generally carry no flow during dry weather, and only partial flows during most rainfalls, storm sewers still must be sized to carry the peak flow from a major storm of specified intensity and duration. Also, unlike sanitary sewers which are placed in relatively deep trenches for service connections, storm sewers can be placed at shallower depths to minimize excavation.

LAYOUT AND DESIGN

A storm sewage collection system consists basically of a network of inlets and pipes located in the street

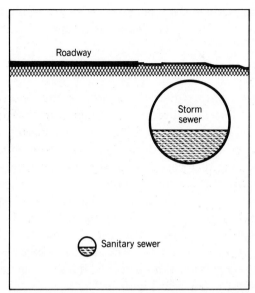

FIGURE 9.8 Sanitary sewers are typically much smaller in diameter and placed in deeper trenches than storm sewers serving the same area.

right-of-way, or in easements along lot lines. A common location within the right-of-way is near the curb or edge of pavement, since inlet boxes can then be connected with fewer manholes and less pipe. Storm drains are located to allow gravity flow to a stream or other body of water, to a storage basin, or in some cases to a special treatment facility. The installation of stormwater pumping stations is especially avoided because of the very large peak flow capacity that would be required.

Modern design practice makes more use of existing natural streams, open channels, or grass-lined swales to carry runoff, before resorting to the use of buried pipes. This is basically to slow down the flow and increase the time of on-site detention (without basin storage), thereby reducing peak flows downstream. Roadway crossings over natural drainage ditches or streams may require the installation of a *culvert*. A culvert is a relatively short section of pipe, or a cast-in-place concrete structure, that carries surface runoff under an obstruction.

Storm sewers are usually built with circular reinforced concrete pipes. Elliptically shaped pipes are sometimes used when the pipe depth is very shallow, to gain leeway for soil cover over the shorter axis of the pipe. Corrugated metal pipe may be used in some cases for storm drains and culverts.

Inlets

A stormwater inlet is a structure that intercepts sheet flow and directs the runoff into the underground pipe system. The location and spacing of inlets are important design factors; there must be sufficient inlet capacity to remove the stormwater from the surface and transfer it into the sewers fast enough to prevent backups. Factors such as hydraulic capacity, clogging, nuisance to traffic, and safety must be evaluated. For example, it is desirable to keep runoff from flooding across street intersections. To accomplish this, inlets are generally located as shown in Figure 9.9.

There are basically three different types of inlet structures: *curb inlets, gutter* or *grate inlets,* and *combined inlets.* A curb inlet has a vertical opening along the curb, into which the runoff flows, as shown in Figure 9.10*a*. A gutter inlet is a horizontal opening in the pavement, covered with a cast-iron grate, as shown in Figure 9.10*b*.

The curb inlet is not an obstruction or nuisance to traffic, but for child safety the opening should be less than 150 mm (6 in.) high; this limits hydraulic capacity of the inlet. Disadvantages of the gutter type inlet are that it can obstruct the smooth flow of traffic (including bicycles) and that the iron grate may become plugged with debris.

A combination inlet has both a curb and a gutter

opening, as shown in Figure 9.10*c*. This inlet may be depressed or lowered for additional hydraulic capacity. The combination inlet is the least subject to clogging.

A cross-sectional view of an inlet basin is shown in Figure 9.11. If the pipe invert were above the bottom of the basin, as shown with the dashed lines, the structure would be called a *catch basin*. A catch basin serves as a trap for grit, sand, and debris which may be washed into the inlet during a storm. Because of the need for periodic cleaning, mosquito control, and odor problems, catch basins are not commonly used in new storm drainage systems. Good inlet design and "self-cleansing" pipe slopes (for adequate flow velocity) would keep the collection system free from debris and blockage problems.

Design Procedure

Using a topographic map of the "drainage district" or area to be sewered, the inlets are first located at street intersections, low points, and at appropriate intervals along the streets. Local engineering departments may specify maximum spacing between inlets. The storm drains are located within the public right-of-way or in drainage easements, following the natural slope of the land for gravity flow.

A storm sewer reach is a section with constant diameter and slope, usually between two inlets. The boundary of the catchment area that is a tributary to each inlet and downstream reach is sketched on the topographic map (as described in Section 3.4). Each of these areas is measured in terms of acres, hectares, or square meters, using a planimeter if necessary.

The Rational Method is commonly used to compute peak design flows in urban or suburban drainage districts. An appropriate runoff coefficient C is selected for each of the inlet catchment areas, based on type of ground cover and land use. The storm recurrence interval for which the system will be designed is selected by the designer, or may be specified in local building and land-use ordinances. Overland flow time, also called *inlet time,* is estimated for each area tributary to an inlet; generally, inlet time should not exceed about 10 min.

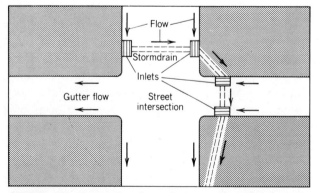

FIGURE 9.9 Proper location of stormwater inlets prevents storm water runoff from flooding a street intersection.

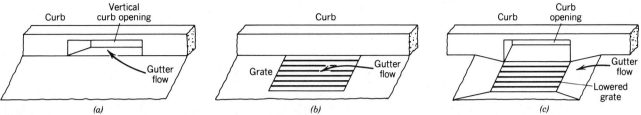

FIGURE 9.10 *There are several different types of stormwater inlets, including (a) the curb inlet, (b) the gutter inlet, and (c) the combined inlet.*

Design begins at the uppermost inlet and proceeds downstream; cumulative areas and composite runoff coefficients are computed for each reach of the system. A new time of concentration must be determined for each section of the system, to account for the gradually increasing travel time to the reach being designed. At any inlet, the longest time of flow for any water entering it must be selected as the time of concentration for that reach, since there must be enough time for the entire basin to contribute runoff in order for the peak flow to develop. In practice, design computations are usually organized in tabular form.

The pipe locations, diameters, and slopes must be shown on the engineering plan drawings; diameter, slope, and invert elevations are determined in much the same way as for sanitary sewers (see Section 8.1). The self-cleansing velocity for storm sewers is about 0.9 m/s (3 ft/s); it is higher than that for sanitary sewers because stormwater tends to carry heavier and more settleable solids than sanitary sewage does.

The following example illustrates the basic procedure for determining design flows and required pipe diameters. We can assume that the given pipe slopes are roughly parallel to the street slope, as determined from a topographic map of the area or street profile. The data for the problem are given in Figure 9.12.

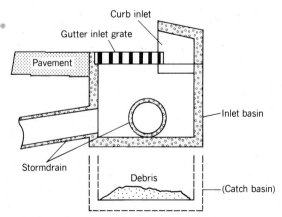

FIGURE 9.11 *A section view of a typical stormwater inlet basin. Catch basins which trap grit, leaves, and debris are not commonly used in new storm sewer systems.*

EXAMPLE 9.7

A storm drainage system has been laid out in a street as shown in Figure 9.12. Individual inlet catchment areas and runoff coefficients have been determined and are shown. Inlet times have been estimated, and the pipe length and slope for each sewer reach is given. Using the Rational Method, compute the design flow and the required pipe diameter for each reach, for a 10-year storm.

Solution:

Reach 1:
First determine the rainfall intensity for area 1, using the inlet time as the time of concentration. Enter Figure 3.5 with T_c = 5 min, and read i = 150 mm/h or 0.15 m/h for the 10-year storm. The catchment area is 1 ha or 10 000 m². Applying the Rational Formula,

$$Q = 0.4 \times 0.15 \text{ m/h} \times 10\,000 \text{ m}^2 = 600 \text{ m}^3/\text{h}$$

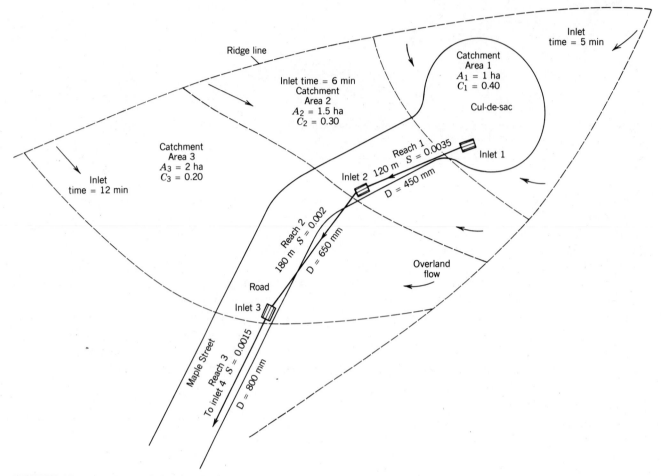

FIGURE 9.12 Illustration for Example 9.7.

and converting to m³/s, we get

$$Q = 600 \text{ m}^3/\text{h} \times 1 \text{ h}/3600 \text{ s} = 0.167 \text{ m}^3/\text{s}$$

To determine the required diameter for reach 1, enter Manning's nomograph with $Q = 0.167$ m³/s and the given slope of 0.0035; read the required diameter $D = 45$ cm and a flow velocity of 1.05 m/s.

Reach 2:

To compute the peak discharge or design flow for the pipeline between inlet 2 and inlet 3, we first compute the composite runoff coefficient for the combined areas 1 and 2, as follows:

$$C = \frac{1}{2.5} \times (1 \times 0.4 + 1.5 \times 0.3)$$

$$= 0.34 \quad \text{(from Equation 9-2)}$$

Next, compute the time of flow in reach 1, using the relationship

$$\text{time} = \text{distance} \div \text{velocity} = 120 \text{ m} \div 1.05 \text{ m/s}$$
$$= 114 \text{ s} = 1.9 \text{ min}$$

The total time of flow to inlet 2 is the inlet time for area 1 plus the channel flow time in reach 1, or $5 + 1.9 = 6.9$ min. This is larger than the inlet time for area 2, therefore use a time of concentration $T_c \approx 7$ min for the composite area draining to inlet 2.

From Figure 3.5, read $i = 145$ mm/h for the 10-year storm.

Applying the Rational Formula, we get

$$Q = 0.34 \times 0.145 \text{ m/h} \times 25\,000 \text{ m}^2 = 1230 \text{ m}^3/\text{h}$$

and

$$Q = 1230 \text{ m}^3/\text{h} \times 1 \text{ h}/3600 \text{ s}$$
$$= 0.34 \text{ m}^3/\text{s for reach 2}$$

Now enter Manning's nomograph with $Q = 0.34$ m³/s and the given slope of $S = 0.002$; read the required diameter $D = 65$ cm and flow velocity $V = 1.02$ m/s.

Reach 3:

The total tributary area to inlet 3 is $1 + 1.5 + 2 = 4.5$ ha, or $45\,000$ m². The composite runoff coefficient is computed to be

$$C = \frac{1}{4.5} \times (0.4 \times 1 + 0.3$$
$$\times 1.5 + 0.2 \times 2) = 0.28$$

The time of flow in reach 2 is 180 m ÷ 1.02 m/s = 176 s = 2.9 min. The total flow time to inlet 3 is then $5 + 1.9 + 2.9 = 9.9$ min (inlet time for area 1 + travel time in reach 1 + travel time in reach 2). But this is less than the individual inlet time for area 3, which is 12 min. Therefore, the 12-min inlet time dominates and is taken as the time of concentration (or storm duration) for the design of reach 3.

Enter Figure 3.5 with $T_c = 12$ min and read $i = 135$ mm/h or 0.135 m/h, for the 10-year storm. Now, applying the Rational Formula, we get

$$Q = 0.28 \times 0.135 \text{ m/h} \times 45\,000 \text{ m}^2 = 1700 \text{ m}^3/\text{h}$$

and

$$Q = 1700 \text{ m}^3/\text{h} \times 1 \text{ h}/3600 \text{ s} = 0.47 \text{ m}^3/\text{s}$$

From Manning's nomograph, with $Q = 0.47$ m³/s and $S = 0.0015$, read $D = 80$ cm and $V = 1$ m/s.

9.3 ON-SITE STORMWATER DETENTION

The present-day approach to stormwater control is increasingly leaning toward temporary storage of the water on-site, in the vicinity where it falls, rather than toward quick discharge to a nearby body of water. To some extent, stormwater storage occurs naturally in most drainage basins. In some cases, intentional ponding of rain water on rooftops or in parking lots for a short period of time can provide enough storage to reduce peak runoff flow rates. And using open grass–swale drainage channels can effectively retard the flow of stormwater runoff.

But the construction of relatively small reservoirs or basins, to hold stormwater after it has been collected from streets, parking lots, and other surfaces, is being required in a growing number of communities undergoing urbanization. These basins store or retain the stormwater and allow it to be released slowly under controlled conditions. They can be effective in controlling relatively short but intense local storms, which tend to cause the most frequent flooding, erosion, and pollution damage in small streams.

Some specific benefits of stormwater storage basins are:

1. Reduction of peak runoff rates.
2. Reduction of the severity and frequency of flooding.
3. Reduction of soil erosion and stream sedimentation.
4. Protection of surface water quality.
5. Groundwater aquifer recharge, if soil conditions permit.

The basic disadvantage of on-site stormwater storage basins is related to the problem of maintenance. The outlet structures are prone to clogging, and the basins themselves often become the depository for sediment and debris. Weed control is a problem, and mosquitos can breed in pools of water

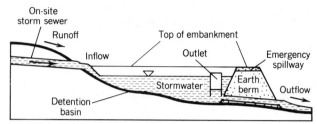

FIGURE 9.13 A section view of a stormwater detention pond. The outlet structure acts as a "bottleneck," restraining the rate of discharge from the pond. The pond or basin is usually empty of water during dry weather, except perhaps for a small stream.

that remain stagnant for a long time. In some cases, safety for children in the area must be considered. Maintenance may be the responsibility of the local municipality, or of nearby property owners.

There are basically three different types of stormwater storage basins. These include *retention basins, detention basins,* and *recharge basins.* A retention basin holds some water all the time, forming a permanent pond, or small lake. In addition to stormwater control, it may also provide aesthetic and recreational benefits on the site. A detention basin, however, only holds the stormwater for a relatively short period of time, during and shortly after periods of rainfall. It is empty and dry most of the time. Sometimes there may be a small stream flowing through the basin, even in dry weather. A section view of a detention basin is shown in Figure 9.13.

The third type, recharge basins, are specifically designed to allow the collected water to percolate into an underlying aquifer. They serve to recharge and replenish groundwater reserves, as well as to control storm runoff. For a recharge basin to be effective, the soils underlying the basin must be permeable to allow relatively rapid infiltration, and the seasonal high water table should be at least 0.5 m below the bottom.

DESIGN PROCEDURE

Local subdivision regulations and municipal land-use ordinances must be reviewed at the very begin-

ning of the project, to determine the specific performance requirements for the basin with regard to storm flow reduction. The computations to determine the *predevelopment peak discharge* (before construction) and the *postdevelopment peak discharge* (after construction) can be made as illustrated in Examples 9.4 and 9.6 in Section 9.1.

Most communities require that a detention basin provide enough storage volume and outflow control to keep postdevelopment runoff equal to or less than predevelopment runoff. In other words, the developer must build a facility that will effectively maintain the rate of runoff from the site just about as it was in its natural condition, before construction.

This requirement usually pertains to the 100-year storm, but some ordinances also specify that the basin should reduce runoff flows from the 2- and 10-year storms as well. If the basin is designed only for the large 100-year storm, it will have no attenuating effect on the smaller but more frequent stormflows. Accommodation of more than one storm return period in the basin is accomplished by proper hydraulic design of the basin volume and outlet structure. A concrete structure with several outlets to handle different size storms is illustrated in Figure 9.14.

Using a topographic map of the site, a suitable location for the detention basin should be established. This should be in the lower part of the trib-

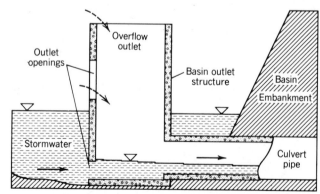

FIGURE 9.14 The outlet structure for a detention basin can be hydraulically designed to reduce peak discharge flows from a range of storm magnitudes or frequencies. Multiple outlet openings, or specially designed "proportional weirs," can be used for this purpose.

utary drainage area. The on-site storm sewer system should be designed, as described in Section 9.2, to direct the runoff into the detention basin. A preliminary estimate of the required basin volume can be made at this point. A simple and quick way of doing this will be described later in this section.

After a preliminary basin size is determined, a grading plan should be sketched on the topographic map. The basin can be constructed by balancing excavation and fill in the low-lying area of the site, forming a confining earth embankment that gradually blends in to the natural topography of the land. A thorough hydraulic analysis of flow through the basin and outlet structure should then be conducted.

If the discharge at the outlet is determined to be equal to or less than the maximum allowable discharge, and if there is enough volume in the basin to prevent flooding of nearby homes, then the basin design is accepted as satisfactory; otherwise, changes would be made and the process would be repeated until an acceptable design was reached. Finally, an emergency spillway is designed to allow flows greater than the design flow to pass through the basin without breaching the embankment.

PRELIMINARY DESIGN COMPUTATIONS

The basic relationship that expresses the function and operation of a stormwater detention basin is that at any given time, the volume of water in storage is equal to the difference between the inflow volume and the outflow volume up to that time, or

$$\text{inflow} - \text{outflow} = \text{storage} \qquad (9\text{-}4)$$

The rate of inflow changes with time. It depends on the intensity and duration of rainfall, as well as on the physical characteristics of the drainage area. The relationship between inflow and time is shown graphically in an *inflow hydrograph.*

The rate of outflow from the basin also changes with time. It depends on the hydraulic characteristics of the basin outlet structure. The outflow is generally a function of the height or depth of water in the basin; the deeper the water, the faster it flows over or through the outlet.

In mathematical terms, we can express the storage equation as

$$I(\Delta t) - O(\Delta t) = \Delta S \qquad (9\text{-}5)$$

where Δt = a small time interval, such as 5 or 10 min (pronounced "delta-t")

I = average inflow rate during Δt

O = average outflow rate during Δt

ΔS = change in storage volume during Δt (pronounced "delta-S")

The solution of this equation leads to the determination of the basin *outflow hydrograph,* which shows the rate of flow out of the basin as a function of time. The procedure for solving the equation and preparing the outflow hydrograph is called *flood routing.*

Typical inflow and outflow hydrographs for a detention basin are shown in Figure 9.15. Initially, the rate of inflow exceeds the rate of outflow and water accumulates in the basin (ΔS is positive). The outlet structure serves, in effect, as a "bottleneck" that prevents the water from flowing out of the basin as fast as it flows in. Eventually, the inflowing stormwater subsides, and the basin gradually empties as the water flows through the outlet

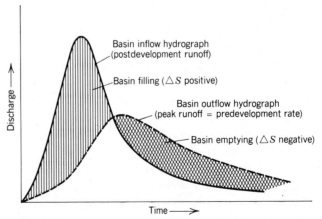

FIGURE 9.15 Typical inflow and outflow hydrographs for a stormwater detention basin. The basin and outlet structure serve to reduce the peak rate of runoff from a developed site.

(ΔS is negative). A comparison of the inflow and outflow hydrographs clearly shows the effect of the basin in attenuating or reducing the peak flow rate.

The flood routing procedure just outlined involves a lot of computation to arrive at a solution of the storage equation and the outflow hydrograph. For our purposes, a simplified procedure will be used in order to be able to illustrate the fundamental concept of stormwater detention without getting bogged down in computations. This procedure is sufficient to provide a preliminary or "ballpark" estimate of the required storage volume or peak outflow rate of a detention basin. It is not an exact method and would not be used for final design or analysis.

In this method, two factors related to the effectiveness of a stormwater detention basin are defined as follows:

$$\text{Storage factor (SF)} = \frac{\text{basin storage volume}}{\text{total rainfall volume}} \quad (9\text{-}6)$$

$$\text{Flow factor (FF)} = \frac{\text{peak outflow rate}}{\text{peak inflow rate}} \quad (9\text{-}7)$$

The relationship between the storage factor and the flow factor can be approximated by a straight line, as shown in Figure 9.16. The line is a graph of the equation

$$\text{FF} = 1.0 - \text{SF} \quad (9\text{-}8)$$

where FF and SF are the flow factor and storage factor, respectively.

If there is no storage volume at all, then SF = 0, and therefore FF = 1.0. In other words, the outflow rate would equal the inflow rate, and there would be no flow attenuation. On the other hand, if a basin big enough to store the total rainfall volume from the storm were provided, then SF = 1.0. Under this circumstance, FF = 0, and there would be no outflow at all. The straight-line relationship approximates what happens in between these two extreme conditions, when some storage volume is provided.

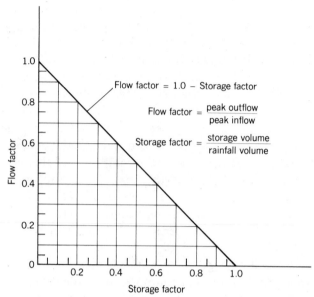

FIGURE 9.16 The relationship between flow factor and storage factor offers a simplified procedure for stormwater detention calculations. (Adapted from Pagan, A. "Flow Factor Line Used in Storage Calculations," *Irrigation Journal*, with permission of the American Society of Civil Engineers.)

Total Rainfall Volume

The total rainfall volume is equal to the area under the inflow (or outflow) hydrograph, since the product of discharge (volume per unit time) and time is equivalent to volume. For practical purposes, it is reasonable to make the simplifying assumption that the inflow hydrograph is triangular in shape.

The peak inflow can be easily computed using the Rational Method. The time for the rising limb of the hydrograph to reach the peak flow value is taken as the time of concentration T_c of the drainage area. The time for the receding limb to reach the base is conservatively taken as twice the time of concentration, or $2T_c$. This triangular hydrograph is illustrated in Figure 9.17.

The total rainfall volume is the area of the triangular inflow hydrograph. The area of a triangle is equal to the product of one half the base times the height ($A = \text{bh}/2$). Since in this case the base $b = 3T_c$ and the height $h = Q_{\max}$, the area is $\frac{1}{2} \times 3T_c \times Q_{\max}$ and,

$$\text{total rainfall volume} = 1.5 \times T_c \times Q_{\max} \quad (9\text{-}9)$$

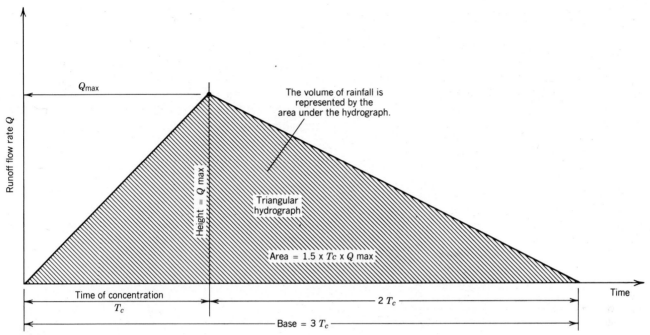

FIGURE 9.17 A triangular hydrograph provides a reasonable estimate of total rainfall volume. The peak flow, or height of the triangle, can be computed using the Rational Formula.

The following examples illustrate the application of this simplified method for preliminary detention basin computations.

EXAMPLE 9.8

A storm causes a peak runoff rate of 5 m³/s in a drainage basin that has a time of concentration of 30 min. A detention basin with 10 000 m³ of volume can be built on site for a residential land subdivision. Estimate the peak outflow from the basin for this storm.

Solution:

$$T_c = 30 \text{ min} \times 60 \text{ s/min} = 1800 \text{ s}$$

Applying Equation 9-9, we get

total rainfall volume
$$= 1.5 \times 1800 \text{ s} \times 5 \text{ m}^3/\text{s} = 13\ 500 \text{ m}^3$$

Now applying Equation 9-6, we get

$$\text{SF} = \frac{10\ 000 \text{ m}^3}{13\ 500 \text{ m}^3} = 0.74$$

From Figure 9.16 (or Equation 9-8), we get FF = 0.26, and from Equation 9-7, we get

$$\text{peak outflow rate} = 0.26 \times 5 \text{ m}^3/\text{s} = 1.3 \text{ m}^3/\text{s}$$

In summary, it can be expected that the 10 000-m³ basin will reduce the peak runoff discharge from 5 m³/s to about 1.3 m³/s.

EXAMPLE 9.9

Estimate the storage volume needed in a detention basin to reduce a peak inflow rate of 150 cfs to 100 cfs, if the total rainfall volume is 300 000 ft³.

Solution:

Applying Equation 9-7, we get

$$\text{FF} = \frac{100}{150} = 0.67$$

From Figure 9.16, we get

$$SF = 0.33$$

Now applying Equation 9-6, we get

$$0.33 = \frac{\text{storage volume}}{300\ 000\ \text{ft}^3}$$

and

basin storage volume

$$= 0.33 \times 300\ 000\ \text{ft}^3 = 100\ 000\ \text{ft}^3$$

EXAMPLE 9.10

Referring to the data given in Examples 9.4 and 9.6 in Section 9.1, assume that the local planning board has required that the land developer provide an on-site stormwater storage basin. The peak runoff after development (Example 9.6) is to be no greater than it was before development (Example 9.4). Assume that the detention basin will have an average water depth of 2.0 m when filled to capacity. How much area of the site, in hectares, will have to be used for the detention basin (a) for a 5-year storm, and (b) for a 100-year storm?

Solution:

(a) 5-year storm:

From Examples 9.4 and 9.6, the predevelopment discharge of 4400 m³/h is set equal to the basin outflow rate, and the postdevelopment discharge of 14 000 m³/h is set equal to the inflow rate. From Equation 9-7, we get FF = 4400/14 000 = 0.31, and from Equation 9-8 we get SF = 0.69.

The time of concentration T_c = 15 min = 0.25 h
Applying Equation 9-9, we get

total rainfall volume
$$= 1.5 \times 0.25\ \text{h} \times 14\ 000\ \text{m}^3/\text{h} = 5300\ \text{m}^3$$

Now, from Equation 9-6, we get 0.69 = storage volume/5300, and

$$\text{storage volume} = 0.69 \times 5300 = 3700\ \text{m}^3$$

Since volume = area × depth, or area = volume/depth, we get

basin area = 3700 m³/2.0 m ≈ 1900 m² or 0.19 ha

(b) 100-year storm

Following the same procedure as in part (a), using the data for the 100-year storm, we get the following results:

$$FF = \frac{9400}{30\ 000} = 0.31$$

and

$$SF = 1.0 - 0.31 = 0.69$$

total rainfall volume = 1.5 × 0.25 h ×
30 000 m³/h = 11 250 m³
storage volume = 0.69 × 11 250 m³ =
7800 m³
basin area = 7800 m³/2.0 m
= 3900 m² or 0.39 ha

This represents only 1.6 percent of the total site area of 0.25 km².

9.4   FLOODPLAINS

Flooding is a natural event that occurs periodically when the water in a stream or river overflows its channel banks and inundates adjacent low-lying land. This land is called the *floodplain*. The portion of the floodplain that is inundated by the 100-year flood is usually called the *flood hazard area*, as il-

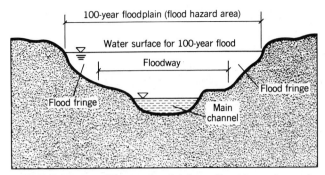

FIGURE 9.18 The 100-year floodplain or *flood hazard area* includes the *floodway* and the *flood fringe* areas.

lustrated in Figure 9.18. The flood hazard area includes the *floodway,* which carries the major portion of the flood at high velocities, and the *flood fringe,* which is covered with shallower, slowly moving water.

Floods are particularly damaging to houses and other structures in the floodway. In addition to economic loss totalling billions of dollars annually in the United States, many lives are lost in severe floods. Further, significant pollution problems, including contamination of drinking water supplies, occur during flood conditions. Flood damages are the result of poor land use and environmental planning, and they can be avoided with proper floodplain management practices.

Improper floodplain management results in the condition illustrated in Figure 9.19. As the floodway is filled in and built upon, the flow path is restricted. This causes an increase in the flood elevation, making the problem even more severe. The basic objective of regulating land use in the flood-

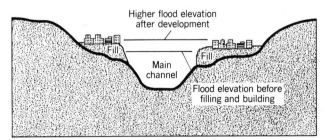

FIGURE 9.19 Urban development in a floodplain causes an increase in flood elevations due to the constricted flow channel.

plain is to reduce the risk of future flood damages. Floodplain regulation is generally a responsibility of local government, subject to state guidelines. Control is also exercised by the federal government under the requirements of the National Flood Insurance Program (NFIP).

Many states have adopted regulations that prohibit certain types of construction or activities in the floodway. These include:

1. New buildings
2. Sanitary landfills
3. On-site sewage disposal systems

Recreational facilities such as playgrounds and picnic areas and farming activities are still permitted because they are unlikely to cause obstructions and increase the potential for flooding or environmental damage. Buildings and sanitary facilities constructed in the flood fringe must be floodproofed to a height above the 100-year flood level, to comply with the requirements of the NFIP. This prevents damage to the structure and contents in the floodplain and reduces pollution problems.

Floodplain management must be preceded by a study that shows the extent of the floodplain on a suitable map. The easiest way to delineate the floodplain is to examine the area topography on available U.S. Geological Survey (USGS) maps. Also, an examination of SCS soil type data can provide useful information; floodplain areas are usually associated with deposits of silty soils laid down by the stream over the years.

The most accurate method to delineate the floodplain is to obtain detailed topographic information from field surveys and to conduct hydrologic and hydraulic analyses of the flood surface profile. A typical flood hazard area map is shown in Figure 9.20.

Structural methods for flood protection include the use of dams and reservoirs, levees or dikes, and channelization. Reservoirs store the excess runoff in upstream areas. Levees and dikes serve to confine floodwaters to a specific channel or flood zone, and channelization increases existing capacity to carry flow. But all structural methods provide only

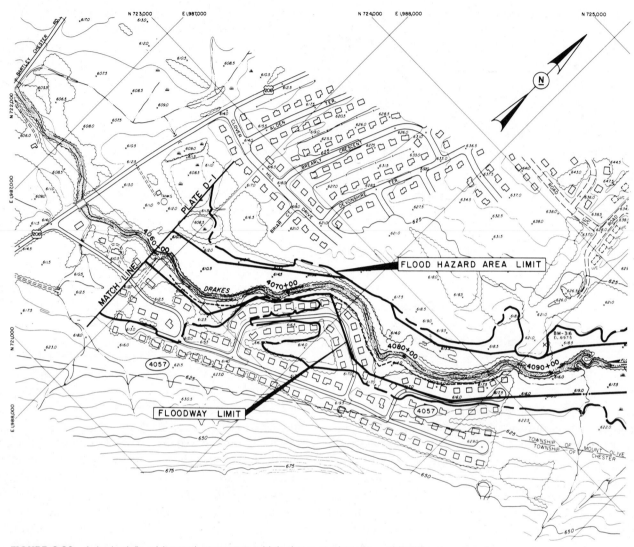

FIGURE 9.20 A typical flood hazard area map which shows a plan view of the floodway and flood hazard area. (Division of Water Resources, N.J. Department of Environmental Protection)

limited protection, up to the storm recurrence interval they are designed for. And they often spur additional development of the floodplain, because of the sense of security they provide.

Reservoirs and levees take up large land areas, and the expense of construction is usually beyond the means of local municipalities. Overall, the preferred method for preventing flood damage is regulation of land use in the floodplain, rather than using structural methods. Local zoning and subdivision ordinances can be effective in this regard. The National Flood Insurance Program provides incentive for proper regulation; federal subsidized flood insurance cannot be purchased unless the municipality participates in the NFIP.

REVIEW QUESTIONS

1. Why is stormwater control an important aspect of environmental technology?

2. Briefly compare the present-day approach toward stormwater control with past practices.

3. Define *time of concentration* of a drainage basin. How is it determined, and what is it used for?

4. What is a *runoff curve number?* What is it used for?

5. What does *csm/in.* stand for?

6. Give two direct effects of urbanization on stormwater runoff. What is the overall effect on a short-term basis, and in the long run?

7. Briefly describe three basic differences between storm sewer systems and sanitary sewer systems.

8. List and briefly describe three types of stormwater inlets.

9. Briefly discuss the design procedure for storm sewers.

10. List five specific benefits of on-site stormwater storage.

11. What is a disadvantage of stormwater storage basins?

12. Briefly describe three different types of stormwater storage basins.

13. Briefly discuss the design procedure for a stormwater detention basin.

14. Sketch a typical inflow and outflow hydrograph for a stormwater detention basin.

15. What is a *floodplain?* What is a *flood hazard area?*

16. Briefly discuss floodplain management measures.

PRACTICE PROBLEMS

1. Estimate the peak rate of runoff on a 15-ac watershed from a storm with rainfall intensity of 2 in./h. Use a runoff coefficient of 0.6.

2. Estimate the peak rate of runoff from a 0.15 km^2 drainage basin that has a runoff coefficient of 0.3, if the rainfall intensity is 60 mm/h.

3. From an air photo of a 22-ac watershed, it is determined that 4.5 ac is flat grassland, 6.0 ac is lightly wooded, 8.5 ac is a suburban residential area with large lots, and 3.0 ac is impervious pavement and roofs. Compute a composite runoff coefficient for the total watershed area.

4. A drainage basin of 180 acres comprises 40 percent wooded areas, 45 percent grassed areas, and 15 percent paved areas. Estimate a conservative value of the composite runoff coefficient for this drainage basin.

5. A 250-ha watershed comprises 80 ha of woods, 120 ha of suburban residential area, 30 ha of business area, and 20 ha of industrial area. Determine a probable range of values for the composite runoff coefficient in this watershed.

6. A 15-ac catchment area has a composite runoff coefficient of 0.5 and a time of concentration of 20 min. Compute the 25-year peak discharge.

7. A 10-ha drainage basin has a composite runoff coefficient of 0.7, and a time of concentration of 30 min. Compute the peak rate of runoff from a 100-year storm.

8. A 0.45-km^2 watershed has a composite runoff coefficient of 0.33. The overland flow distance to the beginning of a stream that drains the watershed is 200 m and the ground slope is 5 percent. The stream is 750 m long, with an average flow velocity of 0.25 m/s. Estimate the peak rate of runoff for the 100-year storm.

9. The watershed described in Problem 8 is to be developed as an industrial area, with a runoff coefficient of 0.9. Part of the stream is channelized, increasing the average flow velocity to 0.4 m/s. Assume the overland flow time remains the same. Determine the 100-year peak discharge from the watershed under developed conditions.

10. An 800-ac forested watershed with Type C soils has been zoned for the following land uses: 300 ac residential, 200 ac business and commercial, and the remaining 300 ac are to be left as undeveloped forest. Assume the time of concentration for the watershed decreases from 60 min to 45 min after development takes place. Using the SCS method, estimate the peak runoff discharge, both before and after development, for a 24-h storm which causes 7 in. of rainfall.

11. A storm drain system is laid out similar to that shown in Figure 9.12, with the following characteristics:

$$\text{catchment area 1: } A_1 = 3 \text{ ac, } C_1 = 0.5, \text{ inlet time} = 5 \text{ min}$$
$$\text{catchment area 2: } A_2 = 4.5 \text{ ac, } C_2 = 0.4, \text{ inlet time} = 8 \text{ min}$$
$$\text{catchment area 3: } A_3 = 6 \text{ ac, } C_3 = 0.3, \text{ inlet time} = 6 \text{ min}$$
$$\text{reach 1 (inlet 1 to inlet 2): } L = 300 \text{ ft, } S = 0.4 \text{ percent}$$
$$\text{reach 2 (inlet 2 to inlet 3): } L = 350 \text{ ft, } S = 0.3 \text{ percent}$$
$$\text{reach 3 (inlet 3 to inlet 4): } S = 0.3 \text{ percent}$$

Using the Rational Method, compute the design flow and required pipe diameter for each reach of the system, for a 25-year storm.

12. A storm drain system is laid out similar to that shown in Figure 9.12, with the following characteristics:

$$\text{catchment area 1: } A_1 = 1.2 \text{ ha, } C_1 = 0.6, \text{ inlet time} = 5 \text{ min}$$
$$\text{catchment area 2: } A_2 = 2.0 \text{ ha, } C_2 = 0.35, \text{ inlet time} = 6 \text{ min}$$
$$\text{catchment area 3: } A_3 = 3.0 \text{ ha, } C_3 = 0.2, \text{ inlet time} = 7 \text{ min}$$
$$\text{reach 1 (inlet 1 to inlet 2): } L = 100 \text{ m, } S = 0.0015$$
$$\text{reach 2 (inlet 2 to inlet 3): } L = 120 \text{ m. } S = 0.003$$
$$\text{reach 3 (inlet 3 to inlet 4): } S = 0.010$$

Using the Rational Method, compute the design flow and required pipe diameter for each reach of the system, for a 5-year storm.

13. A storm causes a peak runoff rate of 8 m^3/s in a drainage basin that has a time of concentration of 20 min. A detention basin with 12 000 m^3 of storage volume is built on site. Estimate the peak outflow rate from the basin for this storm using the simplified storage factor method.

14. Estimate the storage volume needed in a detention basin to reduce a peak inflow rate of 10 m^3/s to 2 m^3/s, when the total rainfall volume is 15 000 m^3.

15. In a land development project, the peak flow before construction is 40 cfs. The peak flow after construction is estimated to be 120 cfs. The local planning board requires that the peak runoff leaving the site after development be no greater than the 40 cfs predevelopment rate, and that on-site storage be provided to accomplish this. If the detention basin will have an average depth of 4 ft at full volume, approximately how many acres of the site will have to be used for the basin? Assume the time of concentration is 30 min.

SELECTED REFERENCES

1. *Design and Construction of Sanitary and Storm Sewers,* American Society of Civil Engineers, Manual of Practice No. 37, New York, 1969.

2. Whipple, W., et al., *Stormwater Management in Urbanizing Areas,* Prentice–Hall, Englewood Cliffs, N.J., 1983.

3. *Residential Stormwater Management/Objectives, Principles, and Design Considerations,* American Society of Civil Engineers, National Association of Home Builders, Urban Land Institute, New York, 1975.

4. McCuen, R., *A Guide to Hydrologic Analysis Using SCS Methods,* Prentice–Hall, Englewood Cliffs, N.J., 1982.

5. *Urban Hydrology for Small Watersheds,* Technical Release No. 55, Soil Conservation Service, U.S. Department of Agriculture, Washington, D.C., 1975.

6. *A Method for Estimating Volume and Rate of Runoff in Small Watersheds,* SCS-TP-149, Soil Conservation Service, U.S. Department of Agriculture, Washington, D.C., 1973.

7. *A Guide to the Environmental Aspects of the Local Planning Process,* Institute for Environmental Studies, Rutgers University, N.J. Dept. of Community Affairs, Trenton, 1976.

8. Debo, T. and H. Ruby, "Detention Basins—An Urban Experience," *Public Works Magazine,* Jan. 1982.

9. Mason, J. and E. Rhomberg, "On-site Detention—The Design Process," *Public Works Magazine,* Dec. 1982.

10. Pagan, A., "Flow Factor Line Used in Storage Calculations," American Society of Civil Engineers, *Journal of Irrigation and Drainage,* Mar. 1977.

10

WASTEWATER TREATMENT AND DISPOSAL

Raw or untreated sewage is mostly pure water. In fact, sanitary wastewater comprises about 99.9 percent water and only about 0.1 percent impurities. In other words, if a 1-L (1 kg) sample of wastewater was allowed to evaporate, only about 1 g or 1000 mg of solids would remain behind.

In contrast to this, seawater is only about 96.5 percent pure water; it contains about 35 000 mg/L or 3.5 percent dissolved impurities. Although seawater contains more impurities than does sanitary sewage, we do not ordinarily consider seawater to be "polluted." The important distinction is not the total concentration but the type of impurities. The impurities in seawater are mostly inorganic salts, but sewage contains biodegradable organic material, and it is very likely to contain pathogenic microorganisms as well.

Actually, sewage can contain so many different substances, both suspended and dissolved, that it is impractical to attempt to identify each specific substance or microorganism. The total amount of organic material is related to the "strength" of the sewage. This is measured by the *biochemical oxygen demand,* or BOD. Another important measure or parameter related to the strength of the sewage is the total amount of suspended solids, or TSS. On the average, untreated domestic sanitary sewage has a BOD of about 200 mg/L and a TSS of about 240 mg/L. Industrial wastewater may have BOD and TSS values much higher than those for sanitary sewage.

Another group of impurities that are typically of major significance in wastewater are the plant nutrients. Specifically, these are compounds of nitrogen, N, and phosphorus, P. On the average, raw sanitary sewage contains about 35 mg/L of N and 10 mg/L of P.

Finally, the amount of pathogens in the wastewater is expected to be proportional to the concentration of fecal coliform bacteria. The coliform concentration in raw sanitary sewage is roughly 1 billion per liter. Coliform concentration, as well as BOD, TSS, N, and P, are parameters of water quality that are discussed in some detail in Chapter 4.

Before discharging wastewater back into the environment and the natural hydrologic cycle, it is necessary to provide some degree of treatment or purification in order to protect public health and environmental quality. The basic purposes of sewage treatment are to destroy pathogenic microorganisms and to remove most of the suspended and dissolved biodegradable organic materials. Sometimes it is also necessary to remove the plant nutrients—nitrogen and phosphorus. Disinfection, usually with chlorine, serves to destroy most pathogens and help prevent the transmission of communicable disease. The removal of organics (BOD) and nutrients helps to protect the quality of aquatic ecosystems.

This chapter describes the most common types of wastewater treatment systems. These treatment methods are grouped into three general categories—preliminary and primary treatment, secondary or biological treatment, and tertiary or advanced treatment. First, the topic of sewage effluent standards is discussed, to put in perspective the overall goals of the various treatment processes. A later section covers on-site subsurface sewage disposal, and the chapter concludes with the topic of sewage sludge management.

10.1 EFFLUENT STANDARDS

In 1972, Congress passed the *Water Pollution Control Act*. One of the specific goals of that law was to clean up the nation's surface waters to the extent that they would be suitable for primary contact recreation as well as for the propagation of fish and wildlife. In other words, all of our streams, rivers, and lakes were to be made "swimmable and fishable" again, as they once were before "progress" and water pollution took their toll.

The *stream standards* that were in effect prior to 1972 were difficult to enforce and administer. In order to supplement those stream standards, which are discussed in Section 5.8, and to facilitate water pollution enforcement efforts, an additional system of *effluent standards* was established under the Water Pollution Control Act. These effluent standards set maximum allowable amounts of specific pollutants that can be discharged from wastewater treatment plants.

Every municipal or industrial treatment facility that discharges wastewater effluent into the environment must have an *NPDES Discharge Permit*. The NPDES (National Pollutant Discharge Elimination System) permit clearly states the allowable amounts of pollutants that a particular facility can discharge, and specifies a compliance schedule if a discharger cannot immediately meet the required limitations.

Under this new system of effluent standards, the focus of pollution control is shifted away from the existing stream standards, which regulate the amount of pollutants in the surface waters, to the amount of pollutants in separate discharges. Effluent standards are easier to enforce than are stream standards.

It is quite clear who is responsible when effluent standards are violated, but the responsibility for violation of stream standards is much more difficult to determine. For example, if the DO level in a stream drops below the required minimum value, it is difficult to prove that a specific polluter is responsible for that problem. On the other hand, it is easier to prove that a specific polluter is violating the conditions of its NPDES permit with regard to excessive BOD discharges. Although effluent standards are now the key to water quality protection, stream standards are still in effect.

The NPDES permit requirements may vary for different treatment plants. They depend on the nature of the wastewater, the classification of the receiving waters, and on the type of treatment. But as a minimum under the Water Pollution Control Act, all water pollution control plants in the United States must attain an 85-percent level of organic and suspended pollutant removal. Wastewater treatment that provides this level of treatment efficiency is called *secondary treatment*. This usually involves the use of biological processes, as described in Section 10.3.

Treatment Efficiency

Treatment efficiency can be defined as the ratio of the amount of pollutants removed to the amount of pollutants in the raw wastewater. In mathematical form, this is

$$\text{efficiency} = \frac{P_{IN} - P_{OUT}}{P_{IN}} \times 100 \qquad (10\text{-}1)$$

where P_{IN} = concentration of pollutant flowing in to the treatment system

P_{OUT} = concentration of pollutant flowing out of the system

EXAMPLE 10.1

Raw sewage flowing into a treatment plant (the plant *influent*) has a BOD$_5$ value of 200 mg/L. What is the maximum concentration of BOD$_5$ allowed in the treated sewage discharge (the plant *effluent*) if the required treatment efficiency is 85 percent? If the flow rate is 5 mgd, how many pounds of BOD will be discharged per day?

Solution:

Applying Equation 10-1, we get

$$85 = \frac{200 - P_{OUT}}{200} \times 100$$

and

$$\frac{200 \times 85}{100} = 200 - P_{OUT}$$

from which we get

$$P_{OUT} = 200 - 170 = 30 \text{ mg/L}$$

Now applying Equation 6-3b (which is not limited to chlorine in its application), we get

$$\begin{aligned}
\text{pounds per day} &= 8.34 \times Q \times C \\
&= 8.34 \times 5 \text{ mgd} \times 30 \text{ ppm} \approx 1300 \text{ lb/d}
\end{aligned}$$

EXAMPLE 10.2

A sewage treatment plant influent has an average TSS concentration of 250 mg/L. If the average effluent TSS concentration is 20 mg/L, what is the removal efficiency for TSS? If the flow rate is 5 ML/d, how many kilograms of suspended solids is discharged in the plant effluent each day?

Solution:

First applying Equation 10-1, we get

$$\begin{aligned}
\text{efficiency} &= \frac{250 - 20}{250} \times 100 \\
&= \frac{230}{250} \times 100 = 92 \text{ percent}
\end{aligned}$$

Now applying Equation 6-3a, kilograms per day = $Q \times C$ = 5 ML/d $\times$ 20 mg/L = 100 kg

NPDES Permit Requirements

Other water quality parameters in addition to BOD and TSS would be limited by permit in a wastewater treatment plant effluent. A typical NPDES permit would include limits as shown in Table 10.1.

The 1972 Water Pollution Control Act actually attempted to set a goal of "zero discharge" of pollutants into the nation's surface waters. It was soon

TABLE 10.1   TYPICAL NPDES EFFLUENT LIMITATIONS

Parameter	Maximum Allowable Concentration
BOD_5	30 mg/L
TSS	30 mg/L
pH	6.0 to 9.0
Fecal coliforms	200 per 100 mL

recognized that this was not only very impractical and expensive, but generally not necessary in many cases. Streams and rivers have a capacity for self-purification that could be used to advantage, without seriously degrading environmental quality or endangering public health.

The Water Pollution Control Act of 1972 was amended in Congress by the Clean Water Act of 1977. Effluent standards are now based upon the best conventional technology, even though this will not achieve the original zero-discharge goal. When secondary treatment levels are not sufficient to achieve stream standards for certain bodies of water, the requirements of the NPDES permit can be made more stringent. Some municipalities and industries have been required to provide *advanced* or *tertiary* treatment levels, which can increase pollutant removal efficiencies to 95 percent or more. Phosphorus and nitrogen can also be removed in order to reduce the rate of eutrophication of lakes, and to reduce the nitrogenous oxygen demand.

Industrial facilities may be served by on-site treatment plants, or they can discharge their wastewater to municipal or publicly owned treatment works (POTWs). But it is usually necessary for industry to provide some degree of *pretreatment* of the industrial wastewater before discharge into a public sewer system, in order to prevent disruption of the treatment process by strong chemical wastes. Regulations have been established for several major industries, including timber, petroleum, mining, textile, pharmaceutical, steel, and chemical production. These regulations limit the effluent concentrations of pollutants such as cyanide, chromium, cadmium, lead, and other toxic substances.

10.2   PRELIMINARY AND PRIMARY TREATMENT

The most basic level of wastewater treatment includes the physical processes of screening and sedimentation, which remove floating objects and settleable solids. This level of treatment removes roughly 60 percent of the suspended solids and about 35 percent of the BOD. Since a minimum of 85 percent of TSS and BOD removal is now required, preliminary and primary treatment alone are no longer adequate. But these processes normally precede other (secondary) treatment processes. In the older "primary" treatment plants, chlorination was the final step for disinfection of the effluent.

The first treatment process for raw wastewater is coarse screening. The *bar screens,* as they are called, are made of long narrow metal bars spaced about 25 mm (1 in.) apart. They retain floating debris such as wood, rags, or other bulky objects that could clog pipes or damage mechanical equipment in the rest of the plant. In most large sewage treatment plants, the bar screens are cleaned automatically by a mechanical device, as shown in Figure 10.1. The collected debris or *screenings* are promptly disposed of, usually by burial on the plant grounds.

In some treatment plants, a mechanical cutting or shredding device, called a *comminutor,* is installed just after the coarse screens. A typical comminutor, shown in Figure 10.2, consists of a slotted cylindrical screen with a moving cutter blade. The comminutor shreds and chops solids or rags that passed through the bar screen. The shredded material is removed from the wastewater by sedimentation or flotation later in the treatment plant. In small sewage treatment plants, a manually cleaned bar screen is generally installed in a channel next to the comminutor, to serve as an emergency bypass when the comminutor needs repair.

Grit Removal

A portion of the suspended solids in raw sewage consists of gritty material such as sand, coffee

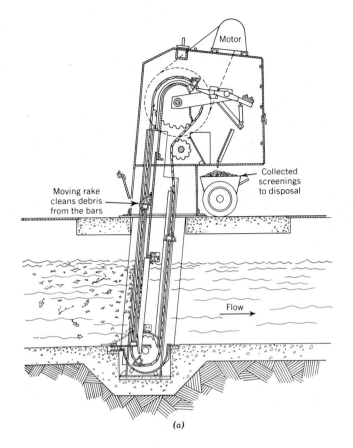

FIGURE 10.2 A typical comminutor installation. (Worthington Pump Division, Dresser Industries, Inc.)

(b)

FIGURE 10.1 Typical bar screen installations, which serve to remove sticks, rags, and other debris from the flow of sewage. *(a)* (Envirex, Inc./a Rexnord Company) *(b)* (FMC Corporation)

grounds, eggshells, and other relatively inert material. In cities with combined sewer systems, large amounts of fine sand and silt may be carried in the sewage. Suspended grit can cause excessive wear-and-tear on pumps and other equipment in the plant. Most of it is nonbiodegradable and will accumulate in treatment tanks. For these reasons, a grit removal process is usually used after screening and/or comminuting.

In the sewers, the flow velocity is generally not less than 0.6 m/s (2 ft/s), the "self-cleansing velocity." Most gritty material is dense enough to settle out of the flow by gravity, if the velocity is reduced to about 0.3 m/s (1 ft/s). Although the grit will settle out at this reduced velocity, the lighter suspended organic solids will still be carried through to the next treatment unit.

The reduction in velocity and the collection of the grit is usually accomplished in long, narrow tanks called *grit chambers*. Even though the sewage flow rate varies throughout the day, the flow velocity can be kept almost constant at 0.3 m/s by the use of a specially shaped outlet weir. Mechanical collectors made of buckets on a continuous

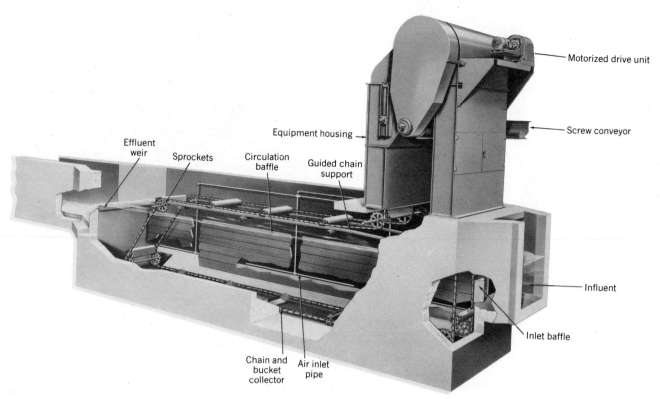

FIGURE 10.3   A mechanically cleaned grit chamber. (Envirex, Inc./a Rexnord Company)

chain serve to remove the grit from the tank; the grit is promptly disposed of by burial on the plant grounds. An aerated grit chamber is shown in Figure 10.3. The rising air bubbles help to keep the organic solids in suspension, while the grit settles to the bottom.

Primary Settling

After preliminary treatment by screening, comminuting, and grit removal, the wastewater still contains suspended organic solids that can be removed by plain sedimentation. The basic principles of sedimentation, or plain gravity settling, have been discussed in Section 6.2 with regard to drinking water purification. For the most part, the same principles apply to sewage treatment as well. There are, of course, differences in the recommended design values for detention time, overflow rate, and tank configuration.

Settling tanks that receive sewage after grit removal are called *primary clarifiers*. They usually provide about 2 hours' detention time; side water depth (SWD) is generally between 2.5 and 5 m (8 and 16 ft). The tanks may be circular or rectangular in shape. In addition to mechanical sludge collectors that continually scrape the settled solids along the bottom to a sludge hopper for removal, a surface skimming device is used to remove grease and other floating materials from the liquid surface.

EXAMPLE 10.3

A primary clarifier has an average influent TSS concentration of 250 mg/L. If its TSS removal efficiency is expected to be 60 percent, what is the expected average effluent TSS concentration?

Solution:

Applying Equation 10-1, we get

$$60 = \frac{250 - P_{\text{OUT}}}{250} \times 100$$

and

$$P_{\text{OUT}} = 250 - \frac{250 \times 60}{100} = 100 \text{ mg/L}$$

The removal efficiencies of primary settling tanks can be increased by the addition of chemical coagulants, although this is not common practice in most modern wastewater treatment plants. In order to achieve BOD and TSS removal efficiencies of at least 85 percent, as required by the Clean Water Act, primary treatment must be followed by an additional treatment process. Generally, this next step is characterized as *secondary treatment*.

10.3 SECONDARY (BIOLOGICAL) TREATMENT

Preliminary and primary treatment processes remove only those pollutants that will either float or settle out by gravity. But about half of the raw pollutant load still remains in the primary effluent. The purpose of *secondary treatment* is to remove the suspended solids that did not settle out in the primary tanks, as well as to remove dissolved BOD that is unaffected by physical treatment. Secondary treatment is generally considered to mean 85 percent BOD and TSS removal efficiency, and now represents the minimum degree of treatment required by law in most cases. In the United States, secondary treatment processes are almost always *biological* systems.

Biological treatment of sewage involves the use of living microscopic organisms. The microbes, including bacteria and protozoa, consume the organic pollutants as food. They metabolize the biodegradable organics, converting them into carbon dioxide, water, and energy for their growth and reproduction. This natural aerobic process requires oxygen, and was previously described under the topic of BOD in Section 4.3.

A biological sewage treatment system must provide the microorganisms with a comfortable "home." In effect, the treatment plant allows the bacteria to stabilize the organic pollutants in a controlled, artificial environment of steel and concrete, rather than in a stream or lake. This helps to protect the dissolved oxygen balance of the natural aquatic environment.

To keep the microbes "happy" and productive in their task of wastewater treatment, they must be provided with enough oxygen, adequate contact with the organic material in the sewage, suitable temperatures, and other favorable conditions. The design and operation of a secondary treatment plant is accomplished with these factors in mind.

Two of the most common biological treatment systems include the *trickling filter* and the *activated sludge process*. The trickling filter is a type of *fixed-growth* system: the microbes remain fixed or attached to a surface while the wastewater flows over that surface to provide contact with the organics. Activated sludge is characterized as a *suspended-growth* system, because the microbes are thoroughly mixed and suspended in the wastewater rather than attached to a particular surface.

TRICKLING FILTERS

A trickling filter consists basically of a layer or bed of crushed rock, about 2 m (6 ft) deep. It is usually circular in shape and may be built as large as 60 m (200 ft) in diameter. Trickling filters are always preceded by primary treatment to remove coarse and settleable solids. The primary effluent is sprayed over the surface of the crushed stone bed and trickles downward through the bed to an *underdrain system*.

A rotary distributor arm with nozzles located along its length is usually used to spray the sewage, although sometimes fixed nozzles are used. The rotary distributor arm is mounted on a center column in the trickling filter; it is driven around by the reaction force or "jet action" of the wastewater that flows through the nozzles.

The underdrain system serves to collect and carry away the wastewater from the bottom of the bed as well as to permit air circulation upward through the stones. As long as topography permits, the sewage flows from the primary tank to the trickling filter by the force of gravity rather than by pumping. A cutaway view of a typical trickling filter unit is shown in Figure 10.4.

As the primary effluent trickles downward through the bed of stones, a biological slime of microbes develops on the surfaces of the rocks. The continuing flow of the wastewater over these fixed biological growths provides the needed contact between the microbes and the organics. The microbes in the thin slime layer absorb the dissolved organics, thus removing "oxygen-demanding" substances from the wastewater. Air circulating through the void spaces in the bed of stones provides the needed oxygen for stabilization of the organics by the microbes.

It should be noted that the so-called trickling filter is not really a "filter" at all, in the true sense of the word. The stones are usually about 75 mm (3 in.) in size, much too large to strain or filter out suspended solids. And by definition, filters have no effect on dissolved solids. The stones in a trickling filter only serve to provide a large amount of surface area for the biological growths, and the large voids allow ample air circulation. Sometimes materials other than rock, such as modules of corrugated plastic or redwood slats, are used to provide the needed surface area and void spaces. But the basic purpose and operation remains the same.

As the microorganisms grow and multiply, the slime layer gets thicker. Eventually it gets so thick that the flowing wastewater washes it off the surfaces of the stones. This is called *sloughing* (pronounced "sluffing"). Since sloughing does occur periodically, there is a need to provide settling time for the trickling filter effluent, in order to remove the sloughed biological solids. These solids consist basically of billions of microorganisms that have absorbed the dissolved organics into their bodies.

The trickling filter effluent is collected in the underdrain system and then conveyed to a sedimentation tank called a *secondary clarifier*. The secondary clarifier, or *final clarifier* as it is sometimes

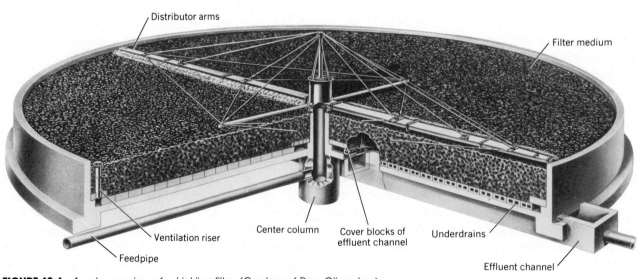

FIGURE 10.4 A cutaway view of a trickling filter. (Courtesy of Dorr–Oliver, Inc.)

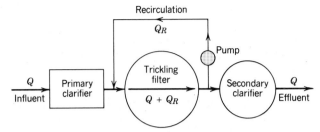

FIGURE 10.5 A flow diagram for recirculation of flow through a trickling filter. The rate of sewage flow applied to the filter is the sum of the influent flow rate and the recirculated flow rate.

called, is similar in most respects to the primary clarifier, although there are differences in detention time, overflow rate, weir loading, and other details.

In order to maintain a relatively uniform flow rate through the trickling filter, and to keep the distributor arm rotating even during periods of low sewage flow, some of the wastewater may be recirculated. In other words, a portion of the effluent is pumped back to the trickling filter inlet so that it will pass through the bed of stones more than once.

Recirculation can also serve to improve the pollutant removal efficiency; it allows the microbes to remove organics that flowed by them during the previous pass through the bed. There are many recirculation patterns and configurations of trickling filter plants. One common pattern, called *direct recirculation,* is shown in a flow diagram in Figure 10.5.

Recirculation

The amount of recirculation can vary. It is characterized by a *recirculation ratio,* which is the ratio of recycled flow to the raw wastewater flow. In formula form, it is:

$$R = \frac{Q_R}{Q} \qquad (10\text{-}2)$$

where R = recirculation ratio
$\quad Q_R$ = recirculated flow rate
$\quad Q$ = raw sewage flow rate

The recirculation ratio, R, is generally in the range of 0.0 to 3.0.

Hydraulic Load

The rate at which the wastewater flow is applied to the trickling filter surface is called the *hydraulic load.* The hydraulic load includes the recirculated flow Q_R; the total flow through the trickling filter is equal to $Q + Q_R$. In formula form, it is:

$$\text{hydraulic load} = \frac{Q + Q_R}{A_S} \qquad (10\text{-}3)$$

where Q = raw sewage flow rate
$\quad Q_R$ = recirculated flow rate
$\quad A_S$ = trickling filter surface area (plan view)

Hydraulic load may be expressed in terms of cubic meters per day per square meter of surface area, or $m^3/m^2 \cdot d$. It also may be expressed in terms of million gallons per acre of surface area per day, or mil gal/ac/d. A typical value for a conventional trickling filter is 20 $m^3/m^2 \cdot d$ (19 mil gal/ac/d).

Organic (BOD) Load

The rate at which organic material is applied to the trickling filter is called the *organic* or *BOD load.* It does not include the BOD added by recirculation. Organic load is expressed in terms of kilograms of BOD per cubic meter of bed volume per day, or $kg/m^3 \cdot d$. It is also expressed in terms of pounds of BOD per thousand cubic feet of bed volume per day, or $lb/1000 \ ft^3/d$. A typical value for organic load on a trickling filter is 0.5 $kg/m^3 \cdot d$, or 30 $lb/1000 \ ft^3/d$. In formula form, the organic load may be expressed as:

$$\text{organic load} = \frac{Q \times \text{BOD}}{V} \qquad (10\text{-}4a)$$
$$\text{(metric units)}$$

or

$$\text{organic load} = 8340 \times \frac{Q \times \text{BOD}}{V} \quad (10\text{-}4b)$$
$$\text{(American units)}$$

where Q = raw wastewater flow, ML/d (mgd)

BOD = BOD_5 in the primary effluent, mg/L (ppm)

V = volume of trickling filter bed, m^3 (ft^3)

EXAMPLE 10.4

A 2-m-deep trickling filter with a diameter of 18 m is operated with a recirculation ratio of 1.5. The raw wastewater flow rate is 2.5 ML/d and the 5-day BOD of the raw sewage is 210 mg/L. Assuming that the primary tank BOD removal efficiency is 30 percent, compute the hydraulic load and the organic load on the trickling filter.

Solution:

First compute the surface area of the trickling filter, as

$$A_S = \frac{\pi \times D^2}{4} = \frac{\pi \times 18^2}{4} = 254.5 \ m^2$$

Since volume = area × depth, we get

$$V = 254.5 \ m^2 \times 2 \ m = 509 \ m^3$$

From Equation 10-2, we can write $Q_R = R \times Q$ and

$$Q_R = 1.5 \times 2.5 = 3.75 \ \text{ML/d}$$

Also

$$Q + Q_R = 2.5 + 3.75 = 6.25 \ \text{ML/d} = 6250 \ m^3/d$$

From Equation 10-3 we get

$$\text{hydraulic load} = \frac{6250 \ m^3/d}{254.5 \ m^2} \approx 25 \ m^3/m^2{\cdot}d$$

The primary effluent BOD can be computed with Equation 10-1:

$$30 = \frac{210 - P_{\text{OUT}}}{210} \times 100$$

and

$$P_{\text{OUT}} = 210 - \frac{210 \times 30}{100} = 147 \ \text{ppm}$$

From Equation 10-4a, we get

$$\text{organic load} = \frac{2.5 \times 147}{509} = 0.72 \ \text{kg/m}^3{\cdot}d$$

Efficiency

The BOD removal efficiency of a trickling filter unit depends primarily on the organic load, the recirculation ratio, and the temperature of the wastewater. Generally, the efficiency increases with decreasing organic load, increasing recirculation, and increasing temperature. For example, with no recirculation ($R = 0$), and a temperature of 20°C, a typical trickling filter would have an efficiency of about 60 percent when the organic load is about 2 $\text{kg/m}^3{\cdot}d$. But if the organic load was 0.5 $\text{kg/m}^3{\cdot}d$, at the same conditions of recirculation and temperature, the efficiency would be 75 percent.

Further, at the 0.5 $\text{kg/m}^3{\cdot}\text{day}$ loading, a recirculation of $R = 1$ instead of $R = 0$ could raise the efficiency to about 80 percent. And if the temperature increased to 22°C, the efficiency would be raised to about 85 percent. Because of the marked effect of temperature on treatment efficiency, trickling filters in northern climates are often enclosed under fiberglass domes to provide protection against wind and snow and to reduce the rate of heat loss from the wastewater.

EXAMPLE 10.5

The BOD removal efficiency of a trickling filter system is 79 percent, and the efficiency of the primary treatment that precedes it is 35 percent. If the raw BOD is 200 mg/L, what is the effluent BOD? Is the treatment plant providing an efficiency that meets the requirement for "secondary" treatment?

Solution:

Although the biological treatment and secondary clarification provide only 79 percent BOD removal, it should be remembered that 35 percent of the raw BOD was already removed by primary settling. Thus 65 percent or $0.65 \times 200 = 130$ mg/L remains as BOD in the primary effluent.

But 79 percent of the BOD entering the trickling filter is removed, leaving 21 percent or $0.21 \times 130 = 27$ mg/L in the secondary effluent. The overall plant efficiency is therefore computed as

$$\text{efficiency} = \frac{200 - 27}{200} \times 100 \approx 87 \text{ percent}$$

This is greater than 85 percent so the treatment plant is providing secondary treatment.

ACTIVATED SLUDGE TREATMENT

The basic components of an activated sludge sewage treatment system include an aeration tank and a secondary settling basin or clarifier. Primary effluent is mixed with settled solids that are recycled from the secondary clarifier and then introduced into the aeration tank. Compressed air is injected continuously into the mixture, through porous diffusers located at the bottom of the tank, along one side. This is illustrated in Figure 10.6.

In the aeration tank, microorganisms consume the dissolved organic pollutants as food. The mi-

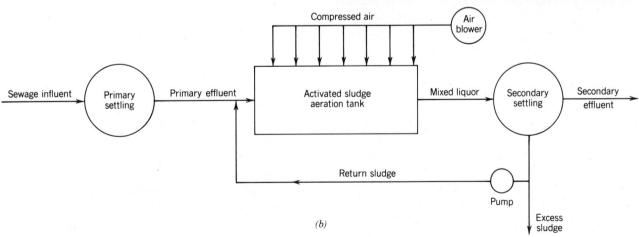

FIGURE 10.6 *(a)* A typical activated sludge sewage treatment plant. The rectangular tanks in the foreground are the aeration tanks; the air diffusers have been raised out of the tank on the right (FMC Corporation) *(b)* A flow diagram for a conventional activated sludge plant.

crobes absorb and aerobically decompose the organics, using oxygen provided in the compressed air; water, carbon dioxide, and other stable compounds are formed. In addition to providing oxygen, the compressed air thoroughly mixes the microbes and wastewater together, as it rapidly bubbles up to the surface from the diffusers. Sometimes mechanical propellerlike mixers, located at the liquid surface, are used instead of compressed air and diffusers. The churning action of the propeller blades mixes air with the wastewater and keeps the contents of the tank in a uniform suspension.

The aerobic microorganisms in the tank grow and multiply, forming an active suspension of biological solids called *activated sludge*. The combination of the activated sludge and the wastewater in the aeration tank is called the *mixed liquor*. In the basic or conventional activated sludge treatment system, a tank detention time of about six hours is required for thorough stabilization of most of the organics in the mixed liquor.

After about six hours of aeration, the mixed liquor flows to the secondary or final clarifier, in which the activated sludge solids settle out by gravity. The clarified water near the surface, called the *supernatant*, is discharged over an effluent weir; the settled sludge is pumped out from a sludge hopper at the bottom of the tank. Recycling a portion of the sludge back to the inlet of the aeration tank is an essential characteristic of this treatment process. The settled sludge is in an "active" state. In other words, the microbes are well acclimated to the wastewater and, given the opportunity, will readily absorb and decompose more organics by their metabolism.

By pumping about 30 percent of the wastewater flow from the bottom of the clarifier back to the head of the aeration tank, the activated sludge process can be maintained continuously. When mixed with the primary effluent, the "hungry" microbes quickly begin to absorb and metabolize the fresh "food" in the form of BOD causing organics. Since the microbes multiply and increase greatly in numbers, it is not possible to recycle or return all the sludge to the aeration tank. The excess sludge, called *waste activated sludge,* must eventually be treated and disposed of (along with sludge from the primary tanks). Sludge management is a major aspect of wastewater treatment, and is discussed in Section 10.6.

F/M Ratio

An important factor used in the design and operation of activated sludge systems is known as the *food-to-microorganism (F/M) ratio*. The "food" is measured in terms of kilograms (pounds) of BOD added to the tank per day. And since the suspended solids in the mixed liquor consist mostly of living microorganisms, the suspended solids concentration is used as a measure of the amount of microorganisms in the tank. This concentration is called the *mixed liquor suspended solids,* or MLSS.

The F/M ratio is an indicator of the organic load on the system, with respect to the amount of biological solids in the tank. For conventional aeration tanks, the ratio is in the range of 0.2 to 0.5. It can be computed from the following formula:

$$\text{F/M} = \frac{Q \times \text{BOD}}{\text{MLSS} \times V} \qquad (10\text{-}5)$$

where F/M = food-to-microorganism ratio, in units of kilograms of BOD per kilogram of MLSS per day

Q = raw sewage flow rate, ML/d (mgd)

BOD = applied 5-day BOD, mg/L (ppm)

MLSS = mixed liquor suspended solids, mg/L

V = volume of aeration tank, ML (million gal)

EXAMPLE 10.6

An activated sludge tank is 30 m long, 10 m wide, and has a SWD of 4 m. The wastewater flow is 4.0 ML/day and the raw 5-day BOD is 200 mg/L. The MLSS concentration is 2000 mg/L. Compute the food-to-microorganism ratio for the system.

Solution:

A conventional activated sludge aeration tank is preceded by primary treatment. Assuming that 35 percent of the raw BOD is removed in the primary clarifier, 65 percent of the BOD would be applied to the aeration tank.

$$0.65 \times 200 = 130 \text{ mg/L}$$

The tank volume is the product of length, width, and depth, or

$$V = 30 \text{ m} \times 10 \text{ m} \times 4 \text{ m} = 1200 \text{ m}^3 \text{ or } 1.2 \text{ ML}$$

Applying Equation 10-5, we get

$$\text{F/M} = \frac{4.0 \times 130}{2000 \times 1.2} = 0.22$$

EXAMPLE 10.7

A conventional aeration tank is to treat a flow of 800 000 gpd of primary effluent with a BOD of 125 ppm. The MLSS concentration is to be maintained at 1800 ppm and a food-to-microorganism ratio of 0.4 is specified. Compute the required volume of the aeration tank. If the side water depth is to be 15 ft and the tank length is to be three times its width, how long should the tank be?

Solution:

Rearranging the terms of Equation 10-5, we get

$$V = \frac{Q \times \text{BOD}}{\text{F/M} \times \text{MLSS}} = \frac{0.8 \times 125}{0.4 \times 1800}$$
$$= 0.139 \text{ million gal}$$

and

$$140\ 000 \text{ gal} \times \frac{1 \text{ ft}^3}{7.5 \text{ gal}} = 18\ 500 \text{ ft}^3$$

Since volume = length × width × depth, or $V = L \times W \times \text{SWD}$, and since the length is to be three times the width, or $L = 3W$, we get $V = 3W \times W \times \text{SWD} = 3 \times W^2 \times \text{SWD}$

We can solve for the tank width as follows:

$$W = \left(\frac{18\ 500 \text{ ft}^3}{3 \times 15 \text{ ft}}\right)^{1/2} \approx 20 \text{ ft}$$

and

$$L = 3 \times 20 = 60 \text{ ft}$$

Sludge Settling

In the activated sludge process, the organic pollutants are absorbed by the billions of microorganisms in an aeration tank. These microorganisms essentially are the "activated sludge." But without proper clarification or separation of the sludge from the liquid portion of the mixed liquor, the treatment process will not be effective at all. For this reason, gravity settling in the secondary clarifier is a most important part of the activated sludge treatment system. If the sludge does not settle fast enough, some of it will be carried over the effluent weirs of the clarifier and cause pollution of the receiving body of water.

Under certain conditions in an activated sludge sewage treatment plant, filamentous or stringy bacteria, usually of the species *Sphaerotilus natans,* grow prolifically in the aeration tank, making the sludge very fluffy and light. Sludge with excessive growths of these filamentous organisms settles very slowly, and a clear supernatant is not formed in the secondary clarifier. Much of the sludge flows out with the effluent. This condition is called *sludge bulking.* Bulking of activated sludge may be controlled or limited by appropriate adjustments in MLSS concentration and F/M ratio; this is accomplished by regulating the rate of sludge return from the clarifier. The amount of aeration may also be a factor; sludge bulking is sometimes associated with too much aeration. Occasionally, adjustments in the mixed liquor pH are made to solve the problem.

A number called the *sludge volume index (SVI)* is used to evaluate the settleability of the activated sludge. It is equal to the volume occupied by 1 g of settled sludge and is expressed in units of milliliters per gram (mL/g).

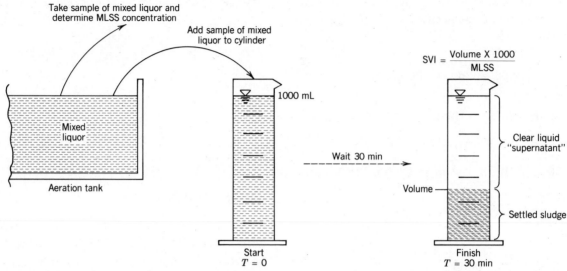

FIGURE 10.7 A schematic illustration of the lab test for *sludge volume index* (SVI), which is used to evaluate sludge settling characteristics.

The determination of SVI involves taking a sample of mixed liquor from the aeration tank and allowing it to settle for 30 min in a 1-L graduated glass cylinder. This is illustrated in Figure 10.7. The volume of settled sludge is read from the markings on the cylinder. The MLSS concentration in the mixed liquor is also measured.

The following formula is used to compute SVI:

$$SVI = \frac{V \times 1000}{MLSS} \qquad (10\text{-}6)$$

where SVI = sludge volume index, mL/g
      V = volume of settled sludge, mL/L
  MLSS = mixed liquor suspended solids, mg/L

EXAMPLE 10.8

An aeration tank has a MLSS concentration of 2000 mg/L. After settling for 30 min in a 1-L graduated cylinder, the sludge volume is measured to be 150 mL. Compute the SVI of the sludge.

Solution:

Applying Equation 10-6, we get

$$SVI = \frac{150 \text{ mL/L} \times 1000}{2000 \text{ mg/L}} = 75 \text{ mL/g}$$

A normal sludge with good settling characteristics generally has an SVI of less than 100. As the SVI increases above 100, sludge settleability decreases and some solids get carried over the effluent weir of the clarifier. Severely bulking sludges would have SVI values of over 200. A very high SVI is an indication to a treatment plant operator that the sludge return rate should be increased, the aeration rate should be decreased, or some other process adjustment should be made. Sometimes chlorine is added to the aeration tank to destroy the filamentous organisms, but this is a last resort to control the problem.

In general, a well-operated activated sludge treatment system can remove about 90 percent of the raw sewage BOD and TSS; sometimes removal efficiency may be as high as 95 percent. In contrast to the simpler trickling filter sewage treatment system, however, an activated sludge plant requires careful operational control. Energy requirements in an activated sludge plant are also high, because of the power consumed for aeration.

MODIFICATIONS OF THE ACTIVATED SLUDGE PROCESS

Several modifications of the conventional activated sludge process have been developed; these serve to increase treatment plant capacity or to reduce the tank volume requirements.

Step Aeration

A process called *step aeration* provides multiple feed points of the primary effluent into the aeration tank, as shown in Figure 10.8. By introducing the organics into the tank in increments or "steps," rather than only once at the head of the tank, the oxygen demand is spread more uniformly over the length of the tank. In this manner, greater treatment plant capacities can be obtained in a given volume of aeration tank than can be obtained using the conventional process.

Extended Aeration

For treating small sewage flow rates from surburban residential developments, hotels, schools, and other relatively isolated wastewater sources, a process called *extended aeration* is often used. These small systems are generally in the form of prefabricated steel tanks and are called "package plants." (Conventional tanks, on the other hand, are usually made with cast-in-place reinforced concrete.) In the extended aeration system, the aeration tank and secondary clarifier are built in a single unit, as illustrated in Figures 10.9 and 10.10.

There are two important distinctions between an extended aeration system and a conventional system. First, screened or comminuted sewage is directed into the extended aeration tank without any primary settling. Second, the detention time or aeration period is about 30 h, whereas the conventional system's detention time is about 6 h.

Another difference is that the extended aeration process operates with F/M ratios as low as 0.05. This means that there is a large population of microorganisms compared to the amount of food (organics). The low F/M ratio and the "extended" period of aeration allow for the stabilization of most of the organics in the wastewater. But eventually

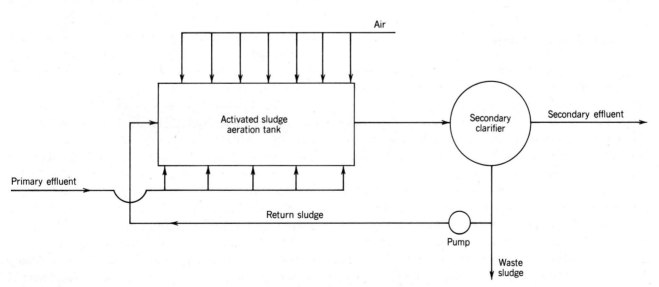

FIGURE 10.8 A flow diagram for the step aeration modification of the activated sludge process.

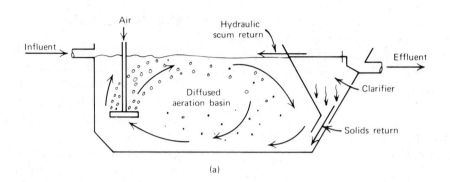

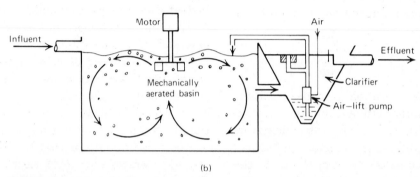

FIGURE 10.9 Two typical small extended aeration treatment systems, using *(a)* diffused aeration and a slotted bottom clarifier for sludge return, and *(b)* mechanical aeration with an air-lift pump for sludge return. (From Hammer, M. J. *Water and Wastewater Technology,* with permission of John Wiley & Sons.)

some sludge has to be removed from the aeration tank for disposal.

Contact Stabilization

In yet another modification of the activated sludge process, the influent sewage is mixed and aerated with return activated sludge for only about 30 min. This process is called *contact stabilization.* The short contact period of 30 min is sufficient for the microorganisms to absorb the organic pollutants, but not to stabilize them.

After the short contact time, the mixed liquor enters a clarifier and the activated sludge settles out; the clarified sewage flows over effluent weirs. The settled sludge is pumped into another aerated tank, called a *reaeration* or *stabilization tank.* The contents of the stabilization tank are aerated for about three hours, allowing the microbes to decompose the

absorbed organic material. The total size of a contact stabilization tank is generally less than that of a conventional plant. This is because the volume of activated sludge being stabilized in the reaeration tank is considerably less than the total wastewater flow. A schematic flow diagram of the contact stabilization process is shown in Figure 10.11.

A typical installation consists of a field-erected circular steel tank, with an inner tank providing a zone for clarification. The annular volume between the inner tank and the outer tank wall provides the room or zones for contact, reaeration, and sludge storage. An "air-lift" type of pump is usually used to transfer sludge between zones of the tank. With some minor modification of piping and baffling arrangements, a contact stabilization system can also be operated in the step aeration or in the extended aeration mode. A circular sewage treatment unit is shown in Figure 10.12.

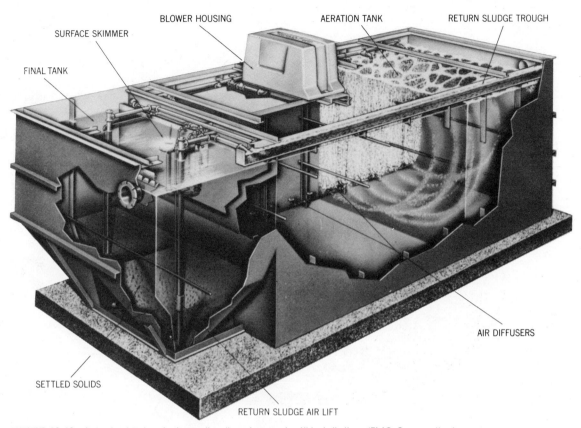

FIGURE 10.10 A typical extended aeration "package plant" installation. (FMC Corporation)

Pure Oxygen Aeration

Air is only 21 percent oxygen. Instead of using air, greater treatment capacities can be achieved by injecting high-purity oxygen into the mixed liquor of an activated sludge sewage treatment plant. A dia-

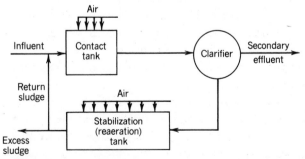

FIGURE 10.11 The contact stabilization modification of the activated sludge process. Organic pollutants are absorbed by the microbes in the contact aeration tank and stabilized in the reaeration tank.

gram of a *high-purity oxygen aeration system* is shown in Figure 10.13. The oxygen is manufactured at the plant site. Primary effluent, return activated sludge, and oxygen are introduced into the first compartment of a multistaged, covered tank. Mechanical agitators mix the oxygen with the wastewater as it flows through the tank. The total aeration period is only about two hours, and the F/M ratio is as high as 1.5. Consequently, the aeration tank volume is considerably less than that required for the conventional system.

OTHER SECONDARY TREATMENT PROCESSES

A biological treatment unit that has been used in Europe, and which in recent years has become an accepted method of sewage treatment in the United States, is the *biodisc* or *rotating biological*

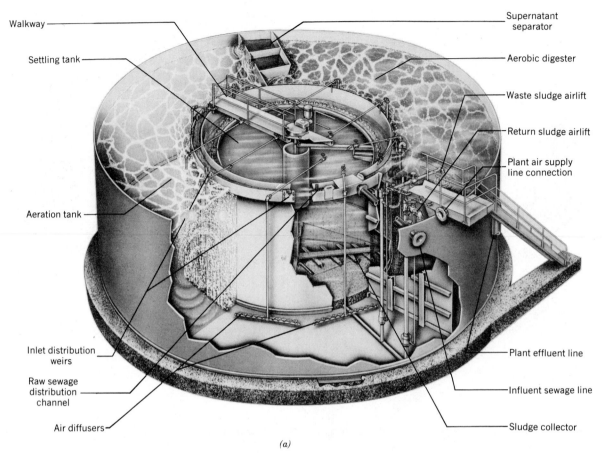

Walkway

Settling tank

Aeration tank

Inlet distribution
weirs

Raw sewage
distribution
channel

Air diffusers

Supernatant
separator

Aerobic digester

Waste sludge airlift

Return sludge airlift

Plant air supply
line connection

Plant effluent line

Influent sewage line

Sludge collector

(a)

FIGURE 10.12   *(a)* Circular prefabricated steel sewage treatment plants are available in diameters up to about 30m (100 ft.) *(b)* They can be used for either step aeration, for contact stabilization, or for extended aeration modes of treatment. (FMC Corporation)

contactor. It consists of a series of large plastic discs mounted on a horizontal shaft. The lightweight discs are about 3 m (10 ft) in diameter, and are spaced about 40 mm (1.5 in.) apart on the shaft. A typical biodisc unit is illustrated in Figure 10.14.

The discs are partially submerged in settled sewage (primary effluent). As the shaft rotates, the disc surfaces are alternately in contact with air and with the wastewater. Consequently, a layer of biological slime grows on each disc, and the attached microbes that form the slime absorb the organic material in the wastewater. This process is similar to the trickling filter system, except that the attached microbial growths are passed through the

wastewater, instead of the wastewater being sprayed over the microorganisms.

The speed of rotation and the number of discs can be varied in order to achieve specific levels of pollutant removal. With several stages of discs, it is possible to remove nitrogenous as well as carbonaceous BOD. This is because growths of nitrifying bacteria predominate in the microbial population on the final disc stages.

In a biodisc system, there is no need to recycle sludge. But a secondary clarifier is needed to settle out the excess biological solids that slough off the discs as the slime layer thickens. Like the trickling filter, the efficiency of the biodisc process is adversely affected by low temperatures. This is be-

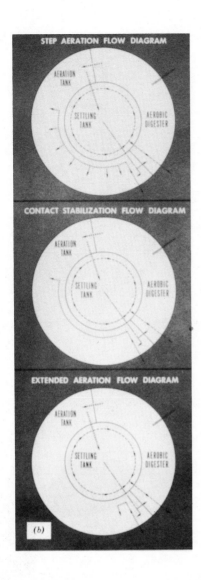

STEP AERATION FLOW DIAGRAM

AERATION TANK

SETTLING TANK

AEROBIC DIGESTER

CONTACT STABILIZATION FLOW DIAGRAM

AERATION TANK

SETTLING TANK

AEROBIC DIGESTER

EXTENDED AERATION FLOW DIAGRAM

AERATION TANK

SETTLING TANK

AEROBIC DIGESTER

(b)

cause the rate of metabolism of the microbes slows down when the temperature drops.

In suburban or rural areas where land is available at relatively low cost, *sewage lagoons* may be used for secondary treatment. They are also called *stabilization* or *oxidation ponds*. The most common type of lagoon used for treating wastewater is the *faculative pond*. In a faculative pond, which is generally about 2 m (6 ft) deep, both aerobic and anaerobic biochemical reactions take place. This is illustrated in Figure 10.15.

Raw wastewater enters the pond, eliminating the need for primary treatment. Organic solids that settle to the bottom decompose anaerobically, producing such substances as methane, hydrogen sulfide, and organic acids. In the liquid above the sludge zone of the pond, incoming organics and the products of anaerobic decomposition are stabilized by *faculative bacteria,* as well as by aerobic microorganisms. Faculative bacteria can grow in either aerobic or anaerobic environments. The average sewage detention time in a faculative pond may be 60 days or more.

Oxygen is added to the wastewater in the pond by wind action and mixing at the surface and from the daylight metabolism of algae. This oxygen supports the aerobic reactions. The mutually dependent relationship between the algae and bacteria in a stabilization pond is very important. Using energy from sunlight, the algae grow and multiply by consuming the carbon dioxide and other inorganic compounds released by the bacteria. The bacteria use both the oxygen released by the algae and the organics from the wastewater.

Although the algae play an important role in the purification process in a sewage lagoon, they can also cause a problem. When these microscopic plants are carried out of the pond in the effluent flow, the allowable levels of TSS for a secondary effluent are usually exceeded. This problem can be particularly severe during the warm summer months. Sometimes, using two or more ponds in series, with careful control of the effluent flow, can eliminate the problem of algae carryover.

Despite the potential difficulty with regard to TSS removal efficiency, sewage lagoons are being used with increasing frequency in areas where land is readily available. The low construction costs, ease of operation and maintenance, and negligible energy costs, offer distinct advantages for this "natural" purification system.

Secondary Effluent Disinfection

The last step in the secondary sewage treatment process is *disinfection*. The purpose of sewage disinfection is to destroy any pathogens in the effluent that may have survived the treatment process, thereby protecting public health. (Removal of BOD

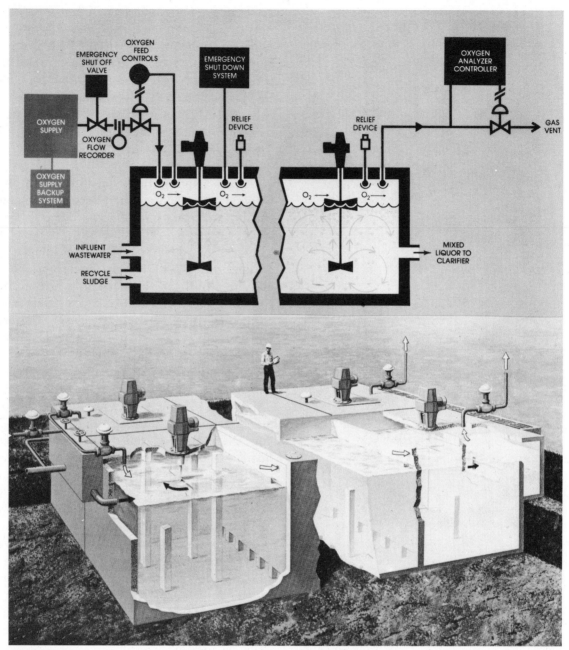

FIGURE 10.13 Pure oxygen can be used in activated sludge wastewater treatment. The tanks are covered to conserve oxygen that is generated on-site. (Air Products and Chemicals, Inc.)

and TSS serves primarily to protect the aquatic environment.) Sewage disinfection is particularly important when the secondary effluent is discharged into a body of water used for swimming or water supply by a downstream community.

Like drinking water, sewage is usually disinfected by *chlorination*. The method of application and the chemistry of this process are discussed in Section 6.5. The chlorine demand of wastewater is relatively high when compared to that of drinking

FIGURE 10.14 A series of rotating biological contactors (RBC), or *biodiscs*, for secondary wastewater treatment. (Bio Systems Division, Autotrol Corp.)

water. A chlorine dosage of about 10 mg/L is required to leave a combined chlorine residual of 0.5 mg/L in the secondary effluent. A residual of 0.5 mg/L is the minimum required by most environmental regulatory agencies for wastewater effluents.

A separate *chlorine contact tank,* with a series of baffles to eliminate short-circuiting of the flow, is used to ensure at least 15 min of contact time between the sewage and the chlorine. Although the presence of a chlorine residual is a good indication of effective disinfection, more than just residual testing is required; specifically, bacteriological testing may be necessary. Most NPDES permits specify a maximum allowable concentration of 200 fecal coliforms per 100 mL in the plant effluent.

It should be noted that excessive chlorination of sewage can have an adverse environmental impact. High chlorine concentrations in the vicinity of sewage treatment plant outfall pipes can kill fish and other aquatic life. The treatment plant operator must carefully control the chlorine dosage to prevent wasting chlorine as well as to prevent fish-kills.

10.4 TERTIARY (ADVANCED) TREATMENT

Secondary treatment can remove between 85 and 95 percent of the BOD and TSS in raw sanitary sewage. Generally, this leaves 30 mg/L or less of BOD and TSS in the secondary effluent. But sometimes this level of sewage treatment is not sufficient to protect the aquatic environment. For example, periodic low flow rates in a trout stream may not provide the amount of dilution of the effluent that is needed to maintain the necessary DO levels for trout survival.

Another limitation of secondary treatment is that it does not significantly reduce the effluent concentrations of nitrogen and phosphorus in the sewage. Nitrogen and phosphorus are important plant nutrients. If they are discharged into a lake, algal blooms and accelerated lake aging or eutrophication may be the result. Also, the nitrogen in the sewage effluent may be present mostly in the form of ammonia compounds. These compounds are

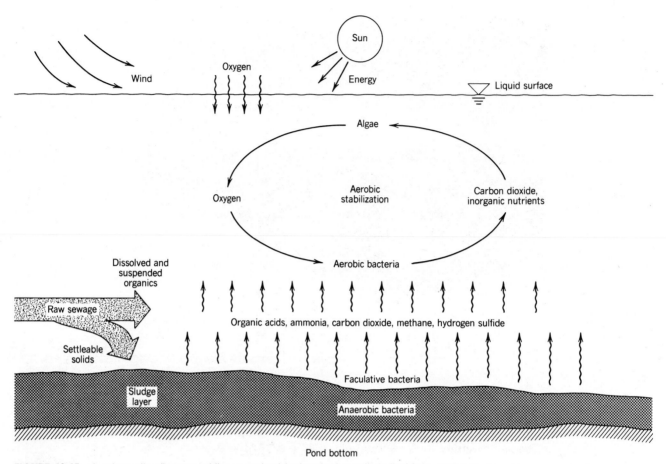

FIGURE 10.15 A schematic diagram of the complex biochemical reactions that take place in a wastewater stabilization pond or lagoon.

toxic to fish if the concentrations are high enough. Yet another problem with the ammonia is that it exerts a "nitrogenous" oxygen demand in the receiving water, as it is converted to nitrates. This process is called nitrification, as discussed in Section 4.3.

When pollutant removal greater than that provided by secondary treatment is required, either to further reduce the BOD and TSS concentrations in the effluent, or to remove plant nutrients, additional or *advanced treatment* steps are required. This is also called *tertiary treatment,* because many of the additional processes follow the primary and secondary processes in sequence.

Tertiary treatment of sewage can remove more than 99 percent of the pollutants from raw sewage, and can produce an effluent of almost drinking wa-

ter quality. But the cost of tertiary treatment, for operation and maintenance as well as for construction, is very high, sometimes doubling the cost of secondary treatment. The "benefit-to-cost ratio" is not always big enough to justify the additional expense. Nevertheless, application of some form of tertiary treatment is not uncommon. Some of the more common tertiary processes are discussed in the following Sections.

EFFLUENT POLISHING

The removal of additional BOD and TSS from secondary effluents is sometimes referred to as *effluent polishing*. It is most often accomplished using a granular-media filter, much like the filters used to purify drinking water. Since the suspended solids consist mostly of organic compounds, filtration removes BOD as well as TSS.

Generally, mixed-media filters are used, in order to achieve in-depth filtration of the effluent. Because of the organic and biodegradable nature of the suspended solids in the secondary effluent, tertiary filters must be backwashed frequently. Otherwise, decomposition would cause septic or anaerobic conditions to develop in the filter bed. In addition to the conventional backwash cycle, an auxiliary surface air-wash is used to thoroughly scour and clean the filter bed. Filtration may be done by gravity in an open tank, or by pressure in closed pressure vessels.

A schematic diagram of an automatic-backwash tertiary filter is shown in Figure 10.16. The filtered water may be stored in an adjacent tank and used for backwash water when the head loss through the filter reaches a predetermined level. Some filter

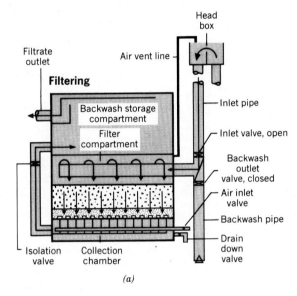

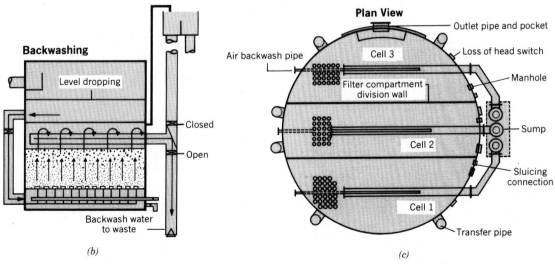

FIGURE 10.16 Auto backwash rapid filters may be used to "polish" the effluent in a tertiary or advanced sewage treatment plant. Diagram *(a)* shows the filtration mode, and diagram *(b)* shows the backwash mode of operation. Three individual filter cells may be constructed in a single prefabricated unit, as shown in diagram *(c)*. (EIMCO Process Equipment Company)

manufacturers mount the backwash storage tank directly above the filter, forming a single self-contained unit.

Another process called *microstraining* also finds application as a tertiary step in wastewater treatment, for suspended solids reduction. The microstrainers, also called *microscreens,* are composed of specially woven steel wire cloth mounted around the perimeter of a large revolving drum. The steel-wire cloth acts as a fine screen, with openings as small as 20 micrometers or millionths of a meter (μm).

The rotating drum is partially submerged in the secondary effluent, which must flow into the drum and then outward through the microscreen. As the drum rotates, captured solids are carried to the top where a high-velocity water spray flushes them into a hopper mounted on the hollow axle of the drum. A typical microscreen installation is illustrated in Figure 10.17.

PHOSPHORUS REMOVAL

As previously discussed, phosphorus is one of the plant nutrients that contribute to the eutrophication of lakes. Raw sewage contains about 10 mg/L of phosphorus, from household detergents as well as from sanitary wastes. The phosphorus in wastewater is primarily in the form of organic phosphorus and as phosphate, PO_4^{3-}, compounds. Only about 30 percent of this phosphorus is removed by the bacteria in a conventional secondary sewage treatment plant, leaving about 7 mg/L of phosphorus in the effluent.

When stream or effluent standards require lower phosphorus concentrations, a tertiary treatment process must be added to the treatment plant. This usually involves *chemical precipitation* of the phosphate ions and coagulation. The organic phosphorus compounds are entrapped in the coagulant flocs that are formed and settle out in a clarifier.

One of the chemicals frequently used in this process is aluminum sulfate, Al_2SO_4. This is called *alum,* the same coagulant chemical used to purify drinking water. The aluminum ions in the alum react with the phosphate ions in the sewage, to form the insoluble precipitate called aluminum phosphate. Other coagulant chemicals that may be used to precipitate the phosphorus include ferric chloride, $FeCl_3$, and lime, CaO.

Adding the coagulant downstream of the second-

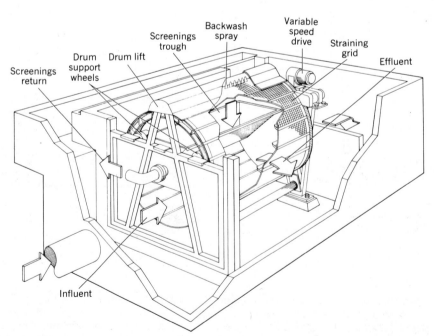

FIGURE 10.17 A perspective view showing the basic components of a microstrainer unit, which may be used for tertiary sewage treatment. (Permutit Co. Inc.)

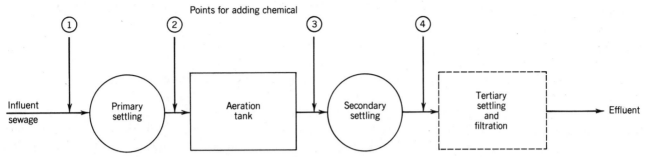

Points for adding chemical

FIGURE 10.18 Phosphorus can be removed from sewage by chemical precipitation; the chemical, usually alum, can be added at one of four different points in the process. Point 2 is the most common point of application.

ary processes provides the greatest overall reliability for phosphorus reduction. It not only removes about 90 percent of the phosphorus, but it removes additional TSS and serves to polish the effluent as well. But when applied in this manner, as a third or tertiary treatment step, additional flocculation and settling tanks must be built. In some cases, even filters may have to be added to remove the nonsettleable floc.

To avoid the need for construction of additional tanks and filters, in most plants requiring phosphorus removal the coagulant is added to the wastewater at some point in the conventional process. For example, alum may be added just before the primary settling tanks. The resulting combination of primary and chemical sludge would be removed from the primary clarifiers.

Or, in activated sludge plants, the coagulant may be added directly into the aeration tanks. In this case, the precipitation and flocculation reactions occur along with the biochemical reactions. Sometimes, the coagulant may be added to the wastewater just before the secondary or final clarifiers. Regardless of the point in the process at which coagulant is added, the total volume and weight of sludge requiring disposal increases significantly. The options for phosphorus removal by chemical precipitation are illustrated in Figure 10.18.

NITROGEN REMOVAL

Nitrogen can exist in wastewater in the form of organic nitrogen, ammonia, or nitrate compounds.

The effluent from a conventional activated sludge plant contains mostly the ammonia nitrogen form, NH_4^+. Effluents from a trickling filter or rotating biodisc may contain more of the nitrate form, NO_3^-. This is because the nitrifying bacteria, those microbes that convert ammonia to nitrate, have a chance to grow and multiply on some of the surfaces in the trickling filter or biodisc units. They do not survive in a mixed-growth aeration tank, where they are crowded out by the faster growing bacteria that consume carbonaceous organics.

Nitrogen in the form of ammonia can be toxic to fish, and it exerts an oxygen demand in receiving waters as it is converted to nitrate. Nitrate nitrogen is one of the major nutrients that cause algal blooms and eutrophication. For these reasons, it is sometimes necessary to remove the nitrogen from the sewage effluent before discharge. This is particularly important if it is discharged directly into a lake.

One of the methods used to remove nitrogen is called *biological nitrification–denitrification*. It consists of two basic steps. First, the *secondary* effluent is introduced into another aeration tank, trickling filter, or biodisc. Since most of the carbonaceous BOD has already been removed, the microorganisms that will now thrive in this tertiary step are the so-called nitrifying bacteria, *Nitrosomoneas* and *Nitrobacter*. In this first step, called *nitrification*, the ammonia nitrogen is converted to nitrate nitrogen, producing a *nitrified effluent*. At this point, the nitrogen has not actually been removed, but only converted to a form that is not toxic to fish and that does not cause an additional oxygen demand.

A second biological treatment step is necessary to actually remove the nitrogen from the wastewater. This step is called *denitrification*. It is an aerobic process in which the organic chemical methanol is added to the nitrified effluent to serve as a source of carbon. The denitrifying bacteria *Pseudomonas* and other groups use the carbon from the methanol and the oxygen from the nitrates in their metabolic processes. One of the products of this biochemical reaction is molecular nitrogen, N_2, which escapes into the atmosphere as a gas. A schematic diagram of this process is shown in Figure 10.19.

Another method for nitrogen removal is called *ammonia stripping*. It is a physical–chemical rather than a biological process, consisting of two basic steps. First, the pH of the wastewater is raised in order to convert the ammonium ions, NH_4^+, to ammonia gas, NH_3. Second, the wastewater is cascaded down through a large tower; this causes turbulence and contact with air, allowing the ammonia to escape as a gas. Large volumes of air are circulated through the tower to carry the gas out of the system. The combination of ammonia stripping with phosphorus removal using lime as a coagulant is advantageous, since the lime can also serve to raise the pH of the wastewater. Ammonia stripping is less expensive than biological nitrification–denitrification, but it does not work very efficiently under cold weather conditions.

LAND TREATMENT OF WASTEWATER

The application of secondary effluent onto the land surface can provide an effective alternative to the expensive and complicated advanced treatment methods previously discussed. A high quality polished effluent can be obtained by the natural processes that occur as the effluent flows over the vegetated ground surface and as it percolates through the soil.

An additional benefit of land treatment is that it can provide the moisture and nutrients needed for vegetation growth, and it can help to recharge groundwater aquifers. In effect, land treatment of wastewater allows a direct recycling of water and nutrients for beneficial use; the sewage becomes a valuable natural resource that is not simply "disposed" of. But relatively large land areas are needed for this kind of treatment. And soil type as well as climate are critical factors in controlling the feasibility and design of a land treatment process.

There are three basic types or modes of land treatment. These include *slow rate, rapid infiltration,* and *overland flow*. The conditions under which they can function and the basic objectives of these types of treatment vary.

In the slow rate system, also called *irrigation*, vegetation is the critical component for the wastewater treatment process. Although the basic objec-

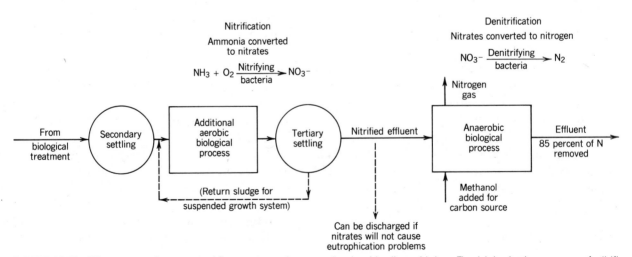

FIGURE 10.19 Nitrogen can be removed from sewage to prevent eutrophication of lakes. The biological processes of *nitrification* and *denitrification* must be carried out after the basic activated sludge process is complete.

tive is wastewater treatment and disposal, another goal is to obtain an economic benefit from the use of the water and nutrients to produce marketable crops (i.e., corn or grain) for animal feed. Another objective might be to conserve potable water by using secondary effluent to irrigate lawns and other landscaped areas.

A schematic diagram of a slow rate land treatment system is shown in Figure 10.20. Wastewater can be applied onto the land by ridge-and-furrow surface spreading or by sprinkler systems. The most common sprinkler system consists of a long spray boom that rotates on wheel supports around a center pivot. In ridge-and-furrow application, the wastewater flows by gravity in small ditches. In either mode of application, most of the water and nutrients are taken up or absorbed through the roots of the growing vegetation.

Not all of the water applied to the land in a slow rate system is absorbed; some of it percolates through the soil to the groundwater zone. Uncontrolled surface runoff of the wastewater is not usually allowed, so the soil must have reasonably good drainage characteristics. The wastewater can be applied to the land at rates of up to 100 mm (4 in.) per week, except during the winter months in northern climates. Of the three basic types of land treatment systems, the slow rate method provides the best results with respect to tertiary treatment levels of pollutant removal. Suspended solids and biochemical oxygen demand are significantly reduced by filtration of the wastewater and biological oxidation of the organics in the top few inches of

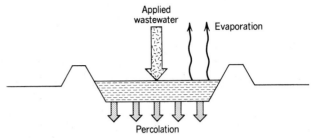

FIGURE 10.21 Rapid infiltration. (EPA)

soil. Nitrogen is removed primarily by crop uptake, and phosphorus is removed by adsorption within the soil.

The rapid infiltration or "infiltration–percolation" mode of land treatment has as basic objectives to recharge groundwater aquifers and to provide advanced treatment of wastewater. A schematic diagram of the rapid infiltration method is shown in Figure 10.21. Most of the secondary effluent percolates to the groundwater; very little of it is absorbed by vegetation. The filtering and adsorption action of the soil removes most of the BOD, TSS, and phosphorus from the effluent, but nitrogen removal is relatively poor. Soils must be highly permeable for the rapid infiltration method to work properly. Usually, the wastewater is applied in large ponds called *recharge basins*.

An overland flow system is illustrated in Figure 10.22. Wastewater is sprayed on a sloped terrace and allowed to flow across the vegetated surface to a runoff collection ditch. Purification is accomplished by physical, chemical, and biological processes as the wastewater flows in a thin film down the relatively impermeable surface. Overland flow

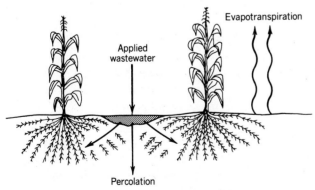

FIGURE 10.20 Slow rate land treatment. (EPA)

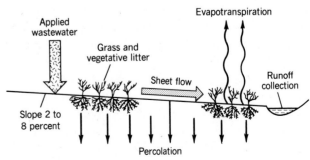

FIGURE 10.22 Overland flow. (EPA)

can be used to achieve removal efficiencies for BOD and nitrogen comparable to other methods of tertiary treatment, but phosphorus removal is somewhat limited. The water collected in the ditch is usually discharged to a nearby body of surface water.

10.5   ON-SITE WASTEWATER DISPOSAL

Several different sewage treatment methods are described in the preceding sections. They are usually applied at a single and centralized location to wastewater that originated from many diverse sources. Very often these centralized facilities are owned and operated by a local municipality, and they are often referred to as *publicly owned treatment works (POTWs)*.

In lightly populated suburban or rural areas, however, it is often uneconomical to build a public sewage collection system and centralized treatment plant. The houses may be spread relatively far apart and the shared cost for each service connection would be excessive. Instead, a system that provides for disposal of wastewater into the ground may be provided for each individual wastewater generator. This is called *on-site subsurface wastewater disposal.*

A secondary benefit of on-site subsurface disposal is that groundwater aquifers are recharged as the sewage effluent reaches the zone of saturation in the ground. The effluent is purified to a large extent by filtration, adsorption, and biochemical oxidation as it percolates through the soil. Aquifer recharge is a particularly important environmental factor in areas where groundwater is a source of potable water supplies.

When sewage is treated at a POTW and discharged into a surface stream, the sewage quickly flows out of the watershed. This contributes to a problem called *groundwater mining* in which the groundwater table drops in elevation. Sometimes the water table drops to the point where many wells go dry, and in areas where groundwater is the source of potable supplies it becomes uneconomical to rely further on the aquifer for public water supply.

In the past, there has been a strong tendency to construct a centralized POTW as soon as possible in a developing area. Subsurface systems were looked upon as a temporary mode of wastewater disposal, and because of frequent failures of these systems, they were considered unreliable and undesirable.

But if properly located, designed, and constructed, an on-site subsurface system can serve effectively for wastewater disposal on a long-term basis. And with realization of the need to avoid mining of groundwater, many environmental agencies are viewing on-site subsurface disposal in a more favorable light, even when population densities may be such that a centralized sewage treatment facility could be used. Presently, about 25 percent of the new homes constructed in the United States make use of on-site subsurface disposal.

SITE EVALUATION

Surface topography as well as subsurface conditions at the proposed site are of importance with regard to planning and design of a subsurface wastewater disposal system. One of the most important factors related to the successful operation of a subsurface system is the *texture of the soil* into which the effluent is discharged. Specifically, the permeability or "hydraulic conductivity" of the soil must be within an acceptable range. If it is too low, the wastewater will not be able to percolate fast enough for effective disposal. But if it is too high, there may not be sufficient time for purification of the effluent before it reaches the water table.

Relatively coarse granular soils that contain a large percentage of sand generally have high hydraulic conductivities. They are usually suitable for subsurface disposal of wastewater, as long as the percolation rate is not excessive. Fine grained soils, such as very fine sands and silts, offer more resistance to the flow of water. Soils containing a large fraction of clay are often unacceptable for subsurface disposal because of their extremely low permeabilities.

In addition to hydraulic conductivity, the *depth to groundwater* and the *depth to bedrock* are important factors with regard to siting and designing a subsurface disposal system. Generally, bedrock and the seasonally high groundwater table must be at least 3 m (10 ft) below the disposal system. These values will vary, depending on local health department and environmental agency regulations.

Before a decision is made regarding the use of subsurface disposal systems, a thorough investigation of site and soil conditions is necessary. Soil survey maps from the local USDA Soil Conservation Office would be studied for preliminary data. In the field, a large *test-pit* would be excavated by a backhoe to a depth of about 4 m (12 ft). Visual observations regarding soil texture, depth to groundwater, and depth to rock would be recorded.

Percolation Test

In order to evaluate the ability of the soil to transmit the flow of water, a *percolation test,* or "perc test," would be conducted. The perc test provides an indirect measure of the soil permeability, or "hydraulic conductivity." Usually, several separate perc tests are required in the vicinity of a proposed subsurface disposal system, particularly if soil conditions are highly variable on the site.

A perc test simply measures the rate at which water seeps into the soil in a *test hole.* It is sometimes called a *falling head perc test,* because it is the rate of drop of the water level in the test hole that is measured. In addition to providing data as to whether or not a conventional subsurface system may be utilized, the perc test provides data for designing the size of the leaching field. Generally, perc tests must be performed by a qualified technician under the supervision of a professional engineer and witnessed by a representative of the local health department.

Specific procedures for conducting perc tests vary somewhat among local environmental agencies and health departments across the country. Although the details vary, including the dimensions of the test hole, most perc test procedures can be summarized as follows:

1. *The test hole.* A test hole of about 200 mm (8 in.) in diameter is dug in the soil to the depth of the proposed leaching field, usually about 0.6 m (2 ft). This may be done on a shallow ledge adjacent to the test pit. The sides of the hole may then be scratched with a sharp tool, and the loose material is removed. An inch or two of coarse gravel may be placed at the bottom of the hole to prevent scour later on.

2. *Soaking the test hole.* The soil is soaked and saturated by filling the hole with clean water. For sandy soils, all of the water is allowed to drain away; for silty soils or soils with a high clay content, the water is allowed to remain in the hole overnight and the test is conducted on the next day. After thorough soaking, the hole is again filled with clean water to a depth of about 200 mm (8 in.). At a uniform time interval of 5 to 30 minutes (depending on the rate of fall), the drop in the water level is measured and recorded. This is repeated until a constant rate of drop is observed.

3. *Measurement of the perc rate.* After the rate of drop becomes constant, the hole is again filled to a depth of 200 mm (8 in.). The time required for the water level to drop 150 mm (6 in.) is recorded. In "tight" or slow draining soils, it may be permissible to measure the drop of water in a 30-min time interval instead of waiting for a 150-mm drop.

4. *Computation of perc rate.* The perc rate for each test hole is computed in terms of minutes per inch of water level drop, or minutes per 25 mm of drop. As long as the perc rates do not vary by more than 20 min, the perc rate for the area is obtained by averaging the rates from each test hole. Otherwise, the variations in soil type would be noted.

EXAMPLE 10.9

A time interval of 45 min is recorded for a 150-mm (6-in.) drop of water level in a test hole. Compute the perc rate.

Solution:

Set up the following ratio, and solve for the perc rate:

$$\frac{45 \text{ min}}{150 \text{ mm}} = \frac{\text{perc rate}}{25 \text{ mm}}$$

$$\text{perc rate} = 25 \times \frac{45 \text{ min}}{150 \text{ mm}} = 7.5 \text{ min/25 mm}$$

(This is equivalent to 18 s/mm.)
In American units, the per rate would be expressed as 45 min/6 in. or 7.5 min/in.

One method for measuring and timing the drop of water in a test hole is illustrated in Figure 10.23. A yardstick or meterstick is simply held against a batter board that is placed across the test hole; the board serves as a fixed reference level for taking the measurement when the bottom of the stick just penetrates the water surface.

Battery operated devices called *percometers* are available commercially. Their operation is based on the fact that water conducts electricity; when the water level drops below the end of a probe placed in the test hole, a gage indicates that the circuit is broken, and the time interval for a predetermined drop in water level can then be measured.

Generally, if the perc rate is slower than 60 min/in. (40 min in some states), the soil is considered unsuitable for conventional septic systems. And if the rate is faster than 1 min/in., subsurface disposal may also be prohibited in order to protect groundwater quality.

SEPTIC SYSTEMS

The most common type of subsurface disposal system includes a *septic tank* and an *absorption* or *leaching field,* as illustrated in Figure 10.24. Briefly, the tank serves to store settled and floating solids and the leaching field serves to distribute the effluent so that it can percolate through the soil. Decomposition of organics takes place under anaerobic conditions, hence the name *septic system.*

The Septic Tank

Settleable and floating solids must be removed from the raw wastewater before it can be applied to the soil in the leaching field. A buried septic tank is used to provide the necessary primary treatment step, as well as to act as a sludge storage tank. Floating solids and grease are trapped by a baffle at the tank outlet and prevented from entering the leaching field. Settleable solids accumulate in the sludge zone at the bottom of the tank, and "primary effluent" flows out into the leaching field.

Some of the sludge decomposes under anaerobic conditions, but eventually the undecomposed solids accumulate and the sludge level rises. To ensure proper operation and a long service life, septic tanks should be pumped clean on a routine basis, every few years. If too much sludge is allowed to accumulate in the tank, solids will eventually be carried out in the effluent and will plug up the absorption field. This is one of the most common causes of failure for septic systems.

Septic tanks are manufactured in a variety of

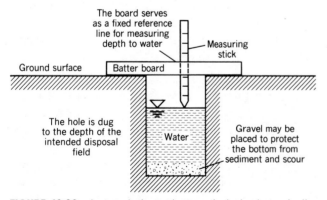

FIGURE 10.23 A perc test can be conducted using a batter board and a meter or yard stick.

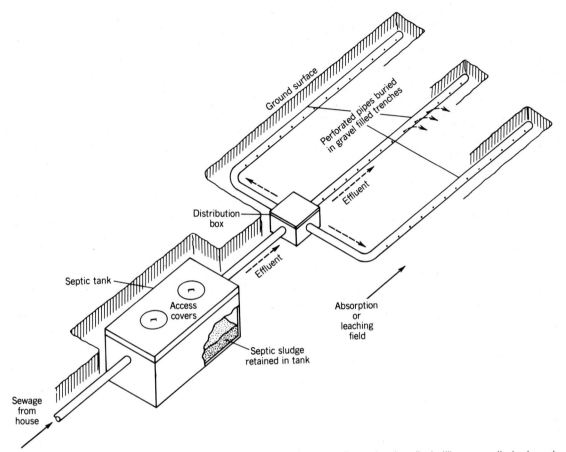

FIGURE 10.24 A perspective view of an on-site subsurface sewage disposal system that utilizes a septic tank and leaching or absorption field.

shapes and sizes. A typical small rectangular concrete tank is shown in Figure 10.25. The minimum size tank allowed for an individual household is generally about 3800 L (1000 gal) of liquid capacity. Septic tanks must be water tight and have adequate access for inspection and cleaning. The top of the tank is usually about 300 mm (1 ft) below the ground surface.

The Leaching Field

The effluent from the septic tank flows into an absorption or leaching field, which distributes the liquid uniformly over a sizable area. From the leaching field, the effluent percolates downward to the water table. As it flows through the soil voids, microorganisms and other pollutants are removed from the effluent. Filtration, adsorption, and biological decomposition each play a role in the purification of the wastewater effluent before it is diluted in the groundwater.

A very common type of leaching field consists of two or more separate trenches with pipes that serve to spread the wastewater. Before reaching the trenches, the effluent flows into a *distribution box,* which serves to evenly divide the flow into each trench. Local health department regulations usually require a minimum of 30 m (100 ft) of separation between the leaching field and a water supply well. A typical layout plan for a residential septic system is illustrated in Figure 10.26.

The disposal trenches in the leaching field are shallow excavations that are a minimum of 0.3 m (1 ft) wide and at least 0.6 m (2 ft) deep. About 150

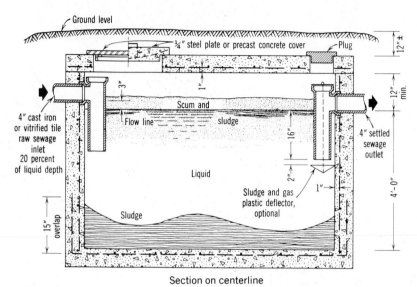

FIGURE 10.25 Cross-section of a typical septic tank. (From Salvato, J., *Environmental Engineering and Sanitation,* with permission of John Wiley & Sons.)

mm (6 in.) of gravel is placed on the bottom of the trench to support the line of perforated effluent distribution pipe. The pipe is covered with more gravel and then with straw or paper to prevent the final layer of backfill soil from penetrating the gravel voids. The slope of the pipe is generally about 0.5 percent or less. On sites with relatively steep slopes, the trenches would run roughly parallel to the ground contours. Individual trenches, or *laterals* as they are sometimes called, should be less

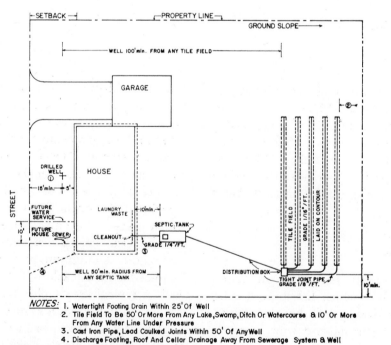

FIGURE 10.26 A plan view of a typical residential on-site subsurface sewage disposal system. (From Salvato, J., *Environmental Engineering and Sanitation,* with permission of John Wiley & Sons.)

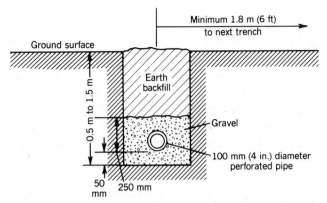

FIGURE 10.27 Cross-section through an absorption field trench.

than 30 m (100 ft) long and should be separated from each other by a distance of at least 1.8 m (6 ft). A typical trench cross-section is illustrated in Figure 10.27.

The design of an absorption field involves the determination of the required number and lengths of the trenches. Data from the percolation test are very important in this regard. For residential systems, a chart such as the one shown in Figure 10.28 may be used by the designer; the requirements differ somewhat for systems serving commercial establishments.

As seen in the graph of Figure 10.28, the required

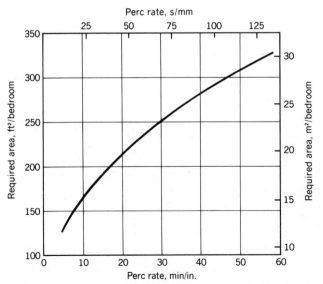

FIGURE 10.28 Typical absorption or leaching field area requirements, for private residences; total required area depends on the perc rate and on the number of bedrooms in the home.

absorption area is related to the soil perc rate. Absorption area can be converted to linear feet of trench, as illustrated in the following example.

EXAMPLE 10.10

A four-bedroom home is situated on a site that has an average perc rate of 30 min/in. How many laterals are required if the trench is 2 ft wide? What are the overall dimensions of the leaching field?

Solution:

From Figure 10.28, we determine that 250 ft² of absorption area is required per bedroom. For a four-bedroom home, a total of 4 × 250 = 1000 ft² of absorption area is needed for the leaching field.

Since the trench width is to be 2 ft, there is a need for 1000 ft² ÷ 2 ft = 500 ft of trench. Also, since the maximum length of an individual lateral is 100 ft, there is a need for 500 ft ÷ 100 ft/lateral, or 5 laterals.

Each lateral will be separated by a distance of 6 ft. Since there are 4 spaces between the 5 laterals, the width of the absorption field will be 6 × 4 = 24 ft. The entire leaching field will occupy 24 × 100 = 2400 ft² of the site.

Seepage Pits

When the site is too small for a conventional leaching field, deeper excavations that take up less area are sometimes used for subsurface disposal. These excavations are called *seepage pits* or *dry wells*. They also may be selected instead of trenches, so as to utilize more favorable deeper soil.

Effluent from the septic tank flows into the pit, where it is stored until it seeps out through the side wall and bottom. A cross-section of a typical seepage pit is shown in Figure 10.29. Although the use of seepage pits is usually discouraged by local health departments, it is acceptable for small wastewater flow rates. It is very important that the high water table be at least 1.2 m (4 ft) below the bottom of the pit, to protect groundwater quality.

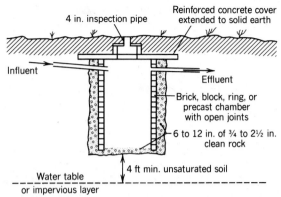

FIGURE 10.29 A cross-section of a typical "seepage pit" for on-site disposal of wastewater. A seepage pit is preceded by a septic tank. (EPA)

The required diameter and depth of a seepage pit depend on the perc rates and thicknesses of the different layers of soil encountered in the excavation. Soil layers that have perc rates slower than 30 min/in. are not included in the design computations. A typical seepage pit is about 3 m (10 ft) in diameter and about 4.5 m (15 ft) deep. When more than one pit is required on a given site, a separation of at least three times the diameter is required. A seepage pit can be dug with conventional excavation machinery, but care must be taken not to excavate when the soil is wet; smearing of the soil would reduce its absorptive capacity.

Mounds

To overcome site restrictions that prohibit the use of either seepage pits or leaching fields, a *mound system* can be used. Typical site restrictions include soils with very slow perc rates, and conditions where bedrock or the water table is close to the ground surface. A mound is an effluent absorption system that is raised above the natural ground surface. It provides a suitable fill material for percolation, as well as adequate separation from bedrock or the water table. A cross-section of a typical mound system is shown in Figure 10.30.

The fill material is usually a medium texture sand from locally available sources. A bed of gravel and distribution laterals are placed in the upper part of the mound, for uniform absorption of the effluent from the septic tank. The effluent must be intermittently applied into the absorption area. The dimensions of a mound system depend on the natural soil conditions, the ground slope, and the depth of fill below the absorption area.

OTHER ON-SITE SYSTEMS

The combination of a septic tank and leaching field is the most common type of system for on-site wastewater treatment and disposal. The septic tank

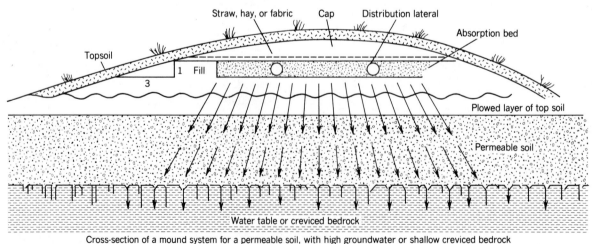

Cross-section of a mound system for a permeable soil, with high groundwater or shallow creviced bedrock

FIGURE 10.30 A *mound system* for on-site sewage disposal. (EPA)

provides minimal treatment, but the quality of the effluent need not be high since the assimilative or purifying capacity of the soil is usually enough to protect the groundwater. The leaching field provides a means of uniform effluent distribution into the soil. As previously discussed, certain restrictions on subsurface disposal into the soil can be overcome with the use of seepage pits or mound systems.

In some instances, however, soil or site conditions are such that subsurface disposal is not feasible at all. Yet it is still necessary to dispose of the wastewater on site because of the lack of a nearby POTW. Several alternatives to disposal by subsurface absorption are available, but they generally are somewhat more expensive to build, operate, and maintain than subsurface systems. These include the use of *evapotranspiration systems* and *intermittent sand filters*. Both of these systems can be used following a septic tank. If a higher level of treatment is required, a small aerobic treatment unit, such as a package extended aeration tank, may be installed on site.

Evapotranspiration Systems

An evapotranspiration (ET) system consists of a sand bed, a network of perforated distribution pipes, and an impermeable liner that prevents the wastewater effluent from reaching the water table. In some cases, the liner may be omitted in order to allow some of the effluent to seep into the soil. But the basic objective of this type of system is to dispose of the wastewater into the atmosphere, and to avoid the need for discharge to either surface or groundwater. A cross-section of a typical ET system is shown in Figure 10.31.

Effluent from a septic tank is distributed throughout the bed in the perforated pipe network. The effluent rises through the sand by capillary action and then evaporates into the air. Grass or other vegetation growing on the top of the bed serves to absorb some of the wastewater in the root zone and transpire it into the air through the leaves. Hence the name, "evapotranspiration" system. One of the most critical factors controlling the use and design of an ET system is the local climate, which affects the rate of evaporation.

Specifically, factors such as average annual rainfall, wind speed, humidity, solar radiation, and temperature are of importance. Evapotranspiration systems operate effectively in areas where the evaporation rate exceeds the rate of precipitation, such as in the southwestern United States. Figure 10.32 illustrates the net relationship between annual precipitation and evaporation in different parts of the country. Transpiration by plants increases the amount of water vapor discharged into the air in soil-covered systems, but only during daylight hours of the growing season.

An important design parameter for an ET system is the hydraulic loading rate, which must be low enough to prevent the bed from filling completely with effluent. A typical loading rate is about 2 L/m^2/d (0.05 gpd/ft^2), in the southwestern part of the country. The hydraulic loading rate is determined from the difference between the rate of evaporation and the rate of precipitation. A typical ET

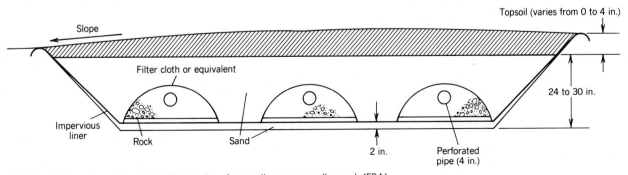

FIGURE 10.31 An *evapotranspiration system* for on-site sewage disposal. (EPA)

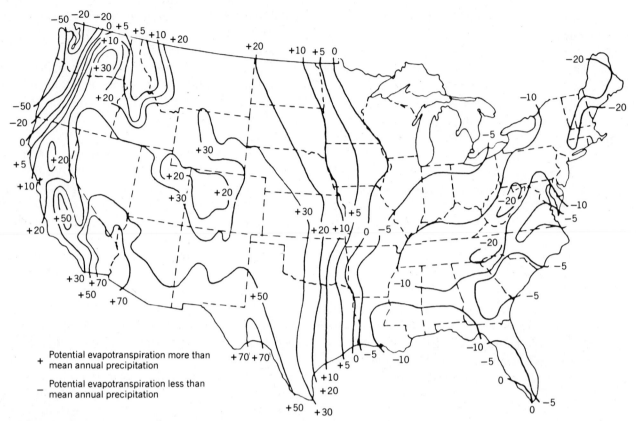

FIGURE 10.32 Evapotranspiration systems are most feasible in areas where the rate of evapotranspiration exceeds the rate of precipitation. (EPA)

system for a single family residence in the southwest would require about 465 m^2 (5000 ft^2) of land area.

Intermittent Sand Filters

One of the oldest methods of wastewater treatment involves intermittent application of settled wastewater to a bed of sand. The sand bed is usually about 1 m (3 ft) deep and is underdrained with gravel and collecting pipes. The collected effluent may be disinfected with chlorine before discharge to land or surface waters.

The filters may be built as open units at ground level, or they may be buried in the ground. They provide efficient treatment while requiring a minimum of maintenance. Many of these systems are in use throughout the United States, providing on-site treatment for homes as well as for small commer-

cial establishments. A profile view of a typical buried filter is shown in Figure 10.33.

The upper perforated distribution lines are level and are spaced about 2 m (6 ft) apart. The underdrain pipes are perforated or open joint lines on a slope of about 0.5 percent. They are spaced about 4 m (12 ft) apart. The hydraulic load on a buried filter is about 0.04 m^3/m^2·d (1 gpd/ft^2). The finished grade of the topsoil over the filter is mounded, to direct runoff away from the bed.

Dosing, or application of the wastewater to the sand bed, is an important factor related to the operation and performance of the filter. Large filters must be dosed intermittently, about two times per day, so as to flood the entire unit. It is important to have a sufficient resting period between doses, to maintain aerobic conditions in the bed.

Small tanks called dosing chambers store the wastewater until it is applied to the filter by a

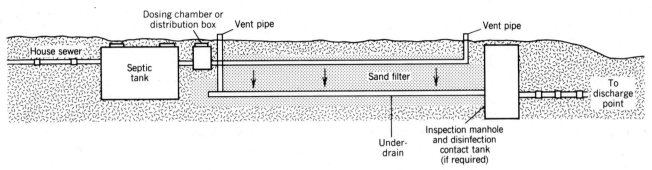

FIGURE 10.33 A cross-section through a typical buried sand filter for on-site sewage disposal. (EPA)

pump or siphon device. A typical siphon dosing chamber is shown in Figure 10.34. The siphon automatically discharges when the wastewater in the tank reaches a level called the *drawing depth;* no mechanical or electrical controls are needed. It is activated by the pressure head of the wastewater. Cast-iron or fiberglass siphons are commercially available in a range of sizes. Siphon chambers require only minimal routine maintenance.

10.6 SLUDGE MANAGEMENT

The pollutants removed from wastewater in a treatment process are concentrated in a form called *sludge,* which is collected and withdrawn from the tanks in which it accumulates. Even in aerobic treatment systems in which it is recycled, eventu-

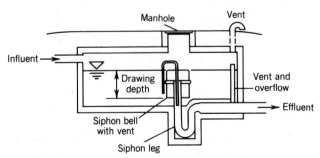

FIGURE 10.34 A section view of a typical dosing chamber and siphon device. (EPA)

ally most of the sludge must be removed from the system. The handling, treatment, and disposal of sewage sludge are major factors in the design and operation of all water pollution control plants.

For many coastal communities, it is economical to dispose of sewage sludge in the ocean. The sludge, in the form of a liquid slurry, may be pumped through a long outfall pipe far enough out so that currents would not carry it back to beaches on the shore. The sludge may also be barged out to sea to selected dumping areas, as discussed in Section 5.7. The practice of sludge disposal in the ocean is under study; many people believe that continued dumping, especially of sludges that contain toxic chemicals, can cause serious environmental damage despite the large dilution factor provided. It is likely that ocean dumping of sludge will eventually be prohibited.

Many inland cities have no alternative but to incinerate or burn the sludge. Incineration of sludge must include the use of costly air pollution control equipment; it also requires fuel and energy consumption to start and maintain the combustion process.

In most cases, the final destination of sewage sludge is the land. Often, it is simply buried in the ground, either separately or in a sanitary landfill along with municipal refuse. (Sanitary landfills are discussed in Section 11.5.) Strict engineering design standards must be followed to prevent groundwater pollution from land disposal of sludge.

Sometimes sanitary sewage sludge is spread on agricultural land, in order to make use of its value as a soil conditioner and fertilizer. Sludge from industrial wastes may contain high concentrations of heavy metals or other toxic chemicals, which would make it unacceptable for land spreading in agricultural areas. The sludge from a water treatment plant consists mostly of precipitated chemicals used in the coagulation or softening processes. It may sometimes be processed in order to reclaim and recycle those chemicals, but when this is uneconomical, water plant sludges are disposed of by on-site burial in the ground.

Land disposal of sludge or sludge incineration is usually preceded by some type of treatment, to reduce its volume as well as to render it easier and less offensive to handle. A discussion of sludge characteristics and quantities, as well as of several of the more common treatment methods, is presented in the following sections.

SLUDGE CHARACTERISTICS

The amount of sludge solids generated in a wastewater (or drinking water) treatment system depends largely on the degree of treatment provided and on the amount of chemicals added. It initially forms at the bottom of a clarifier or settling tank in the form of a concentrated slurry of solids suspended in water. Its volume depends on the relative amounts of solid material and water. The concentration of the sludge is expressed as a percentage by weight or mass. For example, a mass of 100 kg of liquid sludge that contains 3 kg of solids and 97 kg of water, would have a concentration of 3/100 or 3 percent solids. This is equivalent to a solids concentration of 30 000 mg/L.

The concentration of solids has a very significant effect on the total volume occupied by the liquid sludge. *The total sludge volume is inversely proportional to the solids concentration.* For example, if the percentage of solids is doubled, say from 3 to 6 percent, then the total sludge volume would be decreased to half its original volume. This is a very important relationship. By increasing the solids concentration in the slurry a few percentage points, the total sludge volume can be significantly reduced. This reduces the cost of handling, treating, and disposing of the sludge.

For practical purposes, we can assume that the weight or mass per unit volume of liquid sludge is the same as that of pure water. For example, the mass of one cubic meter of water is 1000 kg; the mass of the same volume of a relatively "thick" sludge (such as 10 percent solids) would be only about 1020 kg. The difference of 2 percent can be neglected for design or operational purposes. Based on this, the relationship between sludge concentration and sludge volume can be expressed as follows:

$$S = \frac{M}{V} \times 100 \qquad (10\text{-}7a)$$

or

$$S = \frac{W}{8.34 \times V} \times 100 \qquad (10\text{-}7b)$$

where S = sludge solids concentration, in percent
   M = dry sludge solids, kg (Equation 10-7a)
   W = dry sludge solids, lb (Equation 10-7b)
   V = total sludge volume, L (Equation 10-7a) or gal (Equation 10-7b)
   8.34 = pounds per gallon (Equation 10-7b)

EXAMPLE 10.11

A sludge with a 6-percent solids concentration occupies a total volume of 300 m³. *(a)* What is the water content of the sludge? *(b)* What is the mass of the sludge solids? *(c)* If the sludge was further concentrated (or "dewatered") to a volume of 200 m³, what would the solids concentration be? What would the water content be?

Solution:

(a) If 6 percent of the sludge consists of dry solids, then the water content is simply the difference, or 100 − 6 = 94 percent.
(b) First, V = 300 m³ × 1000 L/m³ = 300 000 L

Applying Equation 10-7a, we get

$$6 = \frac{M}{300\ 000} \times 100$$

and

$$M = \frac{(6 \times 300\ 000)}{100} = 18\ 000 \text{ kg of dry solids}$$

(c) Again applying Equation 10-7a, we get

$$S = \frac{18\ 000}{200\ 000} \times 100 = 9 \text{ percent solids}$$

The water content would be $100 - 9 = 91$ percent.

EXAMPLE 10.12

A volume of 500 000 gal of sludge contains 100 000 lb of dry solids. What is the solids content of the sludge, expressed in percent? If the sludge was concentrated to 4 percent solids, what would its volume be?

Solution:

First, applying Equation 10-7b, we get

$$S = \frac{100\ 000}{8.34 \times 500\ 000} \times 100 = 2.4 \text{ percent}$$

Again applying Equation 10-7b, we get

$$4 = \frac{100\ 000}{8.34 \times V} \times 100$$

and

$$V = \frac{100\ 000}{4 \times 8.34} \times 100 = 300\ 000$$

Primary sludge typically has a solids concentration of about 7 percent. Secondary sludge has a much lower solids concentration because the suspended solids are mostly biological flocs that settle slowly and do not compact to as high a density as primary sludge. Waste activated sludge contains only about 2 percent solids, or less.

Primary and secondary sludge solids are mostly organic materials, with a volatile fraction of up to 0.8. Primary sludge gives off a strong and offensive odor; it can quickly become septic and difficult to handle. In addition to volume reduction, a basic objective of sewage sludge treatment is to stabilize the biodegradable organic solids and to render the sludge unoffensive and easy to handle.

Water treatment plant sludges are mostly inert chemical precipitates that are relatively stable and unoffensive. Quantities of this type of sludge can vary widely depending on the amount and type of chemicals used and on the composition of the raw water.

The quantity of primary sludge produced in a sewage treatment plant depends on the concentration of suspended solids in the raw wastewater and on the TSS removal efficiency. It can be estimated from the following expressions:

$$\text{mass} = E \times \text{TSS} \times Q \qquad \text{(10-8a)}$$
$$\text{weight} = E \times \text{TSS} \times Q \times 8.34 \qquad \text{(10-8b)}$$

where mass = dry sludge solids, kg (Equation 10-8a)
weight = dry sludge solids, lb (Equation 10-8b)
E = TSS removal efficiency, decimal form
Q = sewage flow rate, ML/d (Equation 10-8a), or mgd (Equation 10-8b)

The quantity of secondary sludge produced in a sewage treatment plant depends on the BOD concentration and on the fraction of BOD that is converted to biological solids (microbe cells). It can be estimated by the following expressions:

$$\text{mass} = K \times \text{BOD} \times Q \qquad \text{(10-9a)}$$
$$\text{weight} = K \times \text{BOD} \times Q \times 8.34 \qquad \text{(10-9b)}$$

where K = a coefficient that represents the proportion of BOD converted to biological solids, in decimal form

BOD = applied 5-day BOD, mg/L

The value of K depends on the organic loading or F/M ratio; a typical value for extended aeration or fixed growth systems is $K = 0.25$, and for conventional or step aeration processes, a typical value is $K = 0.35$

EXAMPLE 10.13

A flow of 4 ML/d with raw TSS = 240 mg/L and raw BOD = 220 mg/L enters a trickling filter sewage treatment plant. Removal efficiency in the primary settling tank is 50 percent for TSS and 30 percent for BOD. (a) Compute the mass of primary sludge solids generated per day. (b) Compute the mass of secondary sludge solids generated per day. (c) If the primary and secondary sludges are combined, and the mixture has a solids concentration of 4 percent, what is the total volume of sludge generated per day?

Solution:

(a) Applying Equation 10-8a, we get

mass = $0.50 \times 240 \times 4$
$$= 480 \text{ kg/d of primary sludge}$$

(b) First, compute the BOD applied to the secondary system:

applied BOD = $(1.0 - 0.3) \times 220 \approx 150 \text{ mg/L}$

Using Equation 10-9a, we get

mass = $0.25 \times 154 \times 4$
$$\approx 150 \text{ kg/d of secondary sludge}$$

(c) The total mass of combined primary and secondary sludge is $480 + 154 = 634$ kg/d. Now applying Equation 10-7a, we get

$$4 = \frac{634}{V} \times 100$$

and

$V = 63\,400 \div 4 = 15\,850 \text{ L/d}$
$$\approx 16 \text{ m}^3/\text{d total volume of sludge}$$

SLUDGE TREATMENT

Sludge is treated prior to ultimate disposal for two basic reasons: *volume reduction* and *stabilization of organics*. Stabilized sludge does not have an offensive odor and can be handled without causing a nuisance or health hazard. A reduced sludge volume minimizes pumping and storage requirements and lowers overall sludge handling costs.

There are several processes available to accomplish these two basic objectives. They include sludge *thickening, digestion, dewatering,* and *incineration*. When sludge incineration or ocean disposal is used, there is no need to first digest or stabilize the sludge. Sometimes the sludge is thickened to reduce its volume prior to ocean disposal. Three typical sludge treatment options are shown in Figure 10.35.

Thickening

It is usually impractical to treat "watery" or "thin" sludges that have solids concentrations less than about 4 percent. Waste activated sludge is an example of a thin sludge. Thickening is a physical process that serves to increase the solids concentration of the sludge. As previously mentioned, sludge volume varies inversely with the solids concentration. Doubling the solids content from 3 to 6 percent, for example, would cut the total sludge volume in half.

A treatment unit called a *gravity thickener* is usually used to increase the solids concentration. It resembles a circular sewage sedimentation tank that is equipped with vertical slats or pickets attached to the sludge scraper arm. As the scraper arm slowly rotates, the pickets gently stir the

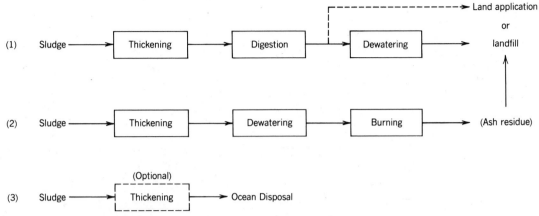

FIGURE 10.35 Three alternatives for sewage sludge management.

sludge. The stirring action serves to release trapped water and gases from the sludge, allowing it to become denser or "thicker." The thickened *underflow* of sludge is withdrawn from the bottom of the tank; the effluent or *supernatant* overflows a weir and is pumped back to the inlet of the treatment plant. Sludge can be thickened to more than 10 percent solids in a gravity thickener.

An alternative to gravity thickening is a process called *dissolved-air flotation,* which is particularly effective for very thin sludges. In this process, air is forced into solution under pressure and then mixed with influent sludge, as shown in Figure 10.36. Air bubbles come out of solution in the open flotation tank, carrying sludge flocs to the liquid surface as they rise. A layer of thickened sludge forms and is skimmed from the surface; the floating sludge layer may be about 400 mm (16 in.) thick. Up to about 6 percent solids concentrations can be obtained in the layer of thickened sludge by dissolved-air flotation.

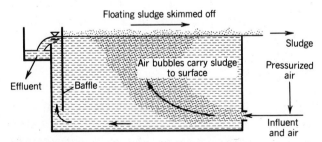

FIGURE 10.36 The dissolved-air sludge flotation process can be used to thicken sewage sludge.

Digestion

Sludge digestion is a process in which biochemical decomposition of the organic solids occurs; in the decomposition process, the organics are converted into simpler and more stable substances. Digestion also reduces the total mass or weight of sludge solids, destroys pathogens, and makes it easier to dry or *dewater* the sludge. Well-digested sludge has the appearance and characteristics of a rich potting soil.

Sludge may be digested under aerobic or anaerobic conditions. Most large municipal sewage treatment plants use anaerobic digestion; aerobic digestion finds application primarily in small, package activated sludge treatment systems. Anaerobic digestion offers an energy-saving advantage over aerobic digestion, because the anaerobic process produces methane gas. The methane may be burned to provide power for other plant processes and equipment, as well as to heat the digester unit. Aerobic systems, on the other hand, are less expensive to build and have fewer operational problems.

In most modern anaerobic systems, digestion takes place in two covered circular tanks. These tanks are typically about 25 m (80 ft) in diameter, and about 15 m (50 ft) deep. The sludge in the first tank is heated to a temperature of about 35°C (95°F) and is thoroughly mixed. A diagram of this type of digestion system is shown in Figure 10.37.

The digestion process is essentially completed in the first tank within about 15 days of detention time. The sludge then flows to a second tank, which

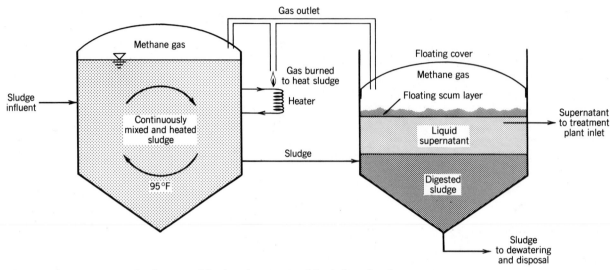

FIGURE 10.37 A schematic diagram of the two-stage anaerobic sludge digestion process.

serves primarily for sludge settling and storage. The digester supernatant, high in BOD and TSS, is pumped back to the inlet of the treatment plant. Digested sludge is removed from the second tank for dewatering and final disposal.

There are many complex biochemical reactions that occur during anaerobic digestion of sewage sludge. Certain species of bacteria first metabolize the complex organic compounds, breaking them down into simpler molecules called organic acids. Then a different species of bacteria metabolizes the organic acids, forming methane. These methane-forming bacteria grow slower, and are much more sensitive to temperature and pH than are the acid formers. The treatment plant operator must maintain careful control over the process to keep the methane-formers alive and healthy; without them, the digestion process cannot continue.

Methane is an important end product of the anaerobic digestion process. As mentioned, its energy value can be utilized in the treatment plant. But even if it is not burned for energy, it serves as an indicator to the plant operator whether or not the process is working properly. The methane content of the sludge digester gas is monitored; if the methane content starts to decrease, corrective action such as adjusting the pH must be taken immediately. Once the methane-forming bacteria die off, the tank must be emptied and the process started up again with fresh sludge.

In aerobic sludge digestion, the alternative to the anaerobic process, the sludge is aerated in an open tank which is similar to an activated sludge tank. Aerobic digestion is usually used in the prefabricated contact stabilization or extended aeration treatment systems. The sludge is aerated for about 30 days, in which time most of the organics are stabilized and the amount of sludge is reduced significantly.

The underlying premise of the aerobic digestion process is that decomposition takes place faster than the rate at which new bacterial cells grow. The BOD loading on the digestion tank is kept very low, so that eventually the microorganisms consume their own cellular mass. This is called *endogenous decay*. But not all of the sludge decomposes, and some of it eventually has to be removed from the tank for disposal. It is a thin sludge that is difficult to thicken or dewater without some type of additional treatment. Although aerobic digestion is more stable in operation than the easily upset anaerobic process, it has the disadvantage of having high energy requirements for aeration.

Dewatering

The process of removing enough water from a liquid sludge in order to change its consistency to that of moist earth is called *sludge dewatering*. Although the process is also called *sludge drying*, the "dry" or dewatered sludge may still contain a significant amount of water, often as much as 70 percent. But at moisture contents of 70 percent or less, the sludge is "liftable," that is, it no longer behaves as a liquid and can be handled manually or mechanically. Sludge is usually dewatered prior to land burial or incineration.

There are several methods available to dewater sludge. The simplest method is to spread the digested sludge on an open bed of sand and allow it to remain there until it dries; drying takes place by a combination of evaporation and gravity drainage. A piping system built under the sand bed collects the water that drains from the sludge. The collected water is pumped back to the inlet of the treatment plant. A cross-section of a typical *sludge drying bed,* as it is called, is shown in Figure 10.38.

Sludge drying beds usually consist of about 200 mm (8 in.) of sand placed on top of a layer of gravel or crushed stone; the pipes that make up the underdrain system are placed in the gravel or crushed stone layer. Since a relatively large amount of land area may be required to construct the sand beds, this method of sludge drying is more common in

rural or suburban communities than in more densely populated urban areas.

Sludge is applied to the sand beds to depths up to 300 mm (12 in.). A typical dewatered sludge "cake," as it is called, has a solids content of about 40 percent; this level of dewatering can be obtained after about 6 weeks of drying. At this point, the sludge can be removed from the sand manually with a pitchfork, or with machinery such as a front-end loader. Sometimes it is necessary to build a glass enclosure, much like a greenhouse, over the sludge beds, in order to protect the sludge from rain and to reduce the drying time in cold weather.

When there is not enough land area available at the plant site to build sludge drying beds, a mechanical system may be used to dewater the sludge. In addition to requiring less space, a mechanical system offers a greater degree of operational control, particularly when thin secondary sludges are being handled. The most commonly used mechanical systems for sludge dewatering include the *rotary-drum vacuum filter* and the *centrifuge.*

The rotary-drum vacuum filter consists of a large cylindrical drum covered with a special filtering fabric. The drum rotates partially submerged in a vat that contains the sludge. A vacuum or suction is applied inside the drum, drawing the sludge up against the fabric cover and extracting the water. A thin layer of dewatered sludge, called *filter cake,* adheres to the fabric. Further rotation of the drum carries the filter cake to a stationary blade that scrapes the sludge into a hopper as the drum moves. The fabric is washed by powerful water sprays before reentering the vat.

A typical vacuum filter installation is shown in Figure 10.39. It is often necessary to add certain chemicals to the sludge in order to coagulate the solids and to improve drainability. This is called *sludge conditioning.* In this process, chemicals such as ferric chloride, lime, or organic polymers are mixed with the sludge before it enters the vat of the vacuum filter.

In the centrifuge dewatering system, the sludge is pumped into a horizontal cylinder, or "bowl," that rotates at high speed. Sludge conditioning chemicals are also injected into the bowl. The solids are forced to the wall of the rotating bowl by centrifugal force; the liquid supernatant is pumped

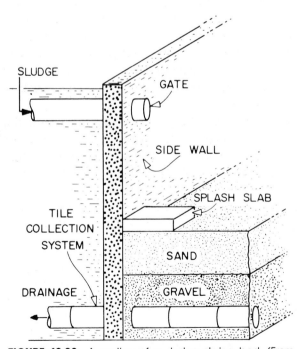

FIGURE 10.38 A section of a sludge drying bed. (From "Operation of Wastewater Treatment Plants," with permission of Water Pollution Control Federation MOP/11, 1976.)

FIGURE 10.39   A sludge vacuum filter. (Envirex, Inc. a Rexnord Company)

back to the head of the plant for treatment. Centrifuges are entirely enclosed, thereby reducing odor problems. But they are sometimes difficult to maintain because of the high speed of the equipment.

Incineration

Sewage sludge can be burned or incinerated to completely evaporate the moisture and to convert the organic solids into an inert ash. The sludge is first dewatered as much as possible in order to minimize the amount of fuel needed. Incineration is used when a suitable site for land disposal is not available, as is usually the case in densely populated cities. Air pollution control becomes an important consideration with this type of sludge processing; the incinerator exhaust gases must be treated to meet EPA air quality standards.

Two common types of sludge incinerators are the *multiple-hearth furnace* and the *fluidized-bed incinerator*. In the multiple-hearth furnace, dewatered sludge enters at the top. As it passes downward through a series of hearths, it is dried and heated to the ignition point. Gas or oil burners furnish the heat for start-up, and the dried sludge itself serves as a fuel to keep the process going. Ash is withdrawn from the bottom hearth; exhaust gases pass through an air scrubbing device in order to control air pollution.

In the fluidized-bed incinerator, an upward flow of air mixed with sludge is forced through a bed of hot sand; the air causes the bed to expand or "fluidize." The sand is preheated to about 800°C (1500°F). The sludge is burned as it passes through the hot sand; the ash, carried out with the exhaust gases, is removed in the final air pollution control equipment.

REVIEW QUESTIONS

1. Compare untreated sanitary sewage and seawater, with respect to the total amount of impurities in each.

2. In raw domestic sewage, what are typical concentrations of BOD, TSS, nitrogen and phosphorus, and coliforms?

3. What are *effluent standards,* and why are they easier to enforce than stream standards? Give an example of typical effluent limitations that would be included on an NPDES permit for a municipal sewage treatment plant.

4. What minimum level of pollutant removal does *secondary treatment* accomplish?

5. What do the terms *influent* and *effluent* refer to?

6. What does *POTW* stand for?

7. Briefly describe preliminary and primary sewage treatment. Approximately what level of BOD and TSS is removed? Is this level of treatment alone adequate by today's standards?

8. What is the difference between a *bar screen* and a *comminutor?*

9. What is the purpose of *grit removal?* How is it accomplished?

10. Give a brief description of a *primary clarifier?*

11. Sketch a flow diagram that shows the overall preliminary and primary sewage treatment processes.

12. What is the purpose of *secondary treatment?* Why is it also called *biological treatment?*

13. What is the difference between a *fixed-growth* secondary treatment system and a *suspended-growth* system? Name one common type of treatment process in each of those two categories.

14. Briefly describe the configuration and operation of a *trickling filter* unit. Why is a *final clarifier* needed?

15. What is the purpose of *recirculation* in a trickling filter?

16. What is meant by *hydraulic load* and *organic load?*

17. How does temperature affect a trickling filter operation?

18. Give a brief description of the configuration and operation of a conventional activated sludge sewage treatment system.

19. What is meant by the term *mixed liquor?*

20. What is *waste activated sludge?*

21. What is the significance of the *F/M ratio?*

22. What does *MLSS* stand for?

23. What is meant by *sludge bulking?* How is the *SVI* related to bulking? How may bulking be controlled?

24. Sketch a flow diagram of an activated sludge treatment process.

25. Briefly describe four different modifications of the conventional activated sludge process. What is a *package plant?*

26. Briefly describe the configuration and operation of a *biodisc* sewage treatment system. In what way is it similar to a trickling filter system?

27. What is a *stabilization pond?* How does it function? In what ways do algae help and hinder the process?

28. What is the purpose of a *chlorine contact tank?* Are there any problems associated with chlorination of secondary effluent?

29. What is the basic purpose of *tertiary treatment* of sewage?

30. What is meant by *effluent polishing?* Briefly describe two different systems used to polish a secondary effluent.

31. Describe a method for removing phosphorus from sewage.

32. Describe two methods for removing nitrogen from sewage.

33. What is a *nitrified effluent?* What is *denitrification?*

34. Describe three types of *land treatment* for sewage.

35. Under what circumstances is on-site sewage disposal warranted? What is a secondary benefit of on-site subsurface disposal?

36. What factors are of importance in planning and designing a subsurface wastewater disposal system?

37. Describe the purpose and procedure of the *perc test*.

38. What are the functions of a *septic tank* in a subsurface disposal system? Is it ever necessary to clean out the tank? Why?

39. What is the function of a *leaching field* in a subsurface disposal system? Briefly describe the configuration of a common type of leaching field.

40. What is a *seepage pit*? When would it be used?

41. Under what circumstances would a *mound system* be used for on-site sewage disposal? Briefly describe this system.

42. Briefly describe the configuration and operation of an *evapotranspiration system*. Under what circumstances would it be used for on-site sewage disposal?

43. Briefly describe the configuration and operation of an *intermittent sand filter system*. Under what conditions would it be used?

44. Briefly discuss some general options for *sludge disposal*.

45. Does sludge volume vary with solids concentration? Describe the relationship in one sentence. Is there a significant difference in volume between a 2-percent sludge and a 4-percent sludge, each of which contains the same mass of solids?

46. Approximately what is the solids content in primary sludge? In secondary sludge? Why is there a difference?

47. What are two basic reasons for treating sewage sludge?

48. Name four different sludge treatment processes.

49. Briefly describe the operation of a *gravity thickener*.

50. Briefly describe the *dissolved-air flotation* process.

51. What is the purpose of *sludge digestion*? Describe the *anaerobic digestion* process. Describe *aerobic digestion*.

52. Describe three methods for drying sewage sludge.

53. What is *filter cake?* Would it be advisable for a treatment plant operator to have filter cake and mixed liquor for lunch?

54. Briefly describe two types of sludge incinerators.

PRACTICE PROBLEMS

1. The influent of a sewage treatment plant has a TSS concentration of 180 mg/L. What is the concentration of suspended solids in the plant effluent, if the plant is achieving a 90-percent TSS removal efficiency? If the flow rate is 10 mgd, how many pounds of solids are discharged per day into the receiving stream?

2. What is the efficiency of a sewage treatment plant that has an influent BOD of 240 mg/L and an effluent BOD of 10 mg/L? If the flow rate is 15 ML/d, how many kilograms of BOD is the plant discharging into the receiving stream?

3. The influent to a primary clarifier has a TSS concentration of 200 mg/L. If the effluent from the tank has a TSS concentration of 60 mg/L, what is the TSS removal efficiency of the clarifier?

4. The influent to a primary settling tank has a BOD of 210 ppm. It is estimated that 33 percent of the BOD will be removed in the tank. What is the expected BOD concentration in the tank effluent?

5. A trickling filter is used to treat a sewage flow of 2 mgd. A direct recirculation ratio of 2 is being utilized. What is the rate at which sewage is applied to the surface of the trickling filter?

6. A trickling filter has a diameter of 20 m and a depth of 2.5 m. It is operated with a direct recirculation ratio of 1.0, and the influent sewage flow rate is 3 ML/d. Influent BOD to the primary tank is 200 mg/L, and the BOD removal efficiency in that tank is 35 percent. Compute both the hydraulic load and the organic load on the trickling filter.

7. The BOD removal efficiency of a trickling filter system is 80 percent and the efficiency of the primary treatment that precedes it is 30 percent. If the raw sewage BOD is 220 mg/L, what is the secondary effluent BOD? What is the overall BOD removal efficiency?

8. An activated sludge tank is 100 ft long, 30 ft wide, and has a SWD of 15 ft. The wastewater flow rate is 1 mgd and the primary effluent BOD is 130 mg/L. The MLSS concentration in the aeration tank is 1800 mg/L. Compute the food-to-microorganism ratio for the system.

9. Sewage is to be aerated in an activated sludge tank; the flow rate is 3 ML/d, and the primary effluent has a BOD of 120 mg/L. The MLSS is to be kept at 2000 mg/L, and the F/M ratio is to be 0.3. If the SWD is 5.0 m, and the tank is to be 20 m long, what should be its width?

10. An aeration tank is operating with a MLSS concentration of 1800 mg/L. After settling for 30 min in a 1-L cylinder, the sludge occupies a volume of 450 mL. What is the SVI of the sludge? Would you expect this sludge to settle satisfactorily in the secondary clarifier?

11. In a time interval of 30 min, the water level in a perc test hole is observed to drop 15 mm. What is the perc rate?

12. A time interval of 2 hours is recorded for a 6-in. drop of water level in a perc test hole. Compute the perc rate.

13. A three-bedroom home is situated on a lot that has a perc rate of 50 min/25 mm. How many laterals are required in the leaching field if the trench width is to be 0.6 m and the maximum trench length is 20 m?

14. A five-bedroom home is situated on a lot that has a perc rate of 17 min/in. How many laterals are required in the leaching field if the trench width is to be 2 ft and the trench length is to be 100 ft? What will be the width of the leaching field?

15. A sludge with a 3-percent solids concentration has a volume of 600 000 gal. How many tons of dry solids are in the sludge?

16. A wastewater treatment plant generates 300 000 L/d of a 2-percent sludge. What is the mass of sludge solids removed from the wastewater each day? If the sludge was thickened to 6-percent solids, what would be the sludge volume and the mass of sludge solids?

17. A flow of 3 ML/d of raw sewage, with TSS = 220 ppm and BOD = 200 ppm, enters a conventional activated sludge treatment plant. Removal efficiencies in the primary clarifier for TSS and BOD are 60 percent and 35 percent, respectively. The primary and secondary sludges are combined and thickened to 6-percent solids. What is the total volume of thickened sludge generated in the treatment plant each day?

SELECTED REFERENCES

1. Parker, H. W., *Wastewater Systems Engineering*, Prentice–Hall, Englewood Cliffs, N.J., 1975.

2. Steel, E. W. and T. J. McGhee, *Water Supply and Sewerage*, 5th ed., McGraw–Hill, New York, 1979.

3. Clark, J. W., et al., *Water Supply and Pollution Control,* 3rd ed., Harper & Row, New York, 1977.

4. Hammer, M. J., *Water and Wastewater Technology,* John Wiley & Sons, New York, 1977.

5. *Design Manual/Onsite Wastewater Treatment and Disposal Systems,* U.S. Environmental Protection Agency, Washington, D.C., Oct. 1980.

6. *Manual of Septic Tank Practice,* U.S. Department of Health, Education, and Welfare, Publication No. 526, Rockville, Maryland, 1969.

11

SOLID AND HAZARDOUS WASTE

11.1 WASTE QUANTITIES AND CHARACTERISTICS

MUNICIPAL REFUSE
HAZARDOUS WASTE

11.2 WASTE COLLECTION AND TRANSPORT

MUNICIPAL REFUSE
HAZARDOUS WASTE

11.3 WASTE PROCESSING AND RESOURCE RECOVERY

INCINERATION
SHREDDING, PULVERIZING, AND BALING
COMPOSTING
RESOURCE RECOVERY AND REUSE

11.4 SANITARY LANDFILLS

SITE SELECTION
CONSTRUCTION AND OPERATION

11.5 HAZARDOUS WASTE MANAGEMENT

DISPOSAL ON LAND
TREATMENT OF HAZARDOUS WASTE
CLEANUP AND CONTAINMENT OF
UNCONTROLLED DUMP SITES

Any material that is "thrown away" or discarded as useless and unwanted is considered *solid waste*. At first glance, the disposal of solid waste may appear to be a very simple and mundane problem. In an age of microcomputers and space flight, it hardly seems possible that "garbage disposal" should present any difficulty. But there are many difficulties that make solid waste disposal a complex technical and environmental problem of huge proportions for our modern (and somewhat wasteful) industrial society.

First, there is no question that improper disposal of solid waste poses a *threat to public health*. Rodents and insects that are sustained by the waste material can act as vectors of disease, transmitting pathogens to human populations. Dozens of communicable diseases can be traced back to the old "garbage dump," including typhoid and plague. Also, certain types of chemical and industrial wastes that are poisonous, flammable, or otherwise very dangerous, can cause more immediate and direct *hazards* to exposed individuals or populations. A more descriptive term, then, for all of society's unwanted and discarded material (which is not disposed of in a sewerage system) is *solid and hazardous waste*.

In addition to the potential for harming public health, improper disposal of solid and hazardous waste can cause serious and perhaps irreversible *environmental damage* in aquatic or terrestial ecosystems. Air pollution can result from incineration and water pollution can result from land burial of waste. This general interrelationship among land, air, and water pollution problems was illustrated in Figure 1.1.

Further complicating the problem are the vast and ever increasing amounts of wastes generated by our "throw-away" society, and the increasingly limited number of options available for their "proper" disposal. For example, one of the most common disposal methods, burial in a *sanitary landfill,* is now criticized by many experts as being unreliable in the long run. And many communities are running out of suitable landfill sites. It is clear that the application of modern engineering and technological methods is necessary, now more than ever, to cope with the solid and hazardous waste disposal problem.

Actually, problems related to solid or hazardous waste go beyond merely their "disposal." In addition to technical and environmental difficulties, there are administrative, economic, and social problems that must be solved. The effort to address all these problems is usually referred to as the practice of *solid and hazardous waste management.*

In this context, the word *management* encompasses planning, design, construction, and operation of facilities for collecting, transporting, processing, and finally "disposing" of the waste material. This is illustrated in Figure 11.1. Overall, solid and hazardous waste management is a multibillion dollar per year industry in the United States. This chapter discusses the fundamentals related to solid and hazardous waste management activities and infrastructure.

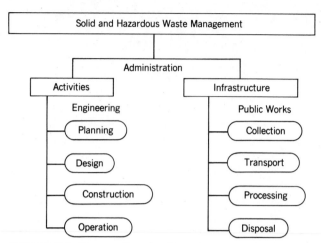

FIGURE 11.1 The practice of *solid and hazardous waste management* encompasses several different technical activities and the operation and maintenance of related infrastructure.

tors related to waste generation, sources, and quantities, and the terminology used in describing waste characteristics, are discussed briefly.

Statistics related to solid and hazardous waste quantities are presented here to illustrate the magnitude of the problem. It should be kept in mind that the numbers are approximate; they are only "ballpark" figures. Average waste generation rates may vary considerably from one location of the country to another.

11.1 WASTE QUANTITIES AND CHARACTERISTICS

Solid and hazardous wastes are grouped or classified in several different ways. These different classifications are necessary in order to be able to address, in an effective manner, the complex problems related to waste management. In this section, fac-

MUNICIPAL REFUSE

The term *refuse* is generally used to describe all of the nonhazardous solid waste from a community that requires collection and hauling (transport) to a processing or disposal site. *Ordinary refuse* includes *garbage* and *rubbish.* Garbage contains putrescible or highly decomposable food waste, such as vegetable and meat scraps. Rubbish contains mostly dry, nonputrescible material, such as glass, rubber, or tin cans, and slowly decomposable or combustible material such as paper, textiles, or wood objects.

Another type of refuse, called *trash,* includes bulky waste material that generally requires special handling, and is therefore not collected on a routine basis. An old couch, mattress, television, or refrigerator, and even a large up-rooted tree stump, are examples of trash items. Community refuse, including garbage, rubbish, and trash, is sometimes referred to as *municipal solid waste,* or simply *MSW.*

A classification scheme for both solid and hazardous waste is illustrated in Figure 11.2. It should be noted that additional classifications may include *agricultural waste, mining waste, construction* or *demolition waste,* and other groups.

On the average, roughly 26 N (7 lb) of ordinary refuse is generated per person each day in the United States. (Keep in mind that this figure includes commercial and industrial refuse as well as residential or household waste.) About 85 percent of it is rubbish, and 15 percent is garbage. Overall, about half of the total amount of refuse is paper. Before compaction, the refuse weighs about 1.5 kN/m³ (250 lb/yd³). Data like these are important for proper planning and design of solid waste collection and disposal facilities. As previously mentioned, actual quantities may vary significantly from these averages, depending on local community characteristics.

On a per-capita basis, the amount of refuse generated may not sound like very much. But when total populations are considered, the amounts add up dramatically. On a nationwide basis, ordinary refuse generation amounts to about 300 million tons per year in the United States. Uncompacted, this amount of refuse would cover roughly 100 square miles of land to a depth of about 10 ft.

HAZARDOUS WASTE

Ordinarily, municipal solid waste that is generated in the average household or commercial establishment is not considered dangerous. But in our modern society, large quantities of dangerous materials are generated by chemical manufacturing companies and related industries. These dangerous or *hazardous wastes* can cause serious illness, injury, or death; they also pose a serious threat to environmental quality if improperly transported or "dumped."

There are hundreds of recent incidents on record in which illegal or improper disposal of hazardous waste caused harm to the public and to the environment. Many of these cases involve contamination of groundwater that was used for public water supply. Indiscriminate handling of hazardous waste can also lead to contamination of rivers and lakes. One such incident, called "Valley of the Drums" by the news media, occurred in Kentucky in the late 1970s. About 6000 damaged and leaky metal drums holding hazardous substances spilled their contents on the ground where they were dumped. Consequently, the soil and surface waters in the area were polluted with about 230 different organic chemicals and metals.

Another infamous hazardous waste incident in the United States is the "Love Canal" episode, which occurred near Niagara Falls, New York, also in the late 1970s. Chemical wastes, buried more than 25 years previously, gradually leaked out of the metal

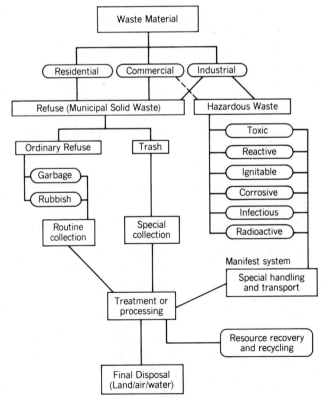

FIGURE 11.2 Classifications of solid and hazardous waste.

drum containers and percolated through the soil into the yards and basements of homes built on the site. Hundreds of families were evacuated after an excessive number of unusual health problems were observed among the residents. Many of the approximately 80 different chemicals identified at the site are suspected of causing cancer in humans.

Hazardous wastes differ from other solid wastes in form as well as in behavior. They often are generated in liquid form, but they may also occur as solids, sludges, or gases. Usually they are contained and confined in metal drums or cylinders. There are four primary classifications or groups of hazardous wastes, based on the properties of *toxicity, reactivity, ignitability,* and *corrosivity.* Two additional types include *infectious* and *radioactive* waste material. These classifications are illustrated in Figure 11.2.

Toxic wastes are poisons, even in very small or *trace* amounts. Some may have an *acute* or immediate effect on humans or animals, causing death or violent illness. Others may have a *chronic* or long-term effect, slowly causing irreparable harm to exposed persons. Some toxic wastes are considered to be *carcinogenic,* causing cancer many years after initial exposure. Others may be *mutagenic,* causing biological changes in the children or offspring of exposed people and animals.

Most toxic wastes are generated by industrial activities, including the manufacture of chemicals, pesticides, paints, petroleum by-products, metals, textiles, and many other products. There are hundreds of different chemicals specifically identified and listed as toxic by the Environmental Protection Agency, and the list is growing. Among the toxic inorganic chemicals are arsenic, asbestos, cadmium, chromate, mercury, and zinc. Some of the toxic organic materials are dioxin, chloroform, ethylene oxide, kepone, methyl parathion, and polychlorinated biphenols (PCBs).

The pesticide dioxin has been categorized by the EPA as "one of the most perplexing and potentially dangerous chemicals ever to pollute the environment." A special control strategy has been adopted for this particular substance, including a systematic investigation to identify the location and extent of dioxin contamination. Health risks will be assessed in order to determine the appropriate response at contaminated sites. Presently, sites that have more than 1 μg/L (one part per billion) of dioxin in the soil would be subject to remedial action in order to protect public health.

Reactive wastes are those that are unstable and tend to react vigorously with air, water, or other substances. The reactions either cause explosions or form toxic vapors and fumes. Ignitable wastes, including organic solvents such as benzene or toluene, burn at relatively low temperatures and present an immediate fire hazard. Corrosive wastes, including strong alkaline or acidic substances, destroy materials and living tissue by chemical reaction. Reactive, ignitable, and corrosive wastes all can cause immediate harmful effects on living organisms or on the physical environment. They present particularly difficult problems related to transport, storage, and disposal, and they must be managed with special care.

Infectious or biological waste material includes human tissue from surgery, used bandages and hypodermic needles, microbiological material, and other substances generated by hospitals and biological research centers. This type of hazardous waste must be handled and disposed of properly to avoid infection and the spread of disease among the general public.

Radioactive waste, particularly from nuclear power plants, is also of special concern. Excessive exposure to ionizing radiation can harm living organisms. And radioactive material may persist in the environment for thousands of years before it decays appreciably. But because of the scope and technical complexity of this problem, radioactive waste disposal is generally considered separately from other forms of hazardous waste.

It has been estimated by the EPA that about 10 percent of all nonradioactive industrial waste in the United States is hazardous. This amounts to roughly 60 million tons per year. Until the mid-1980s, only about 10 percent of this material was disposed of in an environmentally sound manner. Most of it has been "swept under the rug," so to speak, and poses a continuous threat to public health and environmental quality. It has been estimated that there are as many as 40 000 hazardous waste "dumps" in the United States, many of which pose serious health and environmental problems. The cost of cleanup is expected to reach billions of dollars.

11.2 WASTE COLLECTION AND TRANSPORT

At first glance, it may seem that the main problem with respect to solid and hazardous waste is its ultimate disposal. To be sure, disposal is no small problem. But of the billions of dollars spent each year for municipal solid waste management, about 80 percent of it is needed to cover the cost of waste collection and transport. It should also be noted that proper collection and hauling techniques are important to protect public health, safety, and environmental quality. This is particularly true for hazardous waste, where an accidental spill during transport can have tragic consequences.

This section focuses on some of the factors related to the storage, collection, and transport of solid and hazardous waste. Proper storage prior to collection is necessary to protect public health, as well as for aesthetic reasons. Collection and transport are complicated by the fact that house-to-house collection vehicles are generally not suitable for long-distance hauling to a central processing or disposal site. Also, the different characteristics of municipal refuse and hazardous waste require different approaches regarding the storage, collection, and transport of those materials.

MUNICIPAL REFUSE

Proper on-site storage is of particular importance for residential solid waste, which contains a significant amount of putrescible garbage. The use of containers with tight lids reduces the incidence of rodent and insect infestation; also, offensive odors and unsightly conditions can be kept to a minimum if containers and storage areas are washed periodi-

cally, and the waste removed within a week or less. For most individual residences, the 30-gal galvanized metal or plastic container is most effective for temporary storage. In apartment residences, larger portable containers, which can be moved on rollers and then emptied into collection trucks, are commonly used.

Collection

Refuse collection is generally a local municipal responsibility; public employees and equipment are usually assigned to the task. Sometimes it is more economical for private collection companies to do the work under contract to the municipality. And in some communities, private collectors are paid by the individual homeowners. Whatever the actual administrative arrangement, proper planning, operation, and regulation of the collection activity are necessary in order to prevent public health nuisances.

Refuse collection vehicles most commonly used in the United States are of the enclosed, compacting type, with capacities of about 15 m^3 (20 yd^3). Typically, a driver and one or two loaders are needed for each vehicle; solid waste collection is a labor-intensive activity, which accounts for its high relative cost. Since compaction in the collection truck reduces the volume of refuse by about 50 percent, the unit weight of the refuse is doubled to about 3 kN/m^3 (500 lb/yd^3). Typical collection vehicles are shown in Figure 11.3.

Technical decisions must be made with regard to the frequency of collection or pickup from each waste generation site, and the point of pickup (curb, alley, backyard, etc.). These decisions depend primarily on the type of community, population density, and land use in the collection area. Generally, combined collection of garbage and rubbish is more economical than separate collection of each type of waste. But in some communities, certain materials are recycled, so homeowners practice *source separation;* that is, they separate their refuse into two or more types (i.e., paper, glass, food wastes).

The problem of selecting an optimum collection route can be one of the most technically complex

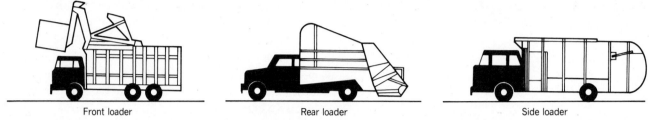

Front loader   Rear loader   Side loader

FIGURE 11.3 Enclosed compaction type refuse collection vehicles reduce the volume of collected waste material by about 50 percent. (New York Department of Environmental Conservation)

problems in solid waste management, particularly for large and densely populated urban areas. An optimum route is one that results in the most efficient use of labor and equipment. Some important characteristics of an optimum route are:

1. Collection vehicles should not travel twice down the same street, that is, collection paths should not overlap.

2. Refuse collection on crowded streets and roads should not occur during morning or afternoon rush hours.

3. Collection should occur in the downhill direction as much as possible, to conserve fuel.

4. The starting point should be close to the collection vehicle garage, and the last collection point should be as close as possible to the destination of a filled vehicle (transfer station, incinerator, processing plant, or sanitary landfill).

In order to satisfy these and other constraints on the collection route, it is sometimes necessary to use mathematical models and digital computers to determine the best route. This sophisticated method of problem solving is called *systems analysis* or *operations research*. It allows the designer to account for all of the many variables in a large and complex collection network.

Transfer Stations

It is not always economically feasible for individual collection trucks to haul refuse to the point of processing or disposal. This is particularly true when the final destination is not in the immediate vicinity of the community in which the waste is collected. To solve this waste transport problem efficiently, one or more *transfer stations* may be used. At a transfer station, the waste from the relatively small collection trucks is consolidated into larger transport vehicles, such as tractor–trailer units. It is more economical for fewer of the large vehicles to transport the waste over the long-haul distance to the processing or disposal location.

There are several types of transfer stations. They may be classified according to capacity, such as *small* (less than 100 tons per day) or *large* (more than 500 tons per day). Or they may be classified as *direct discharge* or *storage discharge*. In a storage discharge transfer station, the refuse is first emptied from collection vehicles into a storage pit or onto a platform. Clamshell equipment or bulldozers are used to hoist or push the waste into large trailer units. In a direct discharge station, the collection vehicles empty directly into the larger transport vehicles. A two-level arrangement is necessary to accomplish this, as illustrated in Figure 11.4.

Innovative Collection–Transport Systems

Modern technology may eventually eliminate the need, in some communities, for above-ground collection and transport of refuse. Pipeline transport of solid waste is feasible, but it is economical only in newly constructed facilities. Pipeline transport may be accomplished by either hydraulic or pneumatic means. In hydraulic systems, shredded waste is mixed with water to form a slurry that can flow through underground pipes. A major disadvantage

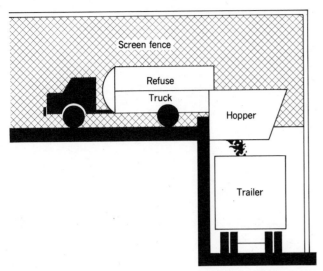

FIGURE 11.4 At a direct-discharge transfer station, several collection trucks deposit refuse into a larger vehicle for hauling to a more distant disposal site. (New York State Department of Environmental Conservation)

is the increased hydraulic and organic load on sewage treatment plants.

In pneumatic systems, the shredded refuse is pulled by suction or vacuum through the underground pipes to a central processing plant. Waste collection at the Walt Disney World amusement park in Florida is accomplished by a system of this type. It eliminates the need for noisy and unsightly garbage trucks. But complex control mechanisms, valves, and high-speed turbines are required for the system to operate, and installation costs are high. For the immediate future, hydraulic and pneumatic solid waste collection and transport systems will probably be feasible only in specialized local situations.

HAZARDOUS WASTE

The storage, collection, and transport of hazardous waste present a special problem. This is because of the potential for serious harm to public health and environmental damage in the event of an accidental spill, as well as because of the unfortunate tendency toward indiscriminate disposal by hazardous waste generators and haulers. The illegal practice of "midnight dumping," as it is called, saves waste generators money, but takes its toll with respect to public and environmental health, as well as public tax dollars. Surface and groundwater contamination are the most common results of midnight dumping.

On-Site Storage

Many generators of hazardous waste store the material on site for varying periods of time. Relatively large quantities may be stored in above-ground basins or lagoons. Above-ground basins may be constructed of steel or concrete, but they are subject to corrosion or cracking and are not suitable for storing reactive or ignitable waste.

The most common type of surface storage impoundment is an open pit or holding pond, called a *lagoon*. The lagoon should be lined with impervious clay soil or a synthetic material to protect the groundwater. But according to EPA estimates, many thousands of such lagoons in the United States do not have liners. And most of them are located above groundwater aquifers used for public water supply.

Some hazardous waste for which no adequate disposal method presently exists is placed in *engineered storage*. Engineered storage serves to hold the material for a relatively long term but temporary basis. The waste may be containerized and buried in a retrievable manner, and is supposed to be subject to "final" disposal when appropriate technology is available for that purpose. High-level radioactive wastes, mercury from batteries, and pesticides such as kepone are examples of substances which are currently managed by engineered storage.

Relatively small amounts of hazardous waste that are generated intermittently may be containerized in 55-gal fiberglass or steel drums, for ease of handling, storage, and transport. Corrosive material would be stored in fiberglass or glass-lined containers to reduce deterioration and leakage. Toxic chemical liquids may be stored in metal drums. A typical steel drum container is illustrated in Figure 11.5a. Unfortunately, too many of these drums have been disposed of improperly, as shown in Figure 11.5b.

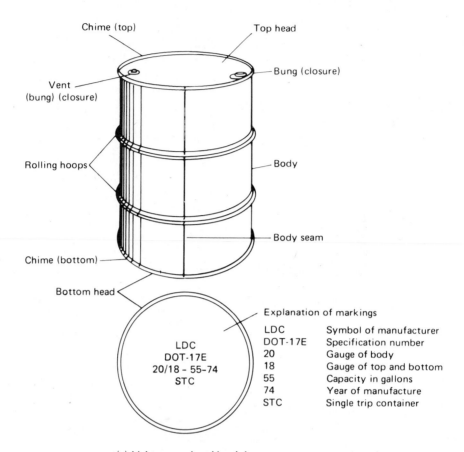

(a) **Light-gauge closed-head drum**

FIGURE 11.5 (a) A typical steel drum container, used for storage and disposal of hazardous waste. (From Tchobanglous, G., *Solid Wastes*, with permission of McGraw–Hill.) (b) Improper storage of hazardous waste material. (Division of Waste Management, N.J. Department of Environmental Protection)

Containers or drums of hazardous waste must be labeled properly before transport to a processing or disposal facility. The label must identify the contents as either explosive, flammable, corrosive, or toxic material. Appropriate signs or placards must be placed on the transport vehicle, to warn the public of potential danger, and to assist emergency response workers if there is an accidental spill along the transport route.

Manifest System

One of the most important laws that regulates the transport and disposal of solid and hazardous waste in the United States is the *Resource Conservation and Recovery Act* (RCRA), of 1976. Under Subtitle C of RCRA, the Environmental Protection Agency is given the responsibility for establishing and regulating a hazardous waste management system. In addition to identifying hazardous waste that will be subject to the regulations, the EPA is charged with the task of administering standards related to hazardous waste storage, transport, and disposal.

One of the key features of RCRA with regard to the transport of hazardous waste is a "cradle-to-grave" *manifest system* for monitoring the journey of the waste from its point of generation to its point of final disposal. It serves primarily to eliminate or reduce problems associated with "midnight dumping" of hazardous waste.

The manifest is a record-keeping document that must be prepared by the generator of the hazardous waste, such as a chemical manufacturing company. The generator has primary responsibility for the ultimate disposal of the waste. The generator gives the manifest to a licensed waste transporter, along with the waste itself. The transporter must comply with the regulations of the U.S. Department of Transportation (DOT) with regard to transport vehicle standards and response to accidental spills.

The manifest must be delivered by the transporter to the recipient of the waste at an authorized off-site storage, processing, or disposal facility. *Each time the waste changes hands, the manifest form must be signed.* Copies of the manifest are kept by each party involved, and copies are sent to the appropriate state environmental agency. Figure 11.6 shows a typical manifest cycle, which is followed by the State of New Jersey.

In addition to curtailing the practice of midnight dumping, the manifest system serves several other purposes. It provides data with regard to sources, types, and quantities of hazardous waste. These data are valuable for future planning and design of hazardous waste management systems. The manifest also ensures that the nature of the wastes is described to the operator of a processing or disposal facility; this prevents accidents due to improper handling or disposal of the material. It also provides information regarding recommended emergency response procedures.

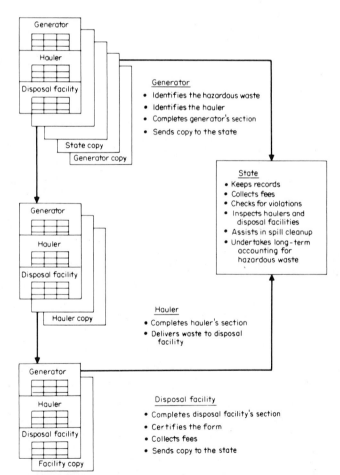

FIGURE 11.6 Hazardous waste *cradle-to-grave* manifest system. (From Bennett G. F., et al., *Hazardous Materials Spills Handbook*, with permission from McGraw–Hill)

11.3 WASTE PROCESSING AND RESOURCE RECOVERY

Solid and hazardous waste may be treated or processed prior to final disposal. Waste treatment or processing offers several advantages. First, it can serve to *reduce the total volume and weight* of material that requires disposal. Volume reduction helps to conserve land resources, since the land is the ultimate "sink" for most waste material. Even after *refuse incineration,* a process that provides the greatest degree of volume reduction, land burial is required for disposal of the ashes and other unburnable residue that remains behind.

Processing can also change the form of the waste

and improve its handling characteristics. Garbage and other organic wastes, for example, can be rendered inoffensive and even useful by a process called *composting*. And certain hazardous wastes can be destroyed, neutralized, or otherwise treated to render them nonhazardous.

Finally, processing can serve to recover natural resources and energy in the waste material, for *recycling* or reuse. Much of the "waste" material can actually be used as raw material for productive purposes. However, a basic disadvantage inherent in any waste processing or recovery system is the additional cost of constructing and operating the facility. This section reviews some of the more common MSW treatment, processing, or recovery methods. Hazardous waste management is discussed in Section 11.5.

INCINERATION

One of the most effective methods to reduce the volume and weight of solid waste is to burn it in a properly designed furnace, under suitable temperature and operating conditions. This process is called *incineration*. It is expensive, and unless appropriate air cleaning devices are provided, atmospheric pollution from the discharge of gaseous and particulate combustion products can occur. It also is a process that requires high-level technical supervision and skilled employees for proper operation and maintenance.

The advantages of incineration, however, often outweigh these disadvantages. Incineration can reduce the total volume of ordinary refuse by more than 80 percent. In densely populated urban areas, where large sites suitable for landfilling are not available within reasonable hauling distances, incineration may be the only economical option for solid waste management. In some cases it is feasible to design and operate the incinerator so that the heat from combustion can be recovered and used to produce steam or electricity. Incineration may also be used to destroy certain types of hazardous waste material.

Incineration is *a chemical process* in which the combustible portion of the waste is combined with

oxygen, forming mostly carbon dioxide and water. This chemical reaction is called *oxidation,* and it results in the release of heat energy. The carbon dioxide and water vapor are released into the atmosphere. For complete oxidation, the waste must be mixed with appropriate volumes of air, and a proper temperature must be maintained for a suitable length of time. Typically, furnace temperatures are about 815°C (1500°F), and the waste must remain in the furnace for about an hour. For some types of refuse, temperatures up to 1400°C (2500°F) may be reached during the primary stages of combustion.

Incineration does not completely destroy all solid waste. *Incinerator residue* includes ash, glass, metal cans, and other unburned substances. The residue from incineration of municipal solid waste is typically about 20 percent (and sometimes as little as 5 percent) of the original waste volume. In addition to some gaseous products of incomplete combustion, one of the products carried in the combustion air stream is *fly ash*. Fly ash consists of finely divided particulate matter, including cinders, mineral dust, and soot. Air pollution control devices installed on municipal incinerators serve primarily to remove the fly ash from the combustion gases, which are then discharged into the atmosphere.

Design and Operation

Most modern municipal solid waste incinerators are designed for *continuous feed* operation, as opposed to the less desirable intermittent or *batch feed* mode of operation. Continuous feed of refuse allows for uniform furnace temperature, which provides more efficient combustion and reduces thermal shock damage to the incinerator components.

A typical incinerator facility includes a below-grade refuse storage pit or "tipping" area, which provides volume for about one day of storage. The refuse is lifted from the pit by a crane with a grab bucket and deposited into a charging hopper and chute. Then it is released from the chute onto a charging grate or stoker. There are various types of mechanical travelling or rocking grates that agitate and move the burning material through the

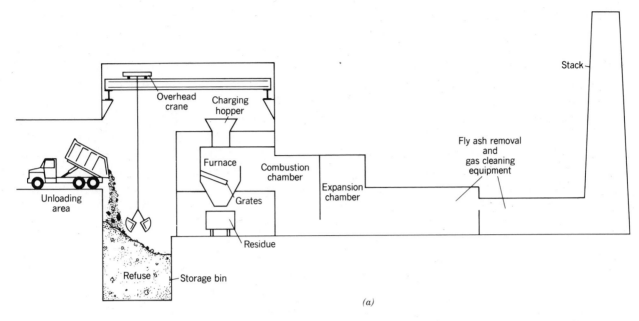

(a)

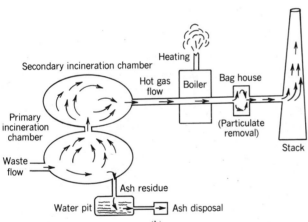

(b)

FIGURE 11.7   *(a)* A simplified section view of a typical municipal incinerator facility. The volume of collected refuse can be reduced by more than 80 percent. Two or more furnaces are usually installed. *(b)* A simplified diagram of a waste-to-energy facility. (Divison of Waste Management, N.J. Department of Environmental Protection)

furnace in a manner that allows a proper draft or flow of air.

Municipal incinerators are built in a variety of configurations, including the *rectangular furnace,* the *rotary kiln furnace,* and the *vertical circular furnace.* A section view of the rectangular type, which is most common for modern municipal incinerators, is shown in Figure 11.7a. Inside the furnace, combustion occurs in two phases: *primary combustion* and *secondary combustion.* In primary combustion, moisture is driven off; the burnable waste is volatilized and then ignited. In secondary combustion the remaining unburned gases and particulates, which are entrained in the airstream after primary combustion, are oxidized. Secondary combustion helps to eliminate odors and reduces the amount of unburned particulates in the exhaust gases.

Sufficient quantities of air must be thoroughly mixed with the burning refuse so that oxygen is available for primary and secondary combustion. Air can be supplied from openings beneath the grates (underfire air) or admitted to the area above the grates (overfire air). The relative amounts of underfire and overfire air must be determined by the operator for best operation of a particular incinerator. The flow of air can be achieved from the natural draft of a chimney or from forced-draft fans. In addition to a controlled air supply, a typical incinerator facility includes air pollution control devices (discussed in Chapter 12). The height of the incinerator stack or chimney depends on several factors, including local topography, land use, climate, average wind conditions, and Federal Aviation Agency regulations.

The furnaces of incinerators that are not used for

energy or heat recovery are typically built with *refractory* material, which resists the damaging effect of very high combustion temperatures. Refractory bricks are made of alumina, magnesia, silica, and a clay mineral called kaolin. Modern incinerators, in which the furnace temperatures may exceed 1100°C (2000°F), have refractory walls of only 225 mm (9 in.) in thickness. Sometimes, suspended refractory walls and roof construction are used, to facilitate repair of damaged refractory material.

Auxiliary gas or fuel oil is sometimes used for furnace warm-up and to initiate primary combustion when the refuse is very wet. Auxiliary fuel also facilitates complete secondary combustion and provides additional smoke and odor control in the exhaust gases.

Energy Recovery

Recovery of the heat given off by burning refuse in an incinerator can be accomplished using a refractory lined furnace, followed by a boiler. This is illustrated schematically in Figure 11.7*b*. The boiler converts the heat from combustion into steam or hot water. In this way, the energy content of the refuse can be recycled and put to beneficial use. Since ordinary municipal refuse in the United States is about 50 percent paper, it has a reasonably high energy content—roughly one third that of coal.

Another type of heat recovery system makes use of a so-called *water-tube wall* furnace. A water-tube furnace is lined with closely spaced welded steel tubes, which are arranged vertically to form continuous sections of wall. Insulation on the outside of the walls reduces heat loss by radiation. Heat is absorbed by the water that circulates through the tubes, and the heated water is used to produce steam. An advantage of this type of system is that the water also serves to control furnace temperature, without the need for excess air. A smaller volume of air flow results in air pollution control costs that are lower than those of a refractory lined furnace.

When raw or unprocessed solid waste is fed as a fuel directly into a heat recovery system, the process is referred to as *mass burning*. In some systems, however, the refuse may be treated or processed by shredding and by separation of noncombustible waste material before being fed into the furnace. In this case, the solid waste is called *refuse derived fuel,* or RDF.

The RDF can be used in loose, shredded form, or it can be compacted and densified into the form of small charcoallike briquettes. RDF can be used as a supplement for an oil or coal-fired furnace, or it can be used in furnaces designed only for the use of RDF as a fuel. As of 1980, there were about 25 RDF heat recovery plants operating (or under construction) in the United States, with capacities ranging between 50 and 2200 tons per day. RDF energy recovery systems are very expensive, since facilities for processing the waste, as well as facilities for burning it, are required. Steady income or revenue from "tipping" fees, as well as from the sale of steam, are necessary to make most RDF projects economically viable. (The tipping fee is paid by the private hauler who unloads waste at the plant.)

Approximately 2.5 lb or 1.1 kg of steam can be produced per 0.5 kilogram or pound of incinerated refuse, but variation in the characteristics of the incoming MSW leads to variations in the rate of steam production. Provision must be made, then, to burn auxiliary fuels when the volume or recoverable heat content of the waste temporarily decreases.

The variation in market demand for the steam can be even more of a problem for a waste heat recovery project. If the demand for steam drops, the revenue from its sale will also drop. This can cause severe economic problems for the owner of the system, since the amortization debt (interest and principal payments on bank financing loans) must be paid off continuously.

Heat recovery and reuse from MSW incineration is an attractive waste management option from an environmental and ecological perspective. But the problems just mentioned, along with the very high costs for equipment and controls, the need for skilled technical personnel, and the need for auxiliary fuel systems, can make it a less attractive option. This is particularly the case for some pragmatic public works officials who are responsible for daily operation and maintenance of municipal waste management systems.

Incineration without heat recovery is simpler to manage and can be less than one third as costly as an RDF project. Nevertheless, plain incineration does result in a total loss of recoverable energy. Because of public and political interest in "recycling," there will probably be an increasing emphasis on the design and construction of MSW heat recovery systems in the coming years. Incineration, with or without heat recovery, is becoming more "attractive" for solid waste management than burial of MSW in a landfill since suitable sites for burial of solid waste are becoming increasingly difficult to find.

Cogeneration

A recent law passed by Congress, called the *Public Utilities Regulatory Policies Act (PURPA)*, requires large power companies to purchase energy from smaller generators of electricity who offer it for sale. Consequently, MSW heat recovery plants that generate more steam than can be used or sold, can convert their excess steam into electricity for sale to the local power company. Of course a turbine would be required at the incinerator facility to convert the steam into electricity.

When both steam and electricity are produced at an MSW plant, the process is called *cogeneration*. Although the process of converting solid waste into electricity is relatively inefficient (compared to producing electricity from fossil fuel, or even compared to producing just steam from solid waste), cogeneration offers some significant advantages to energy recovery from MSW. This is particularly true in areas of the country where the market demand for steam used for heating purposes is seasonably variable. During the warm season, the excess steam can be used to produce electricity, and the MSW energy recovery plant can be economically self-sufficient all year round.

Cogeneration of steam and electricity from solid waste is a relatively new concept in the United States. But it has successful precedents in other in-dustries, and it may greatly enhance the viability of energy recovery as a solid waste management option.

Pyrolysis

Another relatively new development in MSW treatment by thermal–chemical conversion is a process known as *pyrolysis,* which is also called *destructive distillation*. It differs from conventional incineration in that it is an *endothermic* process, that is, it requires continuous input of heat energy to occur. (Incineration, on the other hand, is an *exothermic* process which gives off heat as oxidation occurs.)

Pyrolysis is a high-temperature process (1100°C or 2000°F) which takes place in a low-oxygen or oxygen-free environment. Combustion of natural gas is used to start the process, but if about 70 percent of the gaseous pyrolysis by-products are recycled back to the gas burners, the process can become self-sustaining.

Instead of combustion, pyrolysis involves a complex series of chemical reactions. These reactions decompose or convert the organic carbon components of the solid waste into potentially useful by-products. Pyrolysis also substantially reduces the volume of the solid waste. The gaseous, liquid, and solid by-products of pyrolysis include methane, methanol, tar, and charcoal. They are combustible and can be used as fuels, or they can serve as raw materials for other synthetic chemical products.

Pyrolysis can be particularly useful in converting rubber to fuel oil and combustible gas. Discarded rubber tires are a troublesome form of solid waste, since the tires do not compact or decompose like other waste material, and they must be selectively handled when burned at a conventional incinerator. Millions of tires are discarded each year. Generally they are shredded by special machinery prior to pyrolysis. (Shredded rubber from tires is sometimes also added to asphalt concrete used for road construction.)

The actual composition of pyrolysis end-products may vary, and it is very dependent on the nature of the solid waste as well as on the temperature and pressure under which the process operates. The quality of the by-products can be significantly improved if glass, metal, and other inorganic material is first removed or separated from the solid waste that is fed into the pyrolysis furnace. Although waste separation adds to the expense, the pyrolysis process still has great potential as an effective solid waste management method. In addition to reducing waste volume and producing useful by-products, it poses less of a threat to air quality than does incineration.

SHREDDING, PULVERIZING, AND BALING

Shredding refers to the physical processes of cutting and tearing; whereas *pulverizing* refers to the physical processes of crushing and grinding. Solid waste may be shredded or pulverized in order to reduce its size and increase its uniformity. Since size reduction and uniformity are the objectives of both processes, the terms *shredding* and *pulverizing* are often used synonymously with respect to solid waste treatment and management. *Milling* is yet another term used to describe a mechanical size-reduction process.

Baling refers to a mechanical process in which the raw solid waste is compressed and compacted into rectangular blocks or *bales,* primarily to reduce its volume. It should be noted that the "size reduction" obtained by shredding or pulverizing does not necessarily refer to the total waste volume, but to the individual components of waste material. Baling, though, does achieve significant reductions in total waste volume.

There are many reasons for shredding or pulverizing municipal solid waste. As previously mentioned, the production of refuse derived fuel, or RDF, requires processing of the raw solid waste. This typically includes shredding and/or pulverizing. Composting, which will be discussed in the next section, also first requires some type of size reduction process.

Shredding and pulverizing may first be applied where the basic objective is to recover material from the waste that can be recycled and marketed.

These size reduction and "homogenizing" processes improve the performance of the mechanical separation machinery. Yet another reason for shredding and pulverizing MSW is to prepare it for hydraulic or pneumatic transport in a pipeline. Shredding of refuse prior to land burial can reduce the effort required for proper compaction in the landfill; and finally it reduces the possibility of rodent infestation, since the animals have difficulty in finding food scraps or voids for a habitat in the uniform material.

Hammer Mills

One of the most common types of equipment used for processing MSW into a uniform or homogeneous mass is the *horizontal shaft hammer mill*. A hammer mill is basically an impact device in which the raw solid waste material is hit with a force sufficient to crush or tear the individual components of the waste. Impact is provided by several hammers that rotate at a high speed around the center horizontal shaft.

A section view of a hammer mill is illustrated in Figure 11.8. Cutting bars or breaker plates attached around the periphery of the mill chamber also serve to reduce the size of the waste material. When the size reduction is complete, the processed

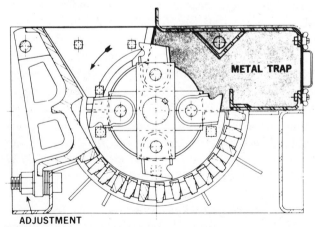

FIGURE 11.8   A section view of a *hammer mill,* used to shred and pulverize solid waste. (Williams Patent Crusher and Pulverizer Company, Inc.)

waste simply falls through the grate at the bottom of the mill. Not unexpectedly, repair and replacement of hammermill components are part of a frequent and common routine, because of the high speeds and impact action of the machinery. In addition to the cost of maintenance, electric power requirements are high.

A hammer mill is a very versatile shredding and pulverizing device because it will accept almost any type of waste material (except, of course, very bulky or dense items such as trees or engine blocks). And it is possible to reduce the size of solid waste material components to uniform fragments between 25 and 50 mm (1 and 2 in.) with proper operation. A typical size hammer mill is a 150-horsepower unit that is capable of processing about 12 tons of solid waste per hour. In addition to the cost of operation and maintenance, though, disadvantages or nuisances of hammer mill operation include noise and dust generation.

High-Pressure Compaction

Compacting solid waste into the form of blocks or bales, roughly 1 to 2 m^3 (yd^3) in size, is called *baling*. When done under high pressures, more than 750 kPa (100 psi), the bales can retain their rectangular shape during handling without the need for wire or strap ties. A high pressure compaction unit is shown in Figure 11.9.

A moisture content of about 40 percent in the refuse helps to hold the compressed pieces of waste together in a single cohesive mass. A compacted bale of refuse can have a unit weight as high as 12 kN/m^3 (2000 lb/yd^3) compared to about 1.5 kN/m^3 (250 lb/yd^3) for the original refuse, because of the elimination of air-filled spaces or voids in the waste material. Volume reduction can be as high as 10:1 or 90 percent of the original waste volume, but it depends to a large extent on the moisture and density characteristics of the raw waste and on the compaction pressures used.

FIGURE 11.9 High pressure compaction units can be used for making rectangular bales or blocks of solid waste. (American Solid Waste Systems—Division of American Hoist and Derrick Company)

As previously indicated, solid waste volume reduction may be expressed in terms of a ratio or in percent. In formula form, this may be expressed as:

percent volume reduction

$$= \frac{Vol_1 - Vol_2}{Vol_1} \times 100 \quad (11\text{-}1)$$

and

$$\text{compaction ratio} = \frac{Vol_1}{Vol_2} \quad (11\text{-}2)$$

where Vol_1 = MSW volume before compaction
Vol_2 = MSW volume after compaction

EXAMPLE 11.1

If the initial volume of a sample of solid waste is 15 m^3, and after compaction the volume is reduced to 3 m^3, what is the percent volume reduction and the compaction ratio? If it is desired to increase the percent reduction to 90 percent, what would the compaction ratio have to be?

Solution:

Applying Equation 11-1, we get

percent volume reduction

$$= \frac{15 - 3}{15} \times 100 = 80 \text{ percent}$$

and from Equation 11-2, we get

$$\text{compaction ratio} = \frac{15}{3} = 5 \text{ or } 5{:}1$$

To obtain the compacted volume at 90-percent reduction, we get:

$$90 = \frac{15 - Vol_2}{15} \times 100 \quad \text{and } Vol_2 = 1.5 \text{ m}$$

At 90-percent volume reduction, then, we get

$$\text{compaction ratio} = \frac{15}{1.5} = 10 \text{ or } 10{:}1$$

From Example 11.1 it can be seen that in order to increase the waste volume reduction by 10 percent, from 80 to 90 percent, it is necessary to double the compaction ratio, from 5:1 to 10:1. An understanding of the relationship between percent volume reduction and compaction ratio is important, particularly when reviewing and interpreting manufacturers' data and selecting or specifying suitable compaction equipment.

The basic advantages of an MSW baling process are the significant decrease in waste volume, the ease of handling the compacted refuse, and the reduction of litter and nuisance potential. The compacted waste can be hauled to a landfill disposal site by conventional vehicles, and the service life of the landfill can be greatly increased because of the smaller volume of waste requiring burial. At the disposal site, the bales can be neatly stacked in place, without the problem of wind-blown debris. The likelihood of animal or insect infestation is decreased, and earth cover requirements are reduced; the need for on-site compaction is eliminated. Sometimes the bales may be left uncovered temporarily, as shown in Figure 11.10.

A unique use of compacted MSW bales has been tried in Japan, where high density bales were encased in asphalt or portland cement concrete, and used for construction purposes. The Japanese process applied pressures in different stages, up to about 35 MPa or 5000 psi. Unfortunately, the problem of anaerobic bacterial decomposition and the production of methane gas can still be troublesome, even in such highly compacted bales of refuse. These bales can pose significant safety hazards if the methane escapes and forms an explosive mixture with air.

It should be noted that one of the reasons that high pressure baling is not practiced more extensively in the United States is that American refuse has a relatively high rubbish content and low mois-

FIGURE 11.10 Uncovered bales of solid waste at a disposal site. (American Solid Waste Systems—Division of American Hoist and Derrick Company)

ture content. This prevents the bales from maintaining their shape as a cohesive mass without ties or other encasement.

COMPOSTING

A process in which the organic portion of municipal solid waste is allowed to decompose biologically, under controlled conditions, is called *composting.* Most large-scale composting processes are carried out in the presence of oxygen, and temperatures in the decomposing waste may reach about 65°C (150°F) because of the aerobic microbial action. The key distinction between composting and other solid waste treatment methods is that composting is basically a *biological process,* rather than a chemical or mechanical process.

The original volume of the organic portion of the waste may be reduced as much as 50 percent. The end product is called *compost* or *humus.* Compost resembles potting soil in its texture and its earthy odor, and it may be used as a soil conditioner or

mulch. In addition to volume reduction, then, composting serves to stabilize the organic waste material, and results in an end product that can be recycled and put to beneficial use. It is an attractive option for solid waste management from an ecological or environmental perspective.

There are several steps involved in a complete MSW composting operation. These include *sorting and separating, size reduction, digestion, product upgrading,* and finally, *marketing.* Initial preparation of the refuse is necessary in order to isolate the organic, decomposable portion of the material from the glass, metal, and other nonbiodegradable substances. This sorting and separating step is usually done mechanically, as will be discussed in more detail in the next section.

Shredding or pulverizing serves to reduce the size of the individual pieces of the organic waste, resulting in a relatively uniform mass of material. This facilitates handling, moisture control, and aeration of the compost pile. It also helps to optimize bacterial activity and increase the rate of decomposition. Waste that has been shredded is ready for the actual composting step, or *digestion* step, as it is also called. The digestion process can be accomplished by the open *windrow method,* or in an enclosed mechanical facility.

A windrow is a long, low pile of the prepared solid waste; it is no more than 3 m (10 ft) wide and 2 m (6 ft) high. The windrows are "turned" or mixed about twice a week, to aerate the decomposing organics; the actual frequency of turning of the pile depends on moisture conditions.

Generally, open-field windrow composting takes about five weeks for digestion or stabilization of the waste material. An additional three weeks may sometimes be required to ensure complete stabilization. As previously mentioned, temperatures in an aerobic compost windrow may reach 65°C (150°F) because of the natural metabolic action of *thermophylic* microbes that thrive at this elevated temperature. This relatively high temperature happens to destroy most of the pathogenic organisms that may be present in the waste.

Open-field windrow composting requires relatively large land areas. A community with a population of 250 000 people, for example, would require about 24 ha (60 ac) of land for an MSW composting facility. There are several types of enclosed mechanical systems that can be used for composting, in lieu of the open windrow method. These systems significantly reduce the land requirements; for example, only about 4 ha (10 ac) of land would be required for a mechanical composting system to serve the community of 250 000 people just mentioned. A schematic diagram of a mechanized compost plant is shown in Figure 11.11.

In addition to reducing land requirements, enclosed digestion of the waste can reduce the time required for stabilization from about five weeks to about one week. A typical mechanical composting system includes one or more enclosed tanks or digesters. They may be equipped with rotating plows, vanes, or augers to mix the shredded waste material and to facilitate aeration. In some systems, mechanized stirring or tumbling is supplemented with compressed air, which is blown into the decomposing waste.

Before the stabilized compost or humus can be sold for use as a mulch or soil conditioner, it must be processed further to improve or upgrade its quality and appearance. Compost upgrading may include drying, screening, and granulating or pelletizing. Sometimes, the compost is placed in bags, although bulk sale is more efficient and economical. The basic purpose of any compost upgrading process is to make the final product more readily marketable than the "raw" or unprocessed compost.

Marketability is the most serious bottleneck or obstacle to the success of composting as a solid waste management option. The demand for compost by farmers is low because of the cost of transporting it and because of the availability and ease of application of inorganic chemical fertilizers. Presently, only a very small fraction (less than 2 percent) of refuse is processed by composting in the United States or in other countries. This is because of the low demand for the end product and the resulting poor economics of the process. But composting may be used more in the future, for several reasons.

It has been found that the addition of sewage

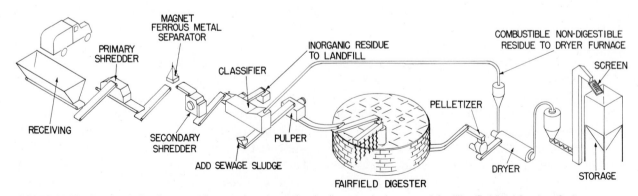

FIGURE 11.11 A schematic diagram of an enclosed, mechanically operated compost facility. (Fairfield Service Co.)

sludge to the organic refuse improves the composting process, as well as the quality of the humus. Composting, then, offers a technically feasible means for processing and recycling both garbage and sewage sludge in a single operation. With regard to final use, the humus can be used for land reclamation as well as for soil conditioner or fertilizer. For example, in areas where top soil was lost because of strip-mining activities, compost can be used to promote plant growth and reduce soil erosion.

As air quality standards become stricter and the cost of air pollution control increases, incineration may tend to lose some of its popularity as a waste management option. And environmental constraints are continuing to diminish the suitability of refuse disposal in landfills. For these reasons, as well as the fact that composting offers an ecologically sound method for recycling resources, the future of the composting process in solid waste management is favorable.

RESOURCE RECOVERY AND REUSE

Domestic, commercial, industrial, and agricultural activity will always result in the generation of solid and hazardous waste material. But our perception of what constitutes "waste," that is, useless and unwanted material, is changing. As we become more aware of ecological or environmental imperatives, the need for recovering and reusing much of what was previously "thrown-away" or "dumped" is becoming more evident. Also, the cost of disposing waste material in an environmentally sound manner makes it more necessary to consider alternative waste management techniques.

The ideal approach to waste management is to reduce the total quantity or volume of material that ultimately requires "disposal." This could be accomplished by *recycling,* that is, separating out and reusing those components of the "waste" that may have economic value. Energy, as well as material, can be recycled. The recovery and reuse of heat energy were already discussed in the section on incineration. And the recovery or recycling of the organic component of solid waste was discussed in the section on composting. In addition to energy

and organic material, there are several other substances that have the potential for recovery, reclamation, and reuse. These, as well as some of the commonly used material separation and sorting procedures will be discussed here.

Separation and Sorting

Before a particular material can be recycled, it must be separated from the bulk of the waste and sorted with other material of the same type or classification. Waste separation can be accomplished at its source or point of generation, or at a central processing facility. At a central facility, the waste separation can be done manually by hand-sorting, mechanically using special types of sorting equipment, or by a combination of manual and mechanical methods.

Source separation, or *curbside separation* as it is also called, is often practiced voluntarily by citizens in many communities. Householders collect newspapers, glass bottles, metal cans, and kitchen or food waste in separate containers. But not all citizens can be relied upon to voluntarily separate their refuse into homogeneous groups, and many are not willing to transport it to an appropriate community "recycling center." Additionally, municipal collection of source-separated refuse would involve separate pick-ups of garbage, paper, glass, and so on. This would raise the cost of collection higher than it already is.

In 1980, New Jersey was the first state in the United States to develop and adopt a statewide strategy for source separation of recycleable material. It is expected that about 25 percent of all MSW in the state will be recycled. The remaining 75 percent is to be incinerated (preferably with heat recovery) or buried in a landfill. A combination of financial incentives, low interest loans, and grants was recommended to help accomplish the plan. It may be necessary, though, to make source separation a legal obligation in order to force public compliance.

Material separation and sorting at a central, mechanized refuse processing plant offers a more efficient and reliable method for waste resource recovery and reclamation than does source separation. The reusable components of the waste can be

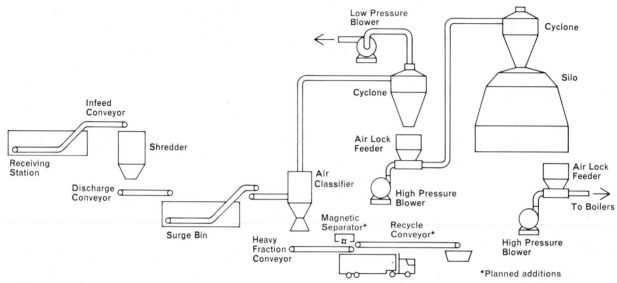

FIGURE 11.12 A simplified flow diagram for a solid waste material recovery plant. (Williams Patent Crusher and Pulverizer Company, Inc.)

separated from the bulk or mixed refuse on the basis of differing physical characteristics, such as size, weight or density, impact resistance, magnetic properties, and even color. A large, centrally located facility serving many communities could accomplish waste separation and sorting on a relatively economical basis.

A schematic diagram of a waste material recovery plant is shown in Figure 11.12. Although there are presently very few such resource recovery plants in operation in the United States, it is expected that in the coming years there will be the construction of several facilities similar in scope to the plant illustrated in Figure 11.12.

After passing through a hammer mill or other shredding device, the bulk waste is processed in an *air classifier,* which serves to remove lightweight material, such as paper, from the refuse. In this device, compressed air is directed upward through a screen that supports the shredded refuse; the light material is carried upward in the air stream, and the heavier material remains behind on the screen.

A *cyclone separator* serves to concentrate the light material, by reducing the volume of the air stream in which it is carried. A *magnetic separator* is used to remove ferrous metal (iron) objects from

the heavier material that was not carried in the air stream. A schematic diagram of a belt type magnetic separator is shown in Figure 11.13.

After magnetic separation, the waste is sometimes processed further in a *rotary drum and trommel screen classifier.* This device operates on the basis of different size and density characteristics of

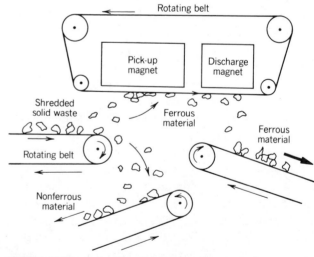

FIGURE 11.13 A simplified diagram of a magnetic separator used to recover ferrous (iron) materials from refuse.

the waste components. It separates out small dense objects, such as pieces of glass, stones, and bottle caps, from the rest of the refuse. An operating trommel screen classifier is illustrated in Figure 11.14. Vibrating screens are also available to separate small dense objects from the waste stream.

Another device that can be used to separate and sort waste material is the *inertial separator*. It operates on the basis of the differing weights of refuse components. Two types of separators are shown schematically in Figure 11.15. In the *ballistic separator*, a rotor throws or ejects the material with a constant force. Since dense objects will travel farther than light objects, because of the effect of air resistance, they can be collected in a separate hopper or bin. In the *inclined separator*, heavy and resilient objects tend to roll downhill against the movement of the conveyor belt, whereas the lighter and less resilient objects are carried upwards.

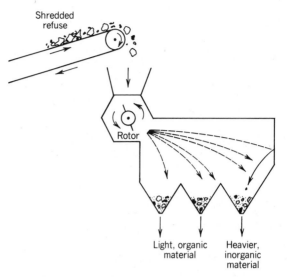

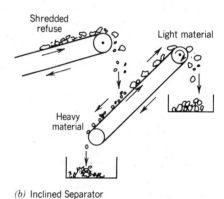

(a) Ballistic Separator

(b) Inclined Separator

FIGURE 11.15 A simplified diagram of two types of inertial separators; material is separated by differences in density and resiliency.

FIGURE 11.14 A trommel screen classifier, used to separate components of solid waste. (Triple/S Dynamics 1031 S. Haskell, Dallas, Texas)

There are several other methods for separating and sorting the different material in refuse, including differential flotation in water and devices that utilize optical sensors to respond to material of specific color, such as light-green glass. Another somewhat sophisticated separation technique, which uses infrared light and computerized comparison of the differing "object signatures" in the refuse, is being developed. This system monitors reflected infrared light for the characteristic absorption spectra of different material. In this way, glass, rubber, paper, plastic, and other substances can be identified and separated from the bulk refuse.

Recycling

As previously discussed, the technology for mechanically separating and sorting various components of solid waste is available. But the goal of recycling and reusing resources that were previously "thrown-away" cannot be accomplished by technology alone. At present, one of the most difficult problems is the *development of a market* for the recovered materials. Until recently, the economics of large-scale recycling has not been very favorable in the United States. For example, there simply has not been sufficient market demand for compost to make composting, or recycling of garbage, a successful waste management option.

Another factor to be considered is that recycling by itself will not eliminate the solid waste problem. There will always be a considerable amount of nonrecycleable residue left for ultimate disposal. Nevertheless, public demand for environmental protection has focused more attention on the need to fully explore the recycling option. As the technology improves, and as the economic situation changes, resource recovery and recycling of many solid waste components may become much more common than it is today.

In addition to the organic component of solid waste (garbage), the materials that may be selected for recycling generally include paper, metal, glass, plastic, and rubber. In the United States, municipal solid waste is mostly paper—roughly 50 percent by weight and 70 percent by volume. It would seem that the best way to cope with this large amount of waste paper is to recycle it. As attractive an option as it sounds, there remain several obstacles or deterrents to large-scale paper recycling.

Recycled paper can be used for several purposes, but it is never "as good as new" after reprocessing. The fibers are weakened, and it is difficult to control the color of the recycled product. Federal law also prohibits the use of recycled paper products for many types of food containers, so as to prevent the possibility of contamination. Yet another limitation to paper recycling is poor economics, including the fact that it often costs less to transport raw paper pulp than recycled paper scrap. Collection, sorting, and transport account for about 90 percent of the cost of paper recycling. And the process of pulping, deinking, and screening wastepaper is generally more expensive than making paper from virgin wood fibers.

It is often noted that paper recycling will help to preserve our forests, since it takes about 17 trees to make 1 ton of paper. But this may be an unrealistic notion. The fact that selective harvesting of mature trees is a necessary forest management technique must be considered. Thinning out timberland by the harvesting or cropping of trees increases the health and productivity of the forest. Therefore, from an ecological perspective, paper recycling does not necessarily offer a benefit or an advantage for preservation of woodlands.

For the purpose of recycling, metals can be classified as being either *ferrous* or *nonferrous*. Ferrous metal contains iron, and is magnetic; nonferrous metal does not contain iron, and is nonmagnetic. Steel is a common ferrous metal, and aluminum is a common nonferrous metal. Most recoverable metal in MSW is contained in cans used for packaging beverages or food; roughly *five billion* metal cans are produced annually in the United States. The so-called "tin can" is actually made of steel, with a coating of tin. The tin coating and the lead solder along the can seam are impurities that hamper the production of high quality steel from the salvaged cans.

Perhaps one of the most successful efforts for reclamation and reuse of discarded metal is the recycling of aluminum cans. About 2 million tons of aluminum are recovered and reprocessed each year. This saves about 96 percent of the energy required to extract aluminum from the raw bauxite ore in which it is originally mined. And aluminum scrap is more easily processed than is other metallic scrap material. It can be magnetically separated from ferrous metal and shredded in a hammer mill, as shown in Figure 11.16. The shredded aluminum can be transported to a smelting facility in carload lots. The impurities it may contain, such as iron and magnesium, actually add strength to the recycled aluminum product. Research is being done now in an attempt to make possible the reclamation of

FIGURE 11.16 *(a)* Aluminum cans are placed in a hopper and carried along a moving belt; then they move through a magnetic separator that removes any steel cans. *(b)* the cans may be crushed and baled or shredded as shown here, prior to reprocessing. (Aluminum Association, Inc.)

all aluminum beverage cans; the most difficult problem is public participation in source separation.

A large source of waste metal comes from abandoned or "junked" automobiles. A modern automobile shredder can demolish a car body in about one minute, but these devices are very expensive. Car bodies can also be compacted by powerful compaction machines. But the cost of collecting and then transporting old automobiles to shredding or compaction plants is very high. Producing high quality

steel from automobile scrap metal is also a problem. For these reasons, a steady market demand for automobile scrap metal has not yet developed in the United States.

Glass is the least troublesome material in municipal solid waste. It is an inert, "nonpolluting" substance, made primarily from silica sand, which is an abundant natural resource. Even though the raw material from which it is made is so readily available, there is still a market demand for waste glass. It can be reused as cullet (broken glass for remelting) to produce a new batch of glass. Other uses of waste glass, such as an aggregate material in asphalt pavement, are being investigated.

With respect to glass beverage containers, though, it is generally cheaper to refill a deposit bottle than to recycle a used no-deposit glass bottle. The legally required use of deposit bottles that is in effect in some areas of the United States is, essentially, a direct form of recycling. An added benefit of the required use of deposit and return glass bottles is the reduction of unsightly litter in the streets and in other community areas.

Reclaimed rubber must be shredded in special shredding machines and broken down chemically before it can be rebonded and remolded. Rubber produced by this process, called *revulcanization,* is not usually as strong as the original product. As an alternative to revulcanization, the use of shredded rubber in asphalt pavements is being investigated by researchers. Another interesting use for discarded automobile tires is the "tire playground" for children, which is becoming a more familiar sight in many communities.

Plastic material is a small but growing component of MSW. Plastic is a nonbiodegradable petroleum-derived substance, composed primarily of carbon, hydrogen, and oxygen. There are many different types of plastic, most of which can be reheated, reformed, and used again. But for this to be done successfully, it is necessary to avoid mixing the different types together. For this reason, it is generally necessary to collect the plastic waste before it is mixed with the bulk refuse. In addition to being remolded, it may be possible to use shredded plastic as an inert fill material, or as an aggregate for lightweight concrete. Plastic bottles are generally not recycled for direct reuse because they tend to absorb hydrocarbons, and the consumer often doubts the cleanliness of a recycled plastic bottle.

11.4 SANITARY LANDFILLS

The oldest method for "getting rid of" solid waste material is *land disposal*. In the past, the waste was simply placed in a heap on top of the ground, and the disposal site or area was called a garbage dump. These open and uncontrolled refuse disposal sites quickly became breeding grounds for vectors of disease, including rats and flies. In addition to posing a direct threat to public health, the dumps were unsightly nuisances; they polluted surface and ground waters, and occasionally they caught on fire. Open dumping of waste material is no longer an acceptable (or legal) disposal method in the United States.

Nevertheless, at the present time most municipal solid waste is still disposed of on land. But the waste is buried in a *sanitary landfill,* not simply deposited in an open dump. A sanitary landfill is an "engineered" waste disposal site. This means that it is designed, constructed, and operated in an environmentally sound manner that does not threaten public health or safety and that minimizes public nuisances (like wind-blown litter and unpleasant odors).

There are three key characteristics of a municipal sanitary landfill that distinguish it from an old-fashioned dump:

1. Solid waste material is placed in a prepared section of the landfill site, in a carefully prescribed manner.

2. Waste material is spread out and compacted in a thin layer, with appropriate heavy machinery.

3. The waste is covered each day with a layer of compacted soil.

Other important aspects of a modern landfill include its initial siting or location (with due regard for environmental, economic, and social factors), initial site preparation (possibly with a bottom liner and drainage or "leachate" collection system), a methane-gas venting or collection system, and a final compacted cover of soil over the entire site, after it is filled to capacity.

In addition to providing an option for waste management, an MSW landfill may also serve to improve or "reclaim" poor quality land. The landfill gradually raises the ground elevation or surface grade of the site. Completed sanitary landfills have been successfully converted into community parks, playgrounds, golf courses, and other beneficial land-use projects. The ultimate use of the site should be decided at the outset of the project, so that the landfill can be constructed and operated keeping that specific goal in mind.

Landfilling is generally the most economic alternative for solid waste disposal, which accounts for its frequent application in most communities. But in recent years, the practice of landfilling has declined in "popularity" as a waste disposal option or choice. It has become increasingly difficult to find suitable landfill sites that are within economical hauling distances for many urban communities. (The suitability of a site is determined by its size as well as by a host of other technical and environmental factors.) Many cities in the United States are presently embarking on plans to construct municipal incinerators (with energy recovery) for solid waste disposal, because of the rapidly declining space for landfill within the city limits.

Another reason for the declining popularity of sanitary landfills is the realization that, in the long run, some environmental damage may occur no matter how well engineered the design, construction, and operation of the site is. It is believed by many people that it is not possible to build a completely safe and secure waste disposal landfill, particularly for hazardous waste. Some of the confined material may eventually escape into the environment, to harm public health and damage the local ecosystem.

Despite these concerns, however, landfilling will continue to be practiced to a significant extent for many years to come, simply because of economic and technical necessity. It is just not possible to reclaim and recycle all solid waste material. In addition, there is no guarantee that other disposal methods are entirely safe. Incineration, for example, will always cause some degree of air pollution, even with use of the most sophisticated air cleaning devices. And there will always be a solid residue from incineration, or any other waste treatment method, which will require ultimate disposal on land or underground.

SITE SELECTION

There are many technical factors involved in selecting a location for a new sanitary landfill. Perhaps the most important include the site's *volume capacity, accessibility,* and *geohydrologic conditions.* Other factors include the climate, as well as local socioeconomic conditions. One of the most difficult and frustrating problems with regard to siting a new landfill, however, is more political than technical in nature.

The general public equates "landfill" with "dump," and takes little note of the fact that a landfill is designed, built, and operated according to up-to-date engineering principles. Citizens, including local politicians, are reluctant to allow construction of a new landfill in their communities, and long drawn out legal battles over proposed sites are not uncommon. A possible way to overcome this problem may be to take the siting approval authority out of the hands of county officials and politicians and leave the final decision up to an appropriate nonpartisan statewide commission.

The total capacity and *design life* of a new landfill depends on the size and topography of the site, the rate of refuse generation, and the degree of refuse compaction. The amount of daily soil cover material adds roughly 20 percent to the overall fill volume, and must be considered when evaluating the capacity of a landfill. There should be enough volume capacity within the working area of the site, so that the landfill will have a design life of about 25 years. The longer the useful life of the site is, the more economical is the overall solid waste

management operation for the communities involved.

A generally accepted rule of thumb regarding municipal landfill capacity is that roughly 1 ha–m (8 ac–ft) of volume is needed each year to serve a population of 10 000 persons. (This can be visualized as 1 ha of land covered to a depth of 1 m, or 1 ac of land covered to a depth of 8 ft, with compacted refuse.) The following examples illustrate typical computations related to landfill capacity.

EXAMPLE 11.2

A rural community of 15 000 persons generates refuse at an average rate of 5 lb/capita/d. A 25-ac landfill site is available, with an average depth of compacted refuse limited to 20 ft by local topography. It is estimated that the compacted refuse will have a unit weight of 1000 lb/yd^3, and that an additional 25 percent of volume will be taken up by the cover material. What is the anticipated useful life of the landfill?

Solution:

The total weight of refuse generated per year is

$$\frac{5 \text{ lb}}{\text{person–d}} \times \frac{365 \text{ d}}{1 \text{ yr}} \times 15\ 000 \text{ persons}$$
$$= 27.4 \times 10^6 \text{ lb/yr}$$

The total yearly volume of refuse is

$$27.4 \times 10^6 \frac{\text{lb}}{\text{yr}} \times \frac{1 \text{ yd}^3}{1000 \text{ lb}} = 27\ 400 \text{ yd}^3/\text{yr}$$

The additional volume for cover material is

$$0.25 \times 27\ 400 = 6850 \text{ yd}^3/\text{yr}$$

Therefore, the total landfill volume required is

$$27\ 400 + 6850 = 34\ 250 \text{ yd}^3/\text{yr}$$

The available volume of the landfill is

$$25 \text{ ac} \times \frac{43\,560 \text{ ft}^2}{1 \text{ ac}} \times 20 \text{ ft} \times \frac{1 \text{ yd}^3}{27 \text{ ft}^3}$$
$$= 8.07 \times 10^5 \text{ yd}^3$$

The useful life of the site is estimated to be

$$8.07 \times 10^5 \text{ yd}^3 \div 34\,250 \text{ yd}^3/\text{yr} \approx 24 \text{ yr}$$

EXAMPLE 11.3

Estimate how many hectares of land would be required for a sanitary landfill, under the following conditions:

Design life of the site	30 years
MSW generation rate	25 N/capita/d
MSW compacted unit weight	5.0 kN/m^3
Average fill depth	10 m
Community population	50 000
MSW-to-cover ratio	5:1 (20 percent of MSW for cover)

Solution:

The quantity of MSW generated each year is

$$25 \frac{\text{N}}{\text{person–d}} \times \frac{365 \text{ d}}{1 \text{ yr}} \times 50\,000 \text{ persons}$$
$$= 4.56 \times 10^5 \frac{\text{kN}}{\text{yr}}$$

The volume of compacted refuse is

$$4.56 \times 10^5 \text{ kN/yr} \div 5.0 \text{ kN/m}^3 = 91\,250 \text{ m}^3/\text{yr}$$

The additional volume required for cover is

$$0.2 \times 91\,250 = 18\,250 \text{ m}^3/\text{yr}$$

The total required volume is 91 250 + 18 250 = 109 500 m^3/yr

The area required is computed as

$$\text{area} = \frac{\text{volume}}{\text{depth}} = \frac{109\,500 \text{ m}^3/\text{yr}}{10 \text{ m}} = 10\,950 \text{ m}^2/\text{yr}$$

Since 1 ha = 10 000 m^2, and the design life is 30 years, we can estimate the total required landfill area as

$$10\,950 \frac{\text{m}^2}{\text{yr}} \times \frac{1 \text{ ha}}{10\,000 \text{ m}^2} \times 30 \text{ yr} \approx 33 \text{ ha}$$

(Additional land would be needed for access roads, buildings, etc.)

In addition to capacity and useful life, *site accessibility* is another important factor related to the location of a sanitary landfill. Accessibility refers to the ease with which refuse collection or transport vehicles can reach the disposal area, on streets and highways, without causing a public nuisance or traffic hazard.

Sites that would require trucks to travel relatively long distances, particularly through residential neighborhoods, are not desirable. It would be best to locate the landfill in a nonresidential area, where the constant truck traffic would not be so much of a nuisance (litter, noise, and odor) or hazard. Since suitable landfill sites are increasingly difficult to find in most urban communities, the landfills are often located well outside city limits. In such cases, centralized *transfer stations* are used to improve the waste hauling operation. The collection trucks first deliver the refuse from the individual neighborhoods to the transfer station, where it is accumulated in large tractor–trailer vehicles for final transport to the landfill.

Capacity and accessibility have always been important factors related to the location of a sanitary landfill. But public concern for environmental health protection is now a more significant force with respect to landfill siting decisions. It is extremely important to minimize contamination from the landfill of the surrounding environment, particularly surface water or groundwater. The local geology and hydrology, or *geohydrology,* of the site has a direct influence on the possibility and the rate of such transport.

Perhaps the key imperative for landfill location and design is that *at no time should the waste be in contact with surface water or groundwater.* Accordingly, landfills are not to be located in low-lying wetland areas such as swamps or marshes, nor are they to be located in the flood plains of streams or rivers. (Location of MSW landfills in such areas would require special engineering design and construction, and would be prohibitively expensive. In addition, landfill in those locations would damage productive ecosystems, or aggravate flooding conditions. There are extra benefits, then, from prohibiting landfills in wetlands or floodplains.)

A thorough investigation of subsurface geologic conditions is of critical importance for landfill siting and design. Considerable attention must be given to the groundwater hydrology of the site, that is, the occurrence and pattern of groundwater flow. This is necessary in order to protect groundwater aquifers from contamination. Data regarding soil gradation, rock types, permeability, and other factors that affect the subsurface flow of water must be obtained.

A minimum separation of about 1.5 m (5 ft) is required between the bottom of an MSW landfill and the seasonally high groundwater table. Relatively impermeable clay soil, or synthetic bottom liners, may be required to ensure that there is no transfer of water between the refuse and the underlying aquifer. The coefficient of permeability must be less than 10^{-5} cm/s. Requirements for landfills that accept hazardous waste are more stringent than those for MSW landfills, as will be discussed in Section 11.5.

CONSTRUCTION AND OPERATION

There are two basic methods for constructing and operating a sanitary landfill: the *trench method,* and the *area method.* A common variation of these two basic modes of operation, called the *ramp method,* is also practiced at some facilities. The method chosen for a particular site depends upon topography, geohydrology, and the availability of suitable cover material.

A feature common to each type of landfill is the *refuse cell.* The cell is the basic "building block" of any landfill. Refuse is spread and compacted in layers in a confined portion of the site; the individual layers are generally about 1 m (3 ft) thick. After compaction, the unit weight of refuse in a layer may be about 6 kN/m^3 (1000 lb/yd^3) (which is roughly four times its initial loose density at the point of generation). Several layers of refuse may be placed on top of each other, up to a depth of about 3 m (10 ft).

At the end of the day's operation, the compacted refuse is covered with a layer of soil, which is also compacted to a thickness of about 150 mm (6 in.). Both the compacted refuse and soil cover make up a single cell of the landfill. Several adjoining cells make up a *lift,* and the completed landfill may consist of several lifts, as illustrated in Figure 11.17. The final soil cover on top of the uppermost lift must be able to support vegetative growth, and is generally about 0.6 m (2 ft) thick.

The trench method of landfilling is used at sites where the groundwater elevation is low and deep excavations are possible. A long trench, wide enough for trucks and compactors to work in, is first excavated and the soil is stockpiled nearby for use as daily cover material. Refuse cells are constructed in the trench and covered with the stock-

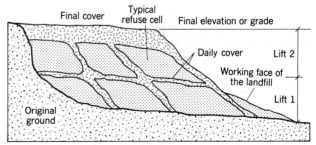

FIGURE 11.17 The basic "building block" of a sanitary landfill is a compacted "cell" of solid waste, which is separated from other cells by a layer of compacted soil.

DAILY EARTH COVER (6 IN.)

EARTH COVER OBTAINED BY EXCAVATION IN TRENCH

ORIGINAL GROUND

COMPACTED SOLID WASTE

FIGURE 11.18   In the trench method of landfill construction, the refuse is deposited and compacted in an excavated trench. The soil from the trench is used as daily cover material. (EPA)

piled soil. A typical trench is about 45 m (150 ft) long and about 3 m (10 ft) deep. The trench method of landfilling is illustrated in Figure 11.18.

The area method of landfilling is generally used at sites with a relatively high groundwater table, where trench excavations are not feasible. Cover material, therefore, has to be hauled in from some other location. It is very important that suitable plans be made to have enough soil available for daily cover. In the ramp method of construction (a variation of the area method), the daily cover material is usually obtained from the original ground just ahead of the working face of the cell. This is illustrated in Figure 11.19.

Landfill Equipment

Various types of heavy machinery are needed at sanitary landfills, to spread and compact the refuse, as well as to haul, spread, and compact the soil cover material. Sometimes traditional earthwork construction equipment can be used. The "bulldozer," for example, is one of the most versatile machines at a landfill. It can be used for various operations, including spreading and compacting.

Extensive landfilling activities may require several additional machines, including large earth-

movers called scrapers or pans. These machines serve to excavate, haul, and spread cover material when the haul distances exceed the range of dozers (about 100 m or 300 ft). Special landfill equipment, such as the steel-wheeled compactor shown in Figure 11.20, may be used to achieve higher density compaction of the refuse. The large-toothed wheels serve as load concentrators, providing more compaction pressure than conventional rubber-tired or crawler tractor equipment.

Gas Production

The organic material in buried solid waste will decompose from microbial action. The rate of decomposition depends primarily on the amount of moisture present in the waste. Decomposition proceeds at a relatively rapid rate when moisture content exceeds 50 percent, but complete breakdown of the waste and landfill stabilization typically takes about 20 years.

At first, the refuse decomposes aerobically, until the oxygen that was present in the freshly placed fill is used up by the aerobic microorganisms. Then the anaerobes take over, producing *methane*, CH_4, and other gases as they metabolize the organics. Methane mixed with air in concentrations of about

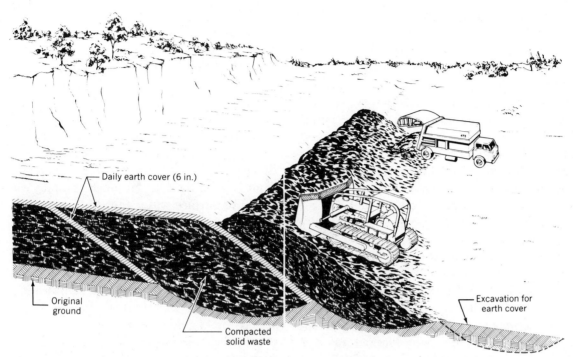

Daily earth cover (6 in.)

Original ground

Compacted solid waste

Excavation for earth cover

FIGURE 11.19 In the ramp method of landfill construction, refuse is placed and compacted on a slope, and daily cover material is obtained directly in front of the working face of the cell. (EPA)

FIGURE 11.20 A steel-wheeled compactor, one of several types of heavy machinery needed for daily operation of a municipal sanitary landfill. (BOMAG: A Unit of AMCA International Corporation)

10 percent is highly explosive. It is also poisonous and can cause death by asphyxiation.

The methane will follow the path of least resistance; it can travel considerable distances horizontally through porous layers of sand and gravel. Hazardous conditions can occur if the gas rises to the surface and accumulates in basements, buildings, or other enclosed areas. The explosions and tragic loss of life that have occurred in the past because of landfill gas, could have been avoided by proper design and construction of the landfill. This is why the possible flow of the gas through the refuse cells and soil must be taken into account during the design stage of an MSW sanitary landfill.

Gas movement can be controlled by providing impermeable barriers in the fill. A 1-m (3-ft) thick layer of moist clay soil, for example, will act as an excellent barrier to the flow of landfill gas at the bottom and periphery of the fill. A venting system can be constructed to collect the blocked gas and vent it to the surface, where it will be safely diluted and dispersed into the atmosphere. Coarse gravel and pipes can be used to ventilate the fill, as illustrated in Figure 11.21.

Instead of venting the gas to the atmosphere, the methane can be collected and recovered for use as a fuel. Actually, methane is one of the most important recoverable resources from MSW landfilling operations. There are several landfills in the United States where recovery of methane is now practiced, and more such gas recovery projects are being planned. Recovery systems that are designed to process the collected landfill gas are commercially available. They use a principle called *membrane permeation* to separate traces of carbon dioxide, nitrogen, and hydrogen sulfide from the gas, leaving a purer methane product for sale as a fuel.

Leachate

In addition to gas, sanitary landfills produce a highly contaminated liquid called *leachate*. Some of the leachate results directly from the decomposition of garbage in the fill. But much of it may come from runoff or surface water that first infiltrates the fill and percolates downward through the refuse cells. Direct contact with the waste material results in the severe contamination of that water. If the leachate then mixes with groundwater, or seeps out of the fill into a nearby stream, significant environmental damage can occur. Perhaps one of the most serious problems related to landfill leachate is the potential contamination of drinking water supplies, particularly "sole source" groundwater aquifers.

There are basically two ways to prevent leachate problems at a sanitary landfill. One is to effectively intercept runoff or surface water and prevent it from entering the landfill. Another is to use a liner on the bottom of the fill to provide an impermeable barrier between the waste and the groundwater. This is illustrated schematically in Figure 11.22.

Leachate control is particularly important at landfills containing industrial chemicals and hazardous waste. New landfills constructed for this purpose are required to install leachate collection and treatment systems and groundwater monitoring wells as well as to employ proper surface grading and a bottom liner. The goal of modern land disposal technology is to make the landfill "secure" from potential leakage of contaminants and environmental damage. This is discussed in more detail in a following section on management of hazardous waste.

Completed Landfills

A problem related to ultimate use of a completed landfill site is that of *settlement*. As decomposition occurs, the finished surface of the landfill will gradually settle to a lower elevation. The amount of set-

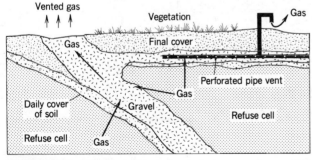

FIGURE 11.21 Gravel and/or perforated pipes can be used to ventilate methane gas from a landfill.

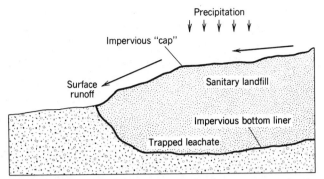

FIGURE 11.22 Leachate can be prevented from contaminating surface or groundwater by diverting runoff away from the site and by constructing a liner to block the underground flow of water.

tlement may be as much as one third of the original height of the fill. This can cause some obvious problems with regard to the construction of buildings or other structures on the site, particularly in the case of uneven or differential settling under a single structure.

Another problem is related to the *bearing capacity* of the completed fill. Bearing capacity is a measure of the ability of the ground to support foundation loads. Bearing capacities for sanitary landfills are roughly 500 lb/ft^2, which is very low in comparison to ordinary construction standards for natural ground surfaces. Obviously, only lightweight structures can be built on sanitary landfills without the need for expensive deep pile foundations. It is basically because of the difficulties associated with excessive settlement and low bearing capacity that completed landfills usually serve as "green acres" or recreation areas. Special consideration must even be given to planted vegetation, so that it will thrive and grow properly.

11.5 HAZARDOUS WASTE MANAGEMENT

Some of the characteristics and problems related to hazardous waste, and with municipal solid waste (MSW), have been discussed briefly in the preced-

ing sections. But the challenge of *hazardous waste management* (HWM) is of such huge proportions in the United States today, that some further discussion is warranted. Hazardous waste is distinguished from municipal solid waste primarily on the basis of its toxicity, reactivity, ignitability, and corrosivity. It includes poisonous chemicals, acids, explosives, and other dangerous substances. Radioactive waste, certainly a dangerous material, is not included in this category, because the problems associated with its disposal are very different from those presented by nonradioactive hazardous waste.

Subtitle C of the Resource Conservation and Recovery Act (RCRA) specifically addresses the problem of hazardous waste management. The RCRA, passed by Congress in 1976, amended and updated the Solid Waste Disposal Act of 1965. It enhanced the legal basis for the control of all solid waste by including hazardous substances as well as MSW in the law, and it promoted resource recovery and reuse as a waste management option. The Environmental Protection Agency is responsible for promulgating specific rules and regulations to carry out the RCRA mandate. The regulations that control hazardous waste are generally more stringent than those for nonhazardous municipal solid waste.

There are several options available for hazardous waste management and disposal. In order of preference, these can be summarized as follows:

1. *Reduce waste quantities* at their source, by modifying the industrial processes that generate the waste material.

2. *Reclaim and recycle the waste* material, after separating and reprocessing it as a resource for some other industrial or manufacturing process.

3. *Stabilize the waste* material, rendering it nonhazardous by appropriate biological, chemical, or physical processes.

4. *Incinerate the waste* at temperatures high enough to destroy or detoxify it.

5. *Apply land disposal methods*, including "secure" landfills (preferably after encapsulating or solidifying the waste to reduce its mobility), or underground (deep-well) injection.

DISPOSAL ON LAND

As previously indicated, the most desirable solutions to the current hazardous waste problem are to reduce the quantity of waste at its source, or to reclaim and recycle the materials for some other productive use. Source reduction is likely to become a top priority of industry because of the rapidly increasing costs of the other options. Reduction can be accomplished by using nontoxic materials and improved process design, and by applying alternative technologies in the manufacturing plants.

Recycling is an increasingly attractive option, particularly for waste that includes high concentrations of metals, oils, acids, and other useful substances. Organizations that serve as so-called *clearinghouses* or *material exchanges* will facilitate recycling efforts. A clearinghouse can help to make arrangements between waste generators and potential users of the waste. A material exchange can serve directly as a transfer agent, purchasing the waste from the generator, reprocessing it if needed, and selling it for reuse by some other industry. It is expected that the function of the clearinghouse or material exchange will play an increasingly important role in hazardous waste management in the coming years.

At present, and in the immediate future, source reduction and recycling cannot be regarded as complete solutions or primary alternatives for managing hazardous waste. Some form of "ultimate disposal" will be required for many years to come. For certain waste materials, such as mercury from the battery industry or kepone from the pesticide industry, no accepted method for "ultimate disposal" exists. For such material, some form of long-term *engineered storage* is required; the waste must be containerized and stored in a retrievable manner, until suitable disposal methods are developed.

Secure Landfills

One method of ultimate disposal of hazardous waste is to put in a landfill. Landfilling of hazardous waste is not an attractive option because of the inherent environmental dangers involved in its practice. But it is the cheapest alternative and may continue to be used until the other preferred hazardous waste management options are perfected and are more fully developed. (In the 1984 amendments to RCRA, dioxins, PCBs, and certain other hazardous wastes were banned from land disposal.)

There are several distinctions between a hazardous waste landfill and a more conventional municipal solid waste landfill, with regard to location, design, construction, and operation. Under the RCRA regulations, it is intended that the hazardous material be deposited in a so-called *secure landfill*, with at least 3 m (10 ft) of height separating the bottom of the fill from bedrock or a groundwater aquifer. Under circumstances where a single site is used for disposal of both MSW and hazardous material, it is necessary to maintain separate disposal operations.

A secure landfill is considered to have four phases in its operation. During the *active phase*, the hazardous wastes are placed in the prepared fill area. Incompatible chemical wastes are placed at separate locations in the fill, so as to avoid explosions or other dangerous reactions. The waste material should first be solidified or containerized in drums, and care must be taken to avoid rupturing the individual containers as they are placed.

During the *closure phase*, an impermeable cap or cover is constructed over the landfill site. A *post-closure* phase, defined as the 30-year period after closure of the site, involves continuous monitoring for chemical leakage from the fill. A fourth and last phase for the landfill, called the *eternity phase*, is expected to involve some leakage of waste material into the environment. It is believed that no landfill can be completely secure forever; the natural degradation of the protective liner or natural geological forces, including the possibility of earthquake, will eventually destroy the structural integrity of the fill.

The key features of a secure landfill include an *impermeable liner*, a *leachate collection system*, an *impermeable cap*, and a *groundwater monitoring system*. As previously discussed, an impermeable liner on the bottom and sides of the landfill serves as a barrier to the flow of water, preventing groundwater from entering or leachate from leaving the fill (see Figure 11.22). The liner of a secure

FIGURE 11.23 A synthetic bottom liner installed at a hazardous waste disposal site; a layer of soil covers the liner to protect it from damage by construction equipment. (Gundle Lining Systems, Inc. 1340 E Richey Rd. Houston, Texas, 77093)

landfill must consist of a synthetic material, with a permeability coefficient equivalent to 10^{-8} cm/s, or less. A typical installation of a synthetic liner is shown in Figure 11.23. Actually, EPA guidelines call for a double liner, for added safety. Two synthetic or plastic liners are preferred, but a layer of compacted clay could also serve as the lower liner. Clay cannot be used alone primarily because certain chemicals may react with the clay minerals, changing its pore structure and allowing leakage.

A double leachate collection system, comprising a network of perforated pipes placed above each liner, is designed to prevent an accumulation of trapped leachate in the fill. The collected leachate can be pumped to a treatment facility for processing. The final cover, or impermeable cap placed over the landfill, is designed to seal the site and to prevent rainwater from infiltrating the surface.

This minimizes the quantity of leachate in the fill, and therefore reduces the potential for environmental pollution. The combination of the double liner, the leachate collection systems, and the cap, serves as a "first line of defense" against leakage during the active and closure phases of the landfill. A schematic diagram illustrating these features of a secure landfill is shown in Figure 11.24.

The third or post-closure phase of a secure landfill operation involves continuous environmental monitoring, to detect any chemical leakage that may occur. This is, in effect, a "second line of defense." For this purpose, a series of deep wells are drilled in and around the site, in specified locations, as illustrated in Figure 11.25. A routine program of sampling and testing must be implemented to detect any plumes of chemical leakage or contaminated groundwater. If leakage does occur, pumps

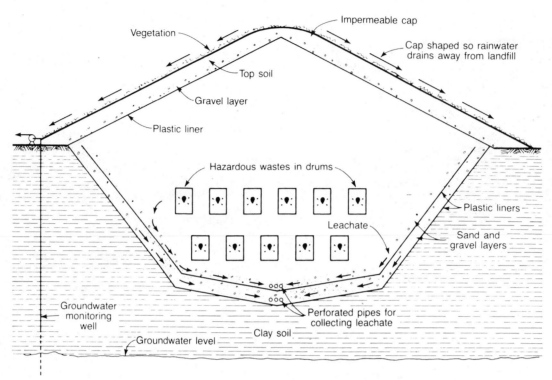

FIGURE 11.24   A cross-section of a typical "secure" landfill. (From Montague, P., "Hazardous Waste Landfills: Some Lessons from New Jersey," Civil Engineering, with permission from the American Society of Civil Engineers)

can be installed in the wells to intercept the polluted water and bring it to the surface for treatment. This "last line of defense," is intended to protect the underlying aquifer from any significant damage due to leakage from the fill.

Underground Injection

Another "land disposal" method for hazardous waste management that should be mentioned here is called *underground or deep well injection*. Unlike landfilling, which is essentially a surface disposal technique, underground injection involves pumping liquid waste down through a drilled well, into a porous layer of rock. The waste is forced into the pores and fissures of the rock under pressure, and is intended to be stored there forever. The layer of rock in which the waste is permanently stored, either limestone or sandstone, must lie below an impervious layer of clay or rock; this injection zone

can be from a hundred meters to a few kilometers below the surface.

Underground injection is relatively inexpensive, takes up little land area, and requires little or no pretreatment. But the most serious objection to this mode of waste disposal is the possibility for eventual leakage and permanent contamination of subsurface groundwater supplies. For this reason, deep well injection is one of the most controversial topics in hazardous waste management; it is not likely that it will ever become a predominant method for disposal. Unfortunately, it has been practiced to a very large extent by many industries, as a means of on-site waste disposal.

TREATMENT OF HAZARDOUS WASTE

Some types of hazardous waste can be detoxified or made less dangerous by either *chemical, biological,*

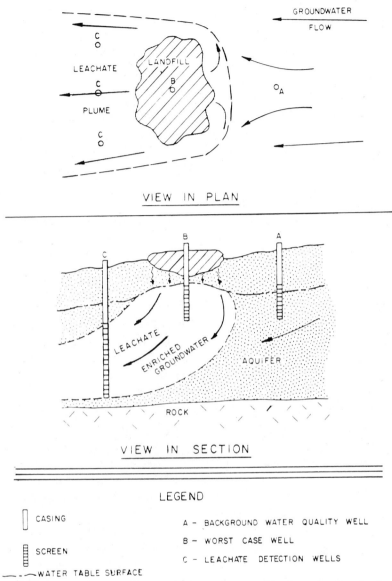

FIGURE 11.25 A groundwater monitoring system consists of a series of wells, strategically located around the waste disposal site. (From Francis, C. W. and S. I. Auerbach, *Environment and Solid Wastes*, Butterworth Publishers)

or *physical* treatment methods. Treatment of hazardous waste may be costly, but it can serve to prepare the material for recycling or for ultimate disposal in a manner safer than disposal without treatment.

Some common chemical treatment processes include *ion exchange, neutralization, oxidation/reduc-* *tion,* and *precipitation.* The process of *incineration* is a high temperature or thermal–chemical process which not only can detoxify certain organic wastes, but can essentially destroy them as well. These processes are discussed briefly in the following paragraphs.

In the ion exchange process, industrial wastewater is passed through a bed of resin that selectively adsorbs charged metal ions. An example of its use in the metal finishing industry is for removal of waste chromic acid from the production rinse water. Neutralization refers to pH adjustment, for reducing the strength and reactivity of acidic or alkaline wastes. Limestone, for example, can be used to neutralize acids, and compressed carbon dioxide may be used to neutralize strong bases.

Oxidation and reduction are complementary chemical reactions that involve the transfer of electrons among ions. Oxidation of waste cyanide by chlorine renders it less hazardous. Precipitation refers to a type of reaction in which certain chemicals are made to settle out of solution as a solid. An example of its application is in the battery industry, where the addition of lime and sodium hydroxide to acidic battery waste causes lead and nickel (toxic heavy metals) to precipitate out of solution.

Incineration, or burning of organic hazardous waste at very high temperatures, can convert the waste to an ash residue and gaseous emissions. The combustion detoxifies the waste material by altering its molecular structure and breaking it down into simpler chemical substances. Although the ash may be treated as a hazardous material, a much smaller volume of waste is left for ultimate disposal.

Not all hazardous waste can be incinerated. Heavy metals, for example, would not be destroyed, but would enter the atmosphere in vapor form. However, incineration has been successfully applied to such potent hazardous waste as chlorinated hydrocarbon pesticides, PCBs, and other organic substances.

Special types of thermal equipment, such as the *fluidized bed incinerator,* the *rotary kiln,* the *multiple-hearth furnace,* and the *liquid injection incinerator,* are available for burning the waste in either solid, liquid, or sludge form. One of the major problems with land-based incineration of hazardous waste is the concern about air pollution from the gaseous emissions. A possible solution that is being explored is the possibility of hazardous waste incineration on a ship at sea, far from population centers.

Biological treatment uses living microorganisms. The microbes utilize the waste material as food and convert it by their natural metabolic processes into simpler substances. It is most commonly used for stabilizing the organic waste in sewage, but certain types of industrial waste can also be treated by this method. Organic waste material from the petrochemical industry, for example, can be treated biologically. It is necessary, though, to utilize bacteria that are acclimated to the waste. In some cases, *genetically engineered* species of bacteria may be used.

In addition to the traditional biological treatment systems, such as activated sludge and the trickling filter, a treatment method called *landfarming* or *land treatment* (which is not the same as landfilling) may be used. The waste is carefully applied to and mixed with the surface soil; microorganisms and nutrients may also be added to the mixture, as needed. The toxic organic material is degraded biologically, whereas inorganics are adsorbed and retained in the soil.

Landfarming can provide a relatively inexpensive method for treatment as well as being a way to ultimately dispose of certain types of hazardous waste. But food or forage crops should not be grown on the same site because they could take up toxic material. A disadvantage of land treatment as a hazardous waste management option, then, is that relatively large tracts of land must be withdrawn from potentially productive agricultural use. Naturally, the surface topography and underground geologic conditions of the site must be suitable, so that erosion or groundwater contamination will not occur as a result of the waste treatment process.

Physical treatment can be used to concentrate, solidify, or reduce the volume of the waste material. The simplest physical process that serves to concentrate and reduce wastewater volume is *evaporation,* which may be facilitated by using mechanical sprayers. Other physical processes that separate the waste from a liquid carrier include *sedimentation, flotation,* and *filtration.* Solidification can be accomplished by encapsulating the waste in concrete, asphalt, or plastic material. This produces a solid mass of material that is resistant to leaching.

CLEANUP AND CONTAINMENT OF UNCONTROLLED DUMP SITES

In addition to using on-site underground injection, many industrial generators of hazardous waste have long been storing or disposing of their dangerous waste material in unlined pits, ponds, or lagoons. The EPA has estimated that almost three-quarters of the approximately 40 000 or so uncontrolled industrial disposal sites in the nation are unlined. Many of these sites pose a serious threat to environmental quality and public health, and efforts to "remediate" or clean them up will continue for years to come. The illegal practice of "midnight dumping" further aggravates the problem (but the "cradle-to-grave" waste tracking provision of RCRA should eventually eliminate this difficulty).

Superfund

In 1980, a federal law called the *Comprehensive Environmental Response, Compensation and Liability Act* (CERCLA) was passed by Congress. This act is commonly referred to as the *Superfund* legislation, because it authorized expenditures of more than $1 billion of federal funds. The money is intended primarily to pay for remedial or cleanup action at the most dangerous hazardous waste dumps in the United States. The original CERCLA authorization was for $1.6 billion, but it is estimated that the ultimate cost for cleanup will eventually reach about $12 billion. It is also expected that the number of sites on the *superfund list* (national priority list) of the most dangerous dumps will climb to about 1800 after thorough investigation.

One of the problems involved in financing the cleanup of abandoned hazardous waste dumps is to identify the company responsible for the dumping. This can be difficult and sometimes impossible. The basic purpose of the Superfund is to raise money for remedial action where the responsible party could not be identified and the government has to oversee the cleanup activity. Most of the Superfund is to be financed by a tax on oil and chemical companies. Unfortunately, the task of CERCLA and Superfund got off to a slow start in the early 1980s, largely because of political difficulties. In the coming years, considerable progress will have to be made, if environmental health and public safety are to be protected.

Remedial action at an abandoned hazardous waste site is quite complicated and cannot be accomplished overnight. There is a significant distinction between an immediate emergency response to an isolated accidental spill of hazardous material and an engineered solution to a long-term "festering" waste dump. The full areal extent of the abandoned dump site and the degree of damage already done to the environment cannot be known until thorough studies have been conducted by scientists, engineers, and technicians. Civil engineers and technicians play a significant role in the site studies and in the design of remedial action plans, particularly with regard to the geotechnical and groundwater contamination aspects of the project. A simplified network diagram used for planning and scheduling a site remediation project is illustrated in Figure 11.26.

Remedial Action Alternatives

There are several alternative courses of action that can be taken to remediate or "clean-up" a hazardous waste dump site. The basic objective, of course, is to eliminate any immediate danger caused by the uncontrolled waste material, as well as to reduce any long-term threat to public health and environmental quality (especially to groundwater). The optimum course of action depends primarily on the type of waste material, the extent of environmental pollution, and the location of the site. Since each site is unique, particularly with regard to hydrogeologic conditions, no two site remediation projects are identical; sound technical judgment must be used in each case.

One possible course of action may be to physically remove the waste material from the site and transport it to some other location for processing, incineration, or ultimate disposal in a landfill. This "off-site" solution may be the most desirable for people living in the vicinity of the dump, but it can be one of the most expensive options. Also, moving the waste from one location to another still involves some risk of environmental pollution.

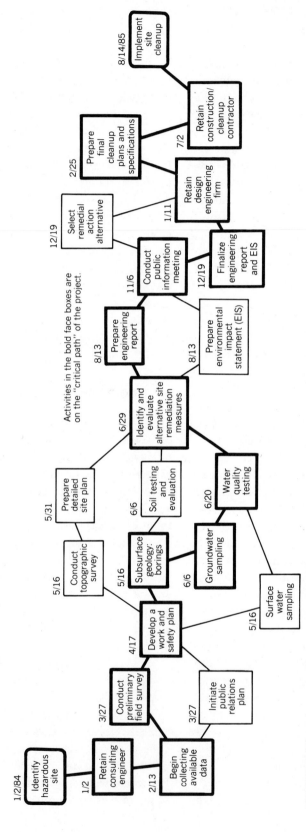

Activities in the bold face boxes are on the "critical path" of the project.

FIGURE 11.26 A network diagram and schedule may be used to manage the many activities for a hazardous site remediation project. More than a year is typically needed for investigation and planning before cleanup and confinement work can actually begin.

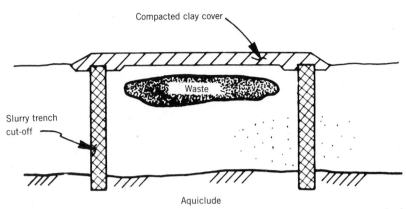

FIGURE 11.27 Slurry trench cutoff walls can be used to prevent the spread of polluted groundwater at an illegal hazardous waste dump site. (From Sweeney, T. L., et al., *Hazardous Waste Management for the 80's*, with permission of Ann Arbor Science Publishers)

On-site remediation, in which the waste is not removed to another location, generally focuses on the need to minimize the production of leachate and to eliminate groundwater pollution. Another primary goal is to contain or prevent the further spread of any groundwater pollution that may have already occurred. This could involve the temporary removal of the waste, construction of a lined or a secure landfill on the same site, and replacement of the waste in the new and "secure" landfill. Or it could involve containment of the waste by constructing an impermeable cover over the site, and by blocking the flow of groundwater with a subsurface cutoff wall.

Subsurface cutoff walls are feasible if there is an *aquiclude,* or naturally occurring impervious layer, below the dump site. The waste can be isolated or contained by constructing a *slurry trench cutoff wall* that penetrates the aquiclude. The trench wall serves as a vertical barrier to the flow of groundwater, and the aquiclude serves as a natural bottom liner. This is illustrated in Figure 11.27.

The slurry trench wall can be excavated from the ground surface with a clamshell or backhoe excavator, as shown in Figure 11.28. It can be built without the need for moving or otherwise disturbing the waste material. Trenches that are about 1 m (3 ft) wide and about 30 m (100 ft) deep can be excavated wihout collapse by filling them temporarily with a bentonite–clay slurry. The dense

FIGURE 11.28 Excavation of a deep slurry trench cutoff wall (Geo-Con, Inc.)

slurry maintains the stability of the trench during excavation, until it is backfilled with a material that forms the vertical barrier. The backfill material may be a mixture of soil and cement, or soil and clay; it must be blended to provide a permeability of 10^{-8} cm/s or less, to block the flow of water. In some cases, a double trench wall, or a trench with an installed synthetic liner sheet, may be built for extra assurance that there will be no groundwater movement into or out of the site.

Field Investigation and Safety

There are many immediate hazards that may be encountered during the field investigation of a waste dump. Naturally, a thorough field study, which includes taking samples to determine the type of the waste material, is necessary before a remedial plan can be designed. Technicians who perform the field work must have a thorough knowledge of the potential dangers, and must proceed with extreme caution. Ideally a field investigation team should consist of three people, for safety purposes. In addition to having a background in the chemical, physical, and biological characteristics of hazardous material, and being trained in hazardous material technology, they should be familiar with first aid techniques. They should also be familiar with the proper use and operation of protective clothing and equipment, as well as sampling and gas detection devices.

One of the primary dangers for personnel at a waste disposal site is the inhalation of toxic gases or vapors. Self-contained breathing apparatus (SCBA) and cartridge respirators or adsorbent gas masks should be available for use. The equipment must be well maintained, and the technician must be thoroughly familiar with its use. It is also necessary that devices for detecting and measuring combustible or toxic gases and oxygen deficient atmospheres be used during the initial investigation of the site. Such devices must be calibrated before use.

In addition to respiratory protection, suitable clothing and equipment for eye and body protection should be available for use. Eye protection includes safety glasses, protective goggles, or faceshields attached to a hard hat, depending on the risk involved. In some cases, fully enclosed "acid suits" may be used for protection. A technician wearing protective equipment while sampling groundwater in the vicinity of a landfill site, is shown in Figure 11.29.

FIGURE 11.29 An example of good safety procedure—a technician wearing protective clothing while collecting samples of groundwater at a waste landfill site.

REVIEW QUESTIONS

1. Give a brief definition of *MSW*, including the meaning of the terms *refuse, garbage, rubbish,* and *trash.* Roughly how much refuse is generated per capita in the United States? What is its unit weight before compaction?

2. What are the major types of *hazardous waste?*

3. Briefly, what does the practice of *solid and hazardous waste "management"* encompass? What is generally the most expensive function or activity?

4. Briefly discuss the basics of MSW collection and transport. What is a *transfer station?*

5. Briefly discuss "cradle-to-grave" monitoring of hazardous waste material. What is the *manifest system?*

6. Briefly discuss three purposes or advantages of solid and hazardous waste processing or treatment.

7. Discuss the process of incineration as a solid waste management option.

8. What is a *water wall furnace?* What does *RDF* mean? Briefly discuss incineration with heat recovery, including advantages and disadvantages. What is *cogeneration?*

9. Briefly discuss the process of *pyrolysis.*

10. What are some reasons for *shredding* or *pulverizing* solid waste? What is a *hammer mill?*

11. What is the purpose of *baling* solid waste? Why is baling not applied very much in the United States?

12. What is *composting?* Briefly describe the process of open-field composting. What are the advantages and/or disadvantages of enclosed digestion? What is the basic reason that composting is not applied very much in the United States?

13. Why would MSW sorting at a central facility be more reliable than source separation, as part of a recycling effort?

14. List four physical characteristics that can be used to separate the different components of municipal solid waste.

15. Make a freehand sketch of a flow diagram showing the various separation processes used at a resource recovery plant.

16. Briefly describe the basic purpose and operating principle of an air classifier and cyclone separator, a magnetic separator, a trommel screen classifier, and an inertial separator.

17. What is one of the most difficult problems with regard to MSW recycling? What types of material are suitable for recycling? Briefly discuss some of the factors related to the recovery and reuse of each type of material. Which materials have been recycled the most successfully?

18. What are three key characteristics of a *sanitary landfill* that distinguish it from an open dump? Why is landfilling declining in popularity as a waste management option? Why will it continue to be used anyway in years to come?

19. Briefly discuss some of the technical factors involved in selecting a new site for a sanitary landfill. What are some of the nontechnical problems? What would be your attitude if a new landfill was scheduled for construction in your neighborhood?

20. What does site *geohydrology* refer to? Why is it of particular importance with regard to landfill siting and construction?

21. What is the basic "building block" of a sanitary landfill? How is it constructed? Make a sketch to illustrate your answer.

22. Briefly describe the difference between the two basic methods for landfilling of MSW. What is the *ramp method?*

23. What type of equipment is typically needed to operate a landfill?

24. Discuss the problem related to gas production at a landfill. What can be done to minimize the problem? Are there any potential advantages or benefits of gas generation?

25. What is *leachate?* How is it generated, and what problems is it likely to cause? What can be done to minimize the problems? Are there any potential advantages or benefits of leachate production at a landfill site?

26. What are two technical problems related to construction of buildings on, and the final use of, a completed landfill site?

27. In order of preference, list five options for the management and control of hazardous waste material. What is *RCRA?*

28. What is a *secure landfill?* Describe four key features regarding its design, construction, and operation. What are the different phases of its "operating life"? Is it really "secure"?

29. Make a sketch showing a cross-section of a secure landfill.

30. What is *underground injection?* Why is it considered by some experts to be an undesirable option for hazardous waste disposal?

31. Briefly discuss some chemical treatment processes that may be applied to hazardous waste. What is the purpose of such treatment? Can all hazardous waste be safely incinerated?

32. What is *land farming* of hazardous waste?

33. What is the purpose of physical treatment of hazardous waste?

34. What is *Superfund?* What are two basic options for hazardous waste dump site remediation or cleanup?

35. What are the primary goals of dump site remediation? How may they be accomplished? Make a sketch showing site containment.

36. Briefly discuss some of the factors related to waste site field investigation and safety. What is one of the primary dangers for personnel at a hazardous waste dump site? Is it advisable for field technicians to be knowledgable about the chemical, physical, and biological properties of hazardous material? Why?

PRACTICE PROBLEMS

1. If the initial volume of a sample of solid waste is 20 yd^3, and after compaction the volume is reduced to 4 yd^3, what is the percent volume reduction and the compaction ratio? If it is desired to increase the percent reduction to 95 percent, what would the compaction ratio have to be?

2. The minimum waste compaction ratio reported by the manufacturer of a high pressure compaction machine is 8:1. What is the corresponding percent volume reduction of the waste? What was the uncompacted waste volume of a 2.5-yd^3 bale?

3. A community of 50 000 people generates refuse at a rate of 8 lb/person/d. It is compacted in a sanitary landfill to a unit weight of 1200 lb/yd^3. After one year of operation, to what depth would a 25-ac landfill be covered? Assume a refuse-to-cover ratio of 4:1.

4.   A community of 15 000 people generates refuse at a rate of 5 lb/person/d. The compacted unit weight of the refuse in the collection vehicles is 500 lb/yd^3. If the capacity of a collection truck is 15 yd^3, how many truckloads of refuse would be unloaded at the landfill each weekday?

5.   A community of 20 000 people uses a 10-ac landfill site that can be filled to an average depth of 50 ft. If refuse is generated at the rate of 6 lb/person/d, its compacted unit weight in the fill is 1000 lb/yd^3, and the refuse-to-cover ratio is 4:1, what is the useful life of the site?

6.   A community of 35 000 people generates refuse at the rate of 30 N/person/d. It is compacted in a sanitary landfill to a unit weight of 6 kN/m^3. To what depth would a 2-ha landfill site be covered after one year of operation? Assume soil cover adds 25 percent to the required volume.

7.   How many acres of land would be required for a sanitary landfill under the following conditions: useful life = 25 yr, waste generation = 5 lb/person/d, compacted weight = 1000 lb/yd^3 in the fill, average fill depth = 35 ft, population = 30 000, and refuse-to-cover ratio = 4:1.

SELECTED REFERENCES

1. Francis, C. W., et al. (eds.), *Environment and Solid Wastes/ Characterization, Treatment, and Disposal,* Butterworth Publishers, Woburn, Mass., 1983.

2. Sweeney, T. L., et al. (eds.), *Hazardous Waste Management for the 80's,* Ann Arbor Science, The Butterworth Group, Ann Arbor, Michigan, 1982.

3. *Recycling in the 1980's,* State Advisory Commission on Recycling, New Jersey Department of Environmental Protection, Trenton, 1980.

4. Tchobanglous, G., *Solid Wastes: Engineering Principles and Management Issues,* McGraw–Hill, New York, 1977.

5. Wilson, D. (ed.), *Handbook of Solid Waste Management,* Van Nostrand Reinhold, New York, 1977.

6. Hagerty, D. J., et al., *Solid Waste Management,* Van Nostrand Reinhold, New York, 1973.

7. Baum, B. et al., *Solid Waste Disposal,* Vol. 1 & 2, Ann Arbor Science, Ann Arbor, Mich., 1973.

8. Brunner, D. and D. Keller, *Sanitary Landfill Design and Operation,* U.S. Environmental Protection Agency Report SW-65ts, Washington, DC, 1972.

9. *Municipal Refuse Collection and Disposal,* Office for Local Government, New York State, Albany, 1964.

10. Sheiman, D., *A Hazardous Waste Primer,* Pub. No. 402, League of Women Voters, Washington, DC., 1980.

11. *The New Jersey Hazardous Waste News,* Association of New Jersey Environmental Commissions, Mendham, New Jersey, Nov./Dec. 1984.

12. *Public Works Manual,* 1984 Edition, Public Works Journal Corp., Ridgewood, New Jersey.

13. Bennet, G. F., et al., *Hazardous Materials Spills Handbook,* McGraw-Hill, New York, 1982.

12

AIR AND NOISE POLLUTION

Air is necessary for the survival of all higher forms of life on earth. On the average, we each need at least 30 lb of air every day to live, but we need only about 3 lb of water and 1.5 lb of food. A person could survive about five weeks without food, five days without water, but only five minutes without air. Nevertheless, many people tend to give little thought to the *atmosphere,* that invisible yet essential ocean of gases in which we live, and upon which we are so dependent.

The atmosphere appears to be so vast in volume, that at first glance it may be difficult to believe that human activity could have any lasting impact on it. Smoke and gases discharged into the air are soon diluted to very low and unnoticeable concentrations, by mixing and dispersion. It seems impossible that human activity could pollute the entire atmosphere, and very unlikely that such small amounts of contaminants could harm public or environmental health.

So it may seem, at first glance. But most of us know better. The atmosphere is extensive, but not infinite. We recall the fact that if the earth was imagined to be about the size of an apple, the depth of the atmosphere would be equivalent to only the thickness of the apple's skin. We know from personal experience the short-term discomfort of being enveloped in the exhaust fumes of vehicles on a congested highway, or when passing by certain manufacturing plants. And we read or hear about the harmful effects of "acid rain," and the claim of many experts that air pollutants from certain regions of the country are causing this problem of acidic precipitation in other areas, several hundred miles away.

Public health studies tend to demonstrate the fact that over the long term, the standard of health for people living and working in highly industrialized urban areas is lower than that for populations in rural areas. There is no direct and conclusive link between air pollution and a specific health problem, such as lung cancer. But there is ample evidence of a distinct relationship between generally "dirty" air and a higher incidence of respiratory disease. Air pollution and atmospheric quality is really an issue of major significance in environmental and public health technology.

The work of civil engineers and technicians often involves air pollution technology, particularly as it relates to transportation or land development projects and their environmental impacts. The objective of this chapter, then, is to present some of the basic facts and concepts regarding air pollution and its control. After a brief consideration of atmospheric factors, the major types, sources, and effects of air pollutants are discussed. Air sampling, testing, pollution control, and air quality standards are also covered. Finally, a section on noise pollution is included. Noise is, after all, a form of air pollution that has public health significance. It also affects the work of civil engineers and technicians, particularly in the areas of construction and transportation.

12.1   ATMOSPHERIC FACTORS

In order to understand topics related to the effects and control of air pollution, it is first necessary to know something about the composition and physical behavior of the atmosphere itself. What does the "pure" atmosphere consist of, and how do meteorological or weather conditions affect the mixing and dispersion of pollutants?

COMPOSITION OF THE ATMOSPHERE

The atmosphere comprises a mixture of many different gases, but mostly it consists of molecular *nitrogen* and *oxygen*. About 78 percent of dry air is nitrogen, and about 21 percent is oxygen. This is expressed on a volume basis. In other words, a container holding 1000 L of air (at standard pressure) would include about 780 L of nitrogen and 210 L of oxygen.

The nitrogen and oxygen add up to only 990/1000 or 99 percent of the total volume. The remaining 10 L, or 1 percent of the "pure" atmosphere, normally includes several other gases. Most of that 1 percent (roughly 0.9 percent) is the inert gas *argon*. The rest includes *carbon dioxide, methane, hydrogen, helium, neon, ozone,* and other gases in trace amounts. Figure 12.1 illustrates the relative amounts of atmospheric gases in graphic form.

The relative amounts or concentrations of gases in air can be expressed in terms of *parts per million* (ppm), as well as in terms of percentage. For example, since 10 000 ppm = 1 percent (see Section 4.1), the oxygen concentration of 21 percent in air can also be expressed as 21 000 ppm. Obviously, it is more convenient to simply express that concentration in percent. On the other hand, the average global concentration of carbon dioxide, 0.0340 percent, may be more conveniently expressed as 340

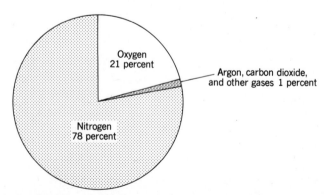

FIGURE 12.1   Molecular nitrogen and oxygen are the main constituents of the atmosphere, but "clean" air also contains argon, carbon dioxide, and trace amounts of several other gases.

ppm. Natural ozone concentrations can be as low as 0.02 ppm.

Water vapor is also a normal component of the atmosphere, but the amount may vary significantly over time and location. Local climate is a major factor that affects the amount of atmospheric moisture. In humid regions, the water or moisture content of air may be as high as 5 percent.

Atmospheric Layers

The full atmosphere extends upward roughly 160 km (100 miles) above the surface of the earth. But the relative composition of gases just outlined pertains only to the *troposphere,* which is the lowermost layer of the atmosphere. The troposphere is only about 12 km (8 miles) thick. It is in this relatively thin layer of air that oxygen-dependent life is sustained, clouds are formed, weather patterns develop, and most of our air pollution problems occur.

The density of air decreases significantly with an increase in altitude or distance above the earth's surface. Consequently, most of the total air mass of the atmosphere is contained within the lower layer or troposphere. The "skin of the apple" mentioned previously refers to this life-supporting layer. Above the troposphere, there is not enough oxygen to support life.

The layer of air above the troposphere, called the *stratosphere,* is a stable layer that extends upward to an altitude of about 30 km (20 miles). Even though it is deeper than the troposphere, the stratosphere contains only a small fraction of the total air mass, because of the lower air density. It does, however, contain much more *ozone,* O_3, than the troposphere.

The ozone in the stratosphere plays an important role in protecting living organisms on the earth from the sun's harmful ultraviolet (UV) radiation. The UV rays are absorbed by the ozone molecules and are then converted into heat energy. The ozone, in effect, acts as a protective filter. It is conceivable that an accumulation of certain pollutants (e.g., freon from aerosol cans) in the stratosphere could react with ozone, diminishing its UV filtering ca-

pacity. There is concern that this may lead to an increase of skin cancer and other health problems in humans.

Layers of the atmosphere above the stratosphere include the *mesosphere,* the *ionosphere,* and the *thermosphere.* These portions of the atmosphere are essentially unaffected by air pollution.

THE EFFECT OF WEATHER

Air pollutants are mixed, dispersed, and diluted in the atmosphere by movement of air masses, both horizontally and vertically. This air movement, and therefore air quality, is very dependent upon local meteorological or weather conditions.

Horizontal dispersion of air pollutants depends upon wind speed and direction. The concentration of pollutants decreases with increasing wind speed, because as the pollutants are discharged from the source, they are more rapidly separated and dispersed by the swiftly moving air. Knowledge of prevailing wind speed and direction in a given locality makes it possible to select sites for new industrial facilities or power plants so as to minimize local air pollution effects. Locating such sites downwind of residential areas is preferable, naturally, to upwind location.

Temperature Inversion

In addition to wind, another important meteorological factor that has a significant effect on the dispersion of pollutants is *atmospheric stability.* The atmosphere is said to be stable when there is little or no vertical movement of air masses. As a consequence, there is little or no mixing of air pollutants in the vertical direction, and pollutants tend to accumulate near the ground. Under such conditions of stability, air pollution problems may become severe.

An unstable atmosphere, on the other hand, is one in which air masses move naturally in a vertical direction, and carry pollutants upward, away from the ground. A condition of instability, then, is preferable to a condition of stability in the atmosphere, with regard to air quality.

Atmospheric stability depends on the *relationship between air temperature and altitude* that prevails at a particular time and location. Normally, in the troposphere, air temperature decreases with increasing altitude—as you go higher, it gets cooler. The lower-most layer of the atmosphere is warmed by heat energy reradiated from the earth's surface. But the relatively warm air near the surface then tends to rise, as it is displaced by cooler and denser air from above. This may result in an unstable condition with constant vertical mixing of air masses, if the rate of temperature decrease with altitude is sufficient to sustain the mixing process.

The rate at which temperature actually changes with increasing altitude at any given time is called the *environmental lapse rate,* or simply, the lapse rate. The specific lapse rate that represents the separation or boundary between a stable and unstable atmosphere is called the *adiabatic lapse rate.* It is equal to $-1°C/100$ m ($-5.4°F/1000$ ft). (The negative sign indicates that air temperature decreases as the altitude increases.)

As long as the environmental lapse rate exceeds the adiabatic lapse rate, the atmosphere will be unstable and vertical mixing of air masses will occur. The colder air from above will descend as the warmed air rises, in a manner similar to the "turnover" of a stratified lake in the fall. This condition is illustrated in Figure 12.2.

Certain weather patterns can cause the environmental lapse rate to be less than, and others make

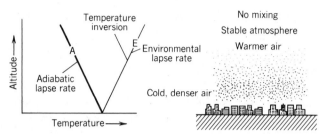

FIGURE 12.3 When local weather conditions temporarily cause air temperatures to increase with altitude, an *inversion* has occurred. The atmosphere is stable during an inversion; air pollutant levels build up because of the lack of mixing and dispersion in the air.

it greater than, the adiabatic lapse rate. In fact, under some circumstances, it is possible for the lapse rate to change direction entirely, that is, to represent an increase rather than a decrease in temperature with altitude. Such a condition is called a *temperature inversion,* and it is a most undesirable condition with respect to air quality.

An inversion is illustrated in Figure 12.3. The denser, colder air is trapped below the warmer air, and vertical motion of air masses is restricted. Since vertical motion is restricted, there is essentially no mixing or dispersion of air pollutants in an upward direction. In an urban area, air quality will decrease rapidly during this period of stability or stagnation, until the weather conditions change and the normal lapse rate is restored.

Temperature inversions can be caused by a variety of local meteorological conditions, and they can occur just about anywhere. But there are certain geographical conditions that can increase the frequency and duration of these inversions. The situation can be particularly severe, for example, for a community located in a valley, which acts as a holding basin or sink for cold, dense air masses near the ground. The surrounding hills also tend to block horizontal air motion, thus adding to the stagnation problem. The city of Los Angeles, for example, lies in a mountain-rimmed "bowl" that traps air pollutants during frequent temperature inversions.

Sometimes a temperature inversion will begin at a certain elevation above the ground surface, leaving a relatively thin layer of unstable air below.

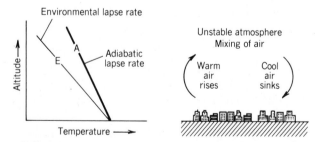

FIGURE 12.2 Line A shows the *adiabatic lapse rate,* and line E shows an *environmental or prevailing lapse rate.* When the air temperature decreases faster than the adiabatic rate, as shown here, air pollutants are dispersed and diluted in the atmosphere.

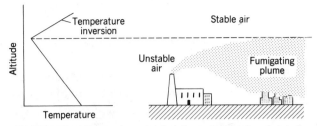

FIGURE 12.4 When a temperature inversion begins above the ground, because of local weather conditions, it acts as a lid or ceiling that prevents further vertical mixing and traps pollutants below it.

Such a condition is illustrated in Figure 12.4. This type of inversion forms a "lid," in effect, that traps pollutants and prevents further vertical mixing. The plume from a smokestack will stop rising when it reaches the inversion altitude. Pollutants in the plume are mixed in the thin but unstable layer near the ground, causing a condition of fumigation for surrounding communities.

12.2 TYPES AND SOURCES OF AIR POLLUTANTS

Air pollution may be simply defined as the presence of "foreign" substances in the atmosphere in high enough concentrations, and for long enough durations, to cause undesirable effects. What are these so-called foreign substances, where do they come from, and what are the undesirable effects? In this section, the nature and sources of common air pollutants will be discussed, and in the following section, some of the most undesirable effects will be considered.

First though, we should make a distinction between so-called *natural air pollution,* and pollution caused by industry, transportation, and other human activities. Not all air pollution is caused by human activity. In fact, at certain times the pollution from natural sources can be far more severe and long lasting than pollution from human activity.

Perhaps the most dramatic and recent example of natural air pollution in the United States was caused by the 1980 eruption of Mount St. Helens, in the state of Washington. Vast quantities of gases and dust were spewed into the atmosphere in a relatively short period of time. Local communities, including the city of Portland, Oregon, were blanketed with volcanic ash for quite a while. In addition to discharges such as those from Mount St. Helens and other active volcanoes around the world, natural air pollutants include smoke and gases from forest fires, windblown dust from deserts, salt seaspray, pollen grains, and other naturally occurring substances.

Those substances that are generally recognized to be of major concern as air pollutants from human activity include the following:

1. Particulates
2. Sulfur dioxide
3. Nitrogen dioxide
4. Carbon monoxide
5. Hydrocarbons
6. Ozone
7. Lead

The principal sources of these air pollutants are considered to be either *mobile* (e.g., automobiles) or *stationary* (e.g., coal fired electric power generating stations). The distinction between mobile and stationary sources of air pollutants is important because of the different dispersion patterns and pollution control technology applied to each type. Chemical manufacturing, fuel combustion for heat, and solid waste incineration are additional stationary sources, but electric power generation is the most significant with respect to total emissions.

Although aircraft, some trains, and ships are also mobile generators of pollution, the automobile and its internal combustion engine contribute most of the contaminants from transportation sources. There are more than 100 million gasoline-powered motor vehicles in the United States. Carbon monoxide, nitrogen dioxide, and hydrocarbons are emitted in the exhaust fumes from the tailpipes of these vehicles; hydrocarbons are also emitted from fuel tank and carburetor evaporation and crankcase blow-by gases. Automobiles account for roughly half of the total emissions of these air pollutants in the United States. Lead is also emitted in automobile exhaust, because of the use of leaded gasoline; this source of atmospheric lead will soon decrease as more and more vehicles use unleaded fuel.

The stationary sources, particularly power plants, are the main contributors of particulates and sulfur dioxide. Nitrogen dioxide is also discharged from power plants. In general, air pollution from stationary, as well as mobile sources, is from the *combustion of fuels*. The contaminants that are emitted directly into the air are called *primary pollutants*. Those that are not directly emitted, but form in the atmosphere through chemical reactions, are called *secondary pollutants*. A brief description of each of the six major air pollutants follows. The effects of these pollutants, and their measurement, will be discussed later on in the chapter.

Particulates

Air pollutants can occur in the form of gases, solids, or liquids. Those that occur as extremely small suspended fragments of solids or liquid droplets are called *particulates*. Particulates are usually distinguished on the basis of particle size and source, rather than on the basis of chemical composition.

Most airborne particulates range in size from 0.1 to 100 μm. (1μm, or one micrometer, is one millionth of a meter; it may also be called a "micron.") Particles that are smaller than 1 μm tend to remain suspended in the air indefinitely, whereas those larger than 1 μm tend to settle out under the force of gravity. Even though the individual particles are so small, the total mass of particulates entering the atmosphere is very large. A recent EPA national emission estimate for *total suspended particulates* (TSP) exceeds 10 million tons per year.

Solid particles larger than 1 μm in size are called *dust* particles, and a suspension of liquid particles of about the same size is called a *mist*. A suspension of solid particles that are less than 1 μm in size is called *smoke;* smoke, of course, is a common product of incomplete combustion. Mechanical

processes that involve grinding or pulverizing of materials are also the source of particulate air pollution, but the particles are generally larger than 10 μm.

Fumes also consist of very small particles, but they are generally formed during high-temperature chemical reactions and vapor condensation. The term *aerosol* generally refers to a quantity of any small particles, liquid or solid, suspended in air. *Soot* is a commonly used term that refers to visible clusters of carbon particles that result from incomplete combustion.

Sulfur Dioxide

Certain fossil fuels, particularly coal, may contain the element sulfur as an impurity. When these fuels are burned for power or heat, the sulfur is also burned or *oxidized*. This chemical reaction can be described by the following equation:

$$S + O_2 \rightarrow SO_2$$
$$\text{sulfur} + \text{oxygen} \rightarrow \text{sulfur dioxide}$$

Sulfur dioxide is a colorless gas with a sharp, choking odor. It is a primary pollutant since it is emitted directly in the form of SO_2. Ordinarily, most people do not think of a gas as having much weight; but when the volumes are large, the total weight can be substantial. Roughly 27 million tons of SO_2 are estimated to be discharged into the atmosphere in the United States, each year.

In the presence of oxygen, water vapor, and sunlight, the SO_2 can be involved in additional chemical reactions. It can react with oxygen to form *sulfur trioxide*, which then can react with water vapor to form a mist of *sulfuric acid*. The chemical reactions are indicated by the following equations:

$$2\,SO_2 + O_2 \rightarrow 2\,SO_3$$
$$SO_3 + H_2O \rightarrow H_2SO_4$$

The sulfuric acid, H_2SO_4, mist is a secondary pollutant, since it is not emitted directly but is formed subsequently in the atmosphere. It is a constituent of "acid rain," an important environmental problem that is currently receiving much attention; this will be discussed again later.

Carbon Monoxide

During complete combustion of fossil fuels, each carbon atom in the fuel combines with two oxygen atoms to form carbon dioxide, CO_2. But the process of combustion is rarely complete: *Incomplete combustion* of the fuel may occur when the oxygen supply is insufficient, when the combustion temperatures are too low, or when residence time in the combustion chamber is too short. *Carbon monoxide,* CO, is one of the primary products of incomplete combustion. It is a colorless and odorless gas. Approximately 100 million tons of CO are emitted in the United States, each year, from both stationary and mobile sources; most of it, though, comes from automobile exhaust gases. The equation that describes the formation of carbon monoxide is written simply as:

$$2\,C + O_2 \rightarrow 2\,CO$$

Nitrogen Oxides

During the combustion of fossil fuel, the temperatures are generally high enough to cause additional reactions between atmospheric nitrogen, N_2, and oxygen, O_2. Nitric oxide, NO, is one of the gases formed. Another is *nitrogen dioxide,* NO_2, a visible red-brown toxic gas with a very sharp odor. More than 20 million tons of nitrogen oxides are discharged each year in the United States.

Hydrocarbons

Compounds that consist of only hydrogen, H, and carbon, C, atoms are called *hydrocarbons.* There are many different types of hydrocarbon compounds, because hydrogen and carbon atoms can combine in an almost unlimited variety of sizes and shapes of molecules.

Most of the hydrocarbons (about 85 percent) that are found in the atmosphere come from natural sources, including forests, and the decomposition of vegetation or other organic matter. Incomplete combustion and evaporation of gasoline from automobiles, trucks, and aircraft, account for most of the emissions caused by human activity. Industrial and chemical manufacturing processes also contribute to hydrocarbon pollution.

Even though most atmospheric hydrocarbon occurs naturally, the concentrations of these highly reactive organic gases in urban areas can cause serious problems, particularly as a result of automobile emissions. Hydrocarbons are involved in atmospheric reactions that form secondary air pollutants, most notably those contained in *photochemical smog.* Ozone is another important atmospheric substance that plays a role in the formation of smog.

Ozone

At very low concentrations, ozone, O_3, is considered to be a natural component of air, particularly in the upper atmosphere where ultraviolet rays from the sun cause a reaction between oxygen molecules. Ozone can also be produced naturally by powerful electric discharges in lightning.

Ozone is not a primary urban pollutant; it can, however, be formed as a secondary pollutant in the troposphere. This occurs in a complex chemical reaction between nitrogen dioxide and hydrocarbons in the presence of sunlight. Ozone formed in this manner in the lower atmosphere becomes a major component of photochemical smog. This type of air pollution is particularly characteristic of urban areas with lots of automobiles and sunshine. Los Angeles, for example, is generally noted for its chronic smog problem.

Lead

This toxic metal is an atmospheric pollutant, in the form of very small particles (less than 0.5 μm) or fumes. The primary sources of lead emissions are automobile exhaust fumes and discharges from some industrial facilities. Most atmospheric lead, though, is caused by the exhaust from automobiles that use leaded gasoline.

People living in urban areas sometimes breathe air with as much as 10 μg/m^3 of lead. This is particularly dangerous for young children, since even slightly elevated levels of lead in their blood cause symptoms of lead poisoning, including intellectual and behavioral deficiencies. Legislation that limits the amount of lead in gasoline, and the gradual change to the use of unleaded gas for all vehicles, will eventually eliminate the automobile as a source of atmospheric lead.

12.3   EFFECTS OF AIR POLLUTION

For discussion, we can classify or group the effects of air pollution into five general categories, according to its effects on:

1.   Human health
2.   Materials
3.   Vegetation, agricultural crops, and livestock
4.   Atmospheric conditions
5.   Aquatic and terrestrial ecosystems

HUMAN HEALTH

Of primary concern are the adverse effects air pollution has on human health. Generally, air pollution is most harmful to the very old and the very young. Many elderly people already suffer from some form of lung or heart disease, and their weakened condition makes them very susceptible to additional harm from pollution. The sensitive respiratory systems of newborn infants are also susceptible to harm from dirty air. But it is not just the elderly or the very young who suffer; healthy people of all ages can be adversely affected by high concentrations of air pollution. In general, major health effects include:

1.   Acute (short-term but severe) illness, or death

2.   Chronic (long-term) respiratory illness, including bronchitis, emphysema, asthma, and possibly lung cancer

3.   Temporary eye and throat irritation, coughing, chest pain, and malaise or general discomfort

The intermittent occurrence of exceptionally high air pollutant concentrations in a community, and the acute public health problems that manifest themselves during the same period, is called an *air pollution episode*. One of the most severe episodes of record occurred in London, in 1952. During a one-week period of very high sulfur dioxide and particulate levels, about 4000 "excess deaths" (more than ordinarily would be expected in that time period) were noted.

In the United States, the first major air pollution episode of record occurred in Donora, Pennsylvania, in 1948. In only a few days during October of that year, 20 excess deaths and about 600 illnesses were attributed to air pollution from local industry. Because of the relatively small population of 14 000 people in Donora, the "per capita death rate" was actually the highest ever recorded during an air pollution episode.

Many other pollution episodes have occurred in the recent past, in many different countries, and we cannot yet rule out the possibility of their recurrence. In general, a typical episode lasts about two to seven days and is characterized primarily by stagnant air and unusually high concentrations of SO_2 and particulates. The stagnant air results from temporary weather conditions, including a temperature inversion and negligible wind speeds. Illnesses and excess deaths occur in all age groups, but mostly the very old, the very young, and previously ill persons are affected.

It is difficult for public health experts to match up any specific air pollutant with a specific disease or health effect, with absolute certainty. But some general conclusions can be drawn from available data. Usually sulfur dioxide, nitrogen dioxide, or ozone cause eye and throat irritation, coughing, and chest pain. These pungent gases can harm lung tissue when inhaled into the respiratory tract, and are associated with bronchitis, emphysema, and other lung diseases.

Inhalation of particulates also affects the breathing process adversely. Although particles larger than about 1 μm tend to be captured by the protective mucus lining and cilia (very small hairs) in the nose and throat, smaller particles can penetrate deep into the lungs. Certain particulates are especially dangerous because of their toxic or carcinogenic properties; lead fumes in automobile exhaust and asbestos fibers are only two such examples.

Carbon monoxide is a colorless and odorless gas that is virtually unnoticeable to our senses. But this makes it all the more dangerous because it can be inhaled without causing irritation or immediate discomfort. It is extremely toxic because it readily combines with hemoglobin in the blood, and takes up the place ordinarily occupied by oxygen, which the body needs continuously. The inhaled CO reduces the ability of the blood to transfer oxygen to body cells, leading to asphyxiation or suffocation.

A CO concentration of about 1000 ppm can cause unconsciousness in a healthy person, in one hour of exposure; death by asphyxiation will occur in about four hours at that concentration. Even much lower concentrations can cause illness or reduced mental awareness. A maximum allowable eight-hour exposure limit for workers in the United States has been set at 50 ppm. Under certain circumstances, particularly in the immediate vicinity of heavily congested highways, atmospheric CO levels may reach one-hour peaks as high as 400 ppm.

MATERIALS

Damage to materials, due to air pollution, occurs continuously in urban areas. It includes the soiling and deterioration of building surfaces, public statues and other outdoor works of art, the corrosion of metals, and the weakening and deterioration of textiles and leather, as well as rubber, nylon, and other synthetic products.

Deposition or settling of particulates on materials is the cause of soiling; the frequent cleaning of soiled surfaces and clothing leads to more rapid deterioration. Abrasion, caused by particulates carried in the wind at high speeds, eventually erodes and wears-away solid surfaces. Examples of direct and irreversible chemical attack include the cracking of rubber that is exposed to ozone, and the severe discoloration of leaded house paint that is exposed to hydrogen sulfide gas. Leather becomes brittle when exposed to sulfur dioxide; the SO_2 is absorbed into the leather material, and is converted to sulfuric acid in the presence of moisture.

The damage of material by air pollution is not merely an aesthetic problem, but also an economic problem of major proportions. Although this is not immediately apparent to the casual observer, it should be noted that the total cost of cleaning and repairing damage caused by air pollution is estimated to exceed $1 billion per year in the United States.

PLANTS, ANIMALS, AND THE ATMOSPHERE

Air pollutants can damage fruits, vegetables, trees, and flowers in various ways. Some pollutants cause collapse of the leaf tissue, others bleach or discolor the leaves. The total cost of air pollution damage to agricultural crops and other vegetation amounts to several hundred million dollars per year in the United States. Certain air pollutants also cause harm to cattle and other livestock, but this is usually a localized problem on farms near specific industrial plants that cause the pollution.

To the general public, the most noticeable effect of air pollution is on the atmosphere itself. Specifically, it is the haze and reduction of visibility due to the scattering of light by suspended particles. Particulates can also affect weather conditions by increasing the frequency of fog formation and rainfall. Perhaps less noticeable, but of far greater significance in the long run, is the fact that an accumulation of suspended particles in the atmosphere can appreciably reduce the amount of solar energy that reaches the earth's surface. This, in effect, would increase the earth's "reflectivity," and possibly lead to a decline in average global temperatures.

GREENHOUSE EFFECT

At the present time, it seems that any increase in the earth's reflectivity is being counterbalanced by a phenomenon called the *greenhouse effect*. The greenhouse effect is caused by carbon dioxide, CO_2, which is not ordinarily considered to be an air pollutant. In fact, it is a normal although minor component of the atmosphere, with an average concentration of about 0.034 percent or 340 ppm. And it does not cause any adverse effects on human health.

But carbon dioxide is released into the atmosphere in vast quantities as a by-product of fossil fuel combustion (coal, oil, gas), which is used in industrial activity and power generation. It is estimated that the average worldwide concentration of carbon dioxide is increasing at a rate of almost 1 ppm per year. This does not cause a public health hazard, nor does it cause damage to plants, animals, or materials. What, then, is the problem with atmospheric carbon dioxide, and why is it called the greenhouse effect?

Carbon dioxide molecules in the air absorb the heat energy reradiated from the earth's surface. The energy coming from the sun is able to penetrate the atmosphere. But when the warmed surface of the earth radiates some of that energy back into space, it is trapped by the carbon dioxide in the troposphere, as if it were a blanket of insulation, or the glass enclosure of a greenhouse. This is illustrated in Figure 12.5. As the CO_2 concentration increases, less heat will escape through the troposphere, and average global temperatures will increase.

The greenhouse effect should not be dismissed as an example of scientific speculation or environmentalists' "doomsday" propaganda. Two independent federal studies published in 1983, one by the Environmental Protection Agency and the other by the National Acadamy of Sciences, concluded that the warming trend is both imminent and inevitable. It is expected that global temperatures will increase

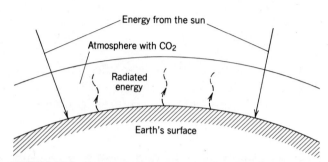

FIGURE 12.5 Energy from the sun can penetrate the atmosphere to warm the earth. But the type of heat energy radiated back from the earth is absorbed by carbon dioxide in the air. Increasing amounts of carbon dioxide from combustion will lead to an increase in atmospheric temperatures, called the *greenhouse effect*.

by about 2°C (3.6°F) within the next 50 years, and by as much as 15°C (27°F) by the year 2100.

Both studies also concluded that even if the use of fossil fuel was banned as of today, the greenhouse effect would not be halted or reversed; there is no known strategy that will mitigate the problem. The only alternative is to plan for ways to cope effectively with the changes in climate that are expected to accompany the warming of the atmosphere. Some of these changes may be beneficial; for example, agricultural production may be improved in certain regions because of a longer growing season and more efficient photosynthesis. On the other hand, the melting of the Arctic ice packs is expected to raise the sea level by about 1 m (3 ft); this will cause extensive economic and social hardship in coastal areas all over the world.

ACID RAIN

A current environmental issue of major public concern is the problem known as *acid rain*. The description "acid rain" refers to the fact that the average pH of rainfall has been decreasing significantly below its normal value, in recent years. The strength of an acidic solution is measured by its pH value (see Section 4.3). Briefly, values of pH range between 0 and 14, with a pH of 7 representing a neutral condition. Values less than 7 indicate acidic conditions; the lower the pH is, the stronger the acid is.

"Pure" rain, in rural areas far removed from human activity, has some natural acidity, with a pH of about 5.5. This is primarily from the formation of carbonic acid, H_2CO_3, by the reaction of moisture and carbon dioxide in the atmosphere. But recent scientific studies show that in urban and industrial areas of the United States, and in other countries, the average pH of rain is less than 4.5. (A pH of 2.2, as acidic as vinegar, was recorded during a rainfall in Scotland in 1974.) On the logarithmic pH scale, a drop of one pH unit represents an increase in acidity by a factor of 10. What is the relationship between acid rain and air quality, and what are the adverse effects of acid rain?

The fact that sulfur dioxide reacts with water vapor to form a mist of sulfuric acid was already discussed. Nitrogen dioxide also reacts with atmospheric moisture to form nitric acid. Oxides of sulfur and nitrogen are among the major air pollutants, and their primary source is power generating stations. The atmospheric mists of sulfuric and nitric acid eventually reach the surface of the earth, in the form of rainfall, dew, or snow. There are several environmental problems attributed to this excessively acidic precipitation, including contamination and damage of:

1. Fresh water lakes
2. Forests
3. Agricultural crops
4. Drinking water
5. Materials

Many species of fish, trees, and agricultural crops are very sensitive to pH values, and do not thrive under acidic conditions. Hundreds of lakes in certain regions of the United States, as well as in many other countries, no longer support fish life; most scientists agree that the "death" of these once productive lakes is directly attributable to acid rainfall.

Acid rain also accelerates the rate at which minerals leach out of the soil. This reduces soil fertility, diminishing the growth and productivity of forests and agricultural crops. Leaching of certain metals from the soil into the groundwater may also contaminate some drinking water supplies. Finally, acid rainfall undoubtedly speeds up the physical deterioration of concrete, metal, and other exposed material.

A factor that complicates the acid rain problem is that most of the sulfur and nitrogen dioxide is emitted from the tall smokestacks or chimneys at power generating plants. The purpose of the tall stacks is to increase the dispersion and dilution of the stack gases and to protect the surrounding community from high levels of air pollution. But discharge from these tall chimneys allows the pollutants to be carried long distances in the atmosphere. The pollution is, in effect, transferred by "air mail" to other regions of the country. For example, most of the acid rain falling in the northeastern region of the United States is believed to be the result of fossil fuel combustion by industries and power plants located in the midwestern section of the nation. About 16 million tons of sulfur emissions each year come from the Midwest. Also, acid rain in Norway is believed to come from industrial areas in England and in continental Europe.

Acid rain is one of the most controversial environmental issues in recent times. In 1984, several northeastern states petitioned the Environmental Protection Agency to order the reduction of emissions from coal-burning power plants in the Midwest. The request was denied by the EPA, on the basis that the existing requirements of the Clean Air Act were not being violated.

The political atmosphere of the early 1980s was one that generally supported further research rather than immediate (and possibly expensive) control of the problem. It was thought by some that acidity in lakes is not simply caused by acid rain. In some cases, though, corrective action was initiated on a statewide basis. For example, in 1984 New York State was the first to require a 30-percent reduction of sulfur emissions by industry and power utilities, specifically to help mitigate the acid rain problem. Finally, in 1986, a $5 billion program to develop cleaner coal-burning technology was endorsed by the federal government; this was part of a joint Canadian-United States effort to control acid rain.

12.4   AIR SAMPLING AND MEASUREMENT

In order to evaluate air quality and to design appropriate air pollution control systems, it is necessary to measure the amount or concentration of the various pollutants. First, of course, an appropriate sample must be collected. There are basically two different approaches for sampling and measuring air pollutants. One involves the sampling and analysis of surrounding "outdoor" or *ambient air quality*. The other involves the sampling and analysis of specific emissions at their point of generation, and may be referred to as *source sampling* or *emissions analysis*.

Ambient Air Quality

Ambient samples are collected from the open atmosphere, after pollutants from various sources have been dispersed and mixed together under natural meteorological conditions. Ambient, or *atmospheric sampling*, as it is sometimes called, serves several purposes. It provides "background" air quality data in urban or rural areas and a basis for developing and updating ambient air quality standards.

Monitoring ambient air quality also provides data to determine if established standards are being met or exceeded. Impending air pollution episodes or emergencies can be predicted in advance, by examining ambient air quality along with meteorological data; this provides time for health officials to warn the public.

Even though samples are taken from the "open air," it is most important that the sampling duration and location be representative of the particular study area and type of pollutant being examined.

Source Sampling

Source or emissions sampling is performed right at the point of pollutant discharge, such as at a vehi-

cle tailpipe or a smokestack. In fact, it is often called *stack sampling* at power plants or industrial facilities where discharge is from a chimney. A basic purpose of source sampling is to evaluate the pollution discharged from a specific generator and to use the results to determine if the so-called *emission standards* are being met or complied with. (Both emission standards and ambient air quality standards will be discussed further in a following section.) Other purposes of emissions sampling are to provide data for designing and operating air cleaning equipment and to measure the working efficiency of that equipment.

For accurate and meaningful results, stack samples must be *isokinetic;* that is, collected by a probe at the same rate at which the gas leaves the stack. The equipment used for this purpose is called a *sampling train,* and it includes several interconnected devices. The basic components are a pitot-tube probe, a vacuum pump to pull the sample out of the stack, a flow meter, and a meter to measure the weight or mass of a specific pollutant in the sample. The temperature of the gas must also be determined. A typical sampling train is illustrated in Figure 12.6.

PARTICULATES

Measurement of particulate air pollutants may be accomplished by several methods, including a *gravity technique,* a *filtration technique,* and an *inertial technique.*

The gravity technique is the simplest method, but it can only measure the amount of *settleable* particulates (dust and fly ash) in the air. A simple device called a *dustfall bucket* has been used for this purpose. The open bucket, containing water to trap and hold the particles, is left exposed in a suitable location, often on a building rooftop. After a collection period of 30 days, the water is evaporated and the dust is weighed.

The measurement results for settleable particulates may be expressed in terms of *grams per square meter per month* ($g/m^2/month$), or more typically, as *tons per square mile per month* (tons/

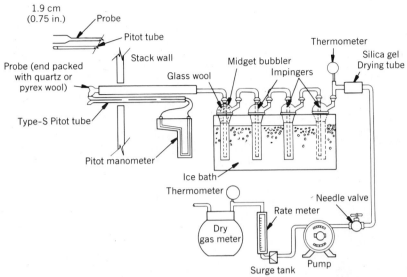

FIGURE 12.6 A stack sampling train for sulfur dioxide. (EPA)

mile²/month), based upon the open top area of the collecting bucket. The total amount of dust that will settle out of the atmosphere in an urban area can be quite high; as much as 50 tons/mile²/month of dustfall have been observed in some cities.

EXAMPLE 12.1

After a 30-day sampling period, 300 mg of dust have settled into a 150-mm-diameter dustfall bucket. Express the dustfall in terms of g/m²/month and tons/mile²/month.

Solution:

The open area of the dustfall bucket is

$$\text{area} = \frac{\pi \times D^2}{4} = \frac{\pi \times (0.15 \text{ m})^2}{4} = 0.0177 \text{ m}^2$$

The monthly rate of dustfall is then

$$0.300 \text{ g} \div 0.0177 \text{ m}^2 = 17 \text{ g/m}^2\text{/month}$$

Converting units to tons/mi²/month, we get

$$\frac{17 \text{ g}}{\text{m}^2 \cdot \text{month}} \times \frac{1 \text{ lb}}{454 \text{ g}} \times \frac{1 \text{ ton}}{2000 \text{ lb}} \times \frac{2.59 \times 10^6 \text{ m}^2}{1 \text{ mi}^2}$$
$$\approx 48 \text{ tons/mile}^2\text{/month}$$

Suspended particles that are too small to settle out of the air by gravity can be collected using the filtration technique. A common filtration apparatus, called the *high-volume sampler,* is shown in Figure 12.7. It acts basically as a vacuum cleaner, except that the air stream first passes through a special leak-proof, glass-fiber filter before it reaches the fan. All the suspended particulates in the air stream are trapped on the filter, which is weighed before and after the sampling period. The difference represents the weight of the *total suspended particulates* (TSP).

The sampling duration is typically 24 hours, in which time about 2000 m³ (70 000 ft³) of air is pulled through the filter. The air flow rate, which gradually decreases as particles accumulate on the filter, is metered and recorded. The measured TSP concentration is typically expressed in terms of mi-

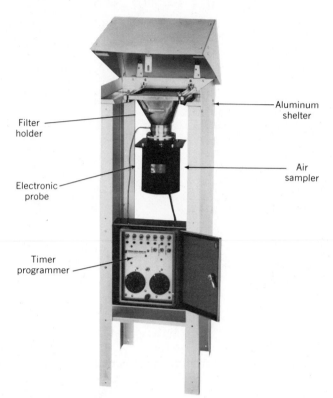

Filter
holder

Electronic
probe

Timer
programmer

Aluminum
shelter

Air
sampler

FIGURE 12.7   A sheltered high-volume (hi-vol) air sampler, used to analyze suspended particulate levels. (General Metal Works, Inc., A Subsidiary of Andersen Samplers, Inc.)

crograms per cubic meter, $\mu g/m^3$. In urban areas, TSP levels are typically around 100 $\mu g/m^3$, although peak values may reach several hundred $\mu g/m^3$; out in the "country," TSP levels are generally about 30 $\mu g/m^3$. Expressing TSP levels in terms of micrograms can give the erroneous impression that the quantities of material are exceedingly small or negligible. It should be noted that a TSP value of 200 $\mu g/m^3$ is roughly equivalent to almost one ton of particles per cubic mile.

EXAMPLE 12.2

The air flow through a high-volume sampler was recorded as 55 ft^3/min at the beginning of sample collection, and 35 ft^3/min after 24 hours of continuous sampling. The filter weighed 10.00 g before,

and 10.20 g after sample collection. What was the TSP level measured in that sample?

Solution:

The average rate of air flow through the filter is simply

$$\frac{55 + 35}{2} = 45 \ ft^3/min$$

The total volume of air passing through the filter in 24 hr is

$$\frac{45 \ ft^3}{min} \times \frac{60 \ min}{hr} \times 24 \ hr = 64 \ 800 \ ft^3$$

and

$$64 \ 800 \ ft^3 \times \frac{0.028 \ 32 \ m^3}{1 \ ft^3} = 1835 \ m^3$$

The weight of particulates is 0.20 g $\times \dfrac{10^6 \ \mu g}{g} =$ 200 $\times 10^3 \ \mu g$. The TSP concentration is computed as

$$\frac{200 \times 10^3 \ \mu g}{1.835 \times 10^3 \ m^3} \approx 110 \ \mu g/m^3$$

Another filtration-type instrument used to collect and measure suspended particulates is called the *paper tape sampler*. Sampling durations with this device are relatively short, typically two hours. A vacuum pump pulls the air stream through a filter tape, which moves automatically on a reel, at preset intervals. The paper tape sampler is illustrated in Figure 12.8.

Trapped particulates form a dark spot on the tape, and the amount of particulates correlates with the darkness of the spot. The relative darkness of the spot is measured by an optical device called a *transmissometer,* which gives a reading in

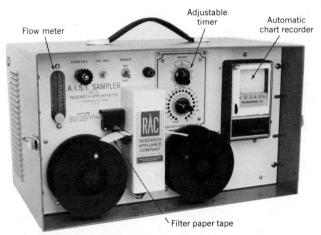

FIGURE 12.8 A paper tape sampler. Air is pulled through a strip of filter paper that traps particulates. Particulate levels are measured by a light transmissometer. (RAC Division, Andersen Samplers, Inc.)

terms of the percentage of light that can pass through the tape. Final results are then expressed in terms of a *coefficient of haze* (COH). A value of 1 COH unit is equivalent to an "optical density" of 0.01.

Say, for example, that after a 2-hr sampling period, the amount of light passing through the clean tape is three times more than the light passing through the spot of trapped particulates. That ratio is called the "opacity" of the spot. The optical density is equal to the logarithm of the opacity. In this example, the logarithm of 3, or 0.477, is the optical density of the spot.

Since a COH of 1 is equal to an optical density of 0.01, the COH of the air sample in this case is 48. This can be converted to COH/1000 linear feet, depending on the area of the spot and the volume of the sample. At a particular location, COH/1000 ft values can be used to monitor hourly fluctuations in particulate air pollution, throughout the day. However, there is no definite relationship between COH/1000 ft and $\mu g/m^3$ of particulates. An advantage of the paper tape sampler is that it is portable and yields quicker results than the high-volume sampler.

The third method of sampling, referred to as the inertial technique, makes use of an obstacle placed in the path of the air stream. The air flows around the obstacle, but because of inertia, the particulates collide with it and become trapped in the device. One of the simplest such devices is the so-called *sticky tape sampler,* illustrated in Figure 12.9. It can be used to collect and measure TSP, as well as to give an indication of prevailing wind direction; the particles collide with and stick on the tape as they are carried by the wind.

Other types of inertial devices are used to collect and analyze specific particulates, such as pollen grains or bacteria. The *cascade impactor,* for example, traps particles on a series of slides that are placed in the air stream. This is illustrated in Figure 12.10. The orifice openings through which the air flows are decreased, thereby increasing the velocity. Particles of different sizes are captured on each slide, because of their inertia, the sudden change in direction of flow, and the different flow velocities. The particulates can be observed on the slides with a microscope.

Smoke Readings

Visual evaluation of smoke plumes that are discharged from a stack or chimney are made with a so-called *Ringlemann Chart,* such as the one illustrated in Figure 12.11. The density or darkness of the smoke is compared to the five standard shades of gray on the chart; Ringlemann smoke readings range from all white (0) to all black (5).

Even though pollutant concentrations are not necessarily correlated exactly with the shade or

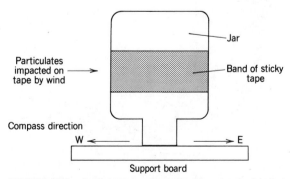

FIGURE 12.9 A sticky tape sampler is a simple inertial device for sampling and measuring particulates and for obtaining results as a function of wind direction.

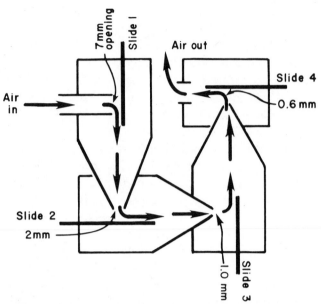

FIGURE 12.10   A cascade impactor for collecting and analyzing particulate air pollutants. (From Vesilind, P. A. and J. J. Peirce, *Environmental Pollution and Control*, with permission of Ann Arbor Science Publishers)

darkness of a smoke plume, Ringlemann readings are of value in monitoring air pollution, and some air quality regulations are still based on smoke density. It should also be noted that the ability to obtain accurate and consistent readings is not an easy task. At least one day of special training is required for a technician to be able to use the Ringlemann chart.

GASEOUS POLLUTANTS

The physical properties and behavior of gases differ markedly from those of particulates. One important example is the fact that gas molecules are small enough to pass through the finest filter.

Two techniques for sampling and measuring the amounts of gases in the atmosphere involve either *absorption* or *adsorption*. The process of *ab*sorption involves the contact and trapping of the gas molecules throughout the volume of a liquid, usually by chemical reaction. The process of *ad*sorption, on the other hand, involves the contact and trapping of the gas molecules on the surface of a solid substance.

Absorption of a specific gas from the air may be accomplished with a simple device called a *bubbler*, as illustrated in Figure 12.12. The air is pumped

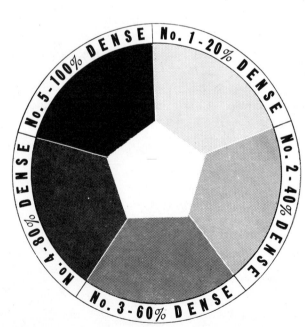

FIGURE 12.11   A Ringlemann type smoke chart. (Plibrico Company, Chicago, Illinois)

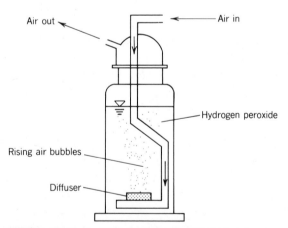

FIGURE 12.12   A glass "bubbler" or absorber may be used for sampling specific gaseous pollutants. For example, hydrogen peroxide will absorb sulfur dioxide from the air, forming sulfuric acid. The level of sulfur dioxide in the air can be computed after measuring the amount of sulfuric acid in the bubbler.

through a small diffuser and bubbled up through a liquid, which will either dissolve the gas under study, or react with it chemically. For example, if a measured volume of air containing sulfur dioxide is bubbled through hydrogen peroxide, H_2O_2, then sulfuric acid is quickly formed, as described by the following chemical equation:

$$H_2O_2 + SO_2 \rightarrow H_2SO_4$$

The amount of sulfuric acid that is formed in the reaction can be measured by standard chemical techniques; from that, the amount and concentration of the sulfur dioxide in the air sample can be computed.

An absorption instrument called a *twenty-four hour bubbler,* shown in Figure 12.13, can be used to test for three different gases at the same time. Separate sampling trains with suitable collecting liquids in the bubbler are connected in parallel to a vacuum pump. The rate of air flow can be con-

trolled and measured. A similar device, called a *sequential sampler,* can be used to collect up to 12 samples in sequence for fixed periods of time, typically 2 hours. The sequential sampler allows peak concentrations of a specific pollutant to be determined on a daily basis.

In adsorption instruments, the gas molecules are attracted to the surface of a solid and held there by molecular bonding forces. Activated carbon is usually used as the adsorbent material; it is a porous solid with a very high surface-area-to-volume ratio. Other materials, such as silica gel and alumina, are also sometimes used as adsorbents. The adsorbent is precoated with a chemical that reacts with and changes color in proportion to the amount of gas adsorbed. For example, the adsorbent in a carbon monoxide detector tube will change from yellow to blue-green, as air containing CO passes through the tube. The CO concentration can be measured by comparing the tube color to a calibrated color chart.

Sometimes it is necessary to collect a small sam-

FIGURE 12.13 *(a)* A three-gas sampler, and *(b)* the sampler in an all-weather shelter. (RAC Division, Andersen Samplers Inc.)

ple of air at a particular location for subsequent analysis in a laboratory. This may have to be done using a minimum of equipment, by an inexperienced technician. One way of collecting a *grab sample,* as it is called, is to utilize an evacuated flask; when the flask is opened at the sampling location, the air sample is drawn into it by the vacuum.

Another type of grab sampling device that is effective if the gas under study is insoluble, is the *liquid-displacement collector,* illustrated in Figure 12.14. An air sample is drawn in at the top of the container, displacing the liquid that is drained out at the bottom. In general, grab samples are not very useful when very small quantities of pollutants are present in the air, since the collected volumes are not large enough for accurate analysis.

On the other end of the spectrum are the modern and sophisticated *continuous monitoring (CM)* instruments, as illustrated in Figure 12.15. These instruments combine collection and automatic analysis for many different air pollutants. Electronic detectors, meters, and recording devices are part of

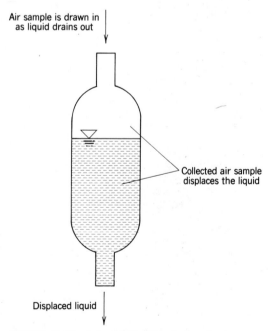

Air sample is drawn in
as liquid drains out

Collected air sample
displaces the liquid

Displaced liquid

FIGURE 12.14   A liquid displacement collector may be used to obtain a grab sample of air for later analysis in a laboratory. The gaseous pollutant to be measured should not react or dissolve in the liquid that is used.

the sampling train of this equipment. Continuous graphs showing the hourly change in pollutant levels or concentrations can be obtained. Expensive CM equipment is used in heavily polluted urban areas, as part of an episode warning system.

12.5   AIR POLLUTION CONTROL

The need to control and minimize air pollution has been evident for several hundred years. One of the earliest attempts to legislate its control was in fourteenth-century England, when King Edward II decreed that anyone found guilty of burning coal while Parliament was in session would be executed. Apparently, the smoke bothered him somewhat.

Today, we know that smoke is more than just a temporary annoyance; smoke and other air pollutants can affect our health and well-being, as well as our aesthetic sensibilities. We also know that threats of medieval-type punishments cannot solve the problem. In our modern industrial society, we cannot avoid generating at least some waste products that will enter the atmosphere, one way or another. Realistic laws and standards are necessary, and the principles of engineering technology must be applied toward the development and use of pollution control equipment.

The first part of this section will briefly address the topic of air quality standards. Air quality criteria and standards serve as the framework within which we can measure the efficacy of air pollution control efforts. They are based on the best available data regarding the health and economic effects of various pollutants. Also discussed in this section are the different types of devices and equipment that are used to reduce pollutant emissions and keep air quality within the legally established standard limits.

AIR QUALITY STANDARDS

In the United States, legislation for air pollution control began with the Clean Air Act of 1963, and

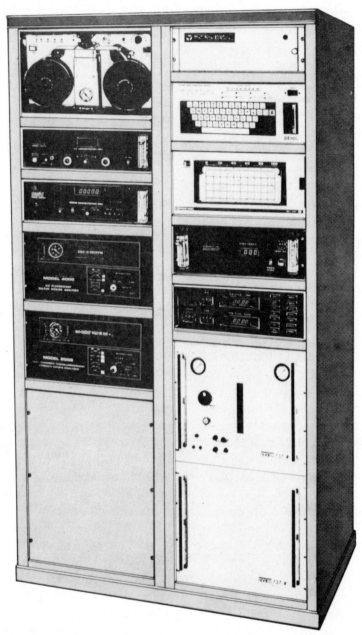

FIGURE 12.15 A continuous ambient monitoring system. (Dasibi Environmental Corp.)

this was supplemented with the Air Quality Act of 1967. Several amendments to the act have been passed since 1967, including those that authorized the EPA to develop air quality criteria and standards. The Occupational Health and Safety Act is another federal law that pertains, in part, to air pollution; it establishes limits of exposure to certain air pollutants for workers in various industries. All air quality standards are under constant review, and may be modified as additional research data become available.

Air quality standards specify maximum allowable concentrations and durations of exposure for specific substances. Three basic types of standards are *threshold limit values (TLV), source performance standards (SPS),* and *ambient air quality standards (AAQS)*.

Threshold limit values focus on specific air contaminants that have well-recognized cause-and-effect health relationships. They serve primarily as occupational standards, limiting the exposure of workers in industrial environments to specific vapors or chemical dust. TLVs are typically established on the basis of exposure for 8 hours per day, 5 days per week.

Source performance (or emission) standards focus on the major municipal and industrial generators of air pollutants, for both stationary and mobile sources. These include power plants, solid waste incinerators, chemical manufacturing plants, oil refineries, passenger cars, and many more.

SPS values are usually expressed in terms of the mass of pollutant emitted per unit of time, of production volume, or of distance. For example, a coal-fired power plant is allowed to discharge no more than 1.2 lb of SO_2 per million BTU (British Thermal Units) of heat input; at a sulfuric acid manufacturing plant, the SPS for SO_2 is 4 lb per ton of acid produced. In 1982, the maximum automobile allowable emission rate for CO was 3.4 g/mi, and 1 g/mi was allowed for nitrogen oxides. In a city where, for example, 50 000 cars travel an average of 25 mi/d, a total of $50\ 000 \times 25 \times 3.4 = 4250$ kg (more than 4 metric tons) of CO would be discharged into the city air (if emission standards were being met).

Ambient air quality standards focus on the allowable levels of "out-of-doors" atmospheric pollutants. They are intended to set limits that will minimize the overall adverse effects on health, comfort, and property. Exposure is assumed to be for 24 hours per day, 7 days per week (in contrast to the more limited exposure times inherent in the TLV standards). Emission standards must be established at levels that will tend to satisfy the ambient air standards in a particular *air quality control region (AQCR)*. For example, if high SO_2 levels are a problem in a particular AQCR, it must be solved by control measures imposed on the SO_2 sources in

that region. The major federal air pollution control regions are shown in Figure 12.16.

Ambient air quality standards for gaseous pollutants may be expressed either in terms of micrograms per cubic meter, $\mu g/m^3$, or parts per million, ppm. The conversion between $\mu g/m^3$ and ppm depends on the molecular weight of the gas as well as on ambient air temperature and pressure.

The National Ambient Air Quality Standards (NAAQS) are presented in Table 12.1. These include limits for the six major air pollutants, as well as for atmospheric lead. Since ambient air quality varies significantly with time, the NAAQS specify certain measuring or averaging times for each pollutant. The *primary standard levels* serve to protect public health, whereas the generally more stringent *secondary standard levels* are meant to protect property and general well-being or comfort. Individual states have been directed to develop *state implementation plans (SIPs),* which will indicate how air quality will be brought up to the NAAQS levels; state air standards can be stricter than the NAAQS levels.

Pollutant Standards Index

In order to be able to provide up-to-date and accurate information about air quality, and to evaluate national air quality trends, a *pollutant standards index (PSI)* has been developed by the EPA. The PSI also serves as a means to provide the general public with information about daily levels of air quality, without having to report exact concentrations of each specific pollutant. This helps to protect public health, particularly for the elderly and other people who should modify their activities (e.g., stay indoors) when the PSI indicates high pollution levels.

The PSI is reported as a single number between 0 and 500. It is computed by a special formula that accounts for the ambient concentrations as well as the potential health effects of five of the "criteria air pollutants": sulfur dioxide, nitrogen dioxide, carbon monoxide, ozone, and total suspended particulates. In effect, the PSI translates ambient concentration data into a single value, and presents information regarding urban air quality across the

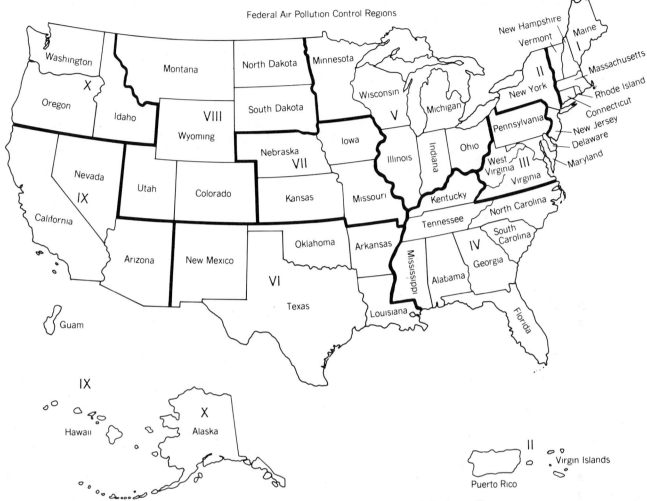

FIGURE 12.16 These areas of the United States are further divided into a total of 247 air quality control regions (AQCR). Ambient pollutant concentrations in an AQCR must be brought into compliance with NAAQS through controls imposed within that AQCR. (Council on Environmental Quality)

nation on a consistent basis. The PSI intervals and corresponding air quality descriptors are summarized as follows:

PSI VALUE	AIR QUALITY DESCRIPTOR
0 to 49	Good
50 to 99	Moderate
100 to 199	Unhealthful
200 to 299	Very unhealthful
300 plus	Hazardous

A PSI value above 100 indicates that at least one of the criteria air pollutants has exceeded its NAAQS level, and people with respiratory or heart ailments should reduce outdoor activity. At PSI values above 200, even healthy people may experience discomfort and other symptoms related to the effects of air pollution.

When the PSI exceeds 300, air quality is generally so poor that death rates for the ill and elderly are likely to increase. Under these hazardous conditions, it is recommended that all persons should minimize physical exertion and avoid traffic or any outdoor activity. Computation of the PSI is based on data from National Air Monitoring Sites (NAMS), which are part of a large standardized

TABLE 12.1 NATIONAL AMBIENT AIR QUALITY STANDARDS

Pollutant	Averaging Time	Primary Standard	Secondary Standard
Particulate matter	Annual (geometric mean)	75 μg/m^3	60 μg/m^3
	24 hr	260 μg/m^3	150 μg/m^3
Sulfur oxides	Annual (arithmetic mean)	80 μg/m^3 (0.03 ppm)	60 μg/m^3 (0.02 ppm)
	24 hr	365 μg/m^3 (0.14 ppm)	260 μg/m^3 (0.1 ppm)
	3 hr	—	1300 μg/m^3 (0.5 ppm)
Carbon monoxides	8 hr	10 mg/m^3 (9 ppm)	10 mg/m^3 (9 ppm)
	1 hr	40 mg/m^3 (35 ppm)	40 mg/m^3 (35 ppm)
Nitrogen dioxide	Annual (arithmetic mean)	100 μg/m^3 (0.05 ppm)	100 μg/m^3 (0.05 ppm)
Ozone	1 hr	240 μg/m^3 (0.12 ppm)	240 μg/m^3 (0.12 ppm)
Hydrocarbons (nonmethane)	3 hr (6 to 9 A.M.)	160 μg/m^3 (0.24 ppm)	160 μg/m^3 (0.24 ppm)
Lead	3 months	1.5 μg/m^3	1.5 μg/m^3

Source: EPA.

network for air quality surveillance throughout the country.

In addition to the PSI, the EPA has published "episodes criteria" for three different air pollution levels or stages. These three stages are *Alert, Warning,* and *Emergency.* They indicate or call for certain control actions that state or local government officials can implement to protect the public. Such actions include the mandatory reduction of manufacturing activities, traffic, and solid waste incineration. Although there is not an exact relationship among the three stages of air pollution and PSI values, the Alert stage generally corresponds to PSI values greater than 200, the Warning stage corresponds to PSI values greater than 300, and the Emergency stage corresponds to PSI values over 400.

Hazardous Air Pollutants

For the most part, air pollution control efforts have focused on the major air pollutants, which we have been discussing in the preceding sections. In recent years, more attention is being given to air pollutants that are particularly hazardous, such as asbestos, mercury, beryllium, vinyl chloride, and others. The EPA has issued emission standards for several sources of these pollutants, as indicated in Table 12.2. Other air pollutants are under consideration for designation as hazardous; those substances suspected of being carcinogenic to humans will be so designated and regulated.

Indoor Air Quality

Up until recently, little attention has been given to air pollutants in the home; discussion of "air quality" has primarily focused on outdoor or workplace conditions, or on specific source emissions. But indoor air quality and the nature of residential pollutants differ from those of major concern in the outdoor air or workplace. Since people spend much of their time at home, indoor air quality can have a significant effect on health.

TABLE 12.2 SOURCES OF HAZARDOUS AIR POLLUTANTS

Pollutant	Source
Asbestos	Asbestos mills, road surfacing with asbestos tailings, manufacturers of asbestos-containing products (fireproofing, etc.), demolition of old buildings, spray insulation
Beryllium	Extraction plants, ceramic manufacturers, foundries, incinerators, rocket motor manufacturing operations
Mercury	Ore processing, chlor-alkalai manufacturing, sludge dryers and incinerators
Vinyl chloride	Ethylene dichloride manufacturers, vinyl chloride manufacturers, polyvinyl chloride manufacturers

Source: EPA.

A factor that makes indoor air pollution a more significant problem now than it was in the past is the recent trend toward complete insulation of buildings and homes for energy conservation. Recently, the ventilation of the average home due to leakage has been reduced from about one air change per hour to about 0.25 changes per hour. This decrease in ventilation means that people are being exposed to staler and possibly more polluted air.

Two of the major ambient pollutants, carbon monoxide and nitrogen dioxide, are also of concern indoors. The highest levels of these pollutants seem to accumulate in the kitchens of well-insulated houses, often exceeding the outdoor concentrations. Two different gaseous substances that particularly affect indoor air quality are radon and formaldehyde. Radon gas is produced continuously from the radioactive decay of radium, which is a natural trace element in most rock and soil. Building materials, therefore, including concrete and brick, as well as the earth foundation itself, are a source of radon gas. Indoor levels of this gas, a suspected carcinogen, are usually higher than outdoor levels.

Formaldehyde is found in foam insulation, carpets, drapes, and other household items. It is also used in plywood and particleboard bonding agents, and is a by-product of natural gas combustion. In addition to causing nausea, respiratory irritation, and other effects, it is suspected of causing more serious long-term problems, such as cancer. In some cases, household formaldehyde levels have been measured in excess of 3 ppm, which is the OSHA standard for the industrial workplace.

Suspended particulates are also an indoor air pollution problem of special concern. Even in enclosed environments where there is no cigarette smoke, the level of respirable suspended particulates (less than 15 μm in size) may be as much as 40 μg/m^3. In environments with cigarette smoke, the levels may be as high as 700 μg/m^3; this exceeds the allowable NAAQS values for particulates by a large margin. It is of special concern to researchers doing epidemiological studies that link respiratory disease to air pollution, and particularly to cigarette smoke.

The OSHA regulations pertaining to air quality focus on the workplace and not on private residences or other nonoccupational environments. It is likely that in the future, EPA authority under the Clean Air Act will extend to indoor air quality, in order to further protect public health. There are methods for controlling indoor air pollution, including the use of activated carbon adsorbers and fiber filters in heating, cooling, and ventilating systems.

POLLUTION CONTROL STRATEGIES

There are several approaches or strategies for air pollution control. The most effective control, of course, would be to prevent the pollution from occurring in the first place. Complete *source shutdown* would accomplish this, but shutdown is only practical under emergency conditions, and even then it causes economic loss. Nevertheless, state public health officials can force the shutdown of industry, and the curtailment of traffic, when an air pollution episode is imminent. But source shutdown can only offer, at best, a very temporary solution to the air pollution problem.

Dilution seems to offer another alternative for air pollution control. Tall smokestacks can be used to reduce the ambient ground-level concentration of pollutants near the source. Taller stacks or chimneys take the gases higher into the atmosphere for greater dilution effects of dispersion and mixing.

Dilution may be effective to some degree, but the use of tall stacks and dispersion for air pollution control has a serious limitation: generally, what goes up must come down. The problem of pollutant transport and acid rain, which was discussed in Section 12.3, demonstrates this weakness. In addition, tall stacks, which at some power plants are as high as 300 m (1000 ft), are very unattractive, and may be hazardous to aircraft. Because of these limitations, the most recent provisions of the Clean Air Act prohibit the use of tall stacks in most cases; smokestacks are supposed to be a last resort for air pollution control.

Another option for air pollution control is to *relocate the source* in order to minimize adverse environmental impacts in a particular locality. Community "air zoning" may be included in the local planning process to require power plants or industrial facilities to be located where fewer people will be affected by the pollutants. The location of these zones can be established on the basis of prevailing wind and weather conditions. But this also has its limitations; although local air quality may be protected, the pollutants can still be "air-mailed" to a neighboring community.

A more positive approach to air pollution control is to make a *fuel substitution and/or process change.* For example, making more use of solar, hydroelectric, and geothermal energy would eliminate some of the pollution caused by fossil fuel combustion. (Nuclear power would do the same, but other problems related to radioactive waste disposal and safety remain to be solved.) Using natural low-sulfur coal and oil would decrease SO_2 emissions from fossil fuel power generating stations. And the technology is available for treating and *desulfurizing* "dirty" fossil fuels prior to combustion, but it is expensive. Completely changing certain industrial manufacturing processes can serve to reduce air pollution; one example of this is to use oxygen or electric furnaces instead of open-hearth furnaces in the steel industry.

Even if the fuel or process is not changed, the *use of correct operation and maintenance practices* can be very important for minimizing air pollution, and should not be overlooked as an effective control strategy. For example, if a power plant operator allows too much excess air into the boiler furnace, fly-ash emissions will increase. And adding too much sulfur at a sulfuric acid manufacturing plant, without providing enough air, can cause excessive sulfur dioxide emissions. Even the failure to properly lubricate a fan motor at an incinerator could lead to unnecessary pollution.

POLLUTION CONTROL EQUIPMENT

The most realistic option for air quality protection is to reduce the amount of pollution at its source. When fuel substitutions or process changes are not possible, or they are simply insufficient to accomplish the goal, then some type of air cleaning equipment must be installed.

There are several types of air pollution control devices that can collect or trap the pollutants before they are emitted into the atmosphere. Some of these devices serve to control only suspended particulates, and others control only gaseous pollutants. The design or selection of a particular type of air cleaning equipment depends on the physical properties of the pollutant to be removed, as well as on the temperature, corrosivity, and other characteristics of the carrier or bulk gas stream.

Control of Particulates

Most ambient particulate air pollution comes from stationary sources, particularly from power plants, industrial processes, and incinerators. The primary factors that determine which type of equipment is best for a specific application are the average particle size and particle density.

The simplest device is the *settling chamber,* which is shown schematically in Figure 12.17. It is basically an enlarged section or compartment in the flue in which the velocity of the carrier gas is reduced. When the air stream velocity is slowed down sufficiently, the coarse particulates (those more than about 40 μm in size) can settle out under

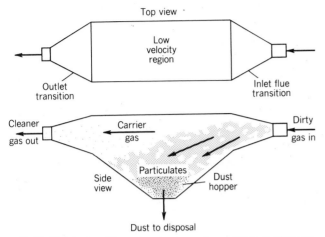

FIGURE 12.17 A settling chamber or enlarged flue section provides a simple way to remove settleable particulates at the source.

the force of gravity. A settling chamber of this type usually serves as a "precleaner," to prevent clogging of the more efficient small-particle collectors that must follow it. Sometimes baffles are added inside the chamber, to increase its efficiency.

A device called a *cyclone* is also used for reducing TSP emissions. Instead of relying only on velocity reduction and gravity settling, the particles are subjected to centrifugal force and friction which separates them from the carrier gas. In a cyclone, which is illustrated in Figure 12.18, the carrier gas enters from a tangential direction at the outer wall of the device and forms a vortex as it swirls around inside a cylindrical and conical shell.

The particulates are forced against the wall by centrifugal force, where they are slowed down by friction, and then they slide down into a dust hopper. The cleaned gas stream then swirls upwards in a narrower spiral, through an inner cylinder and toward the outlet. A cyclone is most efficient when serving to remove coarse particulates, but it is also somewhat effective in removing smaller particles. A typical "cut diameter" for a cyclone is about 15 μm; the cut diameter is that for which 50 percent of the particles are collected, and 50 percent are not. Unlike the settling chamber, the removal efficiency for small particles in a cyclone increases as the velocity increases.

A device called a *wet scrubber* serves to trap sus-

pended particles in a liquid aerosol. In effect, a wet scrubber "washes" the particulates out of the gas stream, as they collide and are intercepted by the countless number of tiny droplets in the aerosol. There are several types of wet scrubbers, the simplest of which is the *spray tower*. In a spray tower, the upward flowing carrier gas stream is washed by water sprayed downward from a series of nozzles. Spray towers are most efficient for removing particles that are 10 μm or greater in size. A schematic diagram of a spray tower scrubber is shown in Figure 12.19.

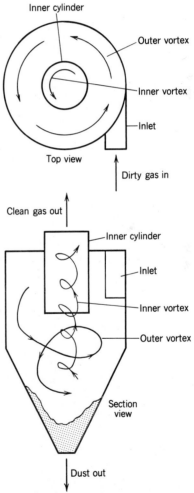

FIGURE 12.18 In a cyclone collector, particulates are spun out toward the outer wall by centrifugal force. They are slowed down by friction and settle to the bottom; clean air flows upward and out the top.

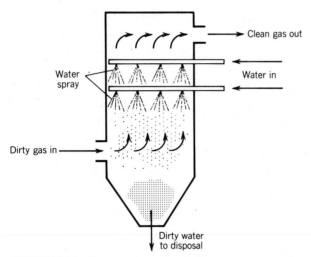

FIGURE 12.19   The spray tower is a type of wet scrubber that removes suspended particulates from the carrier gas.

Another type of wet collector is the so-called *wet centrifugal scrubber*. This device serves to increase the relative velocity between the water droplets and the dirty gas stream, thereby increasing particulate removal efficiency; particles as small as 2 μm can be removed. It should be recognized that in any type of scrubber, the spent wash water that is generated must be collected and treated before discharge into a river or lake. The wet scrubber is a good example of the close interrelationship between air and water pollution control problems, as was illustrated in Figure 1.1.

A relatively large piece of equipment used to remove very tiny particulates from the carrier gas, especially at power generating stations, is the *electrostatic precipitator*. This device applies a high voltage in order to charge the particulates electrically; they can then be attracted in an electric field toward collecting surfaces, on which they become trapped.

The collecting surfaces usually comprise a series of large rectangular metal plates, which are suspended vertically in a parallel arrangement within a boxlike structure. The charged particles that adhere to the plates are removed when the plates are mechanically vibrated or rapped. The falling particles are collected at the bottom of the unit, and disposed of, usually in a landfill. A cut-away view of a

typical electrostatic precipitator is illustrated in Figure 12.20.

An electrostatic precipitator can remove particles as small as 1 μm, with an efficiency exceeding 99 percent. If the TSP level in the carrier gas is very high, a settling chamber or cyclone is generally installed in front of the precipitator, for precleaning of the gas. The removal efficiency of a precipitator is very sensitive to the velocity and distribution of the gas stream flowing across the collecting plates. The flow must be slow and uniform, that is, the same across all of the plates, from top to bottom.

The installation of precipitator units at a power generating facility or at an industrial plant represents a very large dollar investment for the owner. Removal efficiencies must be guaranteed by the equipment manufacturer, so as to ensure the owner that federal and state air quality or source performance standards will not be violated. Because of this, the manufacturers usually build and test exact scale models of each precipitator unit in a laboratory before they are actually constructed in the field. In this way, the required shape or configuration of ductwork and baffles can be predetermined, so as to provide uniform flow velocities across the

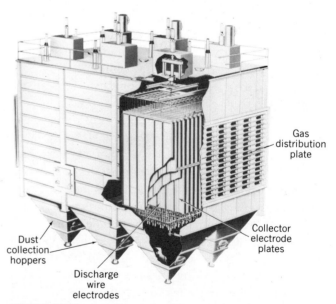

FIGURE 12.20   A cutaway view of an electrostatic precipitator. (Western Precipitation Division, Joy Manufacturing Co.)

FIGURE 12.21 An exact 1/16 scale model of an electrostatic precipitator installation, including the inlet flues. These models are used for testing and evaluation of flow patterns before the actual unit is built in the field. (Research–Cottrell Inc., Air Pollution Control Division)

plates. A typical 1/16 scale model, made of clear Plexiglas, is illustrated in Figure 12.21.

One of the most efficient devices for removing suspended particulates from a carrier gas stream is the *baghouse filter*. It can remove particles as small as 0.01 μm. A typical installation comprises a series of long and narrow filter bags, which are suspended upside down in a large enclosure. It is illustrated schematically in Figure 12.22.

Dirty air is blown through the bottom of the enclosure by fans. The particulates are trapped inside the bags, whereas the clean air passes through the filter fabric and exits at the top of the "baghouse." This equipment is generally preceded by settling chambers or cyclones to reduce the particulate load and the required cleaning frequency for the filter bags. Several compartments of bag filters are often used at a single installation so that any individual

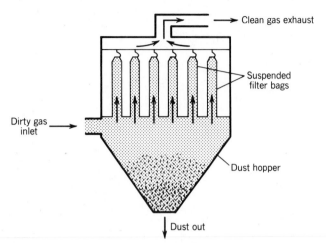

FIGURE 12.22 A section view of a baghouse filter. The filters may be cleaned by mechanical vibrations or by blowing clean air back through the unit.

compartment can be cleaned while the others remain in service.

In cleaning, bag filters are mechanically shaken or vibrated to collect the particulates so that they can be removed for disposal. The air cleaning efficiency of the baghouse filter approaches 100 percent for particles 1 μm or larger in size, and particles as small as 0.01 μm can also be removed to a significant extent. But baghouse filters cause relatively high pressure losses, and they are expensive to maintain. Also, the flue gas or carrier air stream generally must be cooled before passing through the unit; cooling coils needed for this purpose add to the expense.

Control of Gases

Gaseous air pollutants can be controlled using the techniques of absorption or adsorption, similar to the methods used for air sampling, but on a much larger scale. A third method for controlling gaseous pollutants involves combustion. The specific control method selected depends on the nature and properties of the pollutant to be controlled.

A spray tower scrubber, similar to the type used for particulate control, can also be used to control gaseous emissions by absorption. Instead of water, a reactive liquid absorbent may be used to capture the pollutant molecules by chemical reaction. In addition to spray towers, other types of absorbers include packed towers, plate towers, liquid jet scrubbers, or agitated tanks. Each of these devices has a different physical design, but they all operate on the same basic principle; the gaseous pollutant to be removed is put into contact with a liquid, which absorbs it by solution or chemical reaction, and separates it from the flue gas.

Sulfur dioxide, for example, can be removed from the flue gases at coal-fired power plants by scrubbing with a solution of lime, CaO. The SO_2 reacts with the lime, forming an insoluble precipitate of calcium sulfite, $CaSO_3$. The concentrated slurry of insoluble solids that form in this reaction is called *flue gas desulfurization (FGD) sludge*. Application of this process would help to reduce ambient SO_2 levels, and could mitigate the acid rain problem. But in addition to the great expense (which would be passed on directly to the consumer as higher rates for electricity), there remains a major disposal problem for the millions of tons of FGD sludge that would be generated each year.

Gas *ad*sorption, as contrasted with *ab*sorption, is a surface phenomenon: The gas molecules are *sorbed* (attracted and held) on the surface of a solid. Activated carbon is the most common adsorbent material; it is very porous and has an extremely high surface-area-to-volume ratio. A schematic diagram of an activated carbon adsorption unit is shown in Figure 12.23.

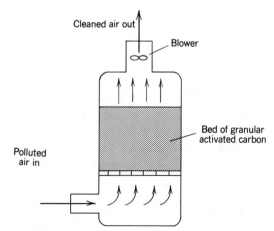

FIGURE 12.23 Activated carbon can be used to adsorb certain gaseous air pollutants.

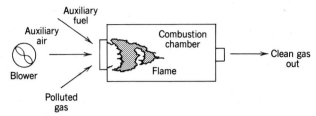

FIGURE 12.24 Direct combustion can be used to control emissions of combustible gases and odors. Natural gas is used as an auxiliary fuel to initiate and maintain a stable flame.

The chemical process of combustion (rapid oxidation) may be used to convert gaseous hydrocarbon pollutants to carbon dioxide and water. The hydrocarbons, usually the product of incomplete combustion, are in effect reburned in a control device, to form the products of complete combustion. The incinerator or afterburner must be designed to provide sufficient oxygen, turbulence, burning time, and temperature, so that complete combustion will occur. A schematic diagram of a combustion device is shown in Figure 12.24.

Certain substances, such as platinum, may act in a manner that will assist the combustion or oxidation reaction. Such substances are called catalysts. A device called a *catalytic converter* is installed in the exhaust system of an automobile. It allows complete oxidation of the combustible gases in the exhaust to occur at relatively low temperatures; in effect, this is a form of "flameless combustion."

The control of gaseous emissions from mobile sources, primarily the automobile, cannot be accomplished by the catalytic converter alone. Certain modifications to the internal combustion engine are needed to first minimize the emission of hydrocarbons, as well as of carbon dioxide and nitrogen dioxide. The use of positive crankcase ventilation (PVC) systems, for example, serves to recycle gases that slip by the piston rings back to the engine intake manifold. A small PVC check valve prevents the buildup of excess pressure in the engine.

Engine operating parameters can also be controlled in order to reduce certain emissions. For example, increasing the air-to-gas ratio of the fuel and air mixture helps combustion and reduces carbon monoxide and hydrocarbon emissions. But the higher combustion temperatures that then occur cause an increase in the emission rate of nitrogen dioxide. Other operating conditions of the engine can be modified, but then there may be a loss of power and fuel economy. Obviously, controlling emissions from the internal combustion engine is a complex problem.

Changing the fuels used in mobile sources can be effective in controlling air pollution. The use of lead-free gasoline, for instance, protects public health from airborne lead; also, lead-free gasoline must be used to prevent damage to catalytic converters. Ultimately, the use of electric vehicles, or other unconventional engines, could be necessary to eliminate serious air pollution problems from mobile sources.

12.6 NOISE POLLUTION

Noise is one of the most undesirable by-products of our modern "mechanized" life-style. It may not seem as insidious as toxic chemical contamination. But it is a pollution problem that affects human health and well-being, and that can contribute to a general deterioration of environmental quality.

The measurement and control of noise in the workplace is the responsibility of acoustical engineers and industrial hygienists. But noise can be a community problem as well as an occupational hazard. Civil engineers and technicians are becoming much more concerned with noise pollution control, particularly in the areas of construction management, land-use planning, and highway design.

More and more communities are imposing noise controls on construction activities, and the environmental impact assessment reports for transportation or land development projects usually must include a section on noise. In order to understand noise regulations, and to plan for effective community noise control, it is necessary to have a basic understanding of sound and its measurement. In this section, an overview of noise pollution, its effects, and its control is presented.

SOUND WAVES

Noise is not a substance that can accumulate in the environment like most other pollutants. Nevertheless, it is a type of waste product; in effect, it is *waste energy,* in the form of sound waves. Sound is caused by mechanical vibrations that transmit pulses of pressure variations or "waves" through the air (or other transmitting media, including liquids or solids.)

Sound pressure waves may be visualized as alternating high and low air density regions, as illustrated in Figure 12.25. The regions of high and low densities or pressures may be represented schematically as the peaks and valleys of a trigonometric "sine curve." The distance between the peaks (or valleys) is called the *wave length,* and the number of waves that will pass a fixed point in one second is called the *frequency* of the wave. The height of the peaks, called the *amplitude* of the wave, represents the pressure intensity and is related to the "loudness" of the perceived sound.

A single wavelength is also called a *cycle,* and the frequency is expressed in terms of *cycles per second (cps).* The term *Hertz,* abbreviated *Hz,* is also used to represent frequency, where 1 Hz = 1 cps. For example, a sound with a frequency of 1000 Hz is one in which the pressure waves pass a given point at a rate of 1000 cps. Frequency should not be confused with the speed of sound, which is constant in a given transmission medium. In air at standard temperature and pressure, sound travels at a constant speed of about 340 m/s (1100 ft/s). The product of the wavelength and the frequency is equal to the speed of the sound wave, as follows:

$$\text{wavelength} \times \text{frequency} = \text{speed}$$
$$\frac{\text{meters}}{\text{cycle}} \times \frac{\text{cycles}}{\text{second}} = \frac{\text{meters}}{\text{second}}$$

Since the speed is constant, there is an inverse relationship between the wavelength and the frequency. In other words, the higher the frequency is, the shorter the wavelength is, and vice versa. This is illustrated in Figure 12.26.

The *pitch* of a sound depends on the frequency of the wave that produces it. High-pitched sounds (e.g., a shrill whistle) have high frequencies. Low-pitched sounds (e.g., a fog horn) have low frequencies. The human ear can detect sounds in the frequency range of about 20 to 20 000 Hz, but for most people, hearing is best in the range of 200 to 10 000 Hz. In normal conversation, the human voice covers a range of 300 to 4000 Hz.

The sine curves or sound waves shown in Figure 12.26 represent pure tones. Noise is typically much more complex than a single pure tone; it usually comprises a combination of many different pressure waves, each with a different frequency and amplitude. The resulting sound wave does not exhibit the neat pattern of a single sine curve, but appears typically as shown in Figure 12.27. A random pattern

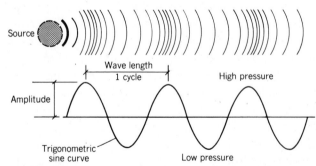

FIGURE 12.25 The alternating high and low pressure regions in a sound wave can be represented by a trigonometric sine curve. The peaks represent high pressure regions and the valleys represent low pressure regions.

(a) Low frequency, long wave, low pitch sound

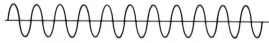

(b) High frequency, short wave, high pitch sound

FIGURE 12.26 The pitch of a sound depends on its frequency. Since a sound wave travels at a constant velocity, the wave length must increase as the frequency decreases.

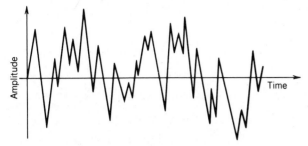

FIGURE 12.27 A sound wave pattern or signal from a typical "noise" is characterized by a random distribution of amplitudes.

or distribution of amplitude and wavelength over time is characteristic of most unwanted sounds or noises.

The average pressure of a cyclical function, such as a sound wave, can be characterized by a mathematical term called the RMS or root-mean-square value. In the following discussion, reference to sound pressure refers to average or RMS pressure.

NOISE MEASUREMENT

As previously discussed, sound is characterized by cyclical changes of air density or pressure, above and below the average atmospheric pressure. Compared to atmospheric pressure, sound pressures are very small, and are expressed in terms of the *microbar,* or μbar. A microbar is approximately one one-millionth (1×10^{-6}) of standard atmospheric pressure at sea level. By definition, 1 bar = 100 kPa (14.7 psi). Therefore, a pressure of 1 μbar = 0.1 Pa (0.000 014 7 psi).

The human ear is a remarkably sensitive organ. Although most sound pressures are extremely small, the average person can detect a sound with an amplitude or pressure intensity as small as 0.0002 μbar. A pressure of 0.0002 μbar is considered to be the lowest audible sound for humans, and is used as a base or reference level in noise measurement. The highest sound pressure that can be perceived without causing pain is about 1000 μbar. This range of pressure, from 0.0002 up to 1000 μbar, covers a very wide spectrum; it is equiv-

alent and comparable to a range of five million to one. $(1000 \div 0.0002 = 5 \times 10^{6})$. Actually, the human ear can perceive sound pressures as high as 10 000 μbar before immediate physical damage occurs to the inner or outer ear organs.

The Decibel Scale

Measuring sound with pressure units that can vary over such a wide range of microbar values is awkward and inconvenient. Another disadvantage of measuring sound in terms of microbars is the fact that the ear responds nonlinearly with respect to pressure. In other words, our perception of "loudness" is not a simple and direct function of sound pressure; a doubling of pressure intensity is not necessarily perceived as a doubling of the loudness of the sound. There are other factors involved in this phenomenon, as will be discussed shortly.

In order to avoid the disadvantages of using pressure directly for sound measurement, a logarithmic relationship called the *decibel scale* is used. The units, called *decibels* (dB), do not represent an absolute physical quantity, such as sound pressure. A noise measurement expressed in terms of decibels represents a *sound pressure level (SPL)*. An SPL is determined with reference to the lowest audible sound pressure of 0.0002 μbar. Mathematically, it is defined as:

$$SPL = 20 \times \log(P/P_0) \qquad (12\text{-}1)$$

where P_0 = 0.0002 μbar, and P is the actual sound pressure

Based on this definition of "sound pressure level," it should be noted that an SPL = 0 dB does not represent the complete absence of sound. Instead, it represents the reference level or least audible sound for most people. This can be seen by substituting P = 0.0002 in Equation 12-1, and computing SPL = 20 × log(0.0002/0.0002) = 20 × log(1) = 20 × 0 = 0. In fact, some people with particularly acute powers of hearing can detect sounds with negative SPL values.

EXAMPLE 12.3

A loud ambulance siren causes a sound pressure of 200 μbar. What is the so-called sound pressure level, or SPL of the siren?

Solution:

$$SPL = 20 \times \log(200/0.0002) = 20 \times \log(10^6)$$
$$= 20 \times 6 = 120 \text{ dB}$$

The threshold of pain for humans, about 1000 μbar, is equivalent to about 134 dB (try the computation yourself, using Equation 12-1). It can be seen, then, that the decibel scale serves to reduce the scope or spread of ordinary sound measurement to a reasonably convenient range of 0 to about 140 dB. To put the decibel scale in perspective with regard to our perception of common sounds and environmental conditions, a list of typical noises and their corresponding SPL values is shown in Figure 12.28.

Combining Noises

In many instances, it is necessary to predict what the combined noise level will be when two or more noise sources act at the same time. For example, it may be important to estimate the noise expected at the boundary of a construction site due to the combined operation of trucks, dozers, pavers, and other machinery. It is important, first of all, to realize that it would be incorrect to add the SPL of each individual noise source, by simple arithmetic.

SPL values cannot be added directly because of the logarithmic nature of the decibel scale. For practical purposes, it is convenient to remember just that the combination of two sounds with equal SPL values always results in only a 3-dB increase over the SPL of one source alone.

For example, if a single siren has an SPL of 120 dB, then two identical sirens sounding simultaneously would result in a noise with an SPL of 123 dB (but not 240 dB!). Likewise, if one backhoe

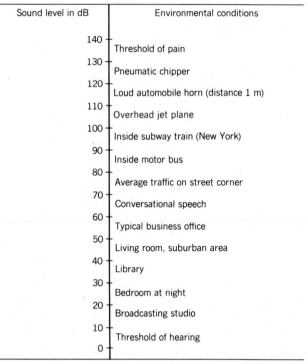

FIGURE 12.28 The decibel, dB, scale is used to measure noise levels. (Bruel & Kjaer Instruments, Inc.)

causes 85 dB of noise, then two backhoes would cause 88 dB of noise. It does not matter what the original SPL value is; the combined SPL will be only 3 dB more than the SPL of the single source. This follows from the mathematical nature of the logarithmic function.

On the other hand, when two sounds that differ by more than 15 dB in SPL are combined, the contribution of the weaker sound is not noticeable to even the most sensitive ear or instrument. For example, if a 95-dB rock drill is operating simultaneously with an 80-dB loader at a construction site, the combined SPL will still be measured as 95 dB. The weaker sound is, in effect, "drowned-out" by the louder sound. This phenomenon is called *masking* of one sound by another.

Sometimes the SPL values of noise sources that differ between 0 and 15 dB must be combined. Instead of using logarithmic computations to add the values, it is more convenient to use a chart, such as the one illustrated in Figure 12.29. The chart is entered first with the numerical difference between

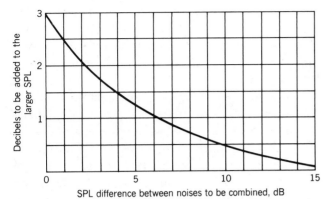

FIGURE 12.29 A chart like this may be used to simplify the addition of decibel values when combining two or more sound levels. The chart is entered on the horizontal axis with the numerical difference between the sound levels to be combined, and the corresponding number of dB to be added to the larger of the two sound levels is read on the vertical axis.

two noise levels to be added. The corresponding number of decibels to be added to the higher SPL value is read from the chart. The following example will demonstrate its use.

EXAMPLE 12.4

Four identical dozers are available for excavation of soil at a construction site. Each dozer has an SPL of 90 dB when operating alone. What is the SPL when the four dozers are operating at the same time? (For simplification, we have been ignoring the effect of distance from the noise source in our discussion of SPL. For this problem, we will assume the dozers are operating together in a confined area.)

Solution:

First, consider what happens when only two of the machines are operating. The numerical difference between the two SPL values is 90 − 90 = 0. Entering the curve in Figure 12.29 with a difference of 0, we read a corresponding 3 dB value on the vertical axis. Therefore, the two dozers operating simultaneously will generate an SPL of 90 + 3 =

93 dB. This is as expected from the previously discussed rule for combining identical SPL values.

With a third dozer operating, we must add another 90 dB to the previous level of 93 dB for two dozers. The difference between the two SPL values is now 93 − 90 = 3 dB. Entering the chart with a 3 dB difference, we find that we must add an additional 1.7 dB to the 93 dB, to obtain the combined SPL. Thus 93 + 1.7 = 94.7 dB from three dozers operating simultaneously. Remember that the SPL increment is always added to the larger of the two SPL values being combined.

Finally, with a fourth dozer, we must combine the 94.7 dB with 90 dB. The difference is 4.7 dB, and the required increment from the chart is 1.3 dB. This results in 94.7 + 1.3 = 96 dB being generated by the four machines. This can be checked by considering that combining two pairs of dozers, at 93 dB a pair, we also arrive at 93 + 3 = 96 dB.

Noise Meters

Many instruments are available for measuring noise. The basic components of a noise meter include a microphone, an amplifier, and a read-out device or scale. Most noise surveys can be conducted with a battery operated, hand-held sound level meter, as illustrated in Figure 12.30. If noise measurements are to be made at one location over a relatively long period of time, such as for a traffic noise survey, the meter can be mounted on a tripod and a recording device and a frequency analyzer can be added to the system. The selection of noise measuring instrumentation depends on the type of noise, the environmental conditions, and the purpose of the survey.

The dB(A) Scale

The human ear responds to sound in a complex way. There is really no simple relationship between the physical measurement of a sound pressure level and an individual's perception regarding the "loudness" of the sound. The relative loudness of sounds depends to an extent on the individual's opinion.

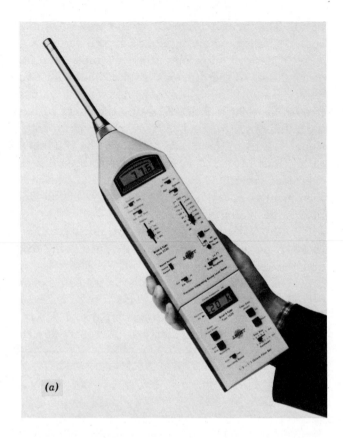

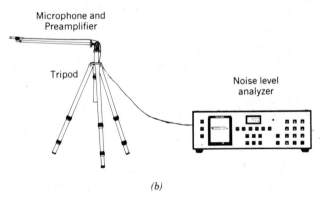

FIGURE 12.30 *(a)* A typical hand-held sound level meter, and *(b)* a typical measuring arrangement used for traffic noise surveys. (Bruel & Kjaer Instruments, Inc.)

One of the factors that affects the apparent loudness of a sound is its frequency. Generally, high pitched sounds are judged by the average person to be louder than low pitched sounds of exactly the same intensity or SPL. For example, a sound of 70 dB at 100 Hz is usually perceived as being quieter than a sound of the same 70 dB, but at a frequency of 1000 Hz. Also, a sound with a frequency of 100 Hz must have an SPL of about 75 dB for it to be judged equally as loud as a 1000 Hz sound with an SPL of 70 dB.

Noise meters are typically supplied with built-in frequency weighting networks, which are labeled either A, B, C, or D. Most regulations and standards for community noise measurement require the use of the so-called A-weighted scale. The A-weighted network automatically adjusts the sound pressure measurements, so as to compensate for the average subjective response of the human ear to ordinary sounds. The other scales are used for more specialized noise measurement surveys. For example, the D-scale is intended primarily to measure aircraft noise at airports.

Noise measurements made using one of the frequency weighting networks are called *sound levels,* instead of sound pressure levels. In order to distinguish them from SPLs, which are expressed in dB, the weighting network used is indicated by adding the corresponding letter to the dB symbol. For example, results of noise measurements made using a meter set on the A-weighted scale are reported in terms of dB(A). The sound levels are read directly from the meter.

EFFECTS OF NOISE

Noise may be simply defined as undesirable and unwanted sound. But the perception of noise is subjective, and may depend to a degree on a person's opinion. What is considered "noise" to one person may literally be "music" to another. But despite differences of opinion, there are some definite harmful effects caused by exposure to high sound levels, whether or not they are called noise. These effects may be physical or emotional, and they can range in severity from being merely annoying to being extremely painful and hazardous.

The most direct harmful effect of noise is physical damage to the ear and the temporary or permanent hearing loss that results from that damage. Temporary hearing loss, called *temporary threshold shift (TTS),* refers to a reduced ability to hear weak sounds, with recovery occurring within one month. Permanent loss, or *noise-induced permanent threshold shift (NIPTS),* as it is called, represents a hearing loss from which there is no recovery.

Below a sound level of 80 dB(A), hearing loss does not usually occur at all. But temporary or TTS effects are noticed at sound levels between 80 and 130 dB(A). About 50 percent of people exposed to 95 dB(A) sound levels at work will develop NIPTS, and most people exposed to more than 105 dB(A) will experience permanent hearing loss to some degree. A level of 150 dB(A) can physically rupture the eardrum.

The degree of hearing loss depends on the duration as well as on the intensity of the noise. For example, about one hour of exposure to a 100 dB(A) noise could produce a TTS that may last for about one day. But in some instances, particularly in factories with noisy machinery, workers are subjected to high noise levels for many hours a day. Exposure to 95 dB(A) for eight hours a day, over a period of ten years may cause about 15 dB(A) of NIPTS.

In addition to hearing loss, excessive noise levels can cause harmful effects on the heart, by raising blood pressure and altering pulse rates. Noise can also cause emotional or psychological effects, such as irritability, anxiety, and stress. Even mental fatigue and lack of concentration are significant health effects of noise. In industry, it results in lowered worker efficiency and productivity, increased sick leave, and higher accident rates. And in the community, high noise levels interfere with sleep, as well as conversations.

In general, then, noise is more than a mere nuisance or annoyance. Excessive noise levels can have a direct effect on public health as well as on aesthetic sensibilities. Noise control is an important aspect of environmental technology.

NOISE CONTROL

There are basically four different ways in which noise levels can be controlled or reduced:

1. Protect the person exposed to the noise.
2. Intercept the noise by blocking its path.

3. Increase the distance from the source.
4. Reduce the sound intensity at the source.

Protection

Ear plugs and ear muffs can be effective in protecting individuals from excessive sound levels. Specially designed muffs can reduce the sound level by about 40 dB(A); these are very useful for protecting industrial workers who are exposed for long periods of time. But it is obviously unrealistic to consider the use of personal ear protectors as a practical solution to general community noise, for the average citizen. Even in occupational settings, they cannot be relied upon as the only solution. Many workers tend not to wear ear muffs, despite company requirements for their use.

Interception

Intercepting or blocking the path of the noise using sound absorbing or reflecting barriers offers a passive means of control. It requires no effort on the part of the potential recipient of the noise. Materials that efficiently absorb or reflect sound energy are called *acoustical materials.* They can be used effectively as covers for noisy machinery, thereby reducing noise levels in the industrial environment. Sometimes acoustical materials are used on the underside of highway bridges, to reduce the reflection of traffic noise from the road to nearby houses.

One of the major sources of community or neighborhood noise is automobile traffic. The path of traffic noise can be blocked by the use of barriers alongside of the highway. A barrier about 4 m (13 ft) high may reduce peak noise levels by about 15 dB(A). These high roadside barriers are usually made of concrete or masonry blocks.

The flat wall surfaces of the barriers must be designed and located properly, so as not to reflect and thereby magnify the noise. The use of an earth berm with dense landscaping may offer a more attractive solution as a barrier than either a concrete or masonry block wall. But this is generally more expensive, and requires the acquisition of additional space or a "right-of-way" along the route.

Most traffic noise comes from the movement of the vehicle tires on the pavement. But noise from the engine and exhaust, from the transmission, and from wind-resistance, add to the problem. Traffic volume and speed have a significant effect on the sound levels: Doubling the speed increases sound levels by about 9 dB(A), and doubling the traffic volume (number of vehicles per hour) increases the sound level by about 3 dB(A). Also, smooth traffic flow causes less noise than does a "stop-and-go" traffic pattern.

Proper highway planning and design are essential for controlling traffic noise. Establishing lower speed limits for highway segments that pass through residential areas, limiting traffic volume, and providing alternate routes for truck traffic, can all be effective noise control measures. The use of a "cut" or depressed section of roadway passing through urban areas is also very effective for reducing traffic noise in the nearby community. This is illustrated in Figure 12.31.

Distance

Land-use planning is one of the most effective methods for highway noise control. Sound levels drop significantly with increasing distance from the noise source. Municipal land-use ordinances pertaining to the location of roadways, and minimum distances between houses and the roads in a residential area, could make use of the attenuating effect of distance on sound levels. Poor land-use planning, on the other hand, can result in situations such as are common in California, where some

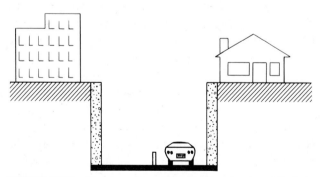

FIGURE 12.31 One way to reduce traffic noise in a city is to build sections of the roadway in a "cut" below street level.

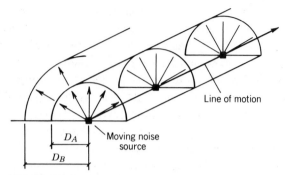

FIGURE 12.32 As a sound wave spreads out from a line source of noise, its sound level decreases at about the rate of 3 dB(A) for each doubling of the distance from the source.

houses have been built as close as 6 m (20 ft) to major highways. At that distance, the sound level inside a house with closed windows, from a passing truck, is about 70 dB(A). At that level, the noise is more than an annoyance; normal speech communication is almost impossible.

The attenuation of sound levels with increasing distance from the source occurs because the fixed amount of sound energy is spread and "diluted" over an increasing area. As the sound wave spreads out, its amplitude and intensity decrease. A highway is considered to be a *line source* of noise, from which sound is propagated in the form of a half-cylinder. This is illustrated in Figure 12.32. The relationship between sound level and distance from a line source can be written as:

$$SL_B = SL_A - 10 \times \log\left(\frac{D_B}{D_A}\right) \qquad (12\text{-}2)$$

where SL_A = sound level at distance D_A from the noise source
SL_B = sound level at distance D_B from the noise source

EXAMPLE 12.5

The sound level measured at a distance of 4 m from a busy highway is 85 dB(A). At what distance from the road would the sound level be reduced to 79 dB(A)?

Solution:

Applying Equation 12-2, we get

$$79 = 85 - 10 \times \log\left(\frac{D_B}{4}\right)$$

and

$$\log\left(\frac{D_B}{4}\right) = \frac{79 - 85}{-10}$$

or

$$\log\left(\frac{D_B}{4}\right) = 0.6$$

From the definition of the log function, we can write

$$\frac{D_B}{4} = 10^{0.6} = 4$$

and

$$D_B = 4 \times 4 = 16 \text{ m}$$

In general, sound level from a line source decreases 3 dB(A) for each doubling of the distance from the source. However, for a single *point source,* from which the sound waves spread out in the form of a sphere instead of a cylinder, sound levels decrease at twice that rate, with increasing distance. Use of a chart as shown in Figure 12.29 for adding sound levels, along with an appropriate distance formula, will allow the prediction of noise levels from transportation, construction, or other noise sources in a community or neighborhood.

The construction industry has long been a focus of complaints related to excessive noise. Construction activities typically require the use of heavy machines, each of which can be a significant source of noise. Relative ranges of noise levels for some common types of heavy construction equipment are r____d in Figure 12.33.

____ls at construction sites can be con-
____ construction management tech-
____lanning and scheduling a job
____ment are operating at
____noise levels. Con-
____e machinery
____gate the

noise problem. Temporary barriers may also be used, to physically block the noise. Construction managers must be aware of any state or local municipal ordinances that focus on noise control, and that may restrict the hours of construction activity.

Source Reduction

Perhaps the most direct approach for noise control is to reduce the sound produced by the source itself. Vehicles and machinery can be effectively muffled to reduce the noise. In industry, occupational noise can be further reduced by using special vibration isolation mountings for the machines and flexible coupling for interior pipelines.

Standards

Maximum permissible workplace noise levels in the United States have been established under the Occupational Health and Safety Act (OSHA). The highest allowable impact noise is 140 dB(A). A sound level of 90 dB(A) is the maximum allowed for an 8-hour exposure. Higher levels are permitted, but for shorter durations; for example, 100 dB(A) would be allowed for up to 2 hours, and 115 dB(A) for 15 minutes.

Under the Noise Control Act, the EPA is authorized to regulate community and construction noise. Other federal noise regulations have been established by the Department of Housing and Urban Development (HUD), and the Department of Transportation (DOT). The Federal Highway Administration has adopted the noise level/land-use relationships summarized in Table 12.3. These design noise levels apply to outdoor areas where noise control is of public benefit. They are designated as L_{10} dB(A) values; the abbreviation L_{10} refers to sound levels that are not to be exceeded more than 10 percent of the time.

Many state and city governments have also enacted noise control laws, and others are in the process of doing so. In addition to establishing maximum ambient noise levels, some ordinances set maximum allowable levels for new vehicles and machinery, thereby requiring design for quieter operation of modern equipment.

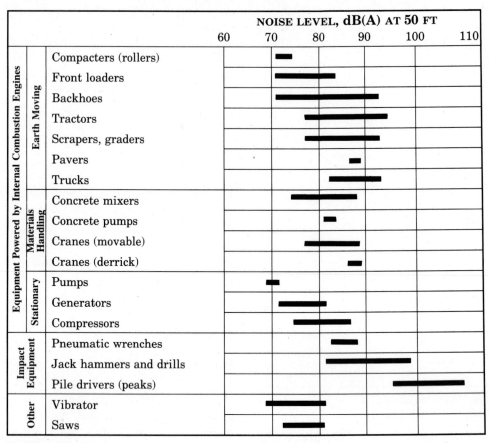

FIGURE 12.33 Relative ranges of noise levels for some common types of heavy construction machinery (EPA)

TABLE 12.3 LAND-USE AND OUTDOOR NOISE LEVELS

Land-use category	Land-use description	L_{10} Noise level
A	Land areas where serenity and quiet serve an important public need, such as amphitheaters or specially designated portions of certain public parks	60
B	Residential areas, motels, hotels, schools, libraries, hospitals, public parks, recreation areas, and meeting rooms	70
C	Developed land not included in category A or B	75

REVIEW QUESTIONS

1. Briefly describe the general composition of the atmosphere.

2. What is the *troposphere?* What is the *stratosphere?*

3. What is meant by *atmospheric stability,* and why is it of significance in air pollution? What is a *temperature inversion?*

4. List three sources and types of natural air pollutants.

5. Briefly discuss the major urban air pollutants: What are they and where do they come from?

6. What is the difference between a *primary* and a *secondary* air pollutant?

7. Briefly discuss the effects of air pollution on human health.

8. What is an air pollution *episode?*

9. What is the *greenhouse effect?*

10. Briefly discuss the problem called *acid rain.*

11. What are the basic purposes of *ambient* and *source* sampling?

12. List and briefly discuss three basic techniques for sampling and measuring particulate air pollution.

13. What is the *Ringlemann Chart* used for?

14. Briefly describe two basic techniques for sampling and measuring gaseous air pollutants.

15. Briefly explain the basic differences among TLV, SPS, and AAQS.

16. What is the *pollutant standards index,* and how is it used?

17. Why is indoor air pollution of particular concern?

18. Briefly discuss five strategies for air pollution control, other than using air cleaning equipment.

19. List and briefly describe the operation of five different devices used to control particulate air pollution from stationary sources.

20. Briefly describe two devices used to control gaseous pollutants.

21. What is FGD sludge?

22. What is the most significant mobile source of air pollution? Briefly describe some methods for its control.

23. Why is noise considered an environmental pollutant?

24. What is sound and how is it characterized? Do both the frequency and speed of a sound wave vary with wave length?

25. What does the term *Hertz* refer to? How is it related to pitch?

26. Why is noise measured in terms of decibels instead of microbars?

27. Would you expect the SPL from two identical noise sources to be twice the SPL from one source alone? Why?

28. What is meant by *masking* of sound?

29. Is it possible to hear a noise with an SPL of 0 dB? Why?

30. At what noise level would most people feel pain?

31. Briefly describe the reason for using units of dB(A).

32. Briefly discuss some of the harmful effects of noise pollution. What is the difference between *TTS* and *NIPTS?*

33. List and discuss four basic strategies for noise pollution control, particularly with regard to traffic and construction noise.

PRACTICE PROBLEMS

1. A 6-in.-diameter dustfall bucket has collected 125 mg of particulates in a 30-day period. Express the dustfall in terms of tons/mile^2/month.

2. The air flow through a hi-vol sampler was recorded as 54 ft^3/min at the beginning, and 48 ft^3/min at the end of a 24-hour sampling period. The filter weighed 9.80 g before, and 10.05 g after sample collection. What was the TSP level?

3. A machine causes a sound pressure of 160 μbar. What *sound pressure level* is this equivalent to?

4. What is the sound pressure, in microbars, caused by a sound with an SPL = 95 dB? By what factor is that pressure higher than the pressure that causes the lowest audible sound?

5. A dozer is being used to push-load a scraper at a construction site. Each machine has an SPL of 90 dB(A) operating alone. What is the SPL of the two machines operating together?

6. What is the combined SPL from three vibratory rollers operating at the same time on a construction site, if each has an SPL of 75 dB(A) when operating alone?

7. What is the combined SPL from a 75 dB(A) roller and a 95 dB(A) scraper at a construction site?

8. The sound level measured at a distance of 25 ft from a road is 76 dB(A). What is the sound level at 100 ft from the road?

SELECTED REFERENCES

1. Bethea, R., *Air Pollution Control Technology,* Van Nostrand Reinhold, New York, 1978.

2. *Environmental Quality,* 11th Annual Report of the Council on Environmental Quality, U.S. Government Printing Office, Washington, D.C., 1980.

3. Stern, A., *Air Pollution,* Vol. 1, 2, 3, Academic Press, New York, 1968.

4. Painter, D. E., *Air Pollution Technology,* Reston Publishing, Reston, Va. 1974.

5. *Air Resource Management Primer,* American Society of Civil Engineers, New York, 1973.

6. Turk, A., et al., *Environmental Science,* W. B. Saunders, Philadelphia, 1974.

7. Vesilind, P.A. and J. J. Peirce, *Environmental Pollution and Control,* 2nd ed., Ann Arbor Science, Ann Arbor, Mich., 1983.

8. Canter, L., *Environmental Impact Assessment,* McGraw–Hill, New York, 1977.

9. *California Tackles Highway Noise,* Civil Engineering-American Society of Civil Engineers, New York, Nov. 1973.

10. Krokosky, E. M., and C. L. Dym, *Noise Control and Civil Engineering,* Civil Engineering-American Society of Civil Engineers, New York, May 1974.

APPENDIX 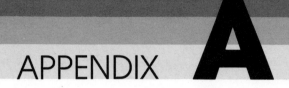 A

ENVIRONMENTAL IMPACT STUDIES

Traditionally, the planning process for civil engineering and construction projects has always included a consideration of the economic as well as the technical problems involved. It was not until 1970, when the National Environmental Policy Act (NEPA) took effect, that *environmental impacts* were also given due consideration in the project planning process.

The NEPA regulations focused on major federal projects that could damage the environment. Shortly after NEPA, many states also adopted similar laws of their own, and eventually, local municipalities began to incorporate thorough environmental planning requirements into land-use or zoning ordinances. Now, even small and privately funded construction projects usually must include an environmental impact study, before approval for the project will be granted. It is important that all civil design and construction professionals have a basic understanding of what an environmental impact study is and how it is used.

A.1   THE ENVIRONMENTAL IMPACT STATEMENT (EIS)

An *Environmental Impact Statement* for a proposed project is a written report that summarizes the findings of a detailed environmental review process. The writing of an EIS is preceded by two steps. First, an *environmental inventory* must be conducted for the vicinity of the proposed project. This inventory includes a thorough description of the existing physical environment, and serves as the basis for evaluating the possible impacts of the project. The second step involves a systematic and comprehensive *environmental assessment*. This assessment, a crucial part of the EIS process, identifies and analyzes the potential adverse environmental consequences of the project. This analysis includes a prediction of all possible environmental changes, as well as a consideration of the magnitude and overall importance of those changes. In many assessments, an attempt is made to measure and describe qualitative environmental impacts in quantitative or numerical terms.

The format of the EIS document or report may vary to some degree, depending on the requirements of the municipality or regulatory agency that will review and approve it. Generally, the following topics or sections are included in a final draft of the EIS:

1. Description of the existing environment
2. Description of the proposed project
3. Environmental assessment
4. Unavoidable adverse environmental impacts
5. Secondary or indirect impacts
6. Methods for reducing adverse impacts
7. Alternatives to the proposed project
8. Irreversible commitments of energy and resources
9. Consideration of public input and review

The EIS is meant to be used as a planning and decision-making tool. It is supposed to be objective and unbiased, and it is not meant to either promote or block the implementation of a proposed project. The greatest benefit of the EIS process is that environmental concerns must be examined thoroughly, and the chances for severe or unexpected damage due to a construction project are significantly minimized.

Unfortunately, EIS reports are sometimes manipulated by developers to promote a construction project, or they are misused by special interest groups to stop a project completely. And a criticism often directed toward the EIS is that it may be imposed upon small projects that might not warrant so much concern. But the role of the EIS as an environmental planning tool is still relatively new, and it is gradually evolving. We can expect that it will soon be "fine tuned" to the point where environmental protection will be achieved in a cost-effective manner. Because it is a permanent and important aspect of civil–construction technology, some of the key features of an EIS are discussed in the following sections.

A.2   DESCRIPTION OF THE EXISTING ENVIRONMENT

A basic objective of an environmental study is to anticipate any potential impacts of a proposed construction project on the environment. It is first necessary to have an accurate and thorough picture of what the predevelopment (existing) environmental conditions are at and near the proposed site. Sometimes an environmental inventory report is already available for an entire city or township. Usually, though, the interdisciplinary team that is preparing the EIS must conduct a more detailed and site specific environmental survey.

The inventory of existing natural resources and urban facilities in the vicinity of the project site typically includes the following categories of data:

Geology, Soils, and Topography.   This would include a description of the types of bedrock that underlie the site, soil types and characteristics, and existing ground slopes or topography. Soil erosion potential is a particularly important factor, as are percolation rates, depth to groundwater, and the location of aquifer recharge areas.

Water Resources. Streams and lakes on or near the project site would be studied and described. Data on existing surface and groundwater quality would be discussed, as would drainage patterns, flood hazards, and streamflow rates. Existing or predevelopment runoff rates would be evaluated so that appropriate measures could be taken to ensure that they are not increased later on.

Vegetation and Wildlife. The extent and type of woodlands or plant growth on the site would be described, and any rare or unique species would be inventoried. Species of animals that use the site as a habitat would also be discussed, and the presence of any endangered species would be determined. Usually, environmental resource data are presented graphically for clarity. An example of a site specific map showing existing vegetation is illustrated in Figure A.2.1.

Air Quality and Noise. Existing air quality data from the nearest state or federal air sampling station would be obtained and discussed. Local meteorological conditions, including average wind speed and direction and the frequency of temperature inversions would be studied and summarized. Noise levels, frequency, and duration in the vicinity of the site would be evaluated.

Transportation. Existing modes of transportation would be described, including automobile, bus, rail, and aircraft. Local traffic volumes, patterns, and existing roadway capacities would be evaluated. An example of a site specific traffic survey is shown in Figure A.2.2.

Public Utilities. The location and service capacities of nearby water supply mains and sewerage systems would be described and shown on a site plan. Gas, telephone, electric, and refuse collection service in the area would be evaluated.

Population, Land Use, and Socioeconomics. Existing population densities and land-use patterns would be studied and described, including residen-

tial, commercial, industrial, and agricultural areas. Local incomes and economic levels, the local tax base, and the capacities of school, fire, and police services in the area would be evaluated.

Historical or Unique Cultural Features. The possibility of an archaeological site existing within the project boundary would be investigated. The location of historical landmarks, museums, or libraries would be described. Any unique aesthetic features, such as a beautiful view, or the last remaining open space in the community, would also be noted.

A.3 DESCRIPTION OF THE PROPOSED PROJECT

In addition to a complete environmental inventory, it is also necessary to have a clear picture of the nature and extent of the proposed project. Although detailed engineering plans are not generally needed for this, a preliminary plan must be made available by the project owner. This plan must be comprehensive enough to allow for a meaningful assessment of environmental impacts.

To illustrate, consider again the fictitious land development project discussed in Section 1.1. An EIS would have to be prepared for this project. The developer's consulting engineer or architect would provide information regarding the total area of the project, the number of building lots, the relative distribution of residential, commercial, and industrial facilities, and other data. A preliminary plan showing the proposed alignment and grading of roadways would be submitted. First-floor elevations of proposed structures, and any changes or regrading of topography would be indicated.

A proposed storm water drainage system would have to be shown, including underground pipelines and any stormwater detention basins. The points of stormwater discharge would be shown. Plans for a proposed water supply and wastewater collection system would be submitted, showing the location and capacities of pipelines and other utilities. In

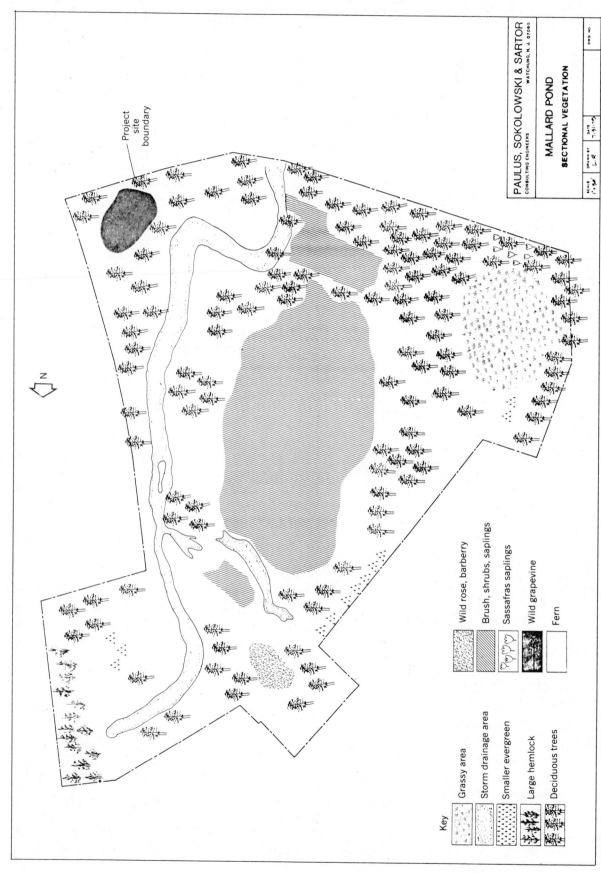

FIGURE A.2.1 A typical project site map showing an inventory of existing environmental conditions—in this case, vegetation. (Paulus, Sokolowski & Sartor, Inc.)

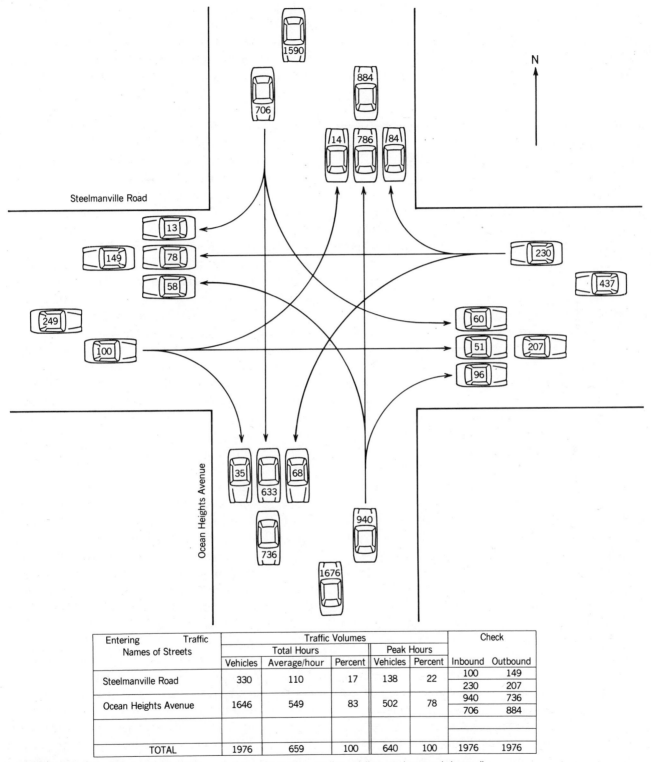

Entering Traffic Names of Streets	Traffic Volumes					Check	
	Total Hours			Peak Hours			
	Vehicles	Average/hour	Percent	Vehicles	Percent	Inbound	Outbound
Steelmanville Road	330	110	17	138	22	100	149
						230	207
Ocean Heights Avenue	1646	549	83	502	78	940	736
						706	884
TOTAL	1976	659	100	640	100	1976	1976

FIGURE A.2.2   A traffic survey is usually part of the EIS section on the existing environmental conditions. (Paulus, Sokolowski & Sartor, Inc.)

some cases, information regarding the type of construction, landscaping, and the expected market value of the constructed facilities would be required.

A.4 ASSESSMENT OF ENVIRONMENTAL IMPACTS

The primary goal of the EIS procedure is to predict any adverse (or beneficial) effects of a proposed project on the natural and urban environment. This is done so that measures can be taken to minimize or eliminate the harmful impacts when the project is implemented. The prediction or assessment of environmental impacts is not an easy task. It must be conducted by an interdisciplinary team including civil engineers and technicians, urban planners, and biologists or ecologists. For large and complex projects, and particularly for sensitive environmental settings, the team may also include geologists, archaeologists, architects, and social scientists.

Certain environmental impacts can be evaluated directly and objectively. They are not really subject to conflicting subjective or personal opinions. For example, the expected increase in stormwater runoff due to the project can be computed and compared to the existing runoff rates and volumes. The effect of the increase on the site and on downstream properties may then be predicted. As discussed in previous chapters, these effects might include flooding, soil erosion, and water pollution.

Air quality impacts can also be assessed using sophisticated mathematical models. Usually, emission of carbon monoxide from cars is of particular significance in land development projects; the increase in automobile traffic would contribute directly to this effect. Basic traffic engineering principles can be applied in order to estimate the increase in traffic, as a function of population density and land use. Using this information, along with data on existing air quality and prevailing weather conditions, the impact of the project on ambient air quality can be anticipated.

Impacts upon vegetation and wildlife are more difficult to evaluate objectively. Although it is relatively easy to estimate how many hectares or acres of woodland will be destroyed by the project, it is much more difficult to agree upon the value or importance of this impact. Of course, if the project site is the last remaining woodland in an urban community, or if some of the trees are among an endangered species, the impacts would be considered more severe than otherwise.

It is important to distinguish between *short-term impacts* and *long-term impacts*. For example, the impacts of construction activities might include a temporary increase in neighborhood noise levels from the heavy machinery. But once the project is completed, these impacts cease; they are therefore considered short term. But the effect of the project on, say, runoff patterns and local aquifer recharge rates would not cease when construction is finished; these would be long-term impacts.

Many procedures for conducting an environmental assessment have been developed over the years. They share the basic goal of providing a comprehensive and systematic environmental evaluation of the project, with the greatest degree of objectivity. These procedures range in complexity from simple checklists to more complex "matrix" methods.

In the checklist method, all potential environmental impacts for the various project alternatives are listed, and the anticipated magnitude of each impact is described qualitatively. For example, negative impacts can be indicated with minus signs. A small or moderate impact could be shown with, say, two minus signs $(--)$, whereas a relatively more severe impact could be shown with three or four minus signs $(---)$. Beneficial or positive impacts can be shown with plus $(+)$ signs. If the environmental impact is not applicable for a particular project alternative, a zero (0) would be shown. Such a list would serve to present a visual overview of the assessment.

In the so-called matrix methods, an attempt is made to quantify or "grade" the relative impacts of the project alternatives and to provide a numerical basis for comparison and evaluation. The anticipated magnitude of each potential impact may be rated on a scale of, say, 0 to 10; the higher numbers may represent severe adverse impacts, whereas the lower numbers represent minor or negligible effects. Zero (0) would indicate no expected impact for a particular activity or environmental component.

Numerical weighting factors are also used in the matrix method, to indicate the relative importance of a particular impact. These weighting factors would be agreed upon by the assessment team, and would be site and project specific. For example, the impacts on groundwater quality may be considered more important in a particular area than impacts on air quality, particularly if the groundwater is a sole source of potable water. Groundwater quality could be assigned a relative importance or weight of, say, 0.5, compared to 0.2 for air quality.

Weighting factors can be multiplied by the respective impact magnitudes, to put each impact in perspective. For example, consider that the impact on groundwater quality has a magnitude of 4 and the impact on air quality has a higher magnitude of 6. But after weighting the impacts (multiplying by the weighting factors), we see that the overall significance of the impact on water quality, $0.5 \times 4 = 2$, is more important or severe than the impact on air quality, $0.2 \times 6 = 1.2$. If the weighted impacts for all the listed items are added together, a composite score or *environmental quality index* can be obtained for each project alternative. The alternative with the lowest index would be the one that would cause the least harmful environmental impact, overall.

A.5 OTHER ASPECTS OF AN EIS

An EIS should include a section in which *mitigating measures* are discussed. These mitigating measures are, in effect, suggested changes or details regarding project design that would tend to reduce or eliminate the adverse impacts. For example, one of the most serious short-term impacts due to construction activities is an increase in soil erosion and sedimentation in local streams; this leads to degradation of surface water quality. Specific measures for controlling erosion and preventing stream sedimentation can be described in the EIS (e.g., using hay bales and temporary seeding). Another example of a mitigating measure would be to relocate where facilities are constructed on the site, if possible, to preserve valuable trees or other vegetation.

An EIS report must also focus upon *unavoidable adverse impacts*—those harmful effects that simply cannot be avoided if the proposed project is implemented. For example, if construction of the project must involve the destruction of a mature stand of beautiful trees, this should be identified as an unavoidable environmental impact. Short of not building the project at all (the *no-action alternative*), and therefore not having to cut down the trees, there is nothing that can be done to mitigate this impact.

All reasonable *project alternatives* should be evaluated and discussed in the EIS. These may include changes in scope or location as well as the so-called *no-project* or *no-action* alternative. The no-action alternative, of course, causes no environmental disruption to the proposed site and environs. But it generally has some adverse socioeconomic side effects. For example, suppose the project involves a residential subdivision; the no-project option would preserve the site in its natural state, but a shortage of available housing in the community could be an unwelcome result.

Many EIS reports must include an evaluation of possible *secondary*, or *indirect*, *impacts* that would be caused by project implementation. Secondary impacts are those that are not immediately apparent and that are not directly caused by the project itself, but probably would not occur if the project were not built.

For example, consider what could happen if a new water main and sewer line were built along a rural road to connect a new subdivision of homes to existing municipal water and sewerage facilities. This is illustrated in Figure A.5.1. It would not be long before new homes were built along that road, causing *strip development*, because water and sewerage utilities were readily available. In effect, construction of the original planned subdivision may indirectly lead to the less desirable future development.

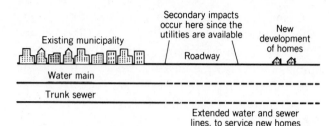

FIGURE A.5.1 Undesirable "strip-development" may be a secondary or indirect environmental impact of utility construction.

Most EIS documents contain a section pertaining to the so-called *irreversible* or *irretrievable* commitments of resources that would result if the proposed project were built. Supplementing the section on unavoidable adverse impacts, this serves to review and focus attention upon energy and material consumption, loss of wildlife habitat, loss of rare or endangered species, and permanent changes in land topography and use.

Finally, a complete EIS would contain a section that responds to public opinion and input. The EIS report is initially prepared in a draft form, which is distributed to the appropriate agency for review as well as to interested citizens and public interest groups. In most cases, a public hearing would be announced and held so that the environmental issues can be discussed openly.

Public input is considered important in the EIS process, because it can provide a perspective or identify an alternative that can otherwise be overlooked by the professionals who prepared the report. Public involvement also serves to resolve disputes early in the planning process. The final copy of the EIS document would reflect this public feedback and input.

ROLE OF THE TECHNICIAN AND TECHNOLOGIST

B.1 EDUCATION
B.2 EMPLOYMENT

The "engineering team" includes *technicians* and *technologists,* as well as *engineers*. It is important for students to have a clear understanding of their future role on this team, and to be aware of the educational requirements necessary to begin a career in the field of civil–environmental technology. It is also helpful for students to be aware of the wide variety of employment opportunities and job responsibilities that exist, as they relate to the different levels of education and training.

B.1 EDUCATION

There are no less than six different levels of education at which a person can begin a career in the field of civil–environmental technology. As would be expected, a higher level of education requires a greater investment of time and stronger academic abilities than does a lower level of education. These educational levels include:

ENGINEERING	TECHNOLOGY	CERTIFICATION
Doctorate degree	Bachelor's degree	Various levels
Master's degree	Associate degree	
Bachelor's degree		

The basic difference between bachelor's-degree programs in *engineering* and in *technology* is in the sequence and level of technical courses in the curriculum. Engineering programs place much more emphasis on math, science, and general analytical abilities than do the technology programs. Specific engineering courses are taken by the student in the junior and senior years of college, after a solid foundation in theoretical principles has been established in the freshman and sophomore years. Most engineering courses rely on a thorough knowledge of calculus.

Engineering is often defined as the *application of science and math* to solving problems for the benefit of people. Technology, on the other hand, can be defined as the *application of engineering principles* for the benefit of people. There is less emphasis on math and theory in the technology programs. Instead, practical applications and "hands-on" skills are stressed. Technology courses usually require knowledge of algebra and trigonometry, but do not rely on calculus, particularly in the freshman and sophomore years. And specific technical subjects may be studied in the freshman year of a technology curriculum.

Generally, a minimum of seven years of full-time university study is required for the doctorate degree (Ph.D.), five years are needed for the master's degree (M.S.), and four years are needed for the bachelor's degree in engineering (B.S. or B.E.). A minimum of four years is required for the bachelor's degree in technology (B.E.T. or B.S.E.T.), and two years are needed for the associate degree in technology (A.A.S.). Some schools offer a master's degree in technology, but this is not very common.

Certification as an operator of a public water supply or sewerage system requires a high school diploma and the passing of a written exam; in many states, several years of operating experience are also required. The levels of certification depend on the type and size of the water or sewerage facility being operated.

Graduates of the bachelor's degree program in engineering technology are called *technologists,* whereas graduates of the associate degree program are called *technicians.* Many employers, however, do not make a distinction between the technologist and the bachelor's degree level engineer; some technologists are hired with a job title that includes the word *engineer.* Most states allow technologists to take the professional engineering (P.E.) licensing exam, but the requirements for years of experience vary. In general, the role of the technician and technologist is that of a liaison between the engineer and builder.

B.2 EMPLOYMENT

For the purpose of discussion, we can categorize employment opportunities into eight different types of activities:

1. *Research and Development* - conducting laboratory and theoretical investigations to further the understanding of environmental processes, and to develop new applications and environmental control equipment.

2. *Teaching* - Instructing and guiding engineering and technology students, developing educational curricula and new courses, writing textbooks, and preparing other instructional material.

3. *Project Planning and Management* - Conducting technical, economic, and environmental feasibility studies, evaluating project alternatives, overseeing the progress of engineering studies and design projects.

4. *Project Design* - Conducting design computations, and preparing detailed plan drawings and specifications to guide the construction of the project.

5. *Construction Management* - Estimating construction costs, scheduling equipment, material delivery, and labor; supervising and coordinating field activities, construction inspection, material testing, quality and safety control.

6. *Facility Operation and Maintenance* - Conducting daily process evaluations and control, water and wastewater testing, supervising maintenance and repair activities.

7. *Regulation and Enforcement* - Monitoring environmental quality, enforcing environmental rules and regulations, reviewing and approving plans for new water supply and waste disposal facilities, and inspecting existing facilities.

8. *Marketing and Sales* - Providing technical support and liaison between manufacturers and users of environmental control products and equipment.

Of course, it is possible to have a career that includes more than one of these eight activities. But the likelihood of working in a specific activity depends somewhat on educational level and training. The likelihood of having a supervisory role in any of those activities is even more dependent on education. This relationship is shown in Figure B.2.1.

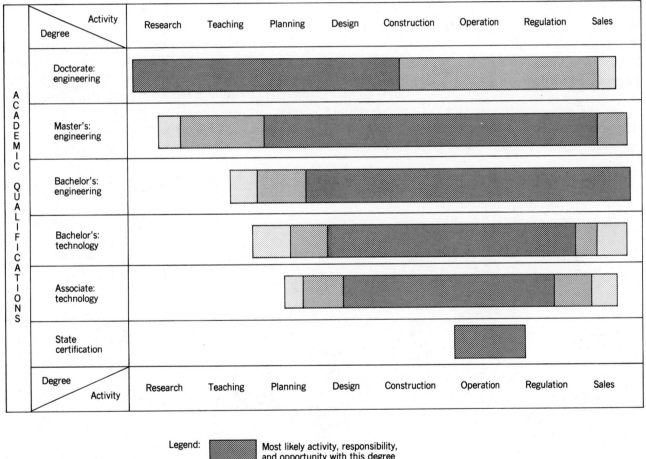

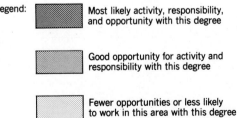

FIGURE B.2.1 Employment opportunities in civil–environmental technology are related to the level of education. In general, greater opportunities for responsibility and advancement are available with higher levels of education.

An engineer with a master's degree, for example, would have opportunities in all eight activities, but engineers at that level are most frequently employed in responsible positions related to project planning and project design. Technicians with associate degrees have significant opportunities to hold responsible positions in construction and facility operation. Technicians may also be employed in planning, design, enforcement, and perhaps research activities, but this is usually under direct supervision of more highly trained and experienced professional engineers. It can be seen from Figure A.2.1 that the range of employment opportunities, and the likelihood of having a position of responsibility in any activity, increases as education and training increase.

Types of Employers

There are several different types of organizations that employ civil–environmental engineers, technologists, and technicians. These include colleges or universities, consulting engineering firms, municipal engineering or public works departments, construction contractors, industries, water and sewerage utilities, and regulatory government agencies. The activities that these different employers are most likely to be engaged in are shown in Figure B.2.2.

Job Activities

Some activities, such as project design, overlap or span the educational spectrum from Ph.D. to the associate degree. How do job duties and responsibilities vary with education in that particular activity? What role do engineers, technologists, and technicians play in project design?

Generally, an engineer with a high level of education or experience would serve as *project manager*. This would involve meeting with the client or project owner, preparing the budget, scheduling and coordinating specific design tasks, and making major decisions about the technical design concepts

Activity \ Employer	College	Firm of Consulting Engineers	Municipal Engineering Department	Contractor	Industry	Government Agency	Water or Sewerage Utility
Research	■ Primary	▒ Secondary			■ Primary	▒ Secondary	
Teaching	■ Primary						
Planning		■ Primary	▒ Secondary			■ Primary	▒ Secondary
Design		■ Primary	■ Primary	▒ Secondary	▒ Secondary	■ Primary	▒ Secondary
Construction		▒ Secondary	▒ Secondary	■ Primary			
Operation		▒ Secondary	■ Primary				■ Primary
Regulation						■ Primary	
Sales					■ Primary		

Legend: ■ Primary activity   ▒ Secondary activity

FIGURE B.2.2 There are many different types of employers in the field of environmental technology. Most employers focus on one or two principal tasks or activities, such as design or construction.

and approach to problem solving. A *project engineer* (with a master's or bachelor's degree in engineering) would work under the direction of the project manager; he or she would assume overall responsibility for the daily design activities involved in preparing plans and specs for that particular project.

If it is a large and complex project—such as the design of a modern water treatment plant—there would be several other engineers, technologists, and technicians working under the direct supervision of the project engineer. Engineers with bachelor's and master's degrees would be involved in the detailed design of specific project components, such as the coagulation or filtration processes. This would include design computations using electronic calculators and microcomputers, preparation of sketches and plans, as well as equipment selection and specification writing. Technologists may also be involved in these detailed design activities.

Technicians with associate degrees would be involved in assisting the design engineers and technologists. Under direct supervision, they would perform specific routine tasks such as making computations, preparing and inputting data for computer analysis, preparing detailed plan drawings, plotting data, and other activities. The technician would also conduct project site surveys, soil or water sampling and testing, and other field investigations. Senior technicians with good technical abilities and several years of experience can assume more design responsibility and supervise other less experienced technicians.

Conclusion

It is not possible to discuss here all the employment opportunities and activities related to environmental engineering and technology. However this discussion should help the student to appreciate the wide range of job activities and types of employers and to understand the general relationship between level of education and opportunities for responsibility and advancement. In particular, it is important that the student understand the distinction between *engineering* and *engineering technology*.

In the coming years, there will be a need in the field of environmental technology for technical personnel at all levels of education and training. Protection of public health and environmental quality is a top-priority goal for most citizens, including politicians and legislators. As researchers and engineers develop new techniques for waste disposal and pollution control, more and more opportunities will be available for technologists and technicians to apply and implement the principles of modern environmental technology.

APPENDIX

COMPUTATIONS

The application of environmental technology to solving practical problems involves mathematical computations. As in any technical field, these computations may range in complexity from the very simple to the very sophisticated and involved. In this introductory text, the example and practice problems do not require mathematics beyond the level of basic algebra. The following sections are meant as a review of the proper use of dimensions and units. A brief list of appropriate conversion factors is also provided.

C.1 UNITS

The *International System of Units,* or simply the *SI metric* system, as it is commonly called, is used in most countries. But in the United States, a voluntary effort to switch from our customary *inch–pound* units to the SI metric system has not yet been entirely successful. The target date of January 1, 1985, set by the Construction Industries Coordinating Committee of the American National Metric Council as the date for an official switch-over for construction activities, has not been met. However, we can still expect to eventually be using SI units rather than the "old" inch–pound units (also called customary English, or American units).

But even after an "official" conversion occurs, it is likely that the use of our customary units will still be necessary to some extent. New projects often have to be coordinated with older plans and specs that were based on the inch–pound system, and many technical documents or reports used for reference will include customary units. For many years to come, it will be necessary for technical personnel to be able to work with both types of units with equal facility. In effect, most engineers and technicians must be bilingual with respect to reading, understanding, and performing technical calculations.

There are several good reasons for "going metric." For example, many of the unwieldy conversions needed to work with American units, such as 7.48 gal = 1 ft^3, 5280 ft = 1 mile, and so on, can be avoided with the SI system. In SI units, all conversions are based on a factor of 10^3 or 1000. Also, it is completely unnecessary to work with fractions in SI units. But the point here is not to justify the switch to SI units as much as it is to emphasize the need to work with both systems.

In this text, data are given in both types of units. Example and practice problems are presented in one or the other type, but no attempt is made to duplicate problems in both systems. This makes it necessary for the student to read all the problems in a particular section, to get full coverage of the topic. In this way, it is less likely for the student (or instructor) to use only one system simply because it is more comfortable to do so. In most cases, though, it is not necessary to convert from one system of units to the other in any given problem.

One of the important differences between the two systems is that *mass (kilogram or kg)* is a basic unit in SI, whereas *force (pound or lb)* is a basic unit in the American system. In customary units, the unit for mass is a derived quantity called a *slug*. In SI, the unit for force, called a *newton (N),* is the derived quantity.

One newton (1 N) is defined as the force required to accelerate a mass of one kilogram (1 kg) at a rate of one meter per second per second (1 m/s^2). It is a good idea to "think metric," rather than to always rely on exact conversion factors when learning to use the SI system. For example, if the student learns and remembers that one newton is roughly equivalent to a quarter of a pound, it will be easier to interpret and have a "feel" for weights and forces expressed in SI units.

In civil-environmental technology, pressure is a physical quantity that is frequently used. Pressure is defined as force per unit area. Most people in the United States are familiar with the customary English term for pressure, which is called a pound per square inch (lb/in.^{2}), or *psi*. But in the SI system, pressure is measured in terms of newtons per square meter (N/m^2), and that unit is called a *pascal (Pa)*. One pascal is a relatively small pressure; it is more common to express pressure in terms of *kilopascals (kPa)*. For example, a typical water main pressure of 60 psi is equivalent to about 400 kPa.

The prefix *kilo* in the word *kilopascal* stands for a multiplier of 1000; 400 kPa is equivalent to 400 000 Pa (a space is used instead of a comma in SI numbers to separate groups of three digits). But it is preferable to write the number in terms of kPa. In general, numbers should be kept within the range of 0.1 to 1000 when presenting numerical data, and the appropriate prefix should be used with the unit. There are several different prefixes used in the SI system.

Those prefixes most commonly used in environmental technology are:

Prefix Name	Symbol	Multiplier
mega	M	10^6
kilo	k	10^3
milli	m	10^{-3}
micro	μ	10^{-6}

In order to solve technical problems, it is necessary to be familiar with many different units and symbols. A selection of those quantities, units, and abbreviations that are frequently used in environmental technology is presented in Table C.1.1.

C.2   UNIT CANCELLATION

One of the best ways to avoid errors in technical computations, and to do conversions, is to write the

TABLE C.1.1   SELECTED QUANTITIES, UNITS, AND SYMBOLS

Quantity	*SI System* Unit (Symbol)	*American System* Unit (Symbol)
length	meter (m) kilometer (km) millimeter (mm) micrometer (μm)	inch (in.) foot (ft) mile (mi)
area	square meter (m^2) square kilometer (km^2) hectare (ha)	square inch (in.^{2}) square foot (ft^2) acre (ac)
volume	cubic meter (m^3) liter (L) megaliter (ML)	cubic foot (ft^3) gallon (gal) acre–foot (ac–ft)
time	second (s) hour (h) annum (a)	second (s) minute (min) hour (hr or h) day (d)
velocity	meters per second (m/s)	feet per second (fps)
flow rate	cubic meters per second (m^3/s) liters per second (L/s) megaliters per day (ML/d)	cubic feet per second (cfs) gallons per minute (gpm) million gallons per day (mgd)
mass	kilogram (kg) milligram (mg) microgram (μg) ton (t)	slug
force	newton (N) kilonewton (kN)	pound (lb) kip (k)
pressure	pascal (Pa) kilopascal (kPa)	pounds per square inch (psi)
energy	joule (J)	foot–pound (ft–lb)
power	watt (W) kilowatt (kW)	horsepower (hp)
temperature	kelvin (K) degree Celsius (^{0}C)	degree Fahrenheit ($^\circ$F)

appropriate units next to each number, and to use *unit cancellation* as a means to ensure that the expression is dimensionally correct. Units can be multiplied, divided, and "cancelled out," just like numbers in arithmetic operations. Several examples are given here to illustrate this procedure.

EXAMPLE C.2.1

Convert 60 in. to an equivalent length expressed in feet.

Solution:

The appropriate conversion, of course, is 1 ft = 12 in., and we simply have to divide the 60 by 12 to obtain the answer, 5 ft. To illustrate unit cancellation, however, consider that the fraction $\frac{1 \text{ ft}}{12 \text{ in.}} = 1$ since 1 ft = 12 in. Therefore, we can multiply the 60-in. quantity by that fraction, without altering its value as a length:

$$60 \text{ in.} \times \frac{1 \text{ ft}}{12 \text{ in.}} = \frac{60 \text{ in.} \times \text{ ft}}{12 \text{ in.}} = 5 \text{ ft}$$

The diagonal slash through the unit indicates cancellation (in./in. = 1). Using unit cancellation helps us avoid the blunder of multiplying the 60 by 12 instead of dividing.

EXAMPLE C.2.2

Convert an area of 350 ha to square kilometers.

Solution:

Since 1 ha = 10 000 m^2, and 1 km^2 = 10^6 m^2, we get

$$350 \text{ ha} \times \frac{10\ 000 \text{ m}^2}{1 \text{ ha}} \times \frac{1 \text{ km}^2}{10^6 \text{ m}^2} = 3.5 \text{ km}^2$$

EXAMPLE C.2.3

Convert a flow rate of 100 gpm to an equivalent flow rate expressed in units of cubic feet per second.

Solution:

$$100 \frac{\text{gal}}{\text{min}} \times \frac{1 \text{ ft}^3}{7.48 \text{ gal}} \times \frac{1 \text{ min}}{60 \text{ s}} = 0.223 \text{ ft}^3/\text{s or cfs}$$

EXAMPLE C.2.4

Convert a flow rate of 5.0 m^3/s to an equivalent flow rate expressed in terms of megaliters per day.

Solution:

$$5.0 \frac{\text{m}^3}{\text{s}} \times \frac{10^3 \text{ L}}{1 \text{ m}^3} \times \frac{1 \text{ ML}}{10^6 \text{ L}} \times \frac{3600 \text{ s}}{1 \text{ h}}$$
$$\times \frac{24 \text{ h}}{1 \text{ d}} = 432 \text{ ML/d}$$

C.3   SELECTED CONVERSION FACTORS

For convenience in solving some of the practice problems in this text, a limited selection of appropriate conversion factors is presented here. A much more thorough listing of conversion factors may be found in the references given at the end of this section.

SI System	American System
1 m = 1000 mm	1 mile = 5280 ft
1 km = 1000 m	1 ft^3 = 7.48 gal
1 m^3 = 1000 L	1 ac = 43 560 ft^2
1 ha = 10 000 m^2	gravity = 32.2 ft/s^2
1 ton = 1000 kg	1 ft^3 of water = 62.4 lb
1 liter of water = 1 kg	1 gal of water = 8.34 lb
gravity = 9.81 m/s^2	1 ton = 2000 lb

In this text, it is generally not necessary to convert from SI to American units, or vice versa; most examples and problems are presented in either one system or the other. But it may be useful to have some basic conversion factors at hand, particularly when first getting used to a new system of units. The following factors are presented for this purpose:

$$1 \text{ m} = 3.28 \text{ ft}$$
$$1 \text{ km} = 0.621 \text{ mi}$$
$$1 \text{ ha} = 2.47 \text{ ac}$$
$$1 \text{ L} = 0.264 \text{ gal}$$
$$1 \text{ kg} = 2.2 \text{ lb}$$
$$1 \text{ kN} = 225 \text{ lb}$$
$$1 \text{ kPa} = 0.147 \text{ psi}$$
$$1 \text{ kW} = 1.34 \text{ hp}$$

Exact or precise conversions from one system to another are generally not used in this text. The student will note, for example, that in many examples a length of 1 in. is taken as 25 mm, instead of the more precise 25.4 mm. For the purpose of this text, and for easier reading, this rounding-off of conversions is sufficient.

C.4 ACCURACY

For most applications of environmental technology, the use of three significant figures in calculations will provide sufficient accuracy. In fact, using too many significant figures can be misleading, since it may imply a degree of accuracy greater than that to which the quantities can actually be measured. Also, much of the raw data used in the computations are only estimates to begin with, and it makes little sense to use a great deal of precision when expressing the calculated results.

Usually, neither the location of the decimal point nor the number of leading or trailing zeros affects the number of significant figures. For example, each of the following numbers has only three significant figures: 1.23, 12.3, 123, 0.123, 12 300, and 0.00123. On the other hand, a number written as 12 300.0 would imply that the zeros to the left of the decimal point are significant, and the number would have five significant figures.

A computed value can be no more accurate than the least accurate number in the original data. For example, if a flow velocity of 3 m/s is multiplied by an area of 0.0706 m^2, the product on a calculator is displayed as 0.2118 m^3/s. But the answer should have no more than one significant figure, since that is the accuracy implied by using 3 instead of 3.00. Therefore, it is best to round off the answer to 0.2 m^3/s. When rounding off, if the digit to be dropped is less than 5, drop it without changing the remaining number. But if the digit to be dropped is 5 or greater, increase the digit to its left by 1. For example, 456.4 would be rounded off to 456, but 456.5 would be rounded off to 457.

SELECTED REFERENCES

1. Lytle, R. J., *American Metric Construction Handbook,* Structures Publishing Company, Farmington, Mich., 1976.

2. Milton, J. M., *Recommended Practice for the Use of Metric (SI) Units in Building Design and Construction,* U.S. Department of Commerce, Washington, D.C., June 1977, NBS 938.

3. Committee Report, *Final Report on Metric Units and Sizes,* Journal American Water Works Association, January 1982.

GLOSSARY AND ABBREVIATIONS

There are many technical terms and abbreviations related to the subject of *environmental technology* that are new to the student. For the most part, these are defined when they are first introduced in the text. In some cases, though, a student may read certain sections or chapters out of sequence. This glossary will then be useful for getting a quick and brief definition of a new term or abbreviation that is unfamiliar. It can also serve as a review and study aid.

The definitions here are intentionally brief; further discussion of each item can be found in the appropriate section of the text. A list of commonly used "environmental" abbreviations follows the glossary definitions.

ABSORPTION A process by which one substance is trapped throughout the volume of another, usually a liquid, by solution or chemical reaction.

ACID A substance that causes an increase in the hydrogen ion concentrations of a solution, and which can react to neutralize a base or alkaline substance.

ACID RAIN Precipitation with higher than normal acidity, caused primarily by sulfur and nitrogen dioxide air pollution.

ACOUSTICAL MATERIAL A material that absorbs or reflects sound energy, and is used for noise control.

ACRE–FEET A unit used for expressing large quantities of water, as in conservation reservoirs, equivalent to the volume that would cover 1 ac of land to a depth of 1 ft.

ACTIVATED CARBON A very porous material which, after being subjected to intense heat to drive off impurities, can then be used to adsorb pollutants from air or water.

ACTIVATED SLUDGE   The suspended solids in an aeration tank or at the bottom of a secondary clarifier in a sewage treatment plant, consisting mostly of living microorganisms.

ACTIVATED SLUDGE PROCESS   A biological sewage treatment system in which living microbes, suspended in a mixture of sewage and air, absorb the organic pollutants and convert them to stable substances.

ADIABATIC LAPSE RATE   The decrease in ambient temperature with elevation, in dry air, that represents the boundary between a stable and an unstable atmosphere; it is equal to $-1°C/100$ m ($-5.4°F/1000$ ft).

ADSORPTION   A physical process involving the contact and trapping of water pollutants or air pollutants on the surface of a solid substance, usually activated carbon.

ADVANCED TREATMENT   Purification processes used after or during secondary wastewater treatment to remove nutrients or additional solids and dissolved organics; also called tertiary treatment.

AERATION   A physical treatment process in which air is thoroughly mixed with water or wastewater for purification.

AEROBE   A microorganism that requires an aerobic environment to live and reproduce.

AEROBIC   In the presence of air or available molecular oxygen.

AEROSOL   A suspension of small solid or liquid particles in air.

AIR CLASSIFIER   A mechanical device used to separate paper and other light materials from solid waste in a resource recovery plant.

ALGAE   Microscopic single-cell plants suspended in water; phytoplankton.

ALGAL BLOOM   Visible overgrowth of algae in a lake, due to eutrophication.

ALUM   Aluminum sulfate, one of the most commonly used chemical coagulants used for water treatment.

AMBIENT SAMPLE   An air sample collected from the surrounding air after pollutants from various sources have been dispersed.

ANAEROBE   A microorganism that lives under anaerobic conditions, without free oxygen.

ANAEROBIC   In the absence of air or available molecular oxygen.

AQUATIC ORGANISM   An organism that lives in water.

AQUICLUDE   An underground layer of relatively impermeable soil or rock that does not yield appreciable quantities of groundwater.

AQUIFER   An underground layer of soil or rock that is porous enough to yield significant amounts of groundwater for public supply.

AREA METHOD   One method by which solid waste is placed, compacted, and covered in a sanitary landfill, without excavation.

ARTESIAN AQUIFER   An aquifer that is enclosed or sandwiched between two impermeable layers of soil or rock, also called a confined aquifer.

ATOM   The smallest part of an element that can exist and still retain the same chemical characteristics.

AUTOTROPHIC ORGANISMS   Self-nourishing green plants that obtain food from photosynthesis; the beginning link of the food chain.

BACKWASH   The washing cycle for a rapid filter in a water treatment plant in which clean water flows up through the filter.

BACTERIA   Microscopic single-celled plants that do not contain chlorophyll and do not nourish themselves by photosynthesis.

BAGHOUSE FILTER   An air cleaning device that removes very small particles from dirty flue gases, as the gas stream passes through a special filter fabric shaped like a long inverted bag.

BALING   A mechanical process in which municipal

solid waste is compressed under high pressure and compacted into rectangular blocks or bales for subsequent land disposal.

BASE   A substance that causes an increase in the hydroxyl radical concentration of a solution, and which can react to neutralize an acid.

BASE FLOW   Dry weather flow in a stream, fed by groundwater seeping out of the ground and into the stream channel.

BATTER BOARDS   Wooden boards placed across a trench during construction to help in establishing line and grade for a sewer pipeline.

BIODEGRADABLE   Readily broken down or decomposed into simpler substances by biological action of microbes.

BOD   Biochemical oxygen demand; the amount of oxygen required by microorganisms to decompose organic waste in water; a measure of the amount of organic pollution.

BUBBLER   A device used to collect gaseous air pollutants for analysis by absorption in an appropriate liquid.

BULKING SLUDGE   Activated sludge that does not settle properly in a secondary clarifier, causing high effluent BOD and TSS.

CARCINOGENIC   Capable of causing cancer.

CATCHMENT AREA   The area that contributes runoff to a stream or urban drainage system; also watershed or drainage basin.

CENTRIFUGAL PUMP   A mechanical device that adds energy to a liquid using a rapidly rotating impeller in a specially shaped casing; most common type of pump used for water treatment and distribution.

CESSPOOL   A covered pit for disposal of sanitary sewage, usually prohibited now.

CHANNEL FLOW TIME   Time of flow in a stream or pipeline to a drainage basin outlet; part of the time of concentration used in stormwater computations.

CHEMICAL OXYGEN DEMAND   The amount of oxygen needed to oxidize all the organics in a wastewater sample, a measure of the level of organic pollution; COD.

CHEMISTRY   The study of the composition and properties of substances—atoms, elements, molecules, and compounds.

CHLORINATION   The process of adding chlorine to water or wastewater, primarily for disinfection.

CHLORINE RESIDUAL   The small amount of chlorine compounds that remain in water or wastewater after disinfection, providing continued sanitary protection in the distribution system.

CLARIFIER   A sedimentation basin or settling tank in which suspended solids settle to the bottom and the clarified water or wastewater is drawn off the top.

COAGULATION   The addition to water or wastewater of certain chemicals that allow very small suspended particles to collide, stick together, and form settleable flocs.

COGENERATION   Energy recovery by production of both steam and electricity at a municipal solid waste incineration facility.

COLIFORMS   A group of mostly harmless bacteria that live in the intestinal tract of warm blooded animals, and which are used as a biological indicator of sewage pollution.

COLLOID   Extremely small particles suspended in water or wastewater, which cannot be removed by plain sedimentation or filtration without coagulation.

COMBINED SEWER   A pipeline that may carry a mixture of sanitary, storm, and industrial sewage, common in older cities, but not used in modern construction.

COMMINUTOR   A mechanical cutting or shredding device often used for preliminary treatment of sewage.

COMMUNICABLE DISEASE   A contagious disease that can be transmitted from person to person in a family or community.

COMPOSITE RUNOFF COEFFICIENT   A weighted average of the runoff coefficients for a large area, used in the Rational Method of stormwater drainage computations; accounts for different types of land use in the watershed.

COMPOSITE SAMPLE   A water or wastewater sample obtained by mixing individual grab samples taken at regular time intervals over the sampling period.

COMPOST   The end product of the composting process, consisting of an inoffensive material resembling potting soil; also called humus.

COMPOSTING   A biological process for treating garbage and/or sewage sludge, involving aerobic decomposition of organic waste under controlled conditions.

COMPOUND   A substance made up of a combination of elements.

CONE OF DEPRESSION   The shape of the groundwater table around a well from which water is being withdrawn.

CONFINED AQUIFER   See *artesian aquifer*

CONSERVATION RESERVOIR   A large open reservoir that serves primarily to store excess wet-weather stream flow for later use during periods of dry weather or drought.

CONTACT STABILIZATION   A variation of the conventional activated sludge process, used for sewage treatment.

COVER MATERIAL   Soil used to cover compacted solid waste in a sanitary landfill waste disposal lift or cell.

CROWN   The top inside wall of a pipe.

CYCLONE   A mechanical air pollution control device to remove particulates.

DARCY'S LAW   A formula that expresses the velocity of groundwater flow as a function of the slope of the water table, and the soil permeability in the aquifer.

DECIBEL   A unit of noise measurement that uses a logarithmic scale, and which is referenced to the lowest audible sound.

DECOMPOSITION   The process by which complex organic and inorganic substances are broken down into simpler substances by biological or physical processes; also called decay.

DEEP WELL   A relatively deep and narrow vertical excavation drilled to penetrate an aquifer for water supply.

DETENTION BASIN   A relatively small reservoir constructed to slow down or temporarily detain surface runoff from a storm.

DETENTION TIME   The average amount of time water or sewage remains in a tank or basin.

DIGESTION   The decomposition of organic waste by microbes under controlled conditions in a sewage treatment plant or garbage compost facility.

DISCHARGE   The volume rate of flow in a stream, river, or pipeline.

DISINFECTION   The destruction of disease causing microbes in water or sewage effluent, usually by the addition of chlorine or ozone.

DISPERSED SOURCE   A broad and unconfined area from which pollutants enter a body of water.

DISTRIBUTION RESERVOIR   A water storage tank connected directly to a distribution system, providing about one day of capacity.

DIVERSITY INDEX   A measure of the variety and population density of different species in an ecosystem.

DRAINAGE BASIN   See *catchment area.*

DRAINAGE DIVIDE   A line sketched on a topographic map that separates adjacent drainage basins; also called a ridgeline.

DRAWDOWN   The vertical distance between the static water level and the pumping water level in a well.

DROUGHT   A long period of dry weather that causes low flows in streams and rivers, and which affects water supplies adversely.

DROUGHT FLOW   Usually the minimum average flow in a stream or river over a seven-consecutive-day period with a recurrence interval of 10 years.

DUG WELL   A shallow excavation that penetrates an unconfined aquifer; usually no longer allowed for drinking water supply.

DUST   Suspended solid particles in air, larger than 1 μm in size.

DUSTFALL BUCKET   A simple device used to collect and measure settleable particulate levels in the atmosphere.

ECOLOGY   The study of living organisms and how they interact with their physical environment.

ECOSYSTEM   An identifiable ecological system containing plants and animals and the air, water, and minerals necessary for their survival.

EFFLUENT   Water or wastewater that flows out from a treatment plant or individual treatment process.

EFFLUENT STANDARDS   Limitations on the maximum amounts of pollutants that can be discharged from a sewage treatment plant.

ELECTROSTATIC PRECIPITATOR   An air cleaning device that removes very small particulates from flue gases in an electric field.

ELEMENT   A substance that cannot be divided into simpler substances by ordinary chemical change.

ENVIRONMENT   Our physical surroundings—air, water, land.

EPHEMERAL STREAM   A stream that becomes completely dry during a drought; also called an intermittent stream.

EPIDEMIC   The temporary but above average occurrence and spread of a particular disease in a community.

EPILIMNION   The uppermost layer of a stratified lake, in which mixing occurs.

EPISODE   The temporary occurrence of high air pollution concentrations and acute public health problems, usually during periods of temperature inversion.

EQUIVALENT PIPE   A computed or "theoretical" diameter and/or length of pipe that would have the same hydraulic characteristics of a series and/or parallel pipe network.

EROSION AND SEDIMENT CONTROL   The use of temporary grass cover, mulch, hay bales, diversion channels, or detention basins at a construction site, to prevent the washing away of exposed soil and the clogging of nearby streams with silt and sand.

EUTROPHICATION   The natural aging of a lake, characterized by high nutrient levels, excessive plant growth, and accumulation of bottom sediments.

EUTROPHIC LAKE   A relatively shallow, warm, and turbid lake, with excessive growths of weeds and algae.

EVAPOTRANSPIRATION   A part of the hydrologic cycle involving the combined processes of evaporation and transpiration of water by vegetation.

EXFILTRATION TEST   A method of testing sewer lines for watertightness, usually as part of an I/I survey.

EXTENDED AERATION   A modification of the conventional activated sludge process for sewage treatment.

FECAL COLIFORMS   Coliform bacteria from the intestines of warm blooded animals.

FECAL STREP   Fecal streptococcus bacteria, which live in the intestines of warm blooded animals; used along with fecal coliforms to determine the source of water pollution—of animal or human origin.

FILTRATION   The removal of suspended particles from water or air using a porous material that allows the fluid to pass through, but traps and retains the particles.

FLOC   A particle large enough to settle out of water or wastewater, formed during the coagulation–flocculation process; also, settleable particles of activated sludge.

FLOCCULATION   Gentle stirring of water or sewage after the addition of coagulation chemicals, which aid in the formation of settleable flocs.

FLOODPLAIN   The land along a river that would be covered by water during a 100-year flood.

FOOD CHAIN   An interrelated series of living organisms that feed on each other in a "one-way" pattern or direction; the producers, the consumers, and the decay organisms.

FOOD WEB   A complex food chain involving interactions among many different species of organisms.

FORCE MAIN   A pipeline through which sewage is pumped under pressure, usually to a sewage treatment plant.

FREEBOARD   The vertical distance between the top of a tank wall and the water surface in the tank.

FREQUENCY ANALYSIS   A statistical method to determine the recurrence interval of storms, floods, or droughts; also, a method for analyzing complex noises.

FUME   Very small solid particles suspended in air, usually formed during high temperature chemical reactions.

GAGE PRESSURE   Pressure measured with reference to atmospheric pressure as a zero or starting point.

GARBAGE   Food wastes in refuse, usually originating in the kitchens of homes or restaurants and in food processing plants.

GRAB SAMPLE   A single sample of water, wastewater, or air collected within a short time span for analysis of pollutants.

GRAVITY FLOW   Open channel flow in a pipe, ditch, or stream bed, characterized by a free liquid surface at atmospheric pressure.

GREENHOUSE EFFECT   The gradual warming of the atmosphere due to increasing levels of carbon dioxide and the trapping of radiated heat energy.

GRIT CHAMBER   One of the preliminary processes in a sewage treatment plant, which serves to remove sand and other inert gritty material from the sewage by gravity settling.

GROUNDWATER   Underground water that occupies the pore spaces in soil or fissures in rock.

GROUNDWATER TABLE   The interface between the zone of aeration and the zone of saturation in the soil, or the level at which the pore spaces in the soil are saturated with water, but are still at atmospheric pressure.

HAMMER MILL   A mechanical device used to shred and pulverize municipal solid waste, prior to composting and/or disposal.

HARDNESS   A property of water characterized by soap curdling and scale deposits in hot water systems, caused primarily by the presence of dissolved calcium and magnesium salts.

HARDY CROSS ANALYSIS   A computational method for analyzing flows and pressures in interconnected water distribution networks.

HAZARDOUS WASTE   Dangerous waste material that can cause serious illness, injury, or death, and environmental damage.

HAZEN–WILLIAMS EQUATION   A formula used to compute major pressure losses in water distribution mains and to design the mains.

HEAVY METALS   Metals such as mercury or lead that have high molecular weights and are toxic to living organisms at trace levels.

HERTZ   A term that indicates the frequency of a wave, such as a sound wave, standing for cycles per second.

HETEROTROPHIC ORGANISM   An organism that cannot manufacture its own food by photosynthesis and must consume plants or animals for energy.

HIGH-VOLUME SAMPLER   A filtration device used for collecting and measuring the amount of suspended particulates in a relatively large sample of air; also called a hi-vol sampler.

HUMUS   The end product of garbage and/or sludge composting.

HYDRAULIC GRADE LINE   A graph of pressure head in a hydraulic system, usually comprising a series of sloping straight lines that show a drop in pressure in the direction of flow.

HYDRAULICS   The study of water at rest and in motion in tanks, reservoirs, pipelines, and pumping systems.

HYDROCARBON   An organic substance that contains only hydrogen and carbon atoms.

HYDROGRAPH   A graph of stream or river discharge versus time.

HYDROLOGIC CYCLE The cycle of water moving through the environment as rainfall, surface and subsurface flow, and vapor.

HYDROLOGY The study of the occurrence and distribution of water on and under the earth's surface.

HYDROSTATIC PRESSURE The force per unit area on the walls of a tank, dam, or pipe caused by the action of a stationary liquid.

HYPOLIMNION The bottom-most layer of a stratified lake, in which little or no mixing occurs.

IGNEOUS ROCK Rock that has cooled and solidified from an original hot, molten condition.

I/I SURVEY A field survey for measuring the extent of infiltration and inflow in a sanitary sewer system.

INCINERATION An engineered process using controlled combustion to burn solid waste and/or sewage sludge, for volume reduction and disposal.

INDUSTRIAL SEWAGE Used water from industrial or manufacturing facilities that carries chemical waste products.

INERTIAL SEPARATOR A device used in a resource recovery plant to separate and sort solid waste based on its relative weight or density.

INFILTRATION In hydrology, a term referring to the penetration of water from precipitation into the ground. In sanitary sewer systems, a term referring to the seepage of groundwater into the sewer line through poorly constructed joints or cracks.

INFLOW Unwanted runoff that gets into a sanitary sewer from illegal connections to roof drains or basement sump pumps.

INFLUENT Liquid that flows into a water or wastewater treatment plant or purification process.

INFRASTRUCTURE Constructed public works facilities that allow human communities to function and thrive productively.

INORGANIC Mineral substances, usually not containing carbon.

INTERCEPTOR A large sewer that collects wastewater from smaller sewer lines and conveys it to a treatment plant or pumping station.

INVERT The bottom inside wall surface of a pipe.

ION An electrically charged fragment of an atom or molecule.

IONIZATION The process by which molecules dissociate into charged fragments called ions or radicals.

ISOKINETIC SAMPLE A sample drawn through a probe in a smokestack at the same velocity of the gas in the stack.

JAR TEST A procedure used to determine the optimum coagulant dose in a water treatment plant.

LAGOON A pond used for biological stabilization of wastewater or for the storage of hazardous waste.

LANDFILL An engineered facility for disposal of solid waste on or in the land, but which does not endanger public health or cause environmental damage.

LAND TREATMENT The controlled spreading of wastewater, sludge, or hazardous waste on selected land parcels for waste treatment and/or disposal.

LAPSE RATE The decrease in air temperature with increasing altitude; also called prevailing or environmental lapse rate.

LASER An instrument that projects an intense narrow beam of light, used to establish line and grade during pipeline construction.

LATERAL A relatively small sewer in a public right-of-way that collects wastewater directly from homes or buildings.

LEACHATE Highly contaminated liquid generated in, and tending to flow out of, a sanitary landfill, thereby causing water pollution.

LEACHING FIELD An area comprising several trenches and buried perforated pipes, for distribution and absorption of septic tank effluent in soil; also, absorption field.

LIFT STATION   A pumping facility for lifting sewage from a low point and moving it in a force main to a higher elevation, usually to a treatment plant.

LOW PRESSURE AIR TEST   A method for determining the degree of watertightness of a sanitary sewer system during an I/I study.

MAJOR LOSSES   Energy loss in a pipeline, seen as a pressure drop due to friction between the layers of flowing water and the pipe wall.

MANHOLE   A structure that provides access to a sewer system for inspection, cleaning, maintenance, sampling, or flow measurement.

MANIFEST SYSTEM   The federally mandated procedure to monitor or track hazardous waste material from "cradle to grave."

MANNING'S FORMULA   An equation used for analyzing and designing gravity-flow storm or sanitary sewer pipelines.

MARSTON'S FORMULA   An equation used to estimate the external load acting on a buried pipe for the purpose of designing its bedding.

MASS BURNING   The incineration of raw or unprocessed municipal solid waste in an energy recovery facility.

MEMBRANE FILTER METHOD   A technique for testing the bacteriological quality of water, using a very fine paper filter to trap and collect the bacteria.

METAMORPHIC ROCK   Rock formed from igneous or sedimentary rock from the action of extreme heat and pressure.

METHANE   A gaseous product from the anaerobic decomposition of garbage or sewage sludge.

MICROBAR   A unit for expressing sound pressures, equal to one one-millionth of standard atmospheric pressure.

MICROBE   A tiny living organism seen with the aid of a microscope.

MICROSTRAINER   A physical treatment device in which water or wastewater flows through a revolving drum that is covered with a finely woven metal fabric that traps suspended solids.

MINOR LOSSES   Energy loss in flowing water manifested as a pressure drop that occurs as the water flows through valves, bends, and other pipeline fittings.

MIST   Very small liquid droplets suspended in the atmosphere.

MIXED LIQUOR SUSPENDED SOLIDS   The contents of an activated sludge aeration tank at a secondary sewage treatment plant.

MOLECULE   The smallest fragment of a compound that can exist and still retain the same chemical properties.

MOST PROBABLE NUMBER   A statistical estimate of coliform bacteria concentration in a sample of water or wastewater, based on the results of the multiple tube fermentation test.

MULTIPLE-TUBE FERMENTATION TEST   A method for estimating the concentration of coliform bacteria in water or wastewater.

MULTIPURPOSE RESERVOIR   A large reservoir built to satisfy two or more needs, including flood control, water supply, power generation, irrigation, and recreation.

MUTAGENIC   Causing harmful health effects in the next generation.

NAPPE   A sheet of water that flows freely over a dam or weir.

NATURAL SUCCESSION   A process by which healthy ecosystems gradually age and change form as time passes.

NITRIFICATION   The conversion of ammonia into nitrates by bacterial action, causing a decrease in dissolved oxygen levels in water.

NOMOGRAPH   A chart used to solve equations graphically.

NONINFECTIOUS DISEASE   A disease that is not transmitted from person to person.

NUTRIENT   A mineral substance that is essential to life.

OLIGOTROPHIC LAKE   A relatively deep, cold, clear young lake, with little aquatic life.

ON-SITE DISPOSAL   Subsurface disposal of sewage at the location where it is generated, usually using a septic tank and leaching field.

OPEN CHANNEL FLOW Gravity flow in a pipe or open conduit with a free surface at atmospheric pressure.

ORGANIC COMPOUND A substance usually made up of complex molecules that comprise carbon with hydrogen, oxygen, and other elements.

OVERLAND FLOW Runoff that has not yet reached a well-defined stream channel or ditch; also called sheetflow.

OVERLAND FLOW TIME The time it takes sheetflow to reach a stream or a stormwater inlet; one part of the time of concentration.

OVERTURN The mixing of water in a stratified lake due to a change of season and air temperature.

OXIDATION A chemical reaction involving combination with oxygen and/or loss of electrons.

OXYGEN SAG CURVE A graph that shows the decrease in dissolved oxygen concentration in a polluted stream, as a function of time or distance; also called the oxygen profile.

PAPER-TAPE SAMPLER A filtration type device used to collect and measure suspended particulates in air.

PARSHALL FLUME A constricted section in an open channel, for the purpose of measuring flow rate.

PARTICULATE A very small fragment of a solid or liquid substance that is suspended in the atmosphere.

PATHOGEN A type of microorganism that can cause disease.

PERCOLATION The flow of water through the pore spaces of soil.

PERC TEST A field test to determine the rate at which water seeps into the ground at a given site, used for septic system design.

PERENNIAL STREAM A stream that has flow all year.

PERMEABILITY The ability of porous soil or rock to allow the flow of water through voids and fissures.

PHOTOSYNTHESIS The natural process by which green plants convert carbon dioxide, water, nu-trients, and sunlight energy into basic food substances.

pH SCALE A logarithmic scale used to indicate the strength of an acidic or basic solution.

PIEZOMETRIC SURFACE An imaginary surface or line that represents the height to which water would rise in an artesian well.

POINT SOURCE A pipe or channel from which pollutants are discharged into a body of water.

POLLUTANT STANDARDS INDEX A computed number related by a formula to the ambient concentrations of five criteria air pollutants, which serves as an overall indicator of air quality.

POROSITY The percentage of rock or soil volume occupied by spaces or voids.

POTABLE WATER Fresh water that is safe and pleasant to drink.

PRESSURE HEAD The height of a column of liquid, usually water, that a given hydrostatic pressure in a system could support.

PRIMARY POLLUTANT A substance that is emitted directly into the environment and that causes harm in its original form.

PRIMARY SLUDGE A concentrated suspension or slurry of solids that accumulate at the bottom of a primary settling tank in a sewage treatment plant.

PRIMARY TREATMENT The removal of floating and settleable solids from wastewater by screening and gravity settling.

PROTOZOA Microscopic single-celled animals that consume bacteria and algae for food.

PUMP HEAD CURVE A graph that shows the relationship between flow rate and pressure head on the discharge side of a pump.

PUTRIFACTION Anaerobic decay of protein compounds.

PYROLYSIS A high-temperature thermal conversion process using little or no oxygen for processing municipal solid waste.

RADICAL An electrically charged group of atoms that act together as a unit in chemical reactions.

RADIUS OF INFLUENCE The horizontal distance from a well to the area where the water table elevation is not affected by pumping.

RAINFALL CURVES A set of graphs that show the relationships among rainfall intensity, duration, and frequency for a given region of the country.

RAINFALL INTENSITY The rate of rainfall expressed in terms of inches per hour or millimeters per hour.

RAMP METHOD One of the methods used for placing, compacting, and covering refuse in a sanitary landfill.

RAPID FILTER A water purification system that removes suspended solids as the water flows through a granular bed of sand or other material, and that is cleaned by backwashing the filter bed.

RATING CURVE A graph that shows the relationship between the stage of a stream and its discharge; also, stage-discharge curve.

RATIONAL METHOD A common procedure for estimating peak stormwater runoff rates.

RAW SEWAGE Wastewater that has not yet been treated to remove pollutants.

REACTIVE WASTE Waste material that is explosive, flammable, or highly corrosive.

REAERATION A natural process that occurs in flowing streams, by which air is mixed in the water, thereby increasing the dissolved oxygen level; also, part of contact stabilization.

RECHARGE AREA A region where water infiltrates the ground surface and percolates to the underlying groundwater aquifer.

RECHARGE BASIN A reservoir built specifically to collect stormwater runoff and allow it to percolate to an underlying aquifer.

RECURRENCE INTERVAL The average number of years between storms of specific intensities and durations; also, return period.

RECYCLING The recovery, reprocessing, and reuse of certain discarded materials as an alternative to final waste disposal.

REDUCTION A chemical reaction involving the removal of oxygen from and/or the addition of electrons to a compound.

REFUSE All of the solid waste from a community that requires collection and hauling to a disposal or processing site, including garbage, rubbish, and trash.

RESPIRATION The process by which organic material is oxidized inside the cells of living organisms, providing energy for growth and reproduction.

RETENTION BASIN A small reservoir holding a permanent pool of water, constructed to retain stormwater runoff.

RETURN PERIOD The average number of years between storms of specific intensities and durations; also, recurrence interval.

REVULCANIZATION The process by which waste rubber is processed for reuse, as part of a solid waste recycling program.

RIDGELINE A line sketched on a topographic map, to show the separation of adjacent watersheds; also, drainage divide line.

RINGLEMANN CHART A set of five standard shades of gray used for visual measurement of smoke plume density.

RIVER BASIN A large watershed encompassing a major river and all of its tributary streams.

SANITARY LANDFILL See *landfill*.

SANITATION The promotion of cleanliness for the prevention of disease and for public health protection.

SCREENING A physical treatment process for water or wastewater in which relatively large floating objects are removed as the liquid passes through a coarse bar or wire mesh screen.

SCRUBBER An air cleaning device that traps particulates or gases in a spray of water; also called a wet collector or spray tower.

SCS METHOD A procedure for estimating the volume and rate of stormwater runoff using soil type as a major criteria.

SECONDARY POLLUTANT A pollutant that is not emitted directly into the atmosphere, but is formed after emission by chemical reactions with other substances.

SECONDARY TREATMENT Biological treatment of wastewater designed to remove at least 85 percent of the suspended solids and biochemical oxygen demand.

SECURE LANDFILL A landfill constructed with an impermeable bottom liner, an impermeable cover or cap, and a groundwater monitoring system for the disposal of hazardous waste.

SEDIMENTARY ROCK Compacted and consolidated soil particles that have become cemented together naturally over a long time.

SEDIMENTATION The slow settling and separation of suspended solids from a liquid under the force of gravity.

SELF-PURIFICATION The processes by which a stream or river assimilates waste and cleanses itself naturally of organic pollutants.

SEPTAGE The contents of a septic tank.

SEPTIC Anaerobic, or without oxygen.

SEPTIC TANK A buried steel or concrete tank that serves for primary settling, sludge digestion, and storage in an on-site sanitary sewage disposal system.

SETTLEABLE SOLIDS The coarser fraction of suspended particles in water, wastewater, or air that settle out because of gravity under relatively quiescent (quiet or still) conditions.

SETTLING CHAMBER An enlarged compartment or section of a flue in which airstream velocity is reduced so as to allow relatively coarse particulates or dust to settle out by gravity.

SETTLING TANK A steel or concrete basin in which settleable solids are allowed to separate from water or wastewater under the force of gravity; also called a clarifier.

SEWAGE Used water from domestic, commercial, or industrial establishments, carrying sanitary or industrial waste material.

SHEETFLOW Runoff that has not yet reached a well-defined stream channel or drainage ditch; also, overland flow.

SHORT CIRCUITING A term referring to the condition in which water or wastewater flows through a treatment tank in less than the theoretical detention time, based on the tank volume and flow rate.

SHUTOFF HEAD The pressure head developed by a centrifugal pump that operates against a closed discharge valve.

SLUDGE A slurry or concentrated suspension of solids that accumulates at the bottom of a settling tank or clarifier.

SLUDGE DEWATERING The process of drying liquid sludge, thereby changing its condition to that resembling potting soil.

SLUDGE DIGESTION Biological stabilization of organic sludge, to reduce its volume, destroy pathogens, and prepare it for drying.

SLUDGE THICKENING A process that increases the solids concentration of sludge in order to reduce its overall volume.

SLUDGE VOLUME INDEX A measure or indicator of the settling behavior of activated sludge in a secondary clarifier.

SLURRY TRENCH A method used to construct subsurface cutoff walls for the purpose of containing buried hazardous waste at an illegal dump site.

SMOKE Very small airborne solid particulates, less than 1 μm in size, formed during incomplete combustion.

SMOKE READING A visual evaluation of a smoke plume emitted from a stack, made by comparing the smoke density to standard shades of gray on a Ringlemann Chart.

SMOKE TESTING A method for testing a sanitary sewer system for watertightness during an I/I survey.

SOFTENING A treatment process that reduces the hardness of water by removing much of the dissolved calcium and magnesium.

SOIL EROSION The wearing away of the land surface, particularly topsoil, by the natural action of wind or water.

SOIL SERIES A group of related soils that have developed from similar rocks and that have similar characteristics.

SOIL SURVEY MAP A map prepared by the Soil Conservation Service showing the soil series in different areas of a county.

SOLID WASTE Useless, unwanted, discarded material including garbage, rubbish, and trash; also called refuse.

SOLID WASTE MANAGEMENT The planning, design, construction, and operation of facilities for the collection, transport, processing, and disposal of solid waste.

SOUND LEVEL A measure of noise, using a meter that weights or adjusts the readings so as to respond approximately as the human ear perceives sound, usually expressed in units of dB(A).

SOUND PRESSURE LEVEL A measure of noise, expressed in decibels, dB.

SOURCE SAMPLING Air samples collected from the pollutant source; also called emissions sampling.

SOURCE SEPARATION The separation of reusable material from refuse at the point of generation, to facilitate recycling efforts.

STACK SAMPLING Samples of flue gas collected by a probe that is inserted directly into the stack, for emission analysis.

STAGE The depth or elevation of water in a stream or lake.

STANDARD METHODS An important reference and guide for water and wastewater sampling and analysis, published jointly by the APHA, AWWA, and the WPCF.

STATIC LEVEL The position or elevation of the water surface in a well that is not being pumped at the time.

STICKY TAPE SAMPLER A simple inertial device for collecting airborne particulates and giving a particle count on the basis of the prevailing wind direction.

STORM SEWAGE Runoff caused by rainfall and collected in a system of surface channels or buried pipes; also, stormwater.

STORM SEWERS Pipelines that are designed to carry only stormwater to a point of storage or disposal.

STORMWATER MANAGEMENT The planned control of surface runoff in natural and urban systems to prevent flooding and pollution.

STRATOSPHERE A stable layer of the atmosphere located just above the troposphere.

SUBMAIN A relatively large sanitary sewer that intercepts the flow from smaller lateral sewers; also, a collector sewer.

SUMMATION HYDROGRAPH A graph that shows cumulative flow versus time and is used to determine required reservoir storage volumes; also, a mass diagram.

SUPERFUND A large fund of money set aside by the federal government to pay for cleaning up illegal hazardous waste dump sites.

SUPERNATANT The water that remains above the sludge layer in a settling tank or digester.

SURCHARGED SEWER A gravity sewerline flowing full with wastewater backed up in manholes and with the hydraulic grade line above the pipe crown.

SUSPENDED SOLIDS Solids carried in water or sewage that would be retained on a glass-fiber filter in a standard lab test.

SYSTEM HEAD CURVE A graph that shows the relationship between the flow rate and the total dynamic head in a water system.

TEMPERATURE INVERSION An atmospheric condition in which air temperature increases with altitude, instead of decreasing as normal.

TERRESTRIAL Growing or living on land.

TERTIARY TREATMENT Processes used after or during secondary wastewater treatment to remove nutrients and/or additional solids and organics; also called advanced treatment.

THERMAL POLLUTION The change in water temperature of a river or lake, usually caused by cooling water discharge from power plants.

THERMAL STRATIFICATION The natural process by which separate layers form in lakes because of water temperature differences.

THERMOCLINE A layer of water that separates the upper epilimnion from the lower hypolimnion layer in a stratified lake.

TIME OF CONCENTRATION The time it takes a drop of water to flow from the most distant point of a watershed to the outlet; also, the storm duration used in the Rational Method.

TOTAL DYNAMIC HEAD The total pressure head in a system against which a centrifugal pump operates.

TOXIC WASTE Poisonous waste that causes acute illness, death, or chronic health problems.

TRANSFER STATION A centrally located facility where refuse from individual collection trucks is transferred into larger vehicles for transport to a more distant disposal site.

TRASH Community refuse that requires special collection.

TROMMEL SCREEN A rotary drum type classifier for sorting solid waste on the basis of size and density characteristics.

TROPOSPHERE The lowermost layer of the atmosphere in which life is sustained and most air pollution problems occur.

TRICKLING FILTER A biological sewage treatment unit in which dissolved organics are absorbed from the settled sewage as it flows over a bed of slime-covered rocks.

TRUNK LINE A large sewer that collects wastewater from submains and conveys it to a treatment plant or pumping station; also called an interceptor.

TUBERCULATION The formation of nodules of rust on the inside wall of a pipe, increasing pressure loss and decreasing flow capacity.

TUBE SETTLER A prefabricated unit of inclined, nested tubes that is installed in a settling tank to increase its efficiency.

TURBIDITY A measure of the light scattering effect caused by finely divided suspended particles in water.

TURNOVER The natural destratification and mixing of the water in a lake or reservoir due to the seasonal change in ambient air temperature during the fall and spring of the year.

UNDERGROUND INJECTION A method for disposing of hazardous waste by pumping it through deep wells into confined porous aquifers.

VACUUM FILTER A mechanical device used at a sewage treatment plant for sludge dewatering.

VECTOR-BORNE A mode of disease transmission through insects, rodents, or other living animal carriers of pathogenic organisms.

VEHICLE-BORNE A mode of disease transmission from contaminated inanimate objects, including food and water.

WATERSHED The land area that contributes runoff to a stream or lake; also, drainage basin and catchment area.

WATER TABLE The top of the zone of saturation, where all soil voids are filled with water at atmospheric pressure.

WEATHERING The gradual process by which solid rock becomes soil, from the action of wind and water, as well as chemical and temperature changes in the environment.

WEIR An obstruction placed in a flowing stream or channel, usually to measure the flow rate.

WINDROW A long narrow pile of garbage and/or sewage sludge in an open-field composting facility.

ZONE OF AERATION The upper layer of soil in which the voids or spaces between the soil particles are not completely filled with groundwater.

ZONE OF SATURATION The layer of soil or rock in which all the voids or fissures are filled with water.

ABBREVIATIONS

AC pipe	asbestos cement pipe
APHA	American Public Health Association
AWWA	American Water Works Association
BOD	biochemical oxygen demand
CAA	Clean Air Act
CEQ	Council on Environmental Quality
CFS	cubic feet per second
CI pipe	cast-iron pipe
COD	chemical oxygen demand
COH	coefficient of haze
CPS	cycles per second
CWA	Clean Water Act
dB	decibel
dB(A)	decibel, A-weighted
DO	dissolved oxygen
EIA	environmental impact assessment
EIS	environmental impact statement
EPA	Environmental Protection Agency
GAC	granular activated carbon
GPCD	gallons per capita per day
FGD	flue-gas desulfurization
F/M	food-to-microorganism ratio
GPG	grains per gallon
GPM	gallons per minute
HGL	hydraulic grade line
HWM	hazardous waste management
Hz	Hertz (cps)
I/I	infiltration and inflow
JTU	Jackson turbidity unit
MA7CD10	minimum average 7-consecutive day 10-year low flow
MCL	maximum contaminant level
MGD	million gallons per day
MLSS	mixed liquor suspended solids
MPN	most probable number
MSW	municipal solid waste
NAAQS	national ambient air quality standards
NEPA	National Environmental Policy Act
NPDES	National Pollutant Discharge Elimination System
NTU	nephelometric turbidity unit
PCB	polychlorinated biphenol
POTW	publically owned treatment works
PPM	parts per million
PSI	Pollutant Standards Index
RBC	rotating biological contactor
RCP	reinforced concrete pipe
RCRA	Resource Conservation and Recovery Act
RDF	refuse derived fuel
SCS	Soil Conservation Service
SDWA	Safe Drinking Water Act
SIP	state implementation plan
SPL	sound pressure level
SPS	source performance standards
SS	suspended solids
STP	sewage treatment plant
SVI	sludge volume index
TDH	total dynamic head
TDS	total dissolved solids
TGIF	Thank Goodness It's Friday
TLV	threshold limit value
TOC	total organic carbon
TS	total solids
TSCA	Toxic Substances Control Act
TSP	total suspended particulates
TSS	total suspended solids
TTHM	total trihalomethanes
VOC	volatile organic carbon
WTP	water treatment plant
WPCF	Water Pollution Control Federation

ANSWERS TO PRACTICE PROBLEMS

Answers have been rounded off to reflect the precision of the data and/or the accuracy of assumed factors in the problem.

CHAPTER 2

1. 22 psi; 8.6 psi   **2.** 5 m   **3.** 290 kPa   **4.** 115 ft; 32 psi   **5.** 40 kPa; 4 m
6. 1.4 m/s   **7.** 12 in.   **8.** 8 m/s   **9.** 0.67 m/s   **10.** 48 psi   **11.** 660 kPa
12. 13 kPa   **13.** 10.5 in.; use 12 in. standard size   **14.** 100 L/s   **15.** 2 cfs
16. 63 L/s   **17.** 560 L/s; 1.17 m/s   **18.** 1800 gpm; 2.3 ft/s   **19.** 450 mm
20. 0.05 percent   **21.** 130 mm; 1.8 m/s   **22.** 1900 gpm; 17 in.   **23.** 1 m/s;
580 L/s; 740 mm   **24.** 440 L/s; 1.3 m/s   **25.** 500 L/s

CHAPTER 3

1. 50 mm/h; 375 ML   **2.** 2 in./h; 8 ac-ft   **3.** 4 in./h; 1.7 in./h; 2 in./h   **4.** 1
percent   **5.** 27 mm/h   **6.** 5 percent   **7.** 35 m^3/s   **8.** See instructor for solution.   **9.** 40×10^6 gal.   **10.** 0.9 mm/h   **11.** sand   **12.** 30 m^3/h; 33 m^3/h

CHAPTER 4

1. 275 mg/L; 16.1 gpg **2.** 68 mg/L **3.** 1250 lb/d **4.** 0.6 mg/L **5.** 10 kg
6. 100 lb/d **7.** 25 ppb **8.** 8 mg/L **9.** > 14 mg/L **10.** 300 mg/L **11.** 270
mg/L; 340 mg/L **12.** 5.8 mg/L **13.** 500 mg/L **14.** 350 mg/L; 43 percent
15. 220/100 mL **16.** 120

CHAPTER 5

1. 180 mg/L **2.** 20 ML/d **3.** 6.4 mg/L **4.** 5.5 mg/L; 6.6 km

CHAPTER 6

1. 8.8 d **2.** 1250 m^3, 5 m **3.** 0.5 in./min **4.** 56.4 ft, 13.4 ft **5.** 20 m by 10
m by 4 m **6.** 4.6 gpm/ft^2, 7.4 in./min **7.** 6 m by 6 m **8.** 2.6 percent **9.** 19
cylinders; yes **10.** 0.6 mg/L **11.** 50 ML/d

CHAPTER 7

1. 35 ML/d or 9.5 mgd **2.** 3200 L/s **3.** 77 L **4.** 0.62 m^2 **5.** 610 gpm at 156 ft **6.** 500 gpm at 180 ft **7.** 615 gpm at 195 ft **8.** 185 L/s at 27 m; Pump b alone, 110 L/s at 30 m **9.** 23 hp; 30 hp **10.** 65 percent **11.** 4100 m^3 **12.** 8 L/s; filling at 34 L/s **13.** 260 mm **14.** 320 mm **15.** 14 in. **16.** AB: 300 L/s; BC: 100 L/s; CA: 200 L/s **17.** 170 kPa **18.** AB: 53 L/s; BC: 11 L/s; DB: 18 L/s; DA: 47 L/s; DC: 29 L/s

CHAPTER 8

1. 830 gpm **2.** 60 L/s **3.** 350 mm @ 2.5 percent; 303.20, 300.70 **4.** 200 mm at 0.0213; 300 mm at 0.00725 **5.** 2000 lb/ft **6.** Class B **7.** Class C **8.** 14 L/min **9.** 870 gal/d/in./mile

CHAPTER 9

1. 18 cfs **2.** 750 L/s **3.** 0.30 **4.** 0.34 **5.** 0.26 to 0.45 **6.** 44 cfs **7.** 3 m^3/s **8.** 4 m^3/s **9.** 15 m^3/s **10.** 1400 cfs; 2400 cfs **11.** 11 cfs, 24 in.; 23 cfs, 30 in.; 35 cfs, 36 in. **12.** 220 L/s, 600 mm; 400 L/s, 650 mm; 600 L/s, 650 mm **13.** 1.3 m^3/s **14.** 12 000 m^3 **15.** 1.2 ac

CHAPTER 10

1. 18 mg/L; 1500 lb/d **2.** 96 percent; 150 kg/d **3.** 70 percent **4.** 140 ppm **5.** 6 mgd **6.** 19 m^3/m^2 d; 0.5 kg/m^3 d **7.** 31 mg/L; 86 percent **8.** 0.21 **9.** 6 m **10.** 250; no **11.** 50 min/25 mm **12.** 20 min/in. **13.** 7 at 20 m long **14.** 5 at 100 ft; 24 ft wide **15.** 75 tons **16.** 6000 kg; 100 000 L **17.** Approximately 9 m^3

CHAPTER 11

1. 80 percent; 5:1; 20:1 **2.** 88 percent; 20 yd^3 **3.** 3.8 ft **4.** 10 loads/d **5.** 15 yr **6.** 4 m **7.** 30 ac

CHAPTER 12

1. 20 tons/mile^2/month **2.** 120 μg/m^3 **3.** 118 dB **4.** 11 μbar **5.** 93 dB(A) **6.** 79.8 dB **7.** 95 dB(A) **8.** 70 dB(A)

INDEX